CELLULOSICS: PULP, FIBRE AND ENVIRONMENTAL ASPECTS

ELLIS HORWOOD SERIES IN POLYMER SCIENCE AND TECHNOLOGY

Series Editors: T. J. KEMP, Department of Chemistry, University of Warwick
J. F. KENNEDY, Director of the Research Laboratory for the Chemistry of Bioactive Carbohydrates and Proteins, Department of Chemistry, University of Birmingham, and Professor of Applied Chemistry, North East Wales Institute of Higher Education.

This series, which covers both natural and synthetic macromolecules, reflects knowledge and experience from research, development and manufacture within both industry and academia. It deals with the general characterization and properties of materials from chemical and engineering viewpoints and will include monographs highlighting polymers of wide economic and industrial significance as well as of particular fields of application.

Barashkov, N. & Gunder, O.	**FLUORESCENT POLYMERS**
Bartenev, G.M.	**MECHANICAL STRENGTH AND FAILURE OF POLYMERS**
Barton, J. & Capek, I.	**RADICAL POLYMERIZATION IN DISPERSE SYSTEMS**
Bershtein, V. & Egorov, V.	**DIFFERENTIAL SCANNING CALORIMETRY OF POLYMERS**
Blazej, A. & Kosík, M.	**PHYTOMASS: A Raw Material for Chemistry and Biotechnology**
Bodor, G.	**STRUCTURAL INVESTIGATION OF POLYMERS**
Bryk, M.T.	**DEGRADATION OF FILLED POLYMERS: High Temperature and Thermal-Oxidative Processes**
Challa, G.	**POLYMER CHEMISTRY: An Introduction**
Crompton, T.R.	**POLYMER LABORATORY INSTRUMENTATION: Vol. 1 & 2**
De. S.K. & Bhowmick, A.K.	**THERMOPLASTIC ELASTOMERS FROM RUBBER–PLASTIC BLENDS**
Gorbatkina, Y.A.	**ADHESIVE STRENGTH IN FIBRE–POLYMER SYSTEMS**
Hearle, J.W.S.	**POLYMERS AND THEIR PROPERTIES: Vol. 1**
Hearle, J.W.S.	**POLYMER MATERIALS** **Vol. 1: Structure, Properties and Performance** **Vol. 2: Complex Systems**
Hongu, T. & Phillips, G.O.	**NEW FIBRES**
Kennedy, J.F., *et al.*	**WOOD AND CELLULOSICS**
Kennedy, J.F., *et al.*	**CELLULOSE: Structural and Functional Aspects**
Kennedy, J.F., *et al.*	**CELLULOSE SOURCES AND EXPLOITATION: Industrial Utilization, Biotechnology and Physico-Chemical Properties**
Kennedy, J.F., *et al.*	**LIGNO-CELLULOSICS: Science, Technology, Development and Use**
Kennedy, J.F., *et al.*	**CELLULOSICS: Pulp, Fibre and Environmental Aspects**
Kennedy, J.F., *et al.*	**CELLULOSICS: Chemical, Biochemical and Material Aspects**
Lazár, M., *et al.*	**CHEMICAL REACTIONS OF NATURAL AND SYNTHETIC POLYMERS**
Nachinkin, O.I.	**POLYMERIC MICROFILTERS**
Nevell, T.J.	**CELLULOSE CHEMISTRY AND ITS APPLICATIONS**
Saifullin, R.S.	**PHYSICAL CHEMISTRY OF INORGANIC POLYMER AND COMPOSITE MATERIALS**
Stepek, J. *et al.*	**POLYMERS AS MATERIALS FOR PACKAGING**
Svec, P. *et al.*	**STYRENE-BASED PLASTICS AND THEIR MODIFICATIONS**
Ulberg, Z.R. & Deinega, Y.F.	**ELECTROPHORETIC COMPOSITE COATINGS**
Vakula, V.L. & Pritykin, L.M.	**POLYMER ADHESION: Physico-Chemical Principles**
Wirpsza, Z.	**POLYURETHANES: Chemistry, Technology and Applications**

ELLIS HORWOOD PTR PRENTICE HALL SERIES IN POLYMER SCIENCE AND TECHNOLOGY

Bartenev, G.M.	**MECHANICAL STRENGTH AND FAILURE OF POLYMERS**
Clarson, S.J.	**SILOXANE POLYMERS**
Galiatsatos, V.	**MOLECULAR CHARACTERIZATION OF NETWORKS**
Shlyapnikov, Y.A.	**ANTIOXIDATIVE STABILIZATION OF POLYMERS**
Vergnaud, J.M.	**LIQUID TRANSPORT PROCESS IN POLYMERIC MATERIALS**

CELLULOSICS: PULP, FIBRE AND ENVIRONMENTAL ASPECTS

Editors:
J. F. KENNEDY B.Sc., Ph.D., D.Sc., C.Biol., C.Chem., F.I.Biol. F.I.F.S.T. F.I. Mgt., F.R.S.C.
Director of the Research Laboratory for the Chemistry of Bioactive Carbohydrates and Proteins, School of Chemistry, University of Birmingham, England, and Professor of Applied Chemistry North East Wales Institute of Higher Education, Wrexham, Wales

G. O. PHILLIPS B.Sc., Ph.D., D.S., F.R.S.C.
Chairman of Newtech Ltd, Deeside, Wales

P. A. WILLIAMS B.Sc., Ph.D., C.Chem. F.R.S.C.
Head of the Polymer and Colloid Chemistry Group North East Wales Institute of Higher Education, Wales

ELLIS HORWOOD
NEW YORK LONDON TORONTO SYDNEY TOKYO SINGAPORE

First published 1993 by
Ellis Horwood Limited
Market Cross House, Cooper Street,
Chichester
West Sussex, PO19 1EB
A division of
Simon & Schuster International Group

Printed and bound in Great Britain
by Bookcraft, Midsomer Norton

British Library Cataloguing in Publication Data

A catalogue record for this book is available from the British Library

ISBN 0–13–053059 X

Library of Congress Cataloging-in-Publication Data

Available from the publisher

1 2 3 4 5 97 96 95 94 93

ACKNOWLEDGEMENTS

This volume arises from the Internatonal Conference on cellulose etc., - Cellulose 91 - which was a joint meeting of the Cellucon Conferences (operated by the Cellucon Trust), the ACS Cellulose, Paper and Textile Division, and the Syracuse Cellulose Conferences.

MEMBERS OF THE ORGANISING COMMITTEE – CELLULOSE 91

This meeting owed its success to the invaluable assistance of the Organising Committee.

Dr. N.R. Bertoniere (Co-Chair)	ACS Cellulose, Paper and Textile Division and Southern Regional Research Center, USDA, USA.
Prof. G.O. Phillips (Co-Chair)	The Cellucon Trust and Newtech Ltd. Clwyd, UK.
Prof. T.E. Timell (Co-Chair)	Syracuse Cellulose Conferences and SUNY College of Environmental Science and Forest, New York, USA.
Dr. H.L. Chum (Programme Chair)	Solar Energy Research Institute, Colorado, USA.
Dr. E.J. Blanchard	Solar Energy Research Institute, Colorado, USA.
Prof. W.G. Glasser	Virginia Tech., Virginia, USA.
Dr. H. Hatakeyama	Industrial Products Research Institute, Tsukuba, Japan.
Prof. J.F. Kennedy	The North East Wales Institute and The University of Birmingham, Birmingham, UK.
Prof. H.L. Needles	University of California, USA.
Dr. R.P. Overend	Solar Energy Research Institute, Colorado, USA.
Prof. A. Sarko	SUNY College of Environmental Science and Forestry, New York, USA.
Dr. P.A. Williams	The North East Wales Institute, Clwyd, UK.
Prof. W.T. Winter	SUNY College of Environmental Science and Forestry, New York, USA.

INDUSTRIAL SPONSORSHIP

The organisers express their grateful thanks to the following for their financial support of Cellulose 91.

U.S. Department of Energy
Conservation and Renewable Energy
Office of Industrial Technologies
Office of Transportation Technologies
Office of Utility Technologies
National Renewal Energy Laboratory
(formerly Solar Energy Research Institute)
U.S. Department of Agriculture
Southern Regional Research Center
Aqualon Co., USA
Asahi Chemical Co., Japan
Daicel Chemical Industries Ltd., Japan
ITT Rayonier, Inc., USA
Kodak/Tennessee Eastman, USA
Monsanto, USA
NOVO Nordisk, Denmark
The Proctor & Gamble Cellulose Co., USA
Westvac, USA

THE CELLUCON TRUST

incorporating

CELLUCON CONFERENCES

International Educational Scientific Meetings
on Cellulose, Cellulosics and Wood

CELLUCON CONFERENCES AND THE CELLUCON TRUST

Cellucon Conferences as an organisation was initiated in 1982, Cellucon '84 was the original Conference, which set out to establish the strength of British expertise in the field of cellulose and its derivatives. This laid the foundation for subsequent conferences in Wales (1986), Japan (1988), Wales (1989), Czecho-Slovakia (1990) and USA (1991). They have had truly international audiences drawn from the major industries involved in the production and use of cellulose pulp and derivatives of cellulose, plus representatives of academic institutions and government research centres. This diverse audience has allowed the cross fertilization of many ideas which has done much to give the cellulose field the higher profile that it rightfully deserves.

Cellucon Conferences are organised by The Cellucon Trust, an official UK charitable trust with worldwide objectives in education in wood and cellulosics. The Cellucon Trust is continuing to extend the knowledge of all aspects of cellulose worldwide. At least one book has been published from each Cellucon Conference as the proceedings thereof. This volume is one of two volumes arising from the 1991 conference held in New Orleans, USA, and the conferences in Wales 1992, Sweden 1993 etc., will generate further useful books in the area.

THE CELLUCON TRUST TRUSTEES

CELLUCON CONFERENCES ORGANISING COMMITTEE

Mr. T. Greenway	Berol Nobel Ltd., UK
Dr. J. Guthrie	University of Leeds, UK
Dr. H. Hatakeyama	Industrial Products Research Institute, Japan
Dr. A. Henderson	DOW Chemicals Europe, Switzerland
Dr. M.B. Huglin	University of Salford, UK
Dr. P. Levison	Whatman Specialty Products Division, UK
Dr. J. Meadows	The North East Wales Institute, UK
Prof. Y. Nakamura	Gunma University, Japan
Mr. W.B. Painting	Hoechst (UK) Ltd., UK
Mr. A. Poyner	Aqualon (UK) Ltd., UK
Mr. R. Price	Shotton Paper Co. Ltd., UK
Prof. J. Roberts	Institute of Science and Technology, University of Manchester, UK
Mr. J.F. Webber	The Forestry Commission, UK
Dr. C.A. White	Fisons Plc, Scientific Equipment Division, UK

Cellucon Conferences are sponsored by:

The Biochemical Society - Chembiotech Ltd., - Hoechst (UK) Ltd., - The North East Wales Institute - USAF European Office of Aerospace Research and Development - US Army Research, Development and Standardisation Group, UK - Welsh Development Agency - Whatman Specialty Products Division.

Cellucon Conferences are supported by:

The American Chemical Society (Cellulose, Paper and Textile Divison) - Aqualon (UK) Ltd., - Berol Nobel Ltd., - Courtaulds Chemicals and Plastics - Ministry of Defence - Crown Berger Europe Ltd., - DOW Chemicals - The Forestry Commission - Shotton Paper Company Ltd., - Syracuse Cellulose Conferences.

Table of contents

Preface 1

PART 1: PULPING, PULP QUALITIES AND PULP BLEACHING 3

1 **Research on the delignification process in pulpings** 5
E.I. Germer

2 **Regulation of crystallinity of solvolysis pulp by delignifcation** 11
Y. Uraki, S. Ogawa, Y. Sano, T. Sasaya and M. Takai

3 **Pulp production by catalytic oxygen-alkali-alcohol (OAA) delignification** 17
E.I. Germer and R.A. Gorenkov

4 **Oxygen-alkali pulping of chips** 25
Yu. S. Ivanov, L.O. Ioffe, Yu. P. Vorobiev, Sh. F. Dvoskin, I.A. Nikiforov and A.B. Broderson

5 **Production of cellulose from Brazilian lignocellulosic residues** 31
S.P.D'O. Barreto, E.H.M. Melo, J.L. Lima Filho, A.C. Duarte Coelho and J.F. Kennedy

6 **Explosion pulping of aspen** 37
X. Hua, B.V. Kokta and S. Kaliaguine

7 **Steam explosion pulping of aspen: the effect of cooking parameters on resulting properties** 45
M. Tomasec and B.V. Kokta

8 **Ultra-high-yield pulping: relation between pulp properties and fibre characteristics by multiple linear regression** 51
F. Carrasco, J.J. Garceau, A. Ahmed and B.V. Kokta

9 **Effect on pulp and paper properties of chemical treatment prior to steam explosion pulping of aspen** 57
Y. Ben, B.V. Kokta and J. Doucet

10 **Papermaking properties of hydrothermally pretreated wood and annual plants** 63
A.J. Zemann, O. Bobleter and P. Krkoska

11 **A new separation and modification system of phytomass constituents: swelling-solvation process** 69
M. Funaoka, S. Fukatsu, J.Matsue and K. Hamaguchi

12 **Anaerobic digestion of cotton and linseed plant stalks for the preparation of pulp and paper** 75
A.J. Shaikh, Y.D. Pai, R.H. Balasubramanya and V.G. Khandeparkar

13 **Possibilities of increasing dissolving pulp reactivity** 83
K. Fischer and I. Schmidt

14 **Bleaching a spruce kraft pulp with light, air and semiconducter** 89
E.L. Springer
15 **Effect of Cr(III) at the ClO_2 stage in the bleaching of kraft (sulfate) paper pulp**.. 95
F. Taner and A. Olcay
16 **Water soluble substances from bleached paper fibres**.. 103
M. Oblak-Rainer, D. Budin, S. Cerne, and B. Lipic

PART 2: PRODUCTION, ANALYSIS, STRUCTURE AND MODIFICATION OF FIBRES.. 109
17 **Do fibre types make paper grades?**.. 111
D.J. Ryder, H.W. Kropholler, C.P. Wilkins and E. Sheffield
18 **Extraction of short staple fibres from flax by an optimized steam explosion technique**.. 117
R.W. Kessler, B. Goth, W. Kessler and M. Tubach
19 **Textile fibers from bagasse pulp**.. 123
D. Paul, H.-J. Gensrich, O. Quintela and E. Ramirez
20 **A novel solvent system for the production of regenerated cellulose fibers**.......... 129
A.A. Sioumis
21 **Using biotechnology to improve cotton fiber quality: progress and perspectives**.. 135
B.A. Triplett
22 **Spectroscopic evaluation of plant fibres and fibre products**................................ 141
I.M. Morrison and D. Stewart
23 **Application of the image analyzer to the analysis of cellulose fibers in newsprint and cotton/bast fiber blends**.. 147
L.T. Hecking
24 **X-Ray diffractometric measurement of cellulose microcrystallite size**............... 153
D.W. Foreman and K.A. Jakes
25 **Inorganic composition of cotton processing dust – should its role in byssinosis be re-examined?**.. 161
J.K.S. Warrier, P.K. Chidambareswaren and V. Sundaram
26 **Lateral order in microfibrils of native and regenerated cellulose**....................... 165
H.-P. Fink, D. Hofmann and H.J. Purz
27 **New findings on the lateral order distribution in cotton fibers**........................... 171
J.K.S. Warrier and P.K. Chidambareswaran
28 **New insight into orientation – property associations in cotton fibers**................. 177
J.K.S. Warrier and P.K. Chidambareswaran
29 **Structural response to tensile deformation in cotton fibers**............................... 181
P.K. Chidambareswaran and J.K.S Warrier
30 **Rate control method for enzyme treatment of cellulosic fiber**............................ 187
M. Ueda, Y. Yoshimura, K. Kumita and M. Morita
31 **Water difffusion in wood pulp fibers**.. 193
T.Q. Li, U. Henriksson and L. Odberg
32 **Potassium hydroxide: a better mercerising agent as revealed by physico-mechanical and structural studies**.. 199
S. Sreenivasan, P. Bhama Iyer and G.S. Patel
33 **Comparison of unique staple-core and conventional cotton-blend fabrics with durable-press finish**.. 205
L.B. Kimmel, A.P.S. Sawhney, G.F. Ruppenicker and R.J. Harper
34 **Core-yarn fabrics with special finishes for military uses**.................................... 213
G.F. Ruppenicker, R.J. Harper, A.P.S. Sawhney and K.Q. Robert

35 **Construction and properties of cotton-rich yarn and fabric, using ARS-patented core-spinning technology** 219
A.P.S. Sawhney, G.F. Ruppenicker and L.B. Kimmel
36 **Kawabata evaluation of cotton/pet blended fabrics after various chemical treatments** 225
D.-H. Shin, L.B. Kimmel and R.J. Harper
37 **Chemical modifications of cotton fibers for improved water absorbency** 231
C. Mizutani, Y. Onogi and H. Inagaki
38 **Fibrets: cellulose acetate microfibers for filtration applications** 237
J.E. Smith
39 **Polarized raman spectra of oriented cellulose triacetate fibres** 243
F. Adar, D. Armellino and H.D. Noether
40 **Resist finishing and dyeing of cationic fabrics** 249
R.J. Harper, Jr. and A.H. Lambert
41 **Photofading characteristics of dyed easy-care finished cotton cellulosic substrate** 255
F.M. Tera, A. Hebeish, N.A. Baset and M.N. Michael
42 **Differences between the properties of cotton fibers treated with urea and carboxylic acid derivatives** 261
S.H. Zeronian, N.R. Bertoniere, K.W. Alger, L.D. Fisher, G. Buschle-Diller, S.A. Holmes and P.A. Slaven
43 **Relationships between properties of crosslinking agents for cellulose and their structures evaluated by molecular modelling methods** 267
H. Petersen, E. Hadicke and S. Brode
44 **Improvements in durable press cotton fabrics from esterification crosslinking.** 283
B.A. Kottes Andrews, B.J. Trask-Morrell and N.M. Morris
45 **Polymerization-crosslinking of cotton fabrics using maleic acid-containing systems** 289
H. Choi

PART 3: PHOTODEGRADATION, WEATHERING, THERMOLYSIS, BIODEGRADATION, PRESERVATION AND POLLUTION ASPECTS OF CELLULOSICS 295
46 **Life cycle analysis: a marketing nightmare** 297
J. Benda and R. Narayan
47 **The environmental aspects of solvent-spun cellulose production and use** 303
C.R. Woodings
48 **Photodegradation of cellulosic cottons by environmental conditions** 315
F.M. Tera and K.E. Shady
49 **Photodegradation of linen by sunlight** 323
N. Kohara, E. Nakayama and H.Toyoda
50 **Effects of air quality and radiation on wood** 329
D.N.-S. Hon
51 **Rapid loss of lignin at wood surfaces during natural weathering** 335
P.D. Evans, K.J. Schmalzl and A.J. Michell
52 **Direct liquefaction of wood extended batch reactions using recycled solvent** 341
T. Ogi, T. Minowa, Y. Dote and S. Yokoyama
53 **Thermal reaction pathways of syringol** 347
C.P. Masuku
54 **Environmentally acceptable conversion technology for the biochemical utilization of lignocellulosics** 355
A. Blazej and M. Kosik

55 **Biological and thermochemical pretreatment of water hyacinth for improved biogas production** 365
V. Patel and D. Madamwar
56 **Cellulases and ligninases of basidiomycetes** 371
Z.R. Ahmedova
57 **Characterization of cellulases activity and electrophoretic analysis from *Ganoderma resinaceum* Boud. Apud Patouillard extracts** 377
M.C. Saagua, Mª M. Carolino and N. Teixeira Rodeia
58 **Complexity of glycoside hydrolases from *Trichoderma viride* A 10** 383
P.-J. Gao, Y.-B. Qu and Z.-N. Wang
59 **Adsorption and interactions of cellobiohydrolase I from *Trichoderma reesei* on microcrystalline cellulose** 391
L. Kuutti, J. Pere, B. Henrissat, H. Chanzy, T. Reinikainen, L. Laaksonen, M. Srisodsuk and T.T. Teeri
60 **Mechanism of enzymic hydrolysis of cellulose and its activation and inhibition by phenolics** 397
A. Blazej and M.Kosik
61 **Microbial degradation of impure cellulosics by *Aspergillus fumigatus* IMI 255091** 409
D.A.J. Wase
62 **Biological degradation of lignocellulose using *Pleurotus ostreatus* strains obtained by genetic manipulation** 415
R. Ramirez, H. Leal, G. Eger, F. Padilla and M. Alcantara
63 **Lignin biodegradation: basic research progress, and applications in soil remediation and biopulping** 421
T.K. Kirk
64 **Mass transfer performance and enzyme production in air-lift and bubble column fermenters** 431
C.K. Erb, W.J. McManamey and D.A.J. Wase
65 **Immobilisation of the cellulolytic fungus *Apergillus fumigatus* IMI 255091** 437
L.A. Hamilton and D.A.J. Wase
66 **The dynamic mechanical behaviour of fungally degraded wood** 445
C.J. McCarthy, C. Birkinshaw, J.T. Pembroke and M. Hale
67 **Products of oxidation of cellulose by hydrogen peroxide during fungal degradation** 451
R.G. Veness and C.S. Evans
68 **Changes in pore structure and cell wall pore volume in wood decayed by brown- and white-rot fungi** 457
D.S. Flournoy and J.A. Paul
69 **The long term effect of surface treatments on the properties of cellulosic and ligneous museum aretefacts** 465
M.A. Robson
70 **Protection of wood against biodegradation by chemical modification** 473
R.M. Rowell
71 **The preservation of sawn Irish sitka spruce (*Picea sitchensis* (Bong.) Carr.)** 479
D.C. Cahill, M. Buggy and E.R. Petty
72 **Pulp and paper mill effluent biobleaching using immobilized *Phanerochaete chrysosporium* BKMF 1767** 485
R.S. Singh, S.S. Marwaha, P.K. Khanna and J.F. Kennedy
73 **Kraft mill effluent: biological treatment** 493
N. Duran, E. Esposito and V.P. Canhos
74 **Mineralization of lignin-bound and free xenobiotics by the white-rot fungus, *Phanerochaete chrysosporium*** 499
H. Sandermann, Jr., D.-H. Pieper and R. Winkler

Preface

Cellulose '91, by any standards, was a monumental International meeting. Three major organisations, with a proven track record in organising cellulose conferences joined forces to celebrate the 50th Anniversary of the Southern Regional Research Center of the United States Department of Agriculture. These were:

- ACS Cellulose, Paper & Textile Division
- Cellucon Conferences
- Syracuse Cellulose Conferences

What could be more appropriate? Celebrating the fifty years of research achievement by the famed Southern Regional in the historic home of cotton - New Orleans. The venue alone ensured a massive turn-out of 500 scientists and camp followers. The meetings were held in the old-style Monteleone Hotel in the French Quarter; dinner was served on the Cajun Queen on the Mississippi River; the daily creole cuisine all provided the stage for a memorable Conference.

The science more than matched the occasion, as this volume testifies. Europe, Japan and the USA combined to provide a high-level cross-section of the world's best cellulose research. The output, set out in this book, confirms the resurgence in cellulose research, development and application. Lignocellulosic sources now offer the most versatile new functional, structural and varied industrial products which are environmentally friendly.

The areas covered in this book support this generalisation:

- Cellulose blends and composites
- Cellulosic enzymes
- New cellulose fibers
- Cellulose fiber-reinforced plastics

- New pulping processes
- High performance derivatives
- New fabrics and finishes
- Biogenesis of cellulose and related polymers
- Cellulosic liquid crystals
- Innovations in cellulose fiber technology
- High performance polymers from lignocellulosics
- Biological degradation

Some years ago I observed the headline in the famous New Orleans daily, the Times Picayune: *KING COTTON RIDES AGAIN*. This would be an appropriate title to this volume, if were not for the fact that a wide variety of cellulosic and lignocellulosic sources are encompassed in the papers. The series of volumes, on cellulosic research, of which this book forms the latest addition, have provided a great stimulus for cellulosic research on the five continents. We all came together in New Orleans in a spirit of friendship and cooperation in the interests of cellulose research. May we thank all who made this memorable meeting possible.

NOELIE R. BERTONIERE, CO-CHAIR CELLULOSE '91

GLYN O.PHILLIPS, CO-CHAIR, CELLULOSE '91

TORE E. TIMELL, CO-CHAIR, CELLULOSE '91

PART 1

1

Research on the delignification process in pulpings

E.I. Germer - VNIIB, Shvernik Ave., 49, St. Petersburg, 194021, Russia.

ABSTRACT

It has been revealed that methods used to state the chemistry of delignification are not sufficient so that out of a great number of reactions of lignin defined while investigating its model compounds to expose those reactions that occur in fact within real pulping and control its stages. To eliminate this defect it has been proposed traditional methods of investigation of the chemistry of delignification supplement with the results of studies of the change of the functional composition of lignin during the real pulping. To obtain representative preparations of lignin it is necessary to use correct methods of extracting it out of liquors and residual wood. Significant differences have been proved in changes of content of the same functional groups in the process of delignification in lignins of liquor and wood residual and also in the lignin extracted in different ways out of liquors in the case of oxygen-alkaline pulping.

INTRODUCTION

Up to now investigations of the chemistry including mechanism of reactions leading to delignification have mainly been based on identification of conversion products of substances modelling lignin main structures "in situ". However the important and necessary information collected only allows one to represent a number of reaction routes illustrating conversion ways of lignin "in situ" that are possible but not obviously realized in real pulping. The information is insufficient to define what of these reactions are determining the separate stages of a real pulping process. Study of products formed during the treatment of protolignin modelling lignin preparations or wood itself also did not confirm the truth of lignin conversion chemistry during pulping because in many cases it was not possible to define lignin structures responsible for either product formation. Besides, the behaviour of lignin preparations during treatment which models pulping, substantially differed from lignin "in situ" behaviour in the pulping. When studying the products of pulping of wood directly, additional problems arise because a wide range of identical aliphatic products form during pulping out of lignin and polysaccharides as well, some products also destruct or form in the pulping liquor, etc.

For the last 10 years the above-mentioned chemical research has included kinetic studies, investigation of reacting particles character, etc. However at present a question on advantages and drawbacks of some (approximately four) usual kinetic models is only at the discussion stage [1]. The use of kinetic research results to determine tne chemistry of delignification is complicated by the fact that delignification is preceded by substantial conversion of lignin in the wood. Kinetic methods can only state the presence of this latent process without any detailed characteristics.

Determination of reacting particle type is very important for defining a mechanism of reaction but gives little information to define delignification chemistry. It relates to the fact that according to the pulping conditions the same reactions of lignin conversion and fragmentation sometimes can be realized both by homolytical and heterolytical mechanisms with the help of different particles.

When a great complexity of the object (lignin), the system where it is located (wood) and the examined process (delignification) overlap, the chemistry of real pulping is more difficult to deterimne than one can imagine even using modern research methods, simulation and computers.

THE CONCEPT AND CONDITIONS OF ITS REALIZATION

However methodology can be suggested which will help to predict from a great number of possible lignin reactions (determined by above-mentioned methods) those ones that practically occur in real pulping and dominate in its separate stages. It can be done if traditional methods of the chemistry will be supplemented by investigation of changes in lignin functional composition during pulping. There are very few papers covering this problem. A great number of research papers are limited to investigation of lignin functional composition at the end of pulping and this is quite insufficient to solve the problem under discussion. At the same time the change in lignin functional composition during the whole pulping process representing the reactions taking place during pulping can make a valuable contribution to the investigation of real pulping chemistry. Such an approach, of course, would be especially successful for those pulping methods where lignin functional composition changes essentially. For example, one of these is oxygen-alkaline wood pulping which will be used to illustrate some aspects of the suggested methodology. This pulping method can be commercially realized due to the invention of Russian original pulping apparatus which successfully solves the problems of chips oxygen-alkaline pulping [2].

There are some conditions that must be fulfilled otherwise the suggested approach can be completely discredeted. Particularly there can be an incomplete or even wrong picture if changes in lignin functional composition will be studied traditionally when only lignin extracted from a liquor is examined. The validity of above-mentioned statements is illustrated by the change of COOH-groups content in the first part of oxygen-alkaline pulping (fig.1). At the final stage of the pulping a different picture is possible: hard extractable lignin of wood residual (LWR) can be converted more deeply than lignin of appropriate liquor. It is illustrated by change of CO-groups content in fig.2.

The mentioned differences in the content of either functional groups in lignin of liquor (LL) and LWR on the same pulping stage may be so great that they will differ not only by value and character of change as it is shown on fig.1 and 2 but even by tendency of change as it is seen from the curves of change in phenolic hydroxyls content (fig.3).

In spite of much information on changes in functional composition of LWR in order to get more complete and true picture of delignification chemistry these data would be better considered together with the results on change in functional composition of LL.

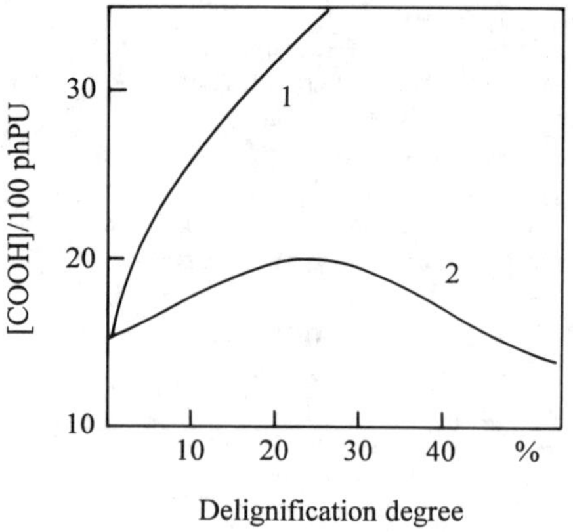

Figure 1. The change of COOH-groups content in liquor (1) and wood residual (2) lignins in the first part of spruce oxygen-alkaline pulping.

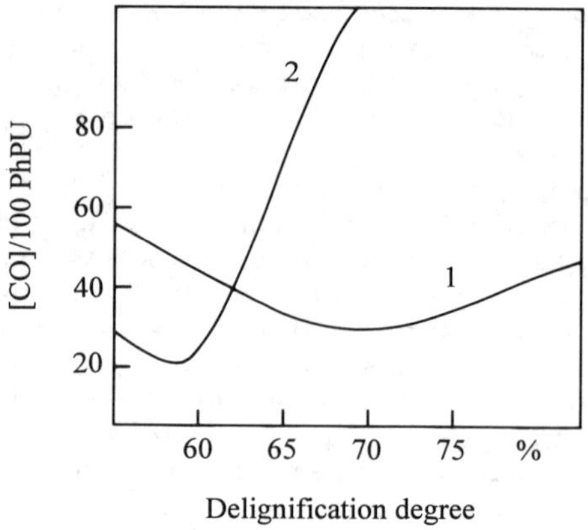

Figure 2. The change of CO-groups content in liquor (1) and wood residual (2) lignins in the second part of aspen oxygen -alkaline pulping.

One more important condition for correct realization of the suggested methodology is a representation of examined preparations of LWR and LL. This condition is mainly guaranteed by correct selection of methods for lignin extraction from wood residual and liquors. Because of partial wood tissue destruction in the wood residual, lignin can be extracted from the latter more completely and in more soft conditions as compared to the extraction from wood. A great part of easy hydrolized links in LWR usually cleaving during lignin preparative extraction is absent in this case because of their cleavage on the earlier stage of pulping.

A separate, very important and often complex problem is to obtain representative preparations of lignin from liquors of different pulping stages. In the case of traditional alkaline pulpings this problem can be easily solved with lignin precipitation by acidification of liquors but for the pulpings causing a substantial destruction of lignin, for example oxygen-alkaline pulping, situation is more complicated. In the latter case a great part of LL accounts for relatively low-molecular water-soluble fraction not precipitatable even in strong acid medium but removable under water washing of lignin from its low-molecular destruction products, carbohydrates and salts. Therefore lignin preparations extracted from such liquors by usual methods will mainly consist of relatively high-molecular fractions with functional composition and other important properties greately differing from real LL composition not only by quantity but by quality as well. The results based on the investigation of these preparations and extrapolated to the whole LL may be unaccurate.

In fig.4 the above statements are illustrated by significant difference not only in methoxyls content but also in tendency of lignin methoxyls changes during oxygen-alkaline pulping when confer acid precipitated lignins and lignins containing low-molecular fractions [3].

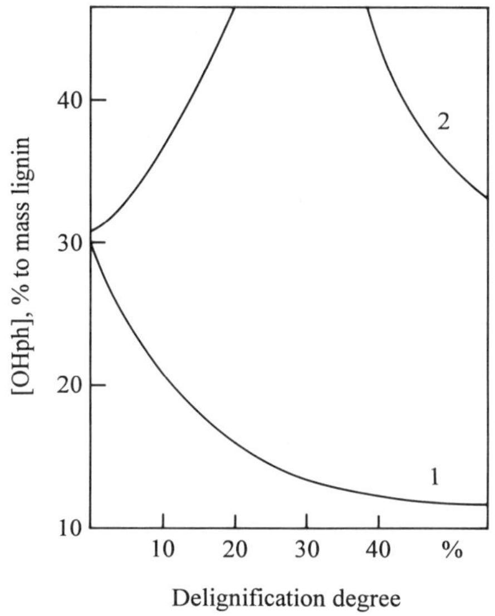

Figure 3. The change of phenolic groups content in liquor (1) and wood residual (2) lignins in the first part of aspen oxygen-alkaline pulping.

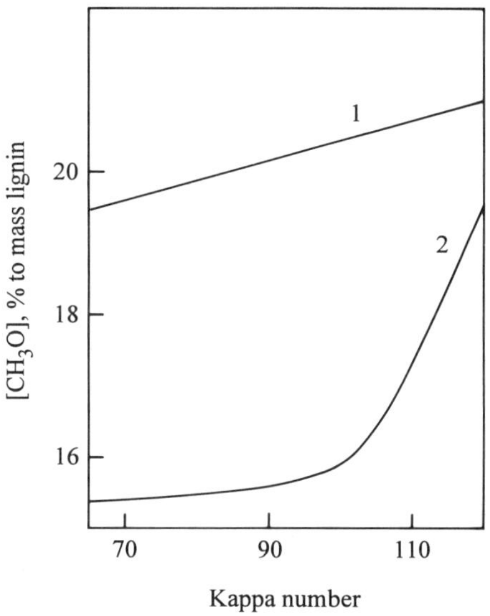

Figure 4. The change of methoxyls content in liquor lignin precipitated mainly as high-molecular fraction (1) and with maintaining low-molecular fractions (2). (Spruce oxygen-alkaline pulping).

Suggested methodology permits not only to define lignin reactions controlling main stages of real oxygen-alkaline pulping but also a number of previously unknown reactions [4], helps with practical recommendations on decrease in oxygen consumption and improvement in yield and some strength properties of pulp [5].

EXPERIMENTAL

Oxygen-alkaline pulpings were performed in batter type swinging system consisting of three steel autoclaves of 860 ml in volume; each was provided by autonomous external electroheating. Wood material - shavings 15*20*(0.5...0.8) mm, original liquor - 10 g/l of sodium carbonate, water to wood ratio - 20, time to cooking temperature 160 C - 40 min, initial oxygen pressure - 1.47 MPa; the oxygen and cooking liquor were added to the autoclaves once before the heating; autoclave free volume to be filled with oxygen - 40%.

Wood and wood residual lignins were extracted by Pepper method in the presence of purified nitrogen. Lignin preparations were isolated from liquors by the developed method which could be used for any pulpings accompanying by appreciable lignin destruction [3].

Methoxyls in lignin preparations were determined according to Ceyzel-Viebock-Shwappach, carbonyls - by oximation at 80 C, carboxyls - by diazomethan methylation following by saponification of methyl ethers, phenolic groups - by differential ionized UV spectroscopy [6] in a slightly modified variant.

REFERENCES

1. Zilberglate M.A., Reznikov V.M. Khimiya drevesiny (Wood Chemistry) 1988, 6, 62.
2. Ioffe L.O., Ivanov Yu.S., Vorobjev Yu.P., Dvoskin Sh.F., Nikiforov I.A. Pulp and Paper International 1989, 12, 50.
3. Shagaev O.V., Germer E.I. Khimiya drevesiny (Wood Chemistry) 1991, 1, 81.
4. Germer E.I., Shagaev O.V. 5-th International Symposium on Wood and Pulping Chemistry. Raleigh, NC, USA. Poster Sessions, 1989, 71.
5. Germer E.I., Shagaev O.V. Bumaghnaya Promyshlennost (Paper Industry) 1989, 2, 8.
6. Zakis G.F. Functional analysis of lignins and their derivatives, Riga, 1987.

2

Regulation of crystallinity of solvolysis pulp by delignification

Y. Yraki, S. Ogawa, Y. Sano, T. Sasaya and M. Takai* - Faculty of Agriculture and *Faculty of Engineering, Hokkaido University, Sapporo 060, Japan.

ABSTRACT

Solvolysis pulps as lignocellulosic material were prepared from conifers with the cresol-water system. X-ray diffractograms showed that the crystalline regions of the pulps maintain the cellulose I structure. The increase in the crystallinity of the pulps resulting from delignification with sodium chlorite was established by X-ray diffraction and ^{13}C-CP/MAS NMR studies. Similar features for commercial KP and SP were observed by the same treatment as that of solvolysis pulps. However, the crystallinity was reduced by the more drastic procedure for delignification. Consequently the delignification conditions seem to be able to regulate the crystallinity.

The reactivity of the pulps was suggested to depend on the crystallinity, because the degree of substitution of the pulp acetates, which were derived with acetic anhydride and perchloric acid under heterogeneous conditions, was closely related to the crystallinity of the pulp.

INTRODUCTION

Organosolv pulping processes as new pulping methods have been investigated to separate wood components efficiently and to eliminate environmental pollution. Solvolysis pulping process with aqueous cresols as a cooking solvent seems to be one of the most suitable methods, since phenol analogs such as cresol are expected to be self-supporting from lignins by pyrolysis, hydrogenolysis and so on, and sulfur compounds and strong acid catalysts is not used.

Physical properties of the solvolysis pulp are slightly weaker than KP owing to the higher content of residual lignin in the solvolysis pulp. However, when a small amount of acetic acid is added to the cooking liquor in order to decrease lignin content, the properties are improved remarkably [1]. Thus the pulps are comparable to KP. The solvolysis pulp is also reported to be readily bleached [2], but the physicochemical properties of the bleached pulp is not clear yet.

In this study, the depolymerization, the crystallinity and the affect of chemical modification on the delignification process with sodium chlorite are examined.

EXPERIMENTAL

200g of chips prepared from fir (*Abies sachalinensis* Masters) and larch (*Larix leptolepis* Gord) were cooked with cresols-water (1:1 and 7:3) mixture in 1.8-L autoclave at 180℃ for 2 h. Heat-up time was 1 h and the ratio for liquor to wood was 5 to 1. After cooking, pulps were filtered and then washed with acetone, water, alcohol and ether followed by drying *in vacuo*. The pulps of *Abies sachalinensis* with cresols-water (1:1) and (7:3) were named A1:1 pulp and A7:3 pulp, respectively. L1:1 pulp and L7:3 pulp were used for the *Larix leptolepis* pulp.

Delignification of the pulp was conducted under two conditions. The procedure under mild conditions was as follows:- 125 mL of 0.1 M aqueous $NaClO_2$ solution (1.13 g) was added to 2.5 g of pulp in 125 mL of 0.1 M acetate buffer (pH4) and then the suspension was left for 2 h at 40 ℃ with occasional shaking, and this process was repeated twice. The other one under severe conditions is according to the Wise method [3]. Briefly, to 2.5 g of pulp in 150 mL of water were added 1 g of $NaClO_2$ and 2 mL of 10 % aqueous acetic acid, and left for 1 h at 70-80 ℃, and this process was repeated 4 times.

X-ray diffractograms of the pulps were obtained by the reflection method. Relative crystallinity was estimated by means of X-ray diffraction based on the peak area of copper as internal standard [4].

Degree of polymerization was measured by a viscometric method [5,6].

RESULTS AND DISCUSSION

Properties of solvolysis pulp

Larch chips were cooked more readily than fir chips by solvolysis with the cresol system, since the residual Klason lignin (KL) content of larch pulp was smaller than that of fir as shown in Table 1. This difference might be attributed to the composition of hemicellulose. Both softwoods contained about 63 % of α-cellulose in holocellulose. When they were cooked, their pulps had higher content of α-cellulose (more than 85 %) and a small amount of KL except for A7:3 pulp delignified under severe conditions. Solvolysis pulp is expected to use as dissolved pulp. Particularly delignified L7:3 pulp might be utilized for the purpose without further purification because of little KL value.

Table 1. Effect of delignification with sodium chlorite on klason lignin and α, β - cellulose of solvolysis pulp.

sample		KL[1] (%)	α, β -cellulose[2] α	β	sample		KL (%)	α, β - cellulose α	β
Abies	wood meal	29.5	–	–	*Larix*	wood meal	29.4	–	–
	holocellulose	3.8	63.2	0.3		holocellulose	4.5	63.1	0.1
A 1:1 pulp	intact pulp	13.6	–	–	L 1:1 pulp	intact pulp	4.1	–	–
	40 ℃ once[3]	8.3	86.8	1.3		40 ℃ once	2.2	89.4	1.8
	40 ℃ twice[4]	7.7	88.1	3.1		40 ℃ twice	1.6	87.7	0.0
	70~80 ℃[5]	1.0	85.0	1.9		70~80 ℃	0.1	85.0	1.7
A 7:3 pulp	intact pulp	15.9	–	–	L 7:3 pulp	intact pulp	4.5	–	–
	40 ℃ once	12.9	84.6	2.6		40 ℃ once	3.2	85.5	0.4
	40 ℃ twice	12.7	85.9	0.1		40 ℃ twice	2.3	83.9	0.2
	70~80 ℃	3.3	78.7	2.4		70~80 ℃	0.0	85.8	4.8

1): % of pulp. 2): % of pulp excluding KL.
3): Pulp was delignified by sodium chlorite in acetate buffer (0.1 M, pH 4) at 40 ℃ (Mild Conditions).
4): Pulp was delignified twice at 40 ℃ (Mild Conditions).
5): Delignification was carried out at 70~80 ℃ (Severe Conditions).

Delignification ratio and degree of polymerization (DP) were plotted against the concentration of sodium chlorite in which the concentration represented summarily stepwise addition of sodium chlorite in two conditions (Fig. 1). The delignification of L1:1 pulp was achieved completely by twice the addition of sodium chlorite under severe conditions ([$NaClO_2$] = 14 g/L). The DP was observed to decrease drastically to 7 g/L followed by a gradual decrease. This indicates that hydrolysis of the pulp is caused by the acidic conditions at elevated temperature together with the proceeding of delignification.

Significant increase in crystallinity of the solvolysis pulp was found by X-ray studies in the early stage of the delignification as shown in Fig. 2. ^{13}C-CP/MAS NMR also suggested the same result (data not shown). But after complete delignification the crystallinity was slightly decreased. A similar feature was observed for commercial unbleached KP and SP.

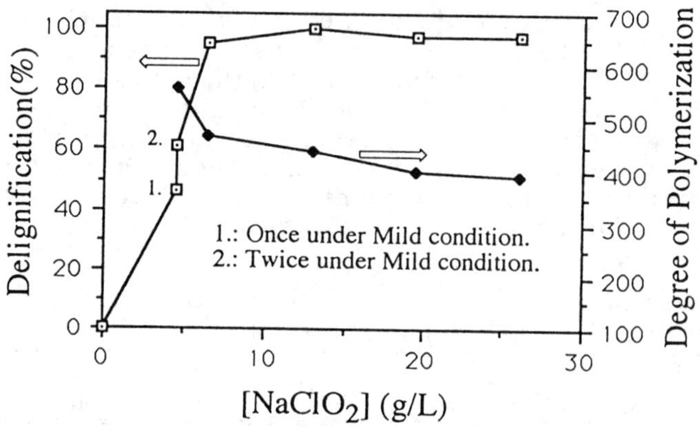

Figure 1. Relationship between delignification and degree of depolymerization of solvolysis pulp (*Larix* 1:1) with sodium chlorite.

Delignification
Degree of Polymerization

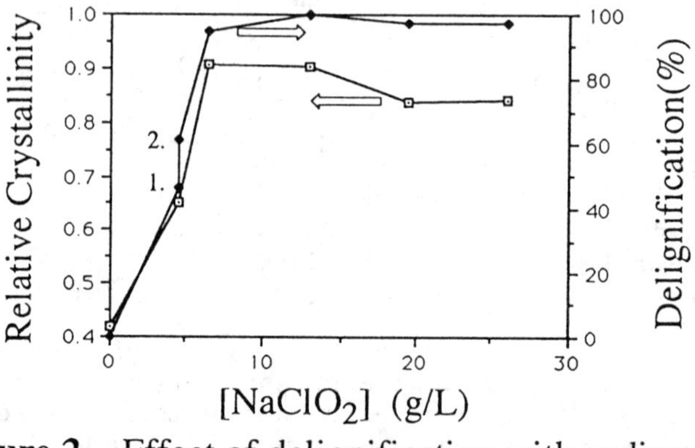

Figure 2. Effect of delignification with sodium chlorite on relative crystallinity of solvolysis pulp (*Larix* 1:1).

Crystallinity of *Larix* Pulp(1:1)
Delignification

From these results the following phenomena for the chemical pulps might be deduced on delignification. In the early stage of the delignification process, amorphous regions of pulp are destroyed by the acidic medium, resulting in an increase in the crystallinity. Subsequently crystalline regions collapsed slightly by excess delignification to decrease DP and crystallinity.

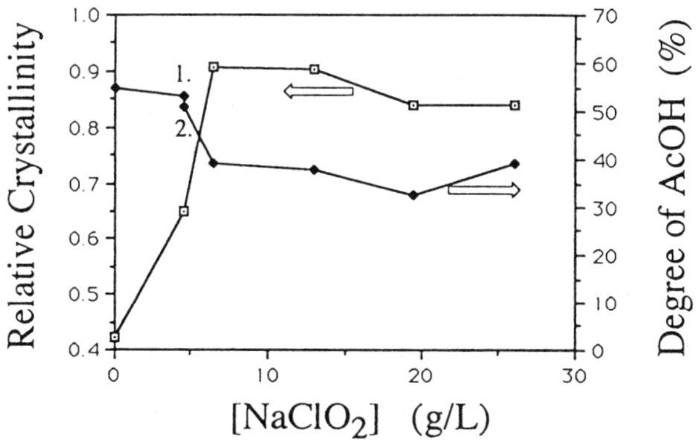

Figure 3. Relationship between relative crystallinity and degree of AcOH of solvolysis pulp (*Larix* 1:1) with sodium chlorite

—□— Crystallinity of *Larix* Pulp(1:1)
—◆— Degree of AcOH

Reactivity of solvolysis pulp

Acetylation of solvolysis pulp was performed to investigate the reactivity of the pulp. Pretreated pulp with acetic acid was suspended in AcOH-toluene-60%$HClO_4$ (in which perchloric acid was a catalyst) followed by reaction with acetic anhydride for 30 min. Fig. 3 shows that the degree of AcOH is high at low crystallinity and declines with elevating crystallinity. When the crystallinity was reduced, the degree of AcOH increased. Consequently the reactivity of the pulp seems to be closely related to its crystallinity.

Completely acetylated pulp (Degree of AcOH = 63.5%) dissolved in methanol-dichloromethane (1:9) mixture even if the pulp had more than 10% KL but did not dissolve in chloroform. The resulting film prepared from the solution indicated characteristics of both cellulose and lignin, for example UV adsorption. It seems that solvolysis pulp may be useful as an industrial material.

These results suggested that it is possible to regulate the crystallinity and reactivity of solvolysis pulp by delignification.

ACKNOWLEDGEMENT

A part of this work was supported by a Grant-in-Aid for Scientific Research (03760106) from the Ministry of Education of Japan.

REFERENCES

(1) Sano, Y.; Endo, M.; Sakashita, Y. *Mokuzai Gakkaishi* 1989, *35*, 807.
(2) Takagi, H.; Kachi, S.; Kawabata, H.; Sandstrom, P. *Japan Tappi Journal* 1989, *43*, 1290.
(3) Wise, L. E.; Murphy, M.; D'Addieco, A. A. *Paper Trade J.* 1946, *2*, 35.
(4) Watanabe, S.; Takai, M. *Kogyo Kagakuzashi* 1968, *71*, 1883.
(5) J. TAPPi No. 44-84.
(6) Wetzel, F. H.; Elliott, J. H.; Martin, A. F. *TAPPI* 1953, *35*,564.

3

Pulp production by catalytic oxygen-alkali-alcohol (OAA) delignification

E.I. Germer and R.A. Gorenkov -VNPObumprom, Shvernik Ave., 49, St. Petersburg, 194021, Russia.

ABSTRACT

Simultaneous use of oxygen, alkali and alcohol (mainly AlkOH:H_2O = 50:50) as pulping reagents permits integration of the advantages of oxygen-alkaline (OA) and alkaline-alcohol (AA) pulping methods into one process while decreasing drawbacks: high alkali discharge in the AA method, high oxygen pressure and hydromodule in OA pulping. Ethanol, propanol and butanol are used as alcohols (methanol can be used as well). The presence of oxygen and its higher solubility in alcohols provide effective use of 1,10-phenanthroline, the catalyst of OA wood delignification. Strength properties of OAA pulp and sulphate pulp are the same. OAA pulping can be successfully realized in a unique pulsating apparatus developed by scientists of Russia for practical realization of OA pulping.

Principal advantages of oxygen-alkaline (OA) pulping are related to the use of oxygen: high delignification selectivity, low concentration of alkali in pulping liquor, high pulp brightness. However the use of oxygen causes difficulties in realization of this method.

Alkaline-alcohol (AA) pulping has great advantages in view of delignification selectivity and strength properties of pulp, the possibility of separate treatment of lignin and carbohydrates and realization at a very low hydromodule. The drawbacks include relatively high NaOH content in liquor, insufficient selectivity, low pulp brightness which requires increase in bleaching stages number, additional consumption of chemicals that result in increase in bleaching effluents load and toxicity.

The comparison of the above-mentioned advantages and drawbacks of examined methods clearly shows that advantages of each method in principle can level the drawbacks of the other one; much higher oxygen solubility in alcohols and water-alcohol mixes than in water. On the basis of these principles we chose the conditions of combined oxygen-alkaline-alcohol (OAA) pulping (see table 1).

Table 1. Comparison of conditions for oxygen-alkaline & alkaline-alcohol pulpings with conditions chosen for oxygen-alkaline-alcohol pulping.

Pulping parameter	Pulping		
	OA	AA	OAA
Alcohol content in pulping liquor, % by vol.	-	30-70	50
NaOH content in pulping liquor, g/l	7-10	20-30	10
Hydromodule (ratio liquor: wood)	20*	10*	10*
Oxygen pressure in digester before pulping, MPa	2*	-	0,5*
Pulping temperature, °C	140-160	170-180	160

* The values are based on insufficient mass transfer in the pulping plant. In the case of more intensive mass transfer these values can be lower.

The results in table 1 show that such important parameters as NaOH content, hydromodule, OAA pulping temperature are lower as compared to the original methods and the oxygen pressure is 4 times lower in comparison with OA pulping.

It will be noted that new variants of oxygen pulping in monobasic alcohol medium (1) and OA pulping in water-alcohol medium (2) do not allow one achievement of the goal because in the first case the pulp lacks strength properties and delignification selectivity is low and in the second case alcohol is not added sufficiently to decrease high oxygen pressure (2 MPa) and only pulping liquor containing a relatively low amount of alcohol (25% by volume) is used. At the same time the increase in alcohol amount (particularly up to 50% by volume) at an oxygen pressure of about 2 MPa does not exclude the possibility of gas phase explosion during pulping.

As can be seen from table 2, under the chosen conditions using pulping liquor containing 10 g/l NaOH the yield of good quality aspen pulp is as high as 59...62% at Kappa number 18...25. The yield of hard pulp (Kappa number 34...39) is of 64...65%. The cooking time for the pulp of average hardness is only 2...3 hours and for hard pulp is 1,5...2 hours. Increase in alkali concentration in pulping liquor to 20 g/l results in significant acceleration of accompanying delignification, but substantial decrease in pulp yield. Consequently, in the case of OAA pulping even the use of low alkali concentration that is typical for AA pulping is not only unsuitable but inadmissible.

The same intensive acceleration of delignification accompanied by unacceptable intensive destruction of polysaccharides is observed during oxygen pulping with water-ethanol solution without alkali, i.e. during oxygen-alcohol pulping mentioned above (1).

The increase in alcohol content from 50 to 70% in oxygen-alcohol pulping without base causes retardation of delignification with some increase in selectivity. In the case of OAA pulping the same increase in alcohol content not only retards the delignification but results in substantial selectivity decrease (cf. pulpings 5 and 9) that is undesirable.

For the pulping 10, oxygen pressure was increased as much as twice. The rate of delignification had increased (cf. pulping 10 and 5) while the selectivity of delignification was maintained or slightly decreased. Therefore a substantial increase in oxygen pressure as compared to the OAA pulping is probably unsuitable.

The pulping 11 is a typical OA cooking performed under conditions chosen for OAA pulping. The comparison of results for this pulping and pulping 4 shows that the rate and selectivity of delignification for the OA pulping is

Table 2. Oxygen-ethanol aspen pulpings with different NaOH contents in pulping liquor

Parameters	NaOH content in water-ethanol pulping liquor, g/l													
	0			10								20		
Pulping NN	1	2	3	4	5	6	7	8	9	10	11	12	13	14
Pulping time, min	69	127	129	79	101	119	150	185	102	96	89	38	60	82
Yield, % o.d. wood	56,3	49,9	60,7	67,1	65,3	64,2	62,5	59,4	59,6	59,9	64,1	54,9	52,6	51,1
Kappa number	28,5	11,9	34,0	47,5	39,0	33,9	25,3	18,3	46,3	22,5	56,4	32,4	31,8	20,9
$\overline{DP}$	...	...	...	...	1070	993	945	...	...	...	...	...	...	...
Liquor pH	2,9	2,8	3,2	6,2	...	6,1	6,0	5,9	6,3	5,9	5,2	9,3	9,2	7,8
Differing parameter	-	-	EtOH 70%	-	-	-	-	-	EtOH 70%	O_2 0,98 MPa	EtOH 0%	-	-	-

The pulpings differ from main pulpings of this series by the value of one parameter.

much lower.

AA pulping when being performed under conditions of OAA pulping (using 10 g/l NaOH) practically is not realized at all. The comparison between ethanol AA and OAA pulpings with a typical for the cooking liquor containing NaOH (20 and 10 g/l respectively) indicates that alkaline-ethanol pulpings are more rapid. However OAA pulping maintains higher selectivity of delignification (the yield is higher in 3...5% o.d. wood), higher brightness and lower alkali consumption. We think that it is advisable to realize OAA pulpings as independent alternatives to environmentally safe pulping or as a component of sulfur free stage (especially considering that to apply oxygen pressure of about 0,5 MPa is much easier than at oxygen pressure of 1...2 MPa in OA pulping).

OAA pulpings using propanol and butanol also allow production of extensively delignified pulp in an acceptable period and with sufficiently high yield. The comparison of OAA pulpings shows that delignification of pulpings using EtOH is more intensive (approximately 1,3 times) but as pulping extends this difference is levelled. The rate of delignification decreases as alcohol molecular mass increases. Selectivity of delignification obviously decreases from EtOH to BuOH and as delignification becomes more extensive the difference is more evident. It relates rather to substantial statistical destruction of cellulose than to its peeling as is indicated by pulp $\overline{DP}$ decrease in 1,5...2,4 times when switching over from ethanol to butanol.

Extrapolation of our results to methanol allows one to assume that the results for OAA pulpings with methanol will be higher compared to the pulpings with ethanol.

In Russia the investigation of delignification chemistry in OA pulping of wood allowed one to discover the most effective catalyst of OA delignification - 1,10-phenanthroline (below phenanthroline) providing the acceleration of OA pulping in 1,8...2 times and over when 0,3...
...0,5% o.d. wood of catalyst is used (with some increase in selectivity of delignification and pulp strength properties (3)).

When using all three alcohols the addition of phenanthroline results in a significant acceleration of OAA pulpings. For OAA pulping using EtOH its average values are 1,4...1,7 with pulp Kappa number of 25...37. Addition of phenanthroline in OAA pulpings using propanol and butanol indicates that in these cases delignification acceleration under phenanthroline action is much higher than 2, i.e. higher than under its action in OA pulping.

When producing pulps within a wide range of Kappa numbers (from 11 to 39) the same selectivity of ethanol OAA pulpings with and without phenanthroline addition is observed. But when switching over to OAA pulpings using propanol and butanol the picture is different. In these cases delignification selectivity of pulpings with phenanthroline addition increase substantially: the yield of pulps having similar Kappa number is higher in 5...7% o.d. wood and $\overline{DP}$ is higher above in 2.

It is interesting to note that phenanthroline addition almost completely levels the difference in delignification rate for pulpings using various alcohols mentioned above for noncatalyzed OAA pulpings.

The addition of phenanthroline in OA pulping performed under the same conditions as OAA pulpings shows much lower effect than, for example, for OAA pulping using EtOH. The selectivity of OAA pulpings with phenanthroline addition is much higher than for comparable OA pulping: the yield is higher by 5...7% and $\overline{DP}$ - in 1,3...1,8 times. In OAA pulpings this was achieved due to the presence of alcohol better dissolving the oxygen.

Table 3. Strength properties of oxygen-alkaline-alcohol (EtOH) aspen pulp.

Pulping method	Pulp yield, % o.d. wood	Kappa number	Breaking length, m	Tearing strength, mN	Bursting strength, kPa
				tearing strength index, mN*m^2/g	bursting strength index, kPa*m^2/g
OAA	55,3	15,2	7700	402/5,74	398/5,69
OAA with 1,10-phenanthroline addition	56,4	18,5	8200	407/5,81	460/6,57

Freeness degree 60...65°SR, basic weight 70 g/m^2.

When producing OAA pulp (with EtOH) for strength test it was decided to use chips of 1...2 mm in thickness instead of shavings as for all pulpings and to increase NaOH content in the pulping liquor from 10 to 15 g/l. Strength properties of the OAA pulp produced are presented in table 3. They are high enough and on the level of aspen kraft pulp. The addition of phenanthroline improved them to some extent. It ought to be considered that chosen pulping conditions and process instrumentation for its realization were not optimized.

The pulpings were performed in swinging steel autoclaves, each of 860 ml volume, provided with electroheating jackets. Aspen shavings of 0,5...0,7 mm in thickness were used. Oxygen was added once before pulping. The autoclave was filled with oxygen in 86%, when producing pulp to determine strength properties - in 59%. Other OAA pulping conditions are presented in table 1. Pulp $\overline{DP}$ was determined on the basis of characteristic viscosity of pulp solutions in ferrumtartaric-sodium complex (FeTNa).

REFERENCES

1. Deineko I.P. Chemical conversion of lignin during wood delignification by oxygen, Dis. Prof. in Chemistry Science, Leningrad, 1990.
2. Ivanov Ju.S., Ioffe L.O., Apushkinsky A.G., Sergeeva V.V., Lukanina L.K. Some characteristic features of oxygen-alkaline pulping in the presence of organic solvents. Chemical processing of wood and wood wastes. Papers on Science, LTA, Leningrad, 1987.
3. Germer E.I. The 6-th Int. Symposium on Wood and Pulping Chemistry. 1991, Melbourn, Australia, Vol.2, 143.

4

Oxygen-alkali pulping of chips

Yu.S. Ivanov, L.O. Ioffe, Yu.P. Vorobiev, Sh.F. Dvoskin, I.A. Nikiforov, A.B. Broderson - VNPOBumprom, Shvernik Ave., 49, St. Petersburg, 194021, Russia.

ABSTRACT

We developed a laboratory and pilot plant process for the oxygen-alkali delignification of wood chips . It was shown that a pulsation apparatus is suitable for oxygen-alkali pulping. After bleaching the quality of oxygen-alkali pulps from both hardwood and softwood moves closer to that of bleached kraft pulps. The production of hardwood bleached pulps is probably chlorine-free.

INTRODUCTION

The aim of the present paper is to obtain information about the realization of an oxygen-alkali pulping process and its potential to enable high-quality bleachable pulps to be produced. The problems of the pulsation apparatus to produce homogeneously cooked pulps from chips are dissoused in a previous paper [1]. We tested two technology alternatives. They are distinguished by the presence in one and the absence in the other of alkaline chips pretreatment. In order to obtain comparable data on pulp quality the kraft and oxygen-alkali pulpings of aspen, spruce and pine were performed. The laboratory plant for oxygen-alkali delignification has been examined earlier [2].

RESULTS AND DISCUSSION

Laboratory study results are presented in Table 1.

Table 1. Characteristics of kraft and oxygen-alkali pulps.

Cook N	Cooking	Wood	Temperature °C	Time min	Yield %	Kappa number	Brightness %
320	O/A 1	Aspen	150	140	64.5	21.4	60.4
352	O/A 2	Aspen	140	180	60.5	18.9	58.0
322	O/A 1	Aspen	150	160	58.7	10.0	66.0
363	O/A 2	Aspen	140	170	58.4	12.5	59.0
1	Kraf	Aspen	165	90	54.1	12.0	34.6
295	O/A 2	Spruce	140	180	48.5	25.7	35.7
329	O/A 2	Pine	150	160	46.1	22.1	36.0
2	Kraft	Spruce	170	90	46.8	25.8	27.1
3	Kraft	Pine	170	90	44.6	26.2	25.8

Note:

1. Oxygen-alkali pulping 2 (O/A 2) - with alkali pretreatment of chips;
oxygen-alkali pulping 1 (O/A 1) - without alkali pretreatment.
2. Pulsation intensity at oxygen-alkali cooking - 2500 mm/min, P_{O2}=1.5 MPa, flow oxygen across autoclav - 1.2 dm^3/min.

From Table 1 can be seen that oxygen-alkali pulp yield is higher than for kraft pulping by 4.3-10.4 %. The aspen oxygen-alkali pulp has the highest yield and alkaline pretreatment decreases pulp yield.

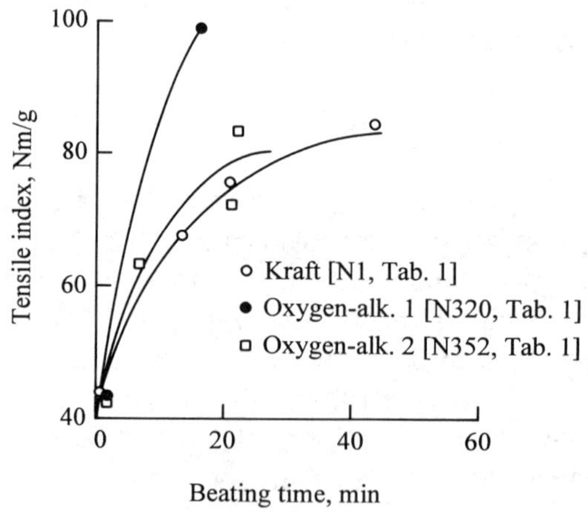

Fig.1. Tensile index VS beating time (aspen pulps)

Fig.1 shows aspen pulp samples breaking strength versus beating time in GRA mill. For the same refining time oxygen-alkali pulp 1 has largest breaking strength.Oxygen-alkali pulp 2 breaking strength at the same beating time is slightly higher than for kraft pulp. Such a relationship is observed also for softwood pulps (Fig.2).

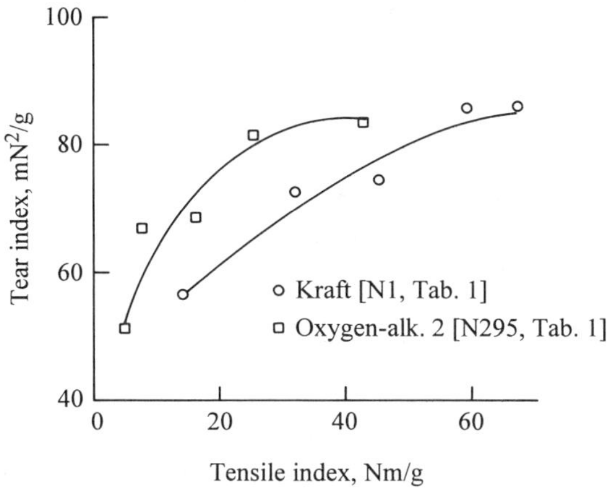

Fig.2. Tensile index VS beating time (spruce pulps)

Fig.3 shows the curves of aspen pulps tearing strength At the same breaking strenght oxygen-alkali pulp 1 has lower tearing strength than kraft pulp.
Alkali pretreatment significantly increases aspen oxygen-alkali pulp 2 strength. However softwood kraft pulp appears to be stronger than oxygen-alkali pulp 2.

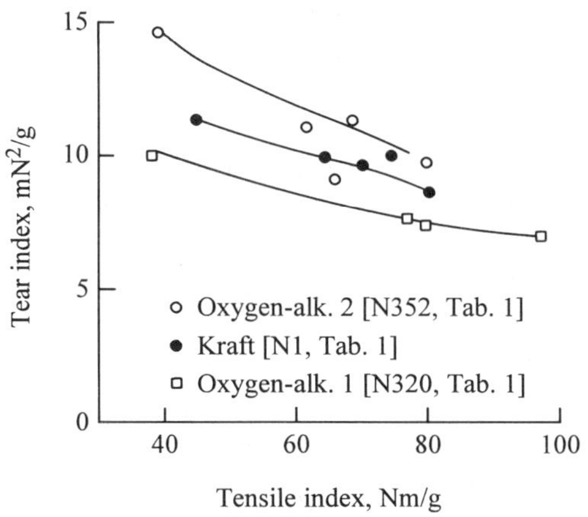

Fig.3. Tear index VS tensile index

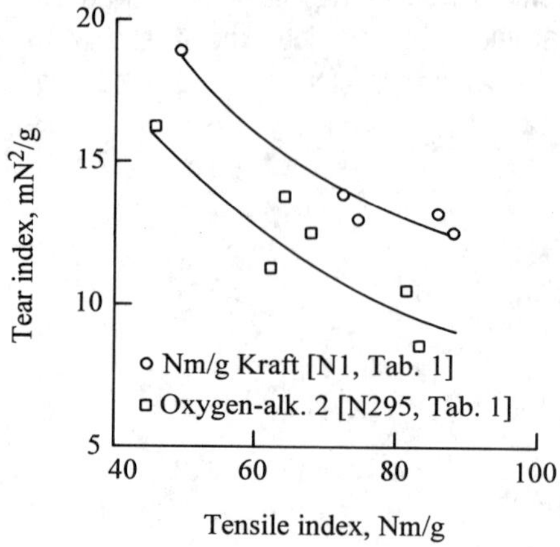

Fig.4. Tear index VS tensile index (spruce pulps).

As seen from Table 1 unbleached oxygen-alkali pulps have higher brightness than kraft pulp by 10-30 % abs. (versus wood species). It causes high oxygen-alkali pulp bleachability .Hardwood pulp is bleached up to 86-88 % without elementary chlorine and its compounds (3). It is greatly important that during bleaching strength does not become impaired.

Softwood oxygen-alkali pulp bleaching in the sequence C/D-Eo-D-E-D (3) gives pulp with brightness of 89 %. The level of strength of bleached pulps is closer to kraft softwood bleached pulp. Nowadays the bleaching sequence of hardwood and softwood pulps with ozone, peracids and hydrogen peroxide are being examined.

CONCLUSION

Due to utilization method and development of oxygen-alkali pulping technology by chips pretreatment and parameters optimization we managed to produce two hardwood pulps. The first of the yield (by 6-8 % higher than a kraft one) has high brighness and satisfactory strength. The second of the yield by 2-4 % higher than for kraft one is closer to kraft pulp in tearing strength. The strength of softwood oxygen-alkali pulp at higher yield is inferior to kraft pulp. Hardwood oxygen-alkali pulp is capable of bleaching without chlorine with its compounds maintaining high strength values. Based on oxygen-alkali chips delignification we may develop ecologically sound sulfur - and chlorine- free bleached pulp production.

REFERENCES

1. Ivanov Yu.S., Ioffe L.O., Vorobiov Yu.P.. Dvoskin Sh.F., Nikiforov I.A. Paperi ja Puu 1991. N10, p.940-945

2. Broderson A.B., Vorobiov Yu.P. The World Pulp and Paper Environmental Conference Leningrad, USSR, 1991, April 16-17 .

3. Ioffe I.O., Ivanov Yu.S. The World Pulp and Paper Environmental Conference Leningrad, USSR, 1991 , April 16-17.

5

Production of cellulose from Brazilian lignocellulosic residues

S.P.D'O. Barreto*, E.H.M. Melo*, J.L. Lima Filho*, A.C. Duarte Coelho and J.F. Kennedy***** - *Laboratorio de Biotecnologia Depto. Bioquimica/LIKA Universdade Federal de Pernambuco (UFPE), Cidade Universitaria, Recife-PE, 50670-901, Brazil. **Depto. Engenharia Quimica, UFPE, Recife-PE, Brazil. ***Research Laboratory for the Chemistry of Bioactive Carbohydrates and Proteins, School of Chemistry, University of Birmingham, Birmingham, B15 2TT, U.K.

INTRODUCTION

Brazilian paper and cellulose industries generally utilize wood (e.g. Eucalyptus and Pine) to produce white cellulose paste which has been used for paper and cellulose derivatives production. Normally, industries that do not utilize white cellulose paste as raw material for industrialized "kraft" paper, have required six tonnes of bagasse to yield one tonne of this paper. The potential of bagasse as waste, considering the updated state of the distilleries technology for fuel alcohol production, are of approximately 7.5 million tonnes/year for the whole of Brazil (Koga, 1988).In the case of Pernambuco State, this amount of bagasse as waste is around 0.6 million tonnes/year. The utilization of regional raw materials, such as sugar cane bagasse, banana tree, sisal, bamboo, etc., as alternative sources for the production of cellulose derivatives, would decrease considerably the use of wood for this purpose and also would diminish tree cut-down.

This work describes a modification of the classic chemical method for pulping and bleaching processes to provide the new and successful processes to the world with a reality of reduction of pollutants in the rivers and preserving the environment.

MATERIAL AND METHODS

Material:

Sugar cane bagasse (*Saccharum officinarum*) and banana tree (*Musa esplendida)* from the North East of Brazil. Cellulose from Sigma (St, Louis, U.S.A.), urea 45% (Nitrofertil, Brazil), sodium hydroxide 50% (w/v), hydrogen peroxide 2% (v/v). All chemicals are commercial grade.

Methodology:

Cellulose Paste Preparation:

Lignocellulosic materials employed in this work for the production of cellulose were dried at 100°C for 24h and had their water content determined. In the next step two types of chemical pulping were carried out: 1 - classical alkaline treatment (D' Almeida, 1988); 2 - modified treatment. In the modified treatment, 400 g of urea were added to 50 l of sodium hydroxide 16% in a 140 l vertical boiler. To this solution 10 kg of either bagasse or banana tree containing 50% of water were added, and cooked at 150°C under 7 atm pressure with gentle stirring for 2 hours.

Bleaching Process:

The classical bleaching process (D' Almeida, 1988) was modified as follows:

First step: Hydrogen peroxide (final concentration 0.5%, 5 ml), was added to a mixture of 1M sodium hydroxide (100ml) and ethanol 92°GL(20ml) and cooked cellulose (1g). The mixture was allowed to react for one hour at 80°C, pH 10-11, under gentle stirring.

Second step: Chlorine dioxide (0.5g) was added to the product (1g) and the mixture was allowed to react at 60°C, pH 5-6, for one hour.

Third step: Sodium hypochlorite (100ml) was added to the product and the mixture was allowed to react at 60°C, pH 8-11 for two hours.

RESULTS AND DISCUSSION

The great amount of pollution caused by the paper and cellulose industry effluents, has directed researches in this field to search for the unique way that will reduce pollution. In this work, two modifications to cellulose production were tried in order to diminish pollutant materials. The first modification was introduced during the pulping process to remove more lignin that can be easily oxidized before its disposal in the rivers. So, with less lignin coating the cellulose, the amount of chlorine-based reagents could be reduced during the bleaching process. The second modification was the substitution of the chlorine normally used in the first step of the process, by hydrogen peroxide. This allowed a reduction on a laboratory scale of the amount of chlorine dioxide from 0.6 g/g of cellulose paste to 0.5 g/g of cellulose paste.

Table 1 shows that pulping of bagasse by the classical chemical method used by the industries in Pernambuco, removed 58 ± 1.2% of lignin. However, when urea was added in the pulping process, approximately 68.0±1.4% of lignin was removed.

Table 1 - Chemical properties of cellulose paste obtained from sugar cane bagasse.

Sample	Lignin klason (%)	Kappa N°	α-cellulose (%)	Pentosan (%)
Native	27.8±0.6	----	----	----
Treated Classical method	11.6±0.6	31.8±0.5	72.9±0.4	10.0±0.3
Treated modified method	8.9±0.2	24.0±0.6	78.0±0.3	7.7±0.1

The same behaviour was observed when trunks of banana trees were treated with and without urea as shown in table 2. The classical pulping treatment in which urea was not added, removed 65.2 ± 1.25%, while an increase to 80.4 ± 1.15% in lignin removal was obtained with the modified pulping treatment in which urea was added. Miyazaki *et al.* (1989) have found similar behaviour with the use of compounds containing amine groups. In both cases the amount of α-cellulose increased, as expected, when urea was used during the pulping process.

The structure of the fibres in native banana tree with and without urea treatment as well as bleached fibres were analyzed by scanning electron microscopy (see figure 1). A comparison between figures 1A and 1B shows that lignin has been removed when urea was not present during pulping. However, the presence of urea in the modified method was found to be more effective than the classical method of removal of lignin (see figure 1C and compare it to figures 1A and 1B).

Figure 2A shows the fibres of cellulose after bleaching, being then very similar to those fibres of commercial cellulose (see figure 2B), except in terms of size of the fibres. The

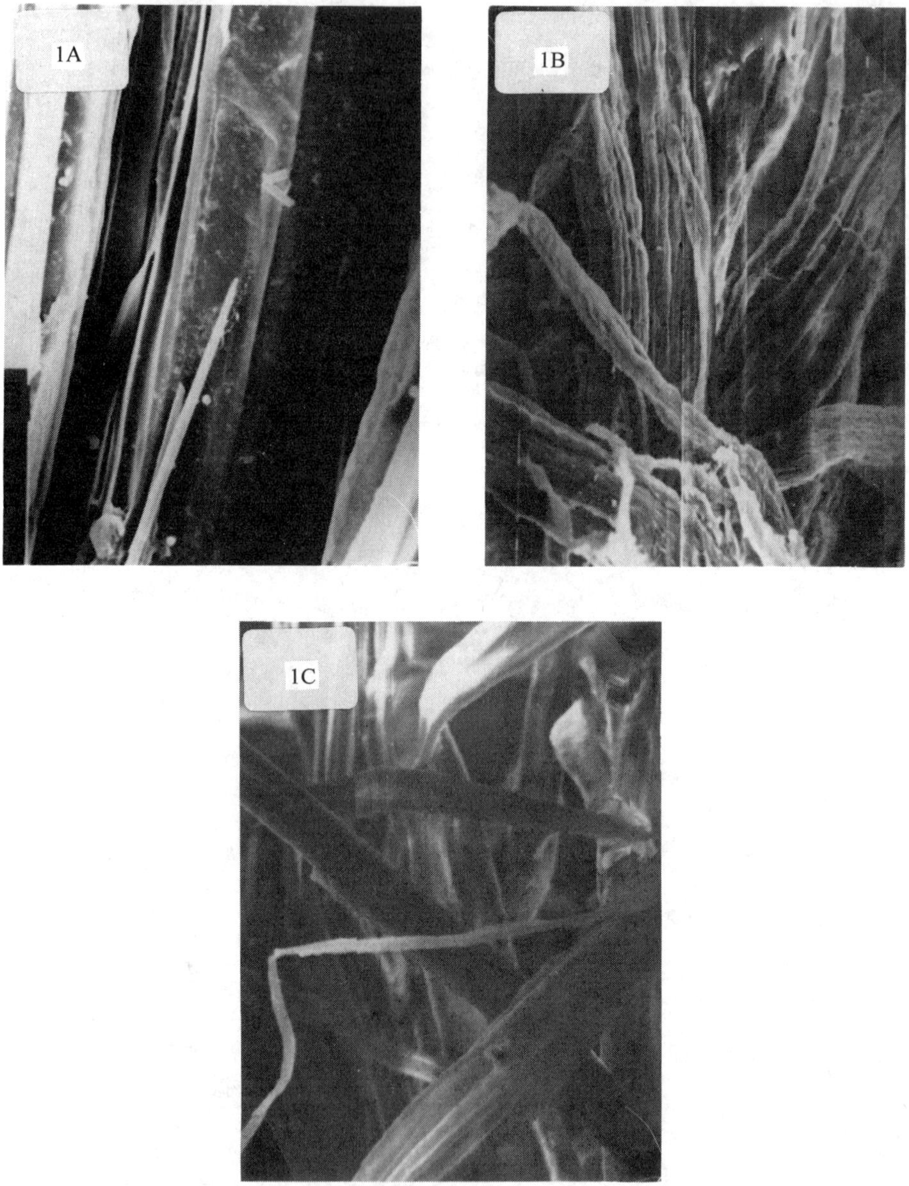

Figure 1 - Scanning electron microscopy (x 750) of a banana tree trunk : A - Native; B - Treated by the classsical method; C - Treated by the modified method.

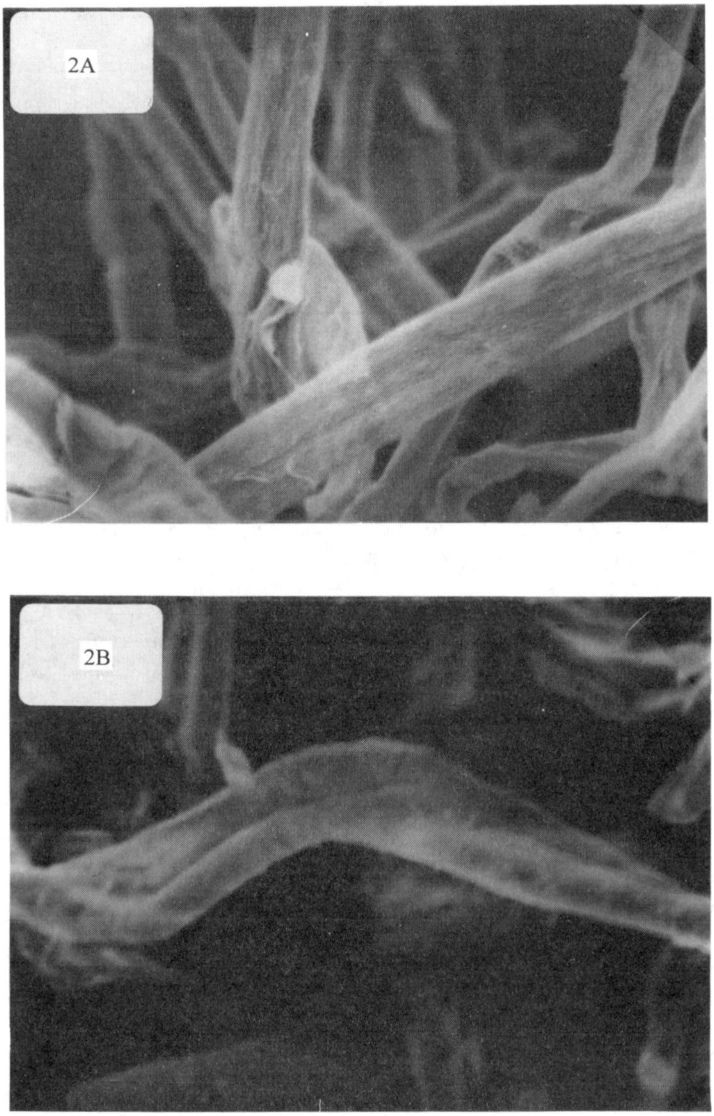

Figure 2 - Scanning electron microscopy (x 750) of: A - Bleached cellulose from banana tree; B - Bleached commercial cellulose.

Table 2 - Chemical properties of cellulose paste obtained from trunk of banana tree.

Sample	Lignin klason (%)	Kappa N^{o}	α-cellulose (%)	Pentosan (%)
Native	18.4±0.4	----	----	----
Treated Classical method	6.4±0.2	33.7±0.3	75.2±0.5	8.4±0.2
Treated modified method	3.4±0.1	23.5±0.1	83.2±0.4	4.8±0.1

fibre sizes of the commercial cellulose seem to be smaller than those of banana tree (see figure 2A).

CONCLUSION:

- The presence of urea in the pulping process seemed to be effective in removing increased amounts of lignin for both sugar cane bagasse and banana tree.

- The substitution of chlorine as the first step of the bleaching process by hydrogen peroxide, and the reduction in the amount of chlorine dioxide in second step did not decrease the cellulose brightening.

REFERENCES:

1 - D' Almeida, M.L.O. (1988). Celulose e Papel. Eds. Instituto de Pesquisas Tecnologicas de Sao Paulo, Sao Paulo, Brasil, p.107-127.

2 - Koga, M.E.T. (1988). Celulose e Papel. Eds. Instituto de Pesquisas Tecnologicas de Sao Paulo, Sao Paulo, Brasil, p.15-40.

3 - Miyazaki, K., Iwanaga, Y. and Tsuchika, H. (1989). In: Wood Processing and Utilization. J.F.Kennedy, G.O. Phillips and P.A.Williams (Eds). Ellis Horwood Ltd., Chichester, p.75-80.

ACKNOWLEDGEMENTS:

This work was supported by FACEPE, JICA, UFPE, Perquimia S.A. and The British Council and forms part of the formal interaction between the Federal University of Pernambuco, Brazil, and the University of Birmingham, U.K.

6

Explosion pulping of aspen

Xujun Hua*, Bohuslav V. Kokta and Serge Kaliaguine*** - Dept. Génie Chimique, Université Laval, St-Foy, Quebec, Canada G1K 7P4. **C.R.P.P., Université du Quebéc à Trois-Rivières, Trois-Rivières, Quebéc., Canada G9A 5H7.

SUMMARY

Explosion pulping of aspen is compared with conventional CTMP and CMP pulping at the same chemical charges. The explosion pulp shows higher physical strength than CMP or CTMP, up to 180% higher in breaking length, depending on freeness required, and saves 14% - 39% refining energy. The high strength of the explosion pulp might be explained by its long fiber length, low content of fine fibers (passing M100). The analysis of ionic contents shows that the explosion pulps cooked with 8% Na_2SO_3 with or without 1% NaOH have a higher sulfonate group content. The surface analysis by ESCA spectroscopy indicates that the explosion pulp has a higher oxygen-to-carbon ratio and less C1 content in C_{1s} peaks. These reveal there are more carbohydrates and less lignin and extractives on the fiber surface of the explosion pulp.

INTRODUCTION

Conventional ultra-high-yield chemithermomechanical or chemimechanical pulping is preferably conducted at a pH level between 4 and 9, and involves either liquid or vapor phase cooking with sodium sulphite-bisulphite solutions for about 10 to 30 minutes at a temperature between 100 and 160 °C[1,2]. It is generally accepted that the chemical treatment is mainly responsible for permanent fiber softening, increase in long fiber content, fiber specific surface area and conformability[3].

There is another softening approach that consists of a steam treatment of chips at high temperatures followed by explosive decompression. The production of pulp using high-pressure and high steam chip softening well above glass transition temperatures of lignin should theoretically lead to lower energy consumption in subsequent refining stages.

It has been suggested by Kokta and Vit[4,5] that the ultra-high-yield (90%+) pulp suitable for papermaking can be produced by vapor phase steam explosion cooking. The initial properties of papers made from exploded softwood chips were similar to those of TMP. However, refining energy was about 20% to 25% lower. Recently, a pulping process entitled "Process for Preparing Pulp for Paper Making"[6-9] referred to as Steam Explosion Pulping Process (SEP) or S-pulping has been proposed both for softwoods and hardwoods. The steam explosion pulping process consists of chemical impregnation of chips, short duration saturated steam cooking at temperatures varying from 170°C to 210°C, pressure release, refining and bleaching (if necessary). In this process, impregnation and cooking conditions were aimed at minimizing yield and brightness loss, maximizing resulting paper properties and decreasing specific refining energy. It has been reported that the preferable conditions for SEP were temperature between 180-195°C, and time 1-4 min. The pulp was easily bleached up to 86% with hydrogen peroxide[6,7].

The objectives of this work are to compare explosion pulping with conventional CMP/CTMP prepared using the same raw materials at the same chemical charges in terms of pulp physical properties and refining energy, and to explain the differences by the fundamental physical and chemical fiber characteristics including the surface chemical properties. Considering the enormous surplus of underutilized aspen in Canada (in Quebec only, about 6 million cubic meters of aspen and birch are left uncut annually[10]), the explosion pulping of aspen was taken as our object of study.

EXPERIMENTAL

1. Raw material and pretreatment

Aspen (Populus tremuloïds Michx) chips were spray impregnated at Hymac Co. in a solution of Na_2SO_3 alone or of Na_2SO_3 and NaOH, before cooking. The ratio of liquor/wood during impregnation was 2:1.

2. Cooking

The explosion pulp was produced with a Stake continuous reactor of 48 tons/day capacity at University of Sherbrooke. Conventional cookings were done under typical CTMP and CMP pulping conditions: 128 °C and 10 min. for CTMP, 150 °C and 30 min. for CMP in pilot plant unit of École Papeterie, CEGEP de Trois-Rivières. The percents of chemical charges during cooking are indicated in the figures below.

3. Refining

Refining was done using a domestic blender OSTERIZER B-8614 at a consistency of 2%. Refining energy was measured with a Digital Power Hi Tester, Model 3181-01. Relative specific refining energy was calculated by subtracting the blending energy of full beaten pulp from the total energy needed to blend the fiber suspension.

4. Pulp physical properties

Handsheets were prepared and tested according to standard CPPA testing methods.

5. Fiber length distribution

The pulp was fractionated into R14, 14/28, 28/48, 48/100, 100/200 and P200 fraction in the Bauer-McNett classifier. The relative fiber length can be calculated from the fractions.

6. Sulfonate and carboxylate groups of pulp

The contents of the sulfonate and carboxylate groups were measured by the conductometric titration method[11].

7. ESCA spectroscopy

The ESCA spectrometer used in this work for surface analysis was an ESCALAB MKII spectrometer fitted on a microlab system from V. G. Scientific and equipped with a non monochromatized dual Mg-Al anode X-ray source. Kinetic energies were measured using a hemispherical electrostatic analyzer with 150 mm radius, working in the constant pass energy mode (20 eV for C_{1s} and O_{1s} peak, 50 eV for S_{2p} and survey spectra). No flood gun was utilized. The vacuum in the test chamber was maintained at the pressure of 10^{-8} Torr. Liquid nitrogen was used for cooling the sample. C_{1s} and O_{1s} peaks were interpreted by a peak synthesis technique and their chemical shifts. The surface atomic ratios were estimated from corresponding peak intensities.

RESULTS AND DISCUSSION

The pulp physical strengths are shown in Fig 1-3, the points in which were obtained by interpolating among many data with different yield and freeness.

Fig 1 shows that the explosion pulp has a considerably higher burst index compared with the conventional CMP and CTMP at the same freeness and yield, from 90-340% dependent on the freeness. The graphs of breaking length and tear index vs refining energy are shown in Fig 2-3. It is indicated that breaking length and tear index of explosion pulp are also higher than those of CMP/CTMP, 50-180% higher in breaking length, 20-100% in tear index. In all the cases, the explosion pulping saves 14-39% refining energy to reach the same freeness. It might be conceived that explosive release of chips after cooking may weaken inter-fiber bonds within the chips to some extent and lead to a lower refining energy consumption.

Explosion pulp has less frac-

Fig 1. BURST INDEX vs. REFINING ENERGY

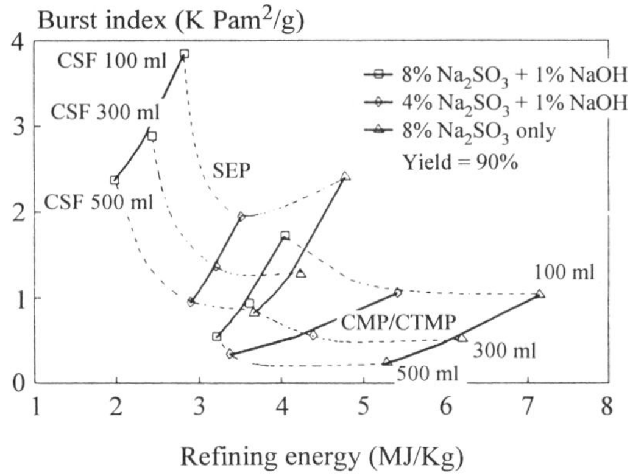

Fig 2. BREAKING LENGTH vs. ENERGY

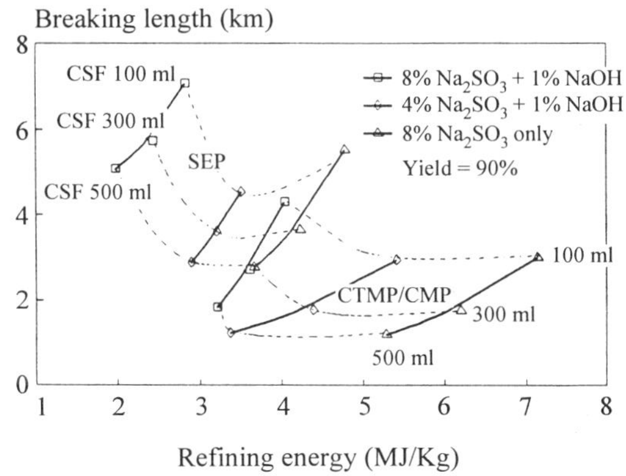

tion retained on 14 mesh which mostly consists of the shives and longer mean fiber length (data not shown here). Fig 4-5 show that SEP has a higher long fiber fraction which is a sum of mesh 14/28 and mesh 28/48 and fine fiber fraction (mesh 100/200 + P200). This means that the fiber length distribution of SEP tends to be more narrow than that of the conventional pulp, and implies that the fibers in the explosion pulp were cut down less and were easier to fibrillate during refining, which improved inter-fiber bonding. Correspondingly, it was found that burst index increases dramatically with an increase of fraction of 14/28+28/48. In contrast to this, the fine fiber fraction has the inverse effect on the strength, as shown in Fig 5. Therefore, the high burst index of explosion pulp might be explained by its lower fine fiber content and higher long fiber content.

Fig 6 shows that SEP has a high level of sulfonation especially those with 8% Na_2SO_3. It has been shown that paper properties improve with an increase of sulfonate content (Fig 7). It seems that the higher sulfonate content is one of the reasons for the higher strength of the explosion pulp due to an increase in fiber hydrophilicity.

ESCA analysis shows that the explosion pulp has a higher oxygen-to-carbon ratio, less C_1 peak area which represents the carbons only linked to hydrogen and/or carbon (Fig 8-9). Since carbohydrates have a higher O/C ratio and are without the C_1 component in the C_{1s} peak com-

Fig 3. TEAR INDEX vs. ENERGY

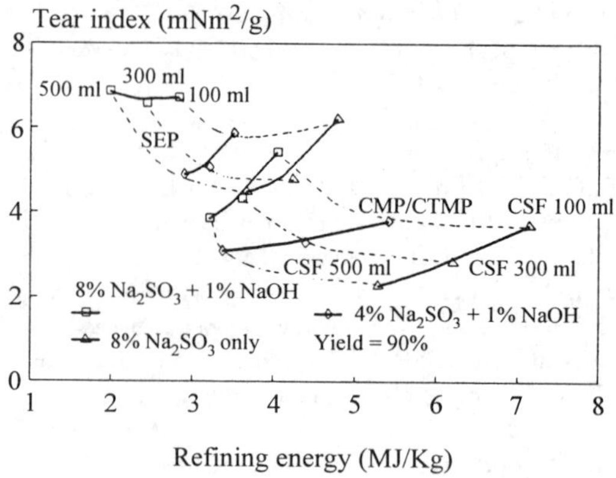

Fig 4. BURST vs. LONG FIBER FRACTION

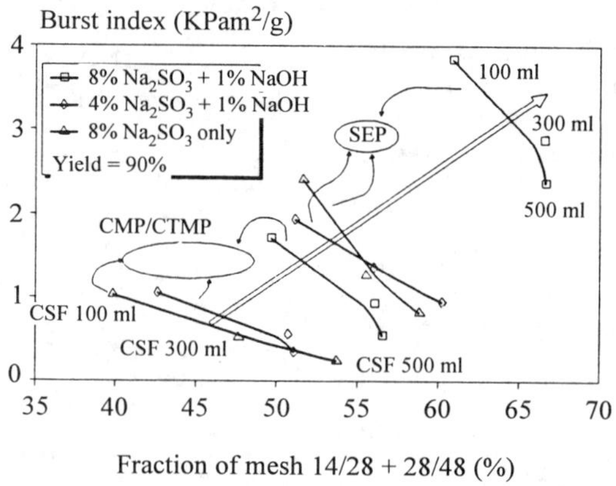

Fig 5. BURST vs. FINE FIBER FRACTION

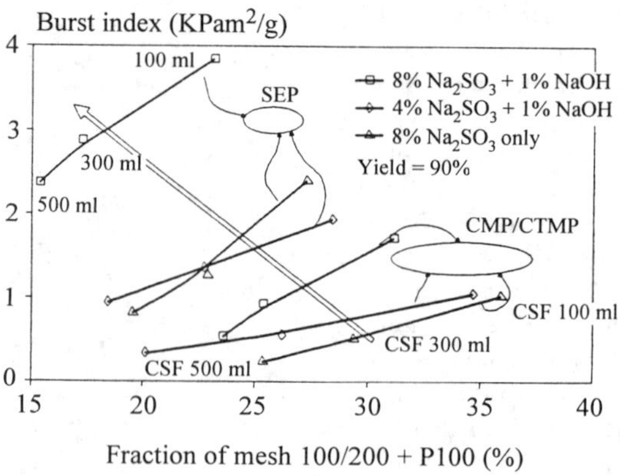

pared with lignin and extractives, it suggests that the explosion pulp has more exposure of carbohydrates on its fiber surface, which has more hydrophilic groups, facilitating formation of hydrogen bonds during paper formation. Based on the theoretical O/C ratios and C1 contents for main components in wood fibers (ie. carbohydrates, lignin and extractives), a triangular graph was established (Fig 10). It illustrates roughly the amount of each component on the surface which confirms that SEP has more carbohydrates on the fiber surface.

Fig 6. SULFONATE GROUP CONTENT

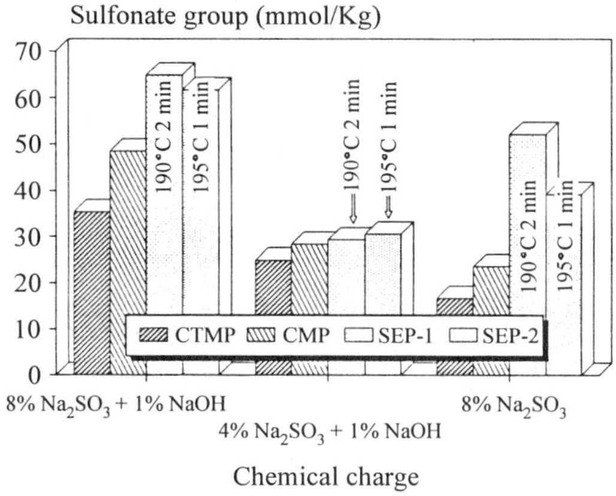

Fig 7. STRENGTH vs. SULFONATE GROUP

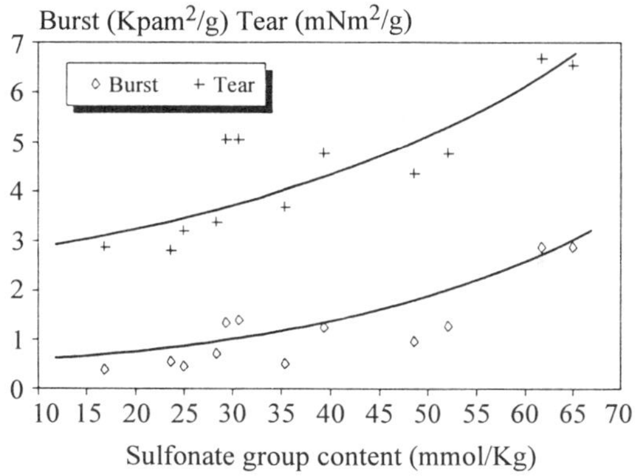

CONCLUSIONS

Compared to conventional CTMP/CMP, the explosion pulp shows superior physical strengths and consumes less refining energy. The superior strengths of the explosion pulp might be explained by its higher content of long fibers and sulfonate groups, lower content of fine fibers, and higher concentration of hydrophilic groups on the fiber surface.

Fig 8. SURFACE O/C ATOMIC RATIO

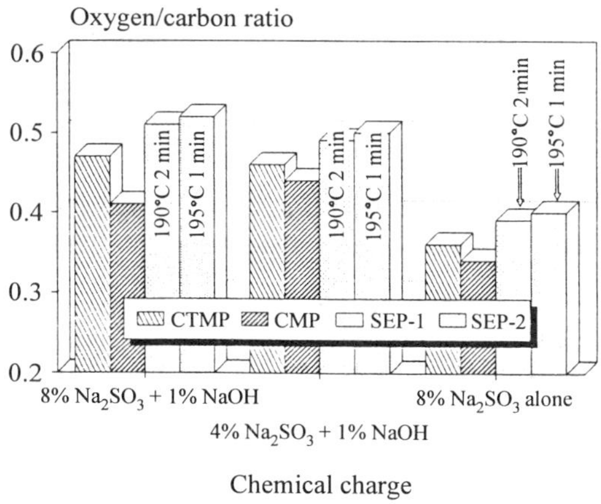

Fig 9. C1 COMPONENT IN C_{1s} PEAK

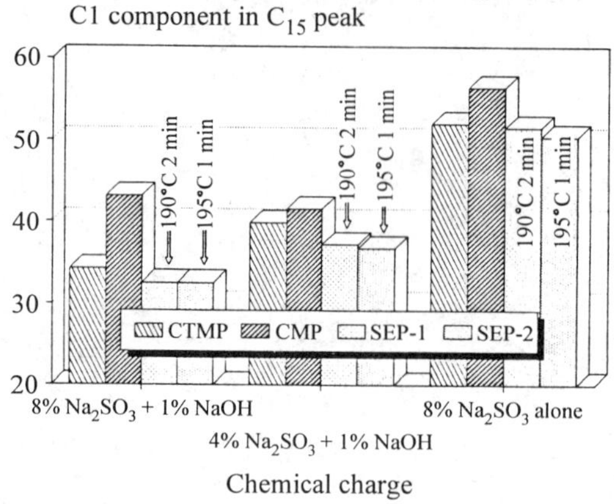

Fig 10. SURFACE COMPONENT FRACTION

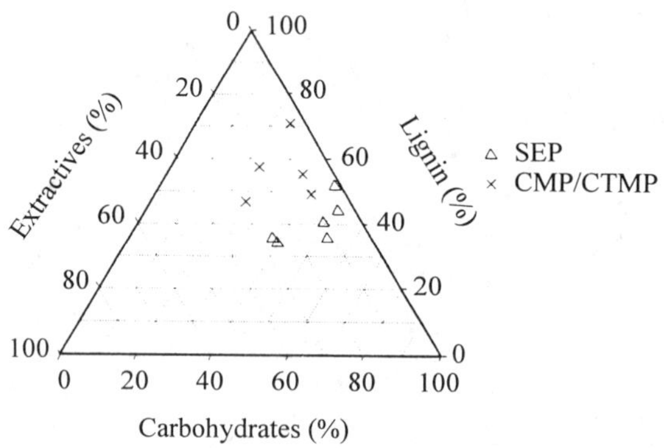

REFERENCES

(1) Leask, R. A., **Pulp and Paper Manufacture, volume 2, Mechanical Pulping,** TAPPI CPPA joint textbook committee, 1987.

(2) Ingruber, O. V., Kocurek, M. J. and Wong, A., **Pulp and Paper Manufacture, volume 4, Sulfite Science and Technology**, TAPPI CPPA joint textbook committee, 1985.

(3) Heitner, C. and Atack, D., **Pulp and Paper Canada**, 84(11), T252-T257 (1983).

(4) B. V. Kokta and R. Vit, "New Ultra-High-Yield V-Pulping Process", **Papir a celluloza**, 44(11), 73-80 (1989).

(5) B. V. Kokta and R. Vit, "Method, Process and Apparatus for Converting Wood, Wood Residue and/or Biomass into Pulp". **Can. Pat.** 1,212,505 (Oct. 14, 1986).

(6) Kokta, B. V., "Process for Preparing Pulp for Paper Making", **Can. Pat.**, #1,230,208 (Dec. 15, 1987).

(7) Kokta, B. V., "Process for Preparing Pulp for Paper Making", **U.S. Pat.**, #4,798,651 (Jan. 17, 1989).

(8) Kokta, B. V., "Improval Process for Preparing Pulp for Paper Making", **Can. Pat. Appl.**, #542,643 (May 1987).
(9) Kokta, B. V., "Process for Preparing Pulp for Papermaking", **Can. Pat.**, #1287 705 (Aug 1991).
(10) B. V. Kokta and Huai-Yu Zhan, **1987 I.M.P.C.** CPPA-TAPPI, EUCEPA, Proceeding, pp193-199, Vancouver, June 2-5, 1987.
(11) Katz, S.; Beatson, R. P. and Scallan, A. M., **Svensk Papperstidning**, 87(6), R48-R53 (1984).

7

Steam explosion pulping of aspen: the effect of cooking parameters on resulting properties

M. Tomasec and B.V. Kokta - Pulp and Paper Research Centre, Université du Québec à Trois-Rivières, C.P. 500, Trois-Rivières, Québec, Canada, G9A 5H7.

ABSTRACT

Explosion pulps, CTMP and CMP were prepared at the same yield values (90% and more). It has been shown that for the same yield and same ionic content values, the explosion pulps have better mechanical properties (expressed as higher tear index at the same breaking length) when compared to CMP and CTMP. The relative specific refining energies of explosion pulps have always been considerably lower than in the case of CMP and CTMP. The pressure increase from 15.5 to 25 atm (achieved by pressurization of the reactor with nitrogen after regular cooking time) increased the mechanical properties and decreased the relative specific refining energy. The positive effect of high temperature and pressure leads to better fiber softening (additional permanent softening of lignin), increases the portion of ordered cellulose I and facilitates fiber separation.

INTRODUCTION

Ultra-high yield pulps are conventionally produced from softwoods and hardwoods by treating wood chips prior to refining, with chemicals that react with but do not solubilize lignin [1,2,3,4,5]. Recently, the "Steam Explosion Pulping for Papermaking" was introduced as an alternative to the classical processes for both the hardwoods and softwoods [6,7,8]. Steam explosion pulping (SEP) process consists of the chemical impregnation of chips, short time cooking at high temperature and high pressure in saturated steam, instantaneous pressure release (explosion), atmospheric refining and bleaching (if necessary). Short cooking time at high temperature and pressure leads to additional permanent lignin softening, ultra-high yields and excellent paper properties [10,11,12]. Explosive pressure release evokes supplementary fiber separation which decreases the refining energy. Best results were achieved in semi-industrial trials, partially as a result of press-impregnation [12,13].

Previous research [12] confirmed lower specific refining energy for explosion pulps (cooked at 180, 190 and 200°C) compared to conventional CMP (170°C) cooked to the same yield value (90%) at the same chemical charge applied during both: impregnation and cooking. This work was carried out to find this relationship at the same conditions for CTMP (128°C), CMP (150 and 160°C) and explosion pulps (190, 195 and 200°C). Another series of experiments was done in order to study the influence of high pressure on resulting pulps.

EXPERIMENTAL

Experimental conditions are described elsewhere [9] and a summary is presented in the Table I:

RESULTS AND DISCUSSION

Experimental design: High yield pulps and even ultra-high yield pulps (90%+) can be prepared under a large range of process conditions, varying such factors as the temperature, chemical charge, etc. Based on the results of our previous research as well as from the theoretical considerations, we assume that increased temperature and pressure are the most important factors contributing to the development of quality parameters in the case of explosion pulps.

Table I: Pulping conditions and some of the characteristics of resulted pulps

#	PULP TYPE	Cooking				IONIC CONTENT [mmol/kg]		
		TIME [min]	TEMP [°C]	PRESSURE [atm]	YIELD [%]	SULFONIC	CARBOXYLIC	TOTAL
1	RMP	-	-	-	98.8	0	77	77
2	CTMP	10	128	1.8	95.3	35	143	178
3	CTMP	60	128	1.8	93.1	39	141	180
4	CMP	55	150	4.0	89.6	37	128	165
5	CMP	30	160	5.2	89.9	38	141	179
6	SEP	1.3	190	11.9	92.6	46	134	180
7	SEP	2	190	11.9	90.4	43	137	180
8	SEP	1.5	195	13.5	90.1	46	134	180
9	SEP	1	200	15.5	89.9	45	121	166
10	N-EXP	2	190	11.9/25	90.0	40	140	180
11	N-EXP	1.5	195	13.6/25	89.7	51	132	183
12	N-EXP	1	200	15.5/25	90.0	42	137	179

In order to examine the influence of the pressure and temperature more thoroughly, we tried to eliminate some of the variables from the experiment.

Firstly, we tried to cook the pulps to the same yield (Table I). In the case of CMP and SEP it was 90%. Explosion pulps were cooked at the usual conditions, while the CMP cooking times were somewhat longer than for usual industrial conditions. In the case of CTMP, the classical conditions gave the yield about 5% higher. However, we succeeded to prepare a CTMP and explosion pulp at comparable yield (about 93%) by manipulating the cooking time.

Secondly, we used the very same degree of chemical treatment for all the pulps with the exception of RMP, where no chemicals were added. The chemical charge was the same during the impregnation as well as during the cook. Even if the total ionic content was similar in the majority of chemically treated pulps, from Table I we can see that the degree of sulfonation is higher in the case of explosion pulps. Ionic content represents one of the most important factors contributing to the mechanical properties. As it was shown before [12], the explosion pulps have better properties at the same value of ionic (sulfonic) content, or, in order to reach the same properties, explosion pulps need lower ionic content in comparison with other processes.

Refining energies: Blender has been reported as the best means for small-scale refining [14] and the properties obtained using the blender as well as the refining energies are well comparable with the semi-industrial results [12,13]. Figure 1 shows the relative specific refining energies for chemically treated pulps at the freeness of 100 ml CSF. The decrease in refining energy in the case of SEP can be attributed to the influence of temperature and explosive discharge. The cooking temperature is in the case of SEP above the glass transition temperature of lignin, which leads to the additional permanent softening of chips. The explosive discharge promotes the separation of well softened chips into individual fibers, as well as to the explosion of fibers themselves [15]. Exploded chips when compared to CMP and CTMP are more flexible and can be easily separated into fibers and refined. Resulting shorter refining times help also to protect the fiber length.

Even if the explosion pulps have higher average fiber length, thanks to the lignin softening, they are more flexible. Density and the light-scattering coefficient are two parameters that reflect the effect of flexibility. The higher is the severity of treatment, the higher is the density. Increased cooking temperature also lowers the LSC of pulp. Higher flexibility of longer SEP fibers dramatically improves the bonding capacity, which explains the superior mechanical properties of explosion pulps [9].

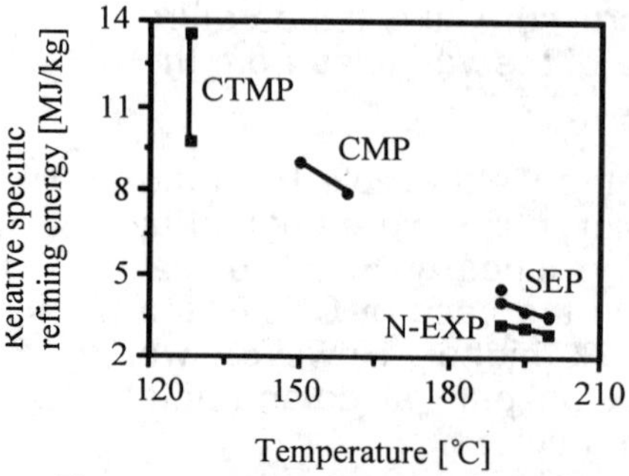

Figure 1: Relative specific refining energy as a function of temperature: as the severity of treatment has increased, the refining energy has diminished almost linearly (100 ml CSF).

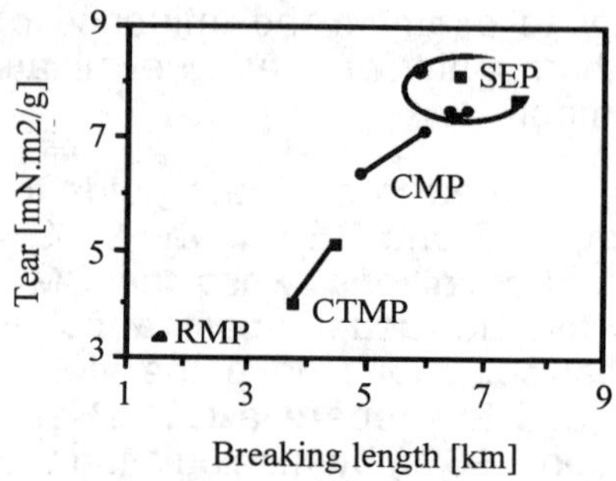

Figure 2: Mechanical resistance of aspen high yield pulps: The tear vs. breaking length relation clearly shows an increase in mechanical properties with increased severity of treatment (100 ml CSF).

The chemical changes that take place during the cook are another important consideration. Tanahashi [15] has recently reported an increase in the crystallinity of cellulose following steam treatment at high temperature and pressure. Using Sukhov's qualitative methods for evaluating cellulosic materials via FTIR (Fourier-transform infrared spectroscopy) and Raman spectroscopy [16], we found [9] that explosion pulps have the highest content of ordered Cellulose I (C-I). Ordered C-I is the highest quality cellulose (highest bonding capacity) and contributes significantly to pulp mechanical properties.

Mechanical properties: The results presented so far show that the physical and chemical changes that take place during the high-temperature-and-pressure cook and subsequent explosive discharge are beneficial and should provide a pulp with superior mechanical properties. This was experimentally confirmed and results for breaking length and tear index are presented in Figure 2.

If we compare CTMP and SEP, the superiority of explosion pulp is obvious in any regard: higher mechanical properties (breaking length, tear and burst), higher density and C-I crystallinity at lower LSC and lower refining energy. This was accomplished either at conventional CTMP conditions (Pulp 2) or at similar yield (Pulp 3) and at the same chemical cooking charge and similar ionic content. The only difference between the CTMP and SEP pulps were the cooking conditions (temperature, pressure and time), which emphasizes the influence of these parameters on pulp quality. While the differences between CMP and SEP were not as dramatic as between CTMP and SEP, Figure 2 shows that the mechanical properties of SEP are superior. This is also apparent in Figures 3 to 5, which correlate mechanical properties to yield, sulfonic content and refining energy.

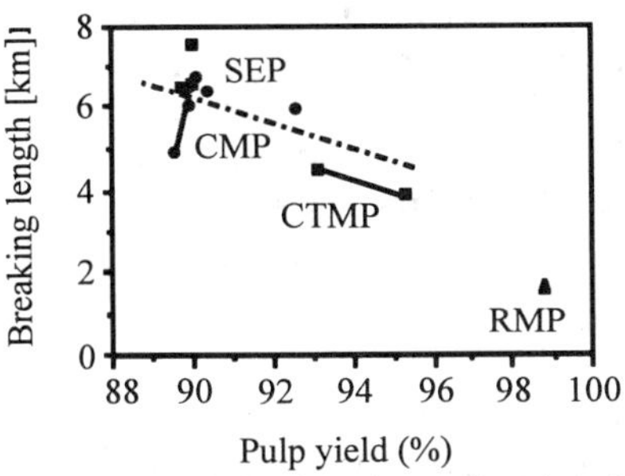

Figure 3: Breaking length as a function of pulp yield: explosion pulps showed best mechanical properties (100 ml CSF).

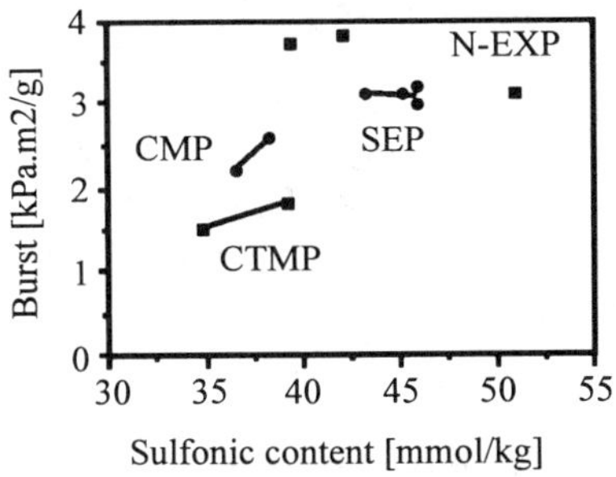

Figure 4: Burst as a function of sulfonic content: even at the same sulfonation, explosion pulps reached higher mechanical properties (100 ml CSF).

Both explosion pulps and conventional mechanical pulps can be prepared at yield values around 90% or even higher (Fig. 3). Our results show that the SEP process reaches higher pulp quality. At the same yield level, breaking length (burst and tear as well) were the highest in the case of SEP. Values of the sulfonic content were also higher in explosion pulps (Fig. 4). Sulfonic content has been proven as one of the key parameters, determining pulp strength. Explosion pulps reach superior resistance even at similar sulfonation, compared to CMP and CTMP (values about 40 mmol/kg).

Figure 5 shows one of the greatest advantages of explosion pulping: energy savings. To refine an explosion pulp to 100 ml CSF value requires at least 50% less energy then in a conventional process. When we refined various pulps to the same properties, energy savings for explosion pulps were even higher, or, at the same amount of applied energy, the properties of SEP were uncomparably higher.

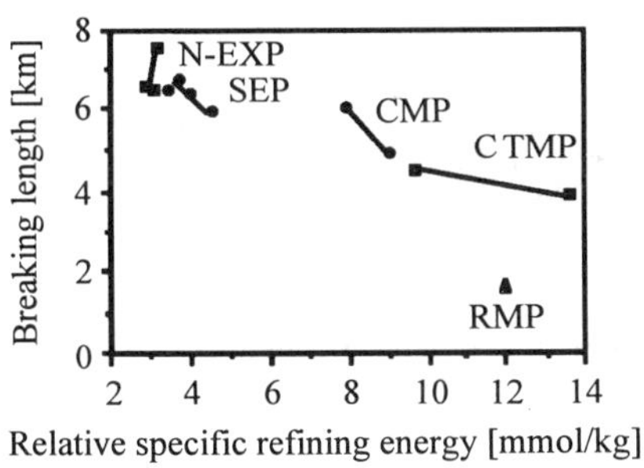

Figure 5: Breaking length as a function of RSRE: explosion pulps save at least 50% of energy, or at the same RSRE have incomparably higher properties (100 ml CSF).

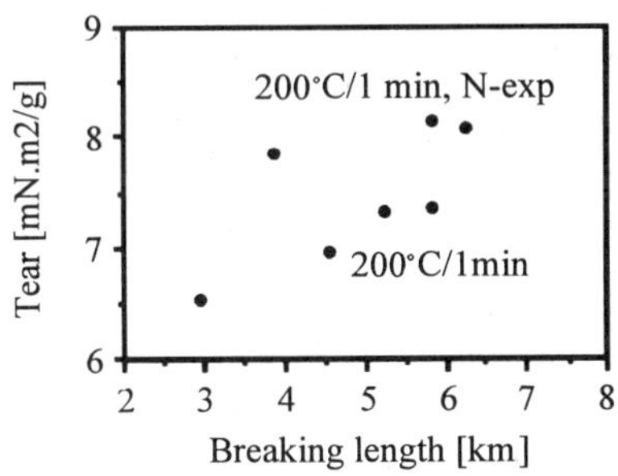

Figure 6: Mechanical strength of explosion pulps prepared at 200°C and 1 minute: pressurizing of the reactor prior to the explosion increased the tear value at constant breaking length.

Influence of pressure: Pressure increase from the value corresponding to the pressure of saturated steam at cooking temperature to 25 atm (prior to the explosion) slightly increased the mechanical properties. Figure 6 shows this differences for the cook at 200°C and 1 minute. Pressure increase also leads to better fiber separation and hence lowers refining energy consumption (by about 20%, Fig. 1).

CONCLUSIONS

Explosion pulping shows promise as a viable ultra-high yield pulping process. Compared with the conventional methods of chemi-mechanical and chemithermomechanical pulping, explosion pulping is very fast (usually 2 min or less) and requires less chemicals. Comparison of high-yield pulps at the same chemical charge, yield and ionic content showed that the explosion pulps had superior mechanical and optical properties and lower refining energy.

The superiority of the explosion pulps can be attributed to the chemical changes (higher crystallinity, more cellulose on the fiber surface, better lignin softening) that occur following the high-temperature-high-pressure cook as well as the physical changes (better and easier fiber separation) that occur following high severity cook and explosive discharge from the digester.

ACKNOWLEDGEMENT

The authors wish to thank NSERC (Canada), Mechanical Pulping Network of Centres of Excellence, FCAR (Quebec), Stake Tech. Co. and CQVB for financial support and Air Liquide Co. for equipment support.

LITERATURE CITED

1. Leask,R.A.: Tappi, 51, #12, 1968
2. Ford,M.J.; Elliot,P.E.: U.S. Patent # 4,116,758 Sept 28,1978
3. Janson,J.; Manstrom,B.: Pulp and Paper Can., 82, #4, p.T 111, (1981)
4. Atack,D.; Heitner,C.; Karnis,A.: Svensk Papperstidn., 83, #5, p.133, 1980
5. Heitner,C.; Atack,D.: Pulp and Paper Can., 84, #11, T 252, Nov. 1983
6. Vit,R.; Kokta,B.V.: Canadian Patent #1,212,505, October 1986
7. Kokta,B.V., Ahmed,A.; Chen,R.; Graczyk,T.; Zhai,D.; Garceau,J.J.: Les Papetières Du Québec, 1, #4, p.17, 1990
8. Barbe,M.C.; Kokta,B.V.; Taylor,J.: Preprints 75th CPPA Annual Meeting, Montreal, 1989, p. A-363
9. Tomasec,M.; Kokta,B.V.: Tappi, 74, #12,p. 119, Dec 1991
10. Kokta,B.V.; Deming,Z.: IPPTA, 3, #2, p.1, 1991
11. Kokta,B.V.; Didwania,D.: Preprints 76th CPPA Annual Meeting, Montreal, 1990, p. B 19
12. In: Kennedy,J.F.; Phillips,G.O.; Williams,P.A.: Lignocellulosics: Science, technology, development and use, chapter 20, p.171-212, Ellis Horwood Limited Publishers, 1992
13. Barbe,M.C.; Kokta,B.V.; Lavallée,H.C.; Taylor,J.: Pulp and Paper Can., 91, #12, p. 142 (T 395), 1990
14. Shaw,A.C.: Pulp and Paper Can., 85, #6, p. T 152, 1984
15. Tanahashi,M.: "Degradation Mechanisms of Wood Components by Steam Explosion", Mitsuhiko Tanahashi, GIFU University, Japan, 144 p., 1989
16. Sukhov,D.A.; Zhilkin,A.N.; Valov,P.M.; Terentiev,O.A.: Tappi, 74, #3, p.201, March 1991

8

Ultra-high-yield pulping: relation between pulp properties and fibre characteristics by multiple linear regression

F. Carrasco*, J.J. Garceau*, A. Ahmed and B.V. Kokta**** - *Department of Chemical Engineering. **Pulp and Paper Research Centre. Université du Québec à Trois-Rivières, Trois-Rivières, Québec, G9A 5H7, Canada.

ABSTRACT

Ultra-high-yield pulps have been obtained by means of steam explosion pulping. This process was carried out as follows: 1) chips soak impregnation (60°C, 24 h, L/C = 6) in solutions containing 8% Na_2SO_3 and a variable concentration of either Na_2CO_3 or $NaHCO_3$, ranging from 0 to 2%; 2) cooking at high temperature (200°C) and short time (1 min); 3) rapid pressure release. This work shows that refining energy requirement was significantly reduced and pulp strength increased when adding Na_2CO_3 or $NaHCO_3$ to an impregnation solution containing 8% Na_2SO_3. However, the higher chemical charge used made the optical properties drop. We have proposed simple mathematical models in order to predict specific refining energy and paper properties from some fibre characteristics. The correctness of these models was checked by multiple linear regression analysis. The best regression coefficients were obtained by using sulfonate content and pulp yield as "key variables". The relative importance of these two variables has been also studied.

INTRODUCTION

Recent work[1] has shown that sulfite treatment of aspen sulfonates lignin and hydrolyzes ester groups in the hemicellulose to carboxylic acids. Moreover, it has been reported that the sulfonic and carboxylic acid content is a major factor in determining the pulp strength and optical properties of conventional chemimechanical pulps[2]. Kokta et al.[3] have demonstrated that aspen chips impregnated with 8% Na_2SO_3 results in relatively high pulp yield and good pulp quality. The combination of caustic and sulfite to achieve chemical pretreatment is known to greatly influence the pulp yield, specific refining energy as well as optical and mechanical properties of chemimechanical pulps[4,5]. In this paper we are going to explain the influence of adding a second chemical (Na_2CO_3 or $NaHCO_3$) to a 8% Na_2SO_3 impregnation solution on the formation of sulfonic acids as well as on the refining energy requirement and pulp properties.

The aim of this work is: 1) to produce and characterize the ultra-high-yield steam explosion pulps. 2) to study the extent of sulfonation as a function of concentration of impregnating chemicals. 3) to establish mathematical correlations which are able to evaluate refining energy consumption and paper properties of explosion pulps by knowing only sulfonate content of pulps and pulp yield, by using multiple linear regression.

EXPERIMENTAL

The average chip size of aspen wood used in this study was the following: length = 2.5-3.7 cm, width = 1-2 cm, thickness = 1-9 mm with maximum distribution at 5 mm. The impregnation of chips (chemical pretreatment) was carried out as follows: 150 g of aspen chips (humidity = 50%) were mixed in plastic bags with solutions made up of 8% Na_2SO_3 along with Na_2CO_3 or $NaHCO_3$ at five different levels: 0, 0.25, 0.5, 1 and 2%. The weight ratio of liquor/chips during impregnation was L/C = 6. The temperature and time of this chemical pretreatment were respectively 60°C and 24 h. Pretreated chips (without liquor) were then cooked in a laboratory batch reactor built by Stake Tech. Ltd. by using saturated steam. The temperature-time sequence used was 200°C-1 min. After cooking, the chips were suddenly decompressed as a consequence of an instantaneous pressure drop from 1.55 MPa (steam pressure corresponding to 200°C) to 101 kPa. Then, the exploded wood chips were thoroughly washed with tap water. The resulting material was then stored in a cold room for further refining and properties measurements. Details concerning yield calculation, refining conditions and properties evaluation are available elsewhere[6].

RESULTS AND DISCUSSION

Figure 1 shows that the steam explosion pulping process leads to the formation of ultra-high-yield pulps (85% +). $NaHCO_3$ is an effective chemical to protect fibres against hydrolysis and degradation. In all the ranges studied, Na_2CO_3 leads to lower pulp yield than $NaHCO_3$. This point has obviously important economic significance. Figure 2 illustrates that the formation of sulfonate groups is very sensitive to the addition of a second impregnation chemical (Na_2CO_3 or $NaHCO_3$), especially at concentrations below 1%. An increase of 24 mmol/kg of sulfonate groups is achieved by adding 1% Na_2CO_3 or $NaHCO_3$ to a 8% Na_2SO_3 solution. This addition has a beneficial effect on refining energy consumption and paper properties.

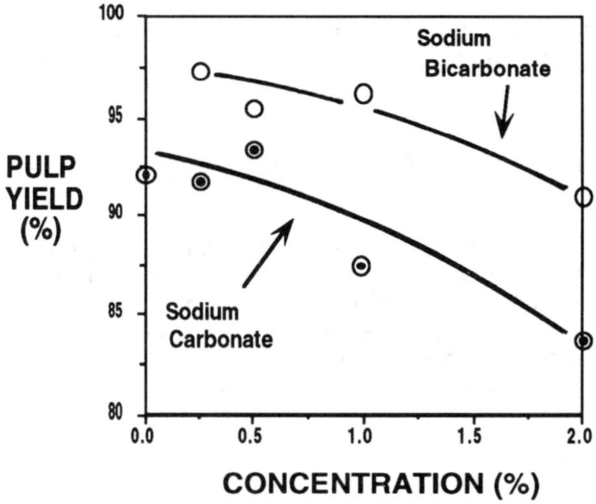

Figure 1: Variation of pulp yield with Na_2CO_3 or $NaHCO_3$ concentration added to a 8% Na_2SO_3 impregnating solution.

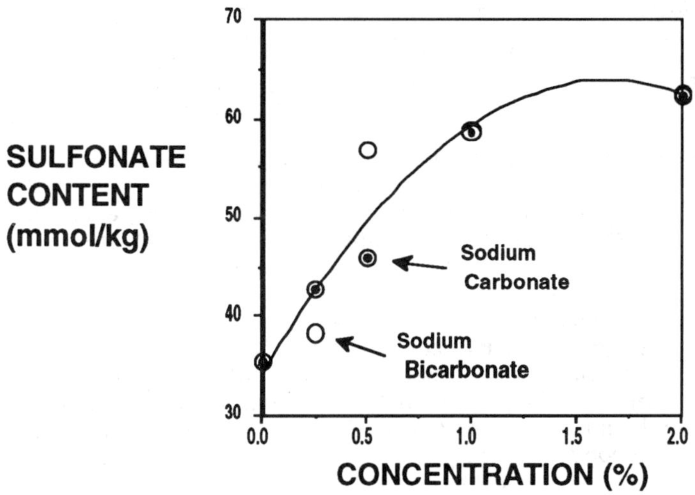

Figure 2: Variation of sulfonate content as a function of Na_2CO_3 or $NaHCO_3$ concentration added to a 8% Na_2SO_3 impregnating solution.

The interdependence between processing conditions and the characteristic properties of fibres and paper is illustrated in Figure 3. It is clear that specific refining energy and paper properties depend on operating conditions (y=f(z)). On the other hand, bulk and fibre characteristics directly influence the paper properties and could explain the energy consumed to attain a certain pulp quality (y=y(x)). Then, it is possible to relate y variables to x variables without taking into consideration the operating conditions (z) because the overall influence of the latter will be included in the intermediate variables x. The variables considered in this study are shown in Figure 3. In order to determine the most significant x variables affecting the values of specific refining energy and paper properties, the following mathematical linear model is proposed:

$$y = a + \sum_{i=1}^{7} b_i . x_i \qquad \text{(a: intercept; } b_i\text{: slopes)}$$

For each product variable (y), a systematic and complete linear regression analysis was performed by considering all the possible combinations between intermediate variables (7 one-x models, 21 two-x models, etc.). The best predictions were achieved when considering sulfonate content and pulp yield as intermediate variables (x). The equations generated by multiple linear regression were the following (see Figure 3 for symbols):

$$y_1 = 5.9 - 0.115\ x_4 + 0.032\ x_7 \qquad (r^2 = 0.91)$$

$$y_2 = 13.8 + 0.078\ x_4 - 0.125\ x_7 \qquad (r^2 = 0.96)$$

$$y_3 = 9.8 + 0.061\ x_4 - 0.103\ x_7 \qquad (r^2 = 0.89)$$

$$y_4 = 5.4 + 0.033\ x_4 - 0.009\ x_7 \qquad (r^2 = 0.71)$$

$$y_5 = 38.7 - 0.264\ x_4 + 0.410\ x_7 \qquad (r^2 = 0.92)$$

$$y_6 = 54.7 - 0.128\ x_4 + 0.417\ x_7 \qquad (r^2 = 0.94)$$

$$y_7 = -139.5 - 4.79\ x_4 + 7.90\ x_7 \qquad (r^2 = 0.95)$$

These equations clearly indicate that an increase in sulfonate content results in an increase in mechanical properties and a decrease in refining energy and optical properties. The opposite effect is observed when raising pulp yield. The relative importance of each intermediate variable can be evaluated by calculating the standardized coefficients corresponding to both slopes. It consists roughly in dividing the actual slope by the mean value of the intermediate variable. Our findings were as follows: Specific refining energy appears to be very sensitive to sulfonate content and depends weakly on pulp yield. Breaking length and burst index (mechanical properties depending on specific surface of fibre) are related to both the sulfonate content and pulp yield, but the former being more significant. Brightness and opacity also depend on both sulfonate content and pulp yield, but being more sensitive to the latter.

Figure 4 provides a visual way to evaluate the correctess of the model proposed. It consists of plotting the predicted values resulting from the mathematical equations above indicated as a function of experimental data. When the model is appropriate, points must be as close as possible to the diagonal line. This type of representation gives more information than just reporting regression coefficient values. It also allows to detect if points are randomly distributed above and below the diagonal line for each series of operating conditions. Figure 4 clearly indicates that specific refining energy is accurately predicted. Breaking length, burst index and optical properties gave also similar results.

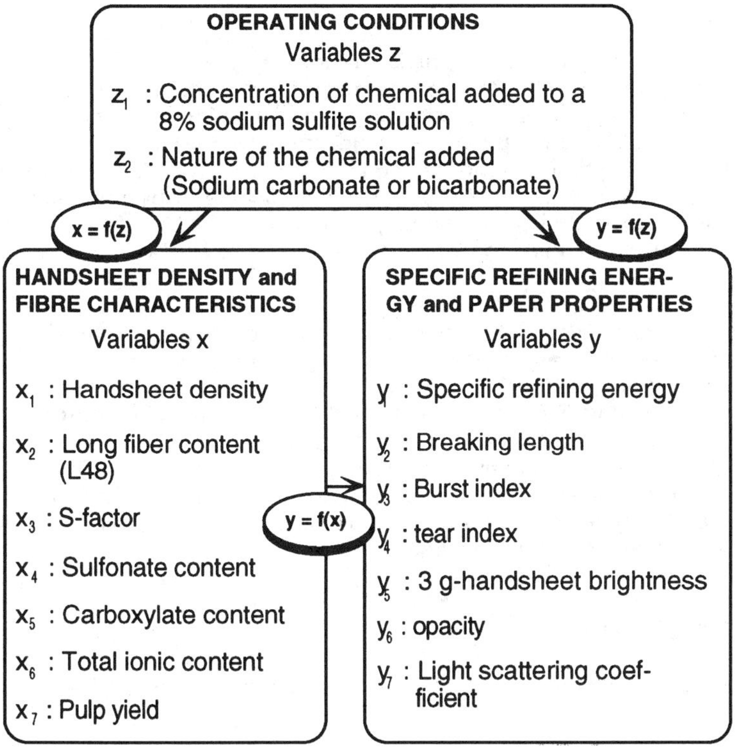

Figure 3: Diagram showing the interrelations between operating conditions and variables representing fibre and paper quality.

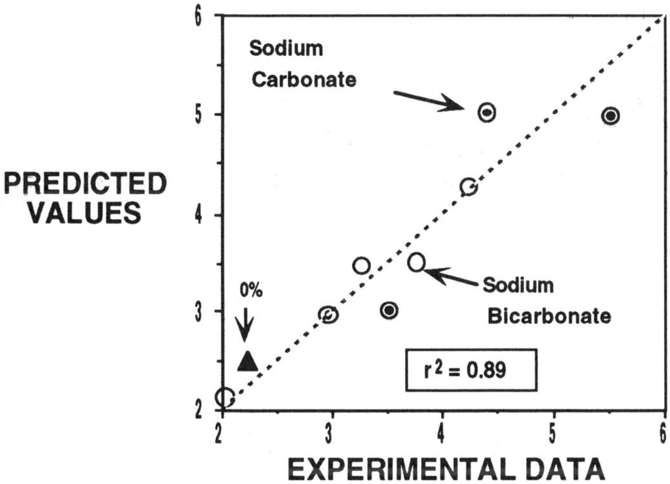

Figure 4: Diagram showing the correctness of the mathematical model: Specific refing energy = f (sulfonate content, pulp yield) by comparing predicted values and experimental data.

CONCLUSIONS

Steam explosion pulping of aspen previously impregnated with sodium salts (sulfite + carbonate or bicarbonate) leads to the manufacture of good quality paper and needs relatively low energy requirements. Moreover, pulp yields are higher than 85% (i.e. ultra-high-yield pulps). Sodium bicarbonate protects fibre against hydrolysis and degradation better than sodium carbonate. The addition of a second chemical (either sodium carbonate or bicarbonate) to a 8% sodium sulfite solution substantially gives rise to the formation of sulfonate groups. The optimum concentration of second chemical appears to be 1%.

It has been demonstrated that it is possible to accurately predict the specific refining energy as well as paper properties by knowing only two fibre parameters (sulfonate content and pulp yield). Apart from tear index, all square regression coefficients were 0.85-0.94. We believe that these correlations will be useful: a) to estimate paper properties before the formation of sheets; b) to compare the properties of pulps resulting from the same pulping process where two or more operating parameters are changed; c) to compare the performance of different chemimechanical pulping processes by taking sulfonate content and pulp yield as a reference.

ACKNOWLEDGEMENTS

This work was financed by grants of NSERC of Canada and FCAR of Québec.

REFERENCES

1. Beatson, R.P.; Heitner, C.; Rivest, M.; Atack, D. *Paperi ja Puu-Papper och Trä* **1985**, 67 (11), 702.
2. Heitner, C.; Atack, D. *Pulp Paper Can.* **1983**, 84 (11), T252.
3. Kokta, B.V.; Ahmed, A.; Zhan, H.Y.; Barbe, M. *Paperi ja Puu - Paper & Timber*, **1989**, 71 (9), 1044.
4. Kokta, B.V.; Ahmed, A.; Zhan, H.Y.; Proulx, R.; Carrasco, F. *Investigacion Técnica Papel*, **1991**, 107, 38.
5. Bengtsson, G.; Simonson, R. *Nordic Pulp Paper Res. J.* **1990**, 1, 16.
6. Carrasco, F.; Garceau, J.J.; Ahmed, A.; Kokta, B.V. *Nordic Pulp Paper Res. J.* (in press).

9

Effect on pulp and paper properties of chemical treatment prior to seam explosion pulping of aspen

Yuxia Ben*, Bohuslav V.Kokta and Jacques Doucet*** - *Départment Sciences du Bois, Université Laval, Ste-Foy, Québec, G1K 7P4, Canada. **Centre de Recherche en Pâtes et Papiers, Université du Québec à Trois-Rivières, Quebéc, G9A 5H7 Canada.

ABSTRACT

The effect of 18 different chemical pretreatments on pulp and paper properties of steam explosion pulping (SEP) of aspen has been studied. The presence of NaOH, $NaHCO_3$ or $MgCO_3$ when compared to Na_2SO_3 alone or other chemical pretreatment systems leads to a lower relative specific refining energy consumption and higher physical strengths. In addition, the presence of $NaHCO_3$ or $MgCO_3$ results also in higher brightness and yield.

INTRODUCTION

Ultra-high-yield chemimechanical (CMP) chemithermomechanical (CTMP) pulps have good pulp and paper properties (1-4): high yield, high opacity and brightness, good bleaching response to hydrogen peroxide, their acceptable physical strengths, low cost and low pollution level and offer an excellent alternative and complement to either rather weak mechanical pulps or rather expensive chemical pulps with low yield (45-50%). However, the major drawback of CMP and CTMP, compared to chemical pulps, is their higher refining energy consumption. Recently, another ultra -high-yield pulping process, called "steam explosion pulping" (SEP) or "S-pulping" was proposed by B. V. Kokta et al for both hardwood and softwood [5,6]. It has been shown (7, 8, 9, 10) that the steam explosion aspen pulps can be prepared to give pulp and paper properties considerably higher than CTMP/CMP pulps prepared with the same chemical cooking charge and similar to that of bleached hardwood kraft pulp. However, refining energy was about 30-50% lower. The SEP process consists of the chemical impregnation pretreatment of chips, short time steam cooking at high temperature, pressure release and refining.

It has been reported (11) that the impregnation solution can protect chips against oxidation during the cooking stage, to develop hydrophilic ionic groups on fiber surfaces during the steam treatment, to make good swelling of chips and to prevent acid hydrolysis leading to yield loss. Therefore, the composition of the impregnation solution is very important during the steam explosion pulping process. Nevertheless only a few kinds of chemicals as impregnation agent have been studied until now. The **OBJECTIVES** of the present study are to study the effect of chemical pretreatment on pulp and paper properties of the steam explosion pulping and to establish the most promising chemical treatment systems.

EXPERIMENTAL

Impregnation of chips

150g aspen chips (50% dryness) were mixed in plastic bags along with 150g of a solution made up of different concentrations of chemical products such as Na_2SO_3 alone or with NaOH, $NaHCO_3$, $MgCO_3$; NaOH with $MgCl_2$ or Na_2S or H_2O_2 alone. The liquor/chips ratio is 3:1. The time of impregnation was 24 hours; the temperature of impregnation was 60 °C. The percentage of chemicals used during solution impregnation appears in the Tables. The impregnation of two other samples consists of two steps. First step: 150g aspen chips (50% dryness) were mixed in plastic bag with 150g of a solution made up of 2% H_2O_2, 1% NaOH, 2.5% Na_2SiO_3, 0.25% $MgSO_3$, 0.025% DTPA. The Time of this pretreatment was 6 or 24 hours; second step: neutralizing samples with $Na_2S_2O_5$ to pH 5.5, washing 3 times with deionized water, then were mixed with 6% Na_2SO_3. The time was 18 or 24 hours. The temperature was 60 °C for both steps. In all cases, 0.5% DTPA was applied. For comparison, only H_2O was used as an impregnation solution in one sample.

Cooking, refining and refining energy

The cooking, refining and refining energy was carried out according to previously published work(11). The temperature of cooking was 190 °C, and time was 4 minutes.

Property evaluation

Paper sheets were prepared and tested according to standard CPPA testing methods. Brightness was evaluated on 3 g paper sheets made from distilled water.

Bleaching

The bleaching was carried out according to a previously published procedure, in one stage, using 2.0% hydrogen peroxide (12). The pulp consistency of the bleaching was 20%, temperature was 80°C, time was 3 hours.

RESULTS AND DISCUSSION

The effect of experimental conditions on pulp and paper properties of the steam explosion aspen pulp has been studied earlier (8, 9, 13). It was shown that the optimum pulp and paper properties and yield were obtained in lab scale with a 16% sodium sulfite impregnation charge at impregnation temperature 60°C for 24 hours and at cooking temperature of 190°C for 4 minutes. The impregnation charge of 16% resulted in approximately 8% chemical uptake.

The variation of the chemical pretreatment system leads to considerable changes of pulp and paper properties. Table I presents the impregnation solutions, relative specific refining energy consumption, pulp and paper properties. Results in Table I show that when the wood chips were impregnated with solution without impregnation chemicals, the yield, brightness and paper physical strengths of the steam explosion pulp were much lower compared with those of pulps impregnated with chemical impregnation solution.

Table I. Pulp and Paper Properties

Impregnation solution	8%Na_2SO_3	8%Na_2SO_3 +1%NaOH	8%Na_2SO_3 +1%$NaHCO_3$	8%Na_2SO_3 +1%$MgCO_3$	8%Na_2SO_3 +1%$MgCl_2$	8%Na_2SO_3 +1%Na_2CO_3	8%Na_2SO_3 +1%$ZnCl_2$	No Chemicals
Yield (%)	90.3	87.0	90.4	93.0	90.4	88.2	91.4	87.9
CSF (ml)	103	110	100	110	125	140	135	105
Refining energy (kJ/kg)	3.68	2.55	2.75	2.73	3.82	2.98	4.14	2.96
Brightness (%)	61.8	58.8	65.5	64.2	63.9	63.1	61.9	34.3
Opacity (%)	91.1	86.9	85.5	90.8	90.5	86.3	91.8	98.9
Porosity (ml)	68	8	16	31	72	25	141	74
Burst index (kPa m^2/g)	3.28	4.91	4.63	4.61	3.33	4.81	2.83	2.09
Tear index (mN m^2/g)	7.16	7.30	7.27	7.26	6.50	7.02	6.09	5.92
Breaking length (km)	5.91	7.10	7.75	7.46	6.27	6.94	5.50	3.31
Light scat. coefficient	488	322	345	411	432	359	469	433

Table I indicates also that the pulp using 8% Na_2SO_3 + 1% $NaHCO_3$ or 8% Na_2SO_3 + 1% $MgCO_3$ as impregnation solution has similar physical strengths and relative specific refining energy consumption as those of the pulp using 8% Na_2SO_3 + 1% NaOH as impregnation solution, but their yields and brightness are much higher, the yield increases from 87% to 90% or 93%, the brightness increases from 58.8% to 65.5% or 64.1%. This is mainly because of the buffering action of the bicarbonate/carbonate as well as due to the protective action of magnesium carbonate.

Moreover, the addition of 1% $MgCl_2$, 1% Na_2CO_3 or 1% $ZnCl_2$ to 8% Na_2SO_3 impregnation solution did not show any distinct promise for explosion pulping under the studied conditions, some pulp and paper properties, either yield or brightness or physical strengths were inferior to those of pulp utilizing 8% Na_2SO_3 + 1% $NaHCO_3$ as impregnation solution.

Table II. Pulp and Paper Properties

Impregnation solution	3%NaOH + 1%$MgCl_2$	2.5%NaOH + 1%Na_2S
Yield (%)	88.4	88.7
CSF (ml)	103	106
Refining energy (kJ/kg)	2.61	2.52
Brightness (%)	45.6	41.5
Opacity (%)	96.8	94.8
Porosity (ml)	21	16
Burst index (kPa m^2/g)	3.34	3.67
Tear index (mN m^2/g)	7.25	7.52
Breaking length (km)	5.43	7.24
Light scat. coefficient	443	373

The 3% NaOH + 1% $MgCl_2$ and 2.5% NaOH + 1% Na_2S are not good impregnation solutions. The product pulp and paper properties, brightness or yields were rather poor (see Table II).

Using H_2O_2 as impregnation solution, the explosion pulp has lower brightness and physical strengths. But if the chips were impregnated thereafter with 6% Na_2SO_3, the brightness and physical properties increase evidently and are superior to those obtained with 8% Na_2SO_3 as impregnation solution (See Table III).

Table III. Pulp and Paper Properties

Impregnation solution	2%H_2O_2	2%H_2O_2 (6 hrs) -- 6%Na_2SO_3 (18 hrs)	2%H_2O_2 (24 hrs) -- 6%Na_2SO_3 (24 hrs)
Yield (%)	92.2	91.0	86.3
CSF (ml)	142	110	110
Refining energy (kJ/kg)	3.25	2.73	2.41
Brightness (%)	47.3	71.1	67.3
Opacity (%)	97.9	84.8	79.7
Porosity (ml)	120	27	7
Burst index (kPa m²/g)	2.64	3.89	4.91
Tear index (mN m²/g)	6.85	7.21	7.50
Breaking length (km)	4.91	7.65	8.66
Light scat. coefficient	487	383	302

The ratio of Na_2SO_3 and $NaHCO_3$ in impregnation solution has been studied (see Table IV). The results show that along with an increase in the amount of $NaHCO_3$ in the impregnation solution, the yield increases slightly, the brightness increases evidently and reaches its peak when the impregnation solution consists of 4% Na_2SO_3 + 4%$NaHCO_3$, then decreases rapidly, the physical strengths go down. The addition of 1-1.5% $NaHCO_3$ to 8% Na_2SO_3 impregnation solution leads to very good pulp and paper properties, but at the expense of opacity and light scattering coefficient.

Table IV. Pulp and Paper Properties

Impregnation solution :	Na_2SO_3 (%) / $NaHCO_3$ (%)	8 / 0	4 / 4	2 / 6	0 / 8	8 / 0.5	8 / 1	8 / 1.5
Yield	(%)	90.3	90.5	91.7	91.8	90.3	90.4	90.3
CSF	(ml)	103	110	129	140	137	100	136
Refining energy	(kJ/kg)	3.68	2.93	2.81	2.68	3.25	2.75	2.68
Brightness	(%)	61.8	67.2	59.8	46.8	92.9	65.5	65.7
Opacity	(%)	91.1	86.6	91.7	96.9	90.3	85.5	86.0
Porosity	(ml)	68	33	55	61	77	16	15
Burst index	(kPa m²/g)	3.28	3.77	3.40	2.80	3.75	4.63	4.60
Tear index	(mN m²/g)	7.16	6.82	7.09	6.99	7.20	7.27	7.37
Breaking length	(km)	5.91	5.83	5.50	4.98	6.54	7.55	7.67
Light scat. coefficient		448	403	420	435	443	345	365

Fig.1 BREAKING LENGTH -- CSF

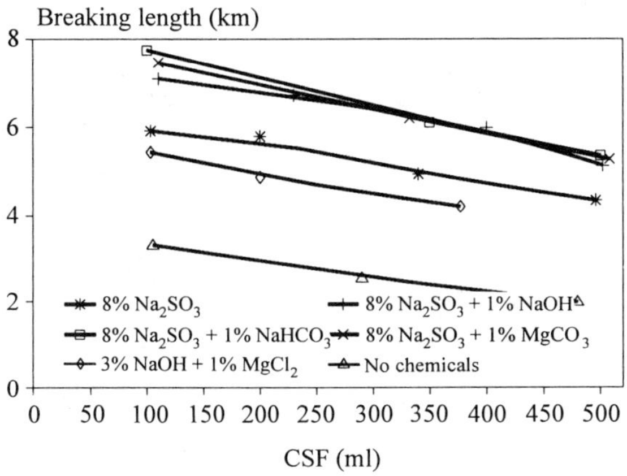

Fig.2 ENERGY -- CSF

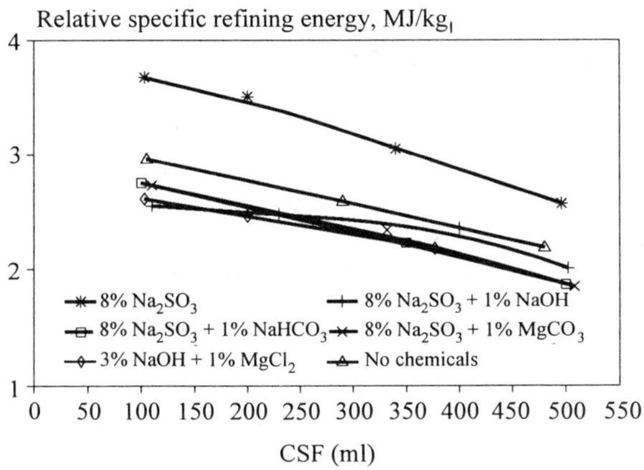

Fig.3 BRIGHTNESS

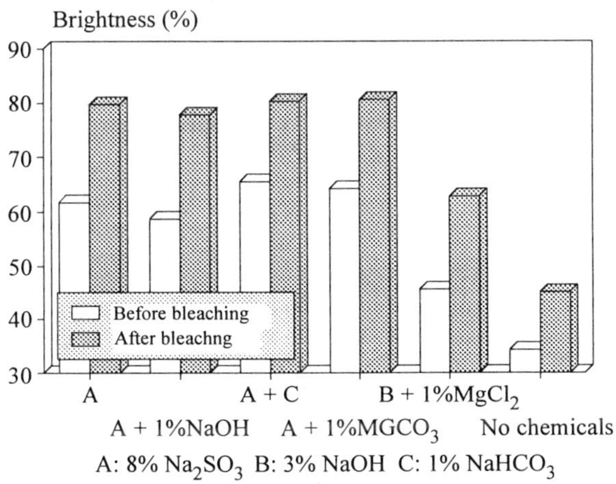

Fig.1-2 presents the breaking length and relative specific refining energy at different CSF levels of some samples. The results show that the pulp using 8% Na_2SO_3 with 1% $NaHCO_3$, 1% $MgCO_3$ or 1% NaOH as impregnation solution has much higher physical strengths and much lower relative specific refining energy consumption at different CSF levels when compared to those of other pulps.

Fig.3 presents the brightness after bleaching with hydrogen peroxide of some samples. The bleaching of the pulp using 8% Na_2SO_3 alone or with 1% $NaHCO_3$, 1% $MgCO_3$ or 1% NaOH as impregnation solution was very easy with 2.0% hydrogen peroxide. The brightness increases to about 80% after bleaching.

In order to explain the effect of the impregnation solution on pulp and paper properties, the physico-chemical characteristics of the explosion pulps have been studied. The results will be published in near future.

CONCLUSION

The presence of NaOH, $NaHCO_3$ or $MgCO_3$ when compared to Na_2SO_3 alone or other chemical systems lead to a lower relative specific refining energy consumption and higher physical strengths.

In addition, the presence of $NaHCO_3$ or $MgCO_3$ when compared to Na_2SO_3 alone or other chemical systems results also in higher brightness and yield.

The other systems (using Na_2SO_3+$MgCl_2$; Na_2SO_3+Na_2CO_3; Na_2SO_3+$ZnCl_2$; NaOH+Na_2S; NaOH+$MgCl_2$; H_2O_2 as impregnation solution) have not shown any distinct promise for explosion pulping under the studied conditions.

REFERENCES

1. Jackson, M., Falk, B. and Akerlund, G. "High yield pulp from North American Aspen(Populous tremuloides)", **Tappi Journal**, vol.68, no.11, 1985, p62-66.
2. Heitner, C., Atack, D., "Dynamic mechanical properties of sulfite treated Aspen", **International symposium on wood pulping chemical**, Japan, vol.2, 1983, p36-41.
3. Franzer, R., Li, K., "Aspen CMP, A supplementary Mechanical pulp", In preprint of **CPPA Annual Meeting** in Montreal, vol.69B, 1983, p163-173.
4. Wegner, T. H., "Improved strength in high yield pulps through chemical pretreatment", **Tappi Journal**, vol.65, no8, 1982, p103-107.
5. Kokta, B. V. Method, "Process for Preparing Pulp for Paper Making", **Can. Pat.**, #1,230,208 (Dec. 15, 1987).
6. Kokta, B. V. Method, "Process for Preparing Pulp for Paper Making", **U.S. Pat.**, #4,798,651 (Jan. 17, 1989).
7. Kokta, B. V. Method, "Improved Process for Preparing Pulp for Paper Making", **Can. Pat.Appl.**, #542643 (may, 1987).
8. Kokta, B. V., Ahemed, A., Zhan, H. and Barbe, M., "Explosion Pulping of Aspen", **Paperi ja puu paper and Timber**, vol. 71, no9 1989, p1044-1055.
9. Kokta, B. V., Ahemed, A., Zhan, H. Y., Barbe, M., "Explosion Pulping of Aspen, II. Effect of Operating Conditions on Resulting Paper Properties", **Proceedings**, 1989 ISWPC, Raleigh, N. C. (May 22-25, 1989).
10. Barbe, M. C., Kokta, B. V., Lavallee, H-C. and Taylor, J., "Aspen pulping: A comparison of Stake explosion and conventional chemimechanical pulping processes", **Pulp and Paper Canada**, vol.91, no.12, 1990, pT395-403.
11. Kokta, B. V., Ahmed, A., Garceau, J. J. and Chen, R., "Progress of Steam Explosion Pulping: An Overview", **Cellucon 90**, Aug. 28, 1990.
12. Kokta, B.V. and Daneault, C., "Brightening Ultra-High-Yield Hardwood Pulps with Hydrogen peroxide and sodium hydrosulfite", **TAPPI**, vol.69, no.9, 1986, p130-133.
13. Zhan, H. Y. and Kokta, B. V. "A study on the explosion pulping of Aspen", **China Pulp and Paper**, no.6, 1988, p29-36.

10

Papermaking properties of hydrothermally pretreated wood and annual plants

A.J. Zemann*, O. Bobleter* and P. Krkoška** - *Institue for Radiochemistry, Leopold-Franzens-University Innsbruck, Innrain 52a, A-6020 Innsbruck, Austria. **Pulp, Paper and Textile Department, Slovak Technical University Bratislava, Radlinskeho 9, 812 37 Bratislava, CSFR.

Abstract

Wood (poplar) and annual plants (sugar cane bagasse, wheat straw) were hydrothermally pretreated.

In the process of hydrothermolysis pure water without any additional chemicals is pumped under elevated temperatures (ca. 200°C) and pressure (20 - 30 bar) through the reaction vessel either in a direct percolation or in a recirculation process. In order to optimize the procedure, the fiber properties of the cellulosic residues were investigated (tensile strength, breaking length, and other papermaking properties).

With hydrothermolysis up to 95% of the lignin can be dissolved in one percolation run whereas during recirculation a certain amount of the dissolved lignin re-precipitates.

Introduction

In recent years some progress has been made in decreasing environmental pollution during pulping processes. Cooking chemicals are recycled from the liquors and new cooking chemicals are developed and used, e.g. in the ASAM process, where anthraquinone and methanol are added to stabilize the carbohydrates from degradation and to avoid recondensation of reactive lignin degradation products[1]. Investigations dealing with the use of enzymes and fungi for pulping and bleaching have been reported[2,3].

Additional processes, which use only little or none of the traditional cooking chemicals or only in order to impregnate wood chips, have been investigated extensively, e.g. the Organocell process[4], steam explosion[5], and hydrothermolysis[6], to name a few.

In the case of hydrothermolysis, pure water without any additional chemicals is pumped through a reaction vessel filled with the fiber raw material. The temperature is kept at approximately 200°C and, to keep the water in a liquid state, a pressure of 20-25 bar is adjusted. As described earlier[6,7], a percolation as well as a recirculation mode can be operated. The first experiments were carried out solely in the percolation mode.

To increase the concentration of dissolved matter in the eluate, the 50 liter pilot plant was equipped with a recirculation pump to lead the eluate back into the reaction vessel. This additional option results in an increase of concentration of the dissolved and extracted material by a factor of 5 to 6. The original motivation for the development of hydrothermolysis was to pretreat plant biomass to higher accessibility for saccharification and fermentation of the solid cellulosic residue[8]. The dissolved part of the biomass, which consists of all the hemicellulose and the larger part of the lignin, was first intended to be used as a feeding stuff. Later on the hemicellulose also became accessible to saccharification and fermentation[9].

Results and Discussion

The materials investigated were poplar wood, wheat straw and sugar cane bagasse. Different parameters and reaction conditions, which are listed in table 1, have to be considered during hydrothermal pretreatment.

Table 1. Reaction Conditions

Material*	P 1	P 2	BA 1	BA 2	S 1	S 2
Temperature (°C)	195	195	195	195	195	195
Processing Mode **	R	R	P	R	P+R	P
Residence Time (min)	40	70	65	50	35	25
Filling-in-ratio (kg/m^3)	188	177	101	130	78	80
KLASON Lignin (%)	38.5	39	13.4	25	9.3	9.5

* P = poplar; BA = bagasse; S = straw

** R = Recirculation; P = Percolation

In the case of hydrothermal pretreated wheat straw, local deviations of the lignin content were observed after percolation. Up to 95% of the original lignin was removed from the material located in the upper part of the reaction vessel, where the fresh water is pumped in.

After pretreatment the fiber residues were stored at room temperature under sterile conditions. For reasons of transportation the samples were air dried before fiber properties were examined.

After determination of the pulp properties, such as tensile strength, zero span breaking length, and shear cohesion, we found that the cellulose samples required at least a slight delignification. Even beating in a laboratory valley hollander was not possible for the unbleached poplar wood and bagasse samples because of the brittleness of the lignin-containing fibers. For this reason a delignification of the poplar samples was carried out in two different 3-stage-procedures:

1. For the P1 residue an acidic delignification sequence with $NaClO_2$/acetic acid, followed by an extraction stage with 3% NaOH was carried out. After a final bleaching step with alkaline H_2O_2 a pulp with a brightness of 82% MgO was obtained.

2. For the P2 sample an alkaline H_2O_2 delignification, followed by an extraction with 3% NaOH and a final alkaline H_2O_2 stage was applied, which rendered a pulp with a brightness of 77% MgO.

Then fiber properties of the bleached and - if possible - of the unbleached samples were determined and are shown in table 2.

Table 2. Fiber Parameters

	unbleached				bleached	
Material	P 1	BA 1	S 1	S 2	P 1	P 2
CSF	260	180	400	370	310	300
l_t (m)	275	660	1,750	2,013	1,990	2,730
l_{to} (m)	915	1,420	4,700	5,200	4,620	4,770
σ_s (kPa)	-	-	556	641	313	548
L_l (mm)	-	-	-	-	0.63	0.63

The fiber properties of the pulp samples, such as fiber length, fiber diameter and coarseness, were determined. To calculate the tensile strength of the paper test sheets by functional parameters an equation, first introduced by Page[10] and modified by Krkoska[11], was applied using these values. In addition, equations developed by Kallmes, Bernier, and Perez[12] were also examined using hydrothermal fiber properties (Table 3).

Table 3. Experimental and Calculated Fiber Properties (meters)

	P 1		P 2	
	l_t	l_{to}	l_t	l_{to}
experimental	1,990	4,620	2,730	4,770
modified Page	2,130	-	2,740	-
Kallmes-Bernier-Perez	2,500	4,670	3,110	5,110

modified Page-equation:

$$\frac{1}{l_t} = \frac{9}{8 \cdot l_{to}} + \frac{0.12 \cdot D \cdot \rho \cdot g}{L_l \cdot \sigma_s \cdot RBA}$$

Kallmes-Bernier-Perez-equation (for RBA > RBA_{cr}):

$$l_t = \frac{1-f_f}{1-f_i} \cdot \varepsilon \cdot \left(1 - \frac{RBA_{cr}}{2 \cdot RBA}\right)$$

$$l_{to} = \frac{1-f_f}{1-f_i} \cdot \frac{9}{8} \cdot E \cdot \varepsilon$$

D = fiber width (mm); ρ = cell wall density (1560 $kg.m^{-3}$); g = gravity constant (9.81 $m.s^{-2}$); L_l = fiber length (mm); σ_s = shear cohesion (Pa); $RBA_{(cr)}$ = (critical) relative bonded area (%/100); $(1\text{-}f_f)$ = proportion of active fibers during the rupture of fibers; $(1\text{-}f_i)$ = proportion of active fibers at the beginning of the deformation; E = module of elasticity (m); ε = total elongation (%/100)

Using the modified Page formula the values for the measured and calculated tensile strengths correspond within a few percent. The equations of Kallmes-Bernier-Perez were applied under the assumption of the term $(1 - f_f)/(1 - f_i)$ to be 1. The equations render higher breaking lengths than the measured values. This led to the assumption that in our pulps this ratio was shifted to a higher portion of active fibers during the rupture of the paper. To calculate the real value of $(1 - f_f)/(1 - f_i)$ the Young's modulus E at different freeness values has to be determined. This work is under way.

In the case of hydrothermolysis, the presence of reactive lignin degradation products mainly influence the fiber properties. As the process of hydrothermolysis uses no chemicals, the solvation mechanisms of traditional cooking procedures, such as the Kraft and sulfite process cannot take place to result in a specific lignin degradation and dissolution of the fragments. If the temperature of the process is kept too high or if recirculation is carried out too long, the possibility of recondensation reactions of reactive lignin fragments increases dramatically. It can be assumed that at least two reaction mechanisms control the hydrothermal degradation of lignin[13,14]:

1. a fast step of degradation with the formation of soluble low molecular weight lignin fragments. These degradation fragments consist mainly of mono- and oligomeric phenolic compounds and are not stable over a longer period of time.

2. a slower step with recondensation reactions of these lignin fragments forming high molecular weight products which reduce the amount of soluble products in the liquid hydrothermal eluates. These polymerisation and recondensation reactions can also take place on the surface areas of the cellulosic fiber, which leads to an increase of the amount of acid insoluble lignin and seem to cause a decrease of the degree of polymerisation of the cellulose molecule.

At temperatures kept below 250°C these reaction rates of formation and recondensation are almost equal. As a consequence, the hydrothermal fiber residues must be delignified and disintegrated shortly after the pretreatment to ensure an optimum of fiber parameters.

Conclusion

Hydrothermolysis is a process which separates the hemicellulose and a large part of the lignin from the cellulosic residue. To optimize the process, combined recirculation-percolation-sequences are investigated in order to minimize the lignin content of the residue. Thereby an economic utilization of the hemicellulose fragments in the liquid phase can be obtained using saccharification and fermentation techniques. Aromatic compounds deriving from degraded lignin can be converted to key organic chemicals.

Acknowledgement

We would like to thank J. Kubikova for her help in calculating the fiber properties.

Literature

1 Patt R., Kordsachia O. and Ghei F. in: G. Grassi, G. Gosse and G. dos Santos, *Biomass for Energy and Industry*, Elsevier Appl. Sci., London, 1990, 2.1056.

2 Trotter P.C.; *Tappi J.* **73/4** (1990) 198.

3 Trotter P.C.; *Tappi J.* **73/5** (1990) 201.

4 Pappens R.; *PPI* **32/3** (1990) 74.

5 Bouchard J., Nguyen T.S., Chornet E. and Overend R.P.; *Biomass* **23** (1990) 243.

6 Zemann A., Bobleter O. and Prutsch W.; *in press.*

7 Bobleter O., Vidotti R., Zemann A. and Prutsch W. in: G. Grassi, G. Gosse and G. dos Santos, *Biomass for Energy and Industry*, Elsevier Appl. Sci., London, 1990, 2.31.

8 Hörmeyer H.F., Schwald W., Bonn G. and Bobleter O.; *Holzforschung* **42** (1988) 282.

9 Walch E., Zemann A., Schinner F., Bonn G. and Bobleter O.; *Bioresource Technol.* **39** (1992) 173.

10 Page D.H.; *Tappi J.* **50** (1967) 449.

11 Krkoška P., Olos P., Mišovec P. and Pirozek J.; *Cellulose Chem. Technol.* **23** (1989) 609.

12 Kallmes O.J., Bernier G. and Perez M.; *Paper Technol. Ind.* **18/7** (1977) 222; **18/8** (1977) 243; **18/9** (1977) 283; **18/10** (1977) 328.

13 Bobleter O. and Concin R.; *Cellulose Chem. Technol.* **13** (1979) 583.

14 Burtscher E., Bobleter O., Schwald W., Concin R. and Binder H.; *J. Chromatogr.* **390** (1987) 401.

11

A new separation and modification system of phytomass constituents: swelling-solvation process

M. Funaoka, S. Fukatsu, J.Matsue, and K. Hamaguchi - Faculty of Bioresources, Mie University, 1515 Kamihama, Tsu, Mie 514, Japan.

ABSTRACT

A new technique (Swelling-Solvation Process) for the separation and modification of phytomass has been developed. This process is a combination of the swelling and dissolution of carbohydrates by concentrated acid and the solvation of lignin with phenol derivatives. The separation of phytomass constituents proceeds quantitatively at room temperature. The resulting lignin (S-S lignin) has several distinctive properties which conventional lignins do not have; it is highly phenolic and scarcely has conjugated systems, showing a white color. S-S lignin can be converted directly to various kinds of products without further modifications. On the other hand, carbohydrates can easily be converted to monosaccharides which are useful for raw materials for chemicals. This process can be applied for the phenol modifications of lignin preparations, giving highly phenolic and light color-lignin derivatives.

PROCESS THEORY

Several wood refining methods have been proposed recently. However, most of those have been developed with the major purpose of manufacturing pulps and/or utilizing carbohydrates. That is, the utilization of lignin has not been taken into account at all at the developing step of the processes and the lignin has been regarded as a barrier for carbohydrate utilization. In these conventional processes, the lignin is subjected to complicated structural modifications and consequently, it is very difficult to reconvert undesirably modified lignins to reactive forms.

A new phytomass refining process (Swelling-Solvation Process, S-S Process) has been designed, aiming at the utilization of both lignin and carbohydrates. It is based on the idea that carbohydrates, not lignin, should be destroyed for the phytomass refining : the quantitative depolymerization of carbohydrates is much easier than that of lignin because of simple interunit linkages, by which the structure of the cell wall would be broken down to give the perfect separation. However, the lignin is also subjected to complicated structural modifications during depolymerizing carbohydrates. Therefore, the key point to achieve this new process is to give lignin new functions as well as to avoid lignin modifications to less reactive forms during the treatment.

Swelling-solvation process is a new technique for rapid and quantitative separation of phytomass constituents and phenol modification of lignin [1,2,3]. This process is due to a combination of the solvation of lignin with phenol derivatives and the swelling and dissolution of carbohydrates by concentrated acid. The characteristic of this process is that carbohydrates and lignin are in the different phases during the treatment. This process includes two steps (Fig. 1):

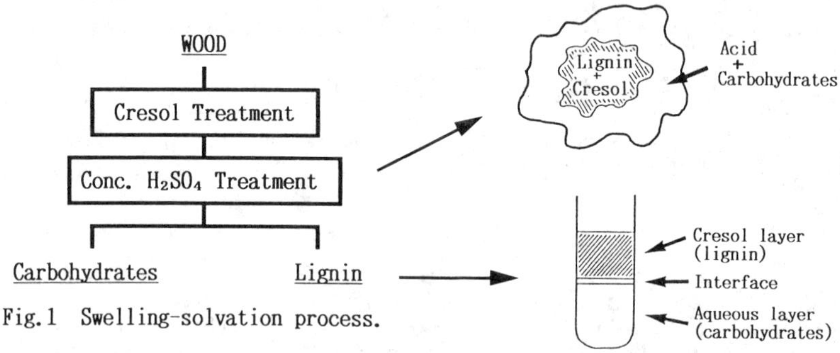

Fig.1 Swelling-solvation process.

● 【First treatment】 .... *p*-Cresol
A phenol derivative, for example *p*-cresol, is a good solvent for lignin, having reactivity similar to that of lignin phenyl nuclei. Furthermore, it does not mix with concentrated sulfuric acid at room temperature.
p-Cresol is added to wood meals with stirring at room temperature. *p*-Cresol has an affinity to lignin and is readily absorbed into wood to solvate with lignin.

● 【Second treatment】 ....Concentrated sulfuric acid
Concentrated sulfuric acid (72%) is added to the mixture of *p*-cresol and wood meals with vigorous stirring at room temperature. Sulfuric acid has a strong affinity with carbohydrates, resulting in the decomposition of tissue structures by swelling cellulose. During this process, the lignin is always surrounded by cresol which is not mixed with sulfuric acid. Consequently, the attack of sulfuric acid to lignin is greatly suppressed. Although lignin contacts with acid in the interface between cresol and acid layers, the resulting carbonium ions (mainly benzyl carbonium ions) are quickly stabilized by cresol, converting lignin to highly phenolic diphenylmethane type derivatives as well as preventing lignin self-condensation.

● 【Separation】
p-Cresol does not mix with concentrated sulfuric acid at room temperature. Furthermore, the specific gravities of cresol and concentrated sulfuric acid are totally different. Therefore, by stopping the stirring, the reaction mixture (a suspension) can be

rapidly separated into cresol and aqueous layers. Lignin is in the cresol layer, whereas carbohydrates in the aqueous layer.

PHYTOMASS REFINING

Separation behavior of lignin and carbohydrates

After the treatment, the reaction mixture was separated into two layers as shown in Fig. 1. White materials (insoluble carbohydrates containing small amount of lignin) were in the interface between aqueous and cresol layers. With the treatment of 20 min at room temperature, about 80 % of lignin were separated into the cresol layer and about 87 % of carbohydrates into the aqueous layer. With extended time, insoluble carbohydrates became soluble and moved to the aqueous layer, whereas lignin incorporated into carbohydrates moved to the cresol layer (Tables 1 and 2).

Table 1 Distribution of lignin*1 in S-S treatment*2.

Treatment time(min)	Aqueous layer	Interface	Cresol layer
20	2.1	17.6	80.3
60	1.7	8.7	89.6
120	1.6	2.7	95.7

*1 cresol free, % of original lignin
*2 Yezo spruce, 10 ml of *p*-cresol/g of wood, 20 ml of 72% H_2SO_4/g of wood.

Table 2 Distribution of carbohydrates*1 in S-S treatment*2.

Treatment time(min)	Aqueous layer	Interface	Cresol layer
20	86.6	11.5	1.9
60	91.4	7.2	1.4
120	95.3	3.1	1.7

*1 % of total carbohydrates
*2 Yezo spruce, 10 ml of *p*-cresol/g of wood, 20 ml of 72% H_2SO_4/g of wood.

The separation of hardwoods proceeded more rapidly than softwoods. This difference would be due to the difference not in tissue structures between softwood and hardwood but in lignin structures such as the distribution and quantity of ether linkages and the degree of condensation.

The separations of various wood species have been carried out. However, there was no difference in the separation behavior between wood species. It is because the wood tissue structures are broken down completely by the swelling and dissolution of carbohydrates during the treatment.

Properties of separated lignins (S-S lignins)

Compared to the milled wood lignin, S-S lignins were 5 % higher in carbon and 5 % lower in oxygen, due to the elimination of oxygen containing functional groups and the phenolation in those places. The amounts of combined cresol were about 25 % (about 0.65 mol/C_9) in softwood lignins and about 30 % (about 0.9 mol/C_9) in hardwood lignins (Table 3).

In general, the lignins isolated under acidic conditions have a dark brown color. Also the lignins phenolated using H_2SO_4 or HCl catalyst at high temperature were black. However, S-S lignins were white, having a slightly pinkish color, in spite of the isolation with concentrated sulfuric acid and a large amount of combined cresol in the side chains. This means that protolignins were not subjected to complicated structural modifications other than the phenolation in the side chains, due to the solvation with cresol. S-S lignins were readily dissolved in various solvents such as methanol, ethanol, acetone, etc.

The S-S lignin had a sharp peak only at 280 nm, having no other peak or shoulder, in UV spectrum (Fig. 2). ΔEi spectrum of S-S lignin had one sharp peak only at 300 nm (Fig. 3). These facts indicate that the S-S lignin is highly phenolic and has almost no conjugated systems in the side chains. This is one of the distinctive characteristics of S-S lignins.

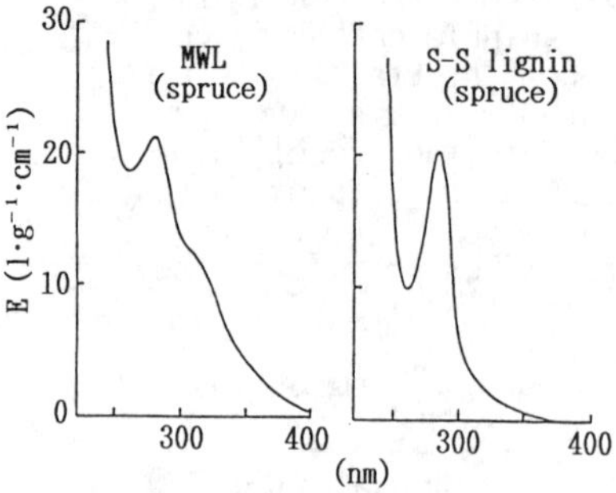

Fig. 2 UV spectra of lignins.
S-S treatment time: 20 min

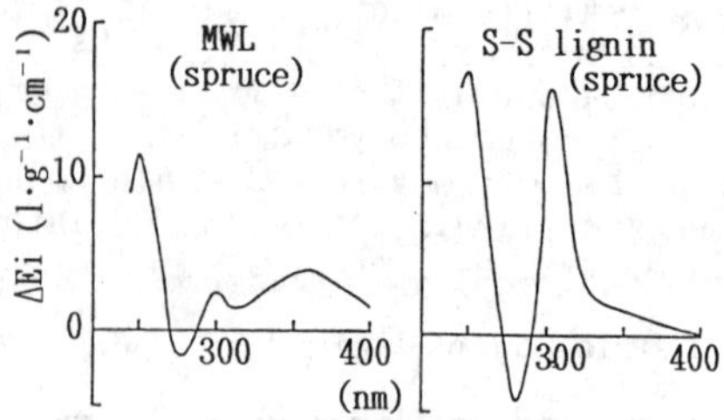

Fig. 3 ΔEi spectra of lignins.
S-S treatment time: 20 min

IR spectrum of S-S lignin was very sharp, having almost no absorption around 1650 cm^{-1} assigned to conjugated carbonyl groups and a sharp absorption around 800 cm^{-1} due to combined cresol (Fig. 4). These are in agreement with the characteristics of S-S lignin shown in UV spectra.

S-S lignins had no benzyl alcohol and the quantity of Cγ-hydroxyl groups was comparable to that of the milled wood lignin. Phenolic hydroxyl contents of S-S lignins (except for cresolic OH), which were higher by about 0.25 mol/C_9 than that of the milled wood lignin, remained almost unchanged, independent of the treatment time, implying that β-O-4 linkages were stable during the treatment (Table 3).

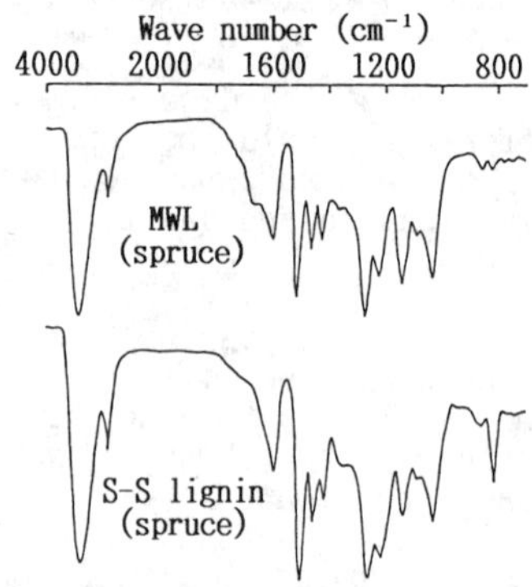

Fig. 4 IR spectra of lignins.
S-S treatment time: 20 min

From the above results, the modification of lignin during the treatment may be summarized as follows: The lignin is greatly protected from the attack of

Table 3 Properties of separated lignins (S-S lignins*1).

Species	Elemental composition(%) C	H	O	Combined cresol(%)	OH group (mol/C_9) Cα	Cγ	Phenolic
Spruce MWL	61.5	5.8	32.7	-	0.35	0.80	0.35
Yezo spruce	66.8	6.0	27.2	25.9	Trace	0.79	1.26*2
Slash pine	66.9	5.9	27.2	25.7	Trace	0.87	1.27*2
Birch MWL	59.7	6.1	34.2	-	0.53	0.82	0.32
Japanese birch	64.3	6.0	29.7	30.9	Trace	0.80	1.51*2
Kapur	65.7	6.0	28.3	30.7	Trace	0.91	1.56*2

*1 S-S treatment: 10 ml of *p*-cresol/g of wood, 20 ml of 72% H_2SO_4/g of wood, room temperature, 20 min. *2 Lignin Ph(OH) + Cresolic OH

sulfuric acid by solvating with cresol. The lignin contacts with acid for a very short time only in the interface between cresol and acid phases, and at that time, reactive benzyl alcohols and ethers mainly react with cresol to convert lignin to highly phenolic diphenylmethane type derivatives. Furthermore, cresol attacks also the α-β double bonds and α-carbonyl groups, resulting in no conjugated system in the side chains.

PHENOL MODIFICATION OF INDUSTRIAL LIGNIN

The S-S system used for phytomass refining, a heterogeneous reaction system consisting of a mixture of cresol and concentrated acid at room temperature, was applied for the phenol modification of industrial lignin (kraft lignin, KL). The results were compared with those by conventional phenolation processes. Phenolation conditions are shown in Table 4.

Table 4 Phenolation conditions of kraft lignin.

Reaction system	Acid (mol/g lignin)	*p*-Cresol (ml/g lignin)	Temp. (℃)	Time (min)	Vessel	Abbreviation of phenolated KL*
S-S (Heterogeneous)	H_2SO_4 (0.2410)	10	25	20	Open	SS-KL
Homogeneous	HCl (0.0012)	2	130	360	Sealed	Homo-KL(HCl)
Homogeneous	H_2SO_4 (0.0015)	2	170	180	Open	Homo-KL(H_2SO_4)

* Acetone sol.- ether insol. fraction

In contrast to the conventional phenolations in which the reaction mixture forms the homogeneous system at high temperature, it is expected that in the heterogeneous reaction system (S-S system) at room temperature, consisting of the cresol layer with lignin and the aqueous layer with a catalyst, the phenolation in the lignin side chains would proceed exclusively without complicated structural modifications.

The kraft lignin phenolated in the S-S system (SS-KL) had the highest solubilities toward acetone and alcohols, compared with the lignins by conventional processes (Homo-KLs). Homo-KLs had a black color, whereas SS-KL was grey, lighter than original kraft lignin (Fig. 5, Tables 5 and 6). SS-KL had combined cresols of 0.24 mol/C_9 which was, although more than that of Homo-KL(HCl), one half as much as that of Homo-KL(H_2SO_4) (Table 6). However, ΔEi spectrum of SS-KL had a sharp peak at 300 nm, showing highly phenolic property (Fig. 6). Furthermore, the broad peak around 350-400 nm for the original kraft lignin was not observed in SS-KL, due to the introduction of cresol to

Table 5 Yields of phenolated kraft lignins.

Phenolation system	Yield (%) Acetone insol.	Acetone sol. Ether insol.
S-S (Heterogeneous)	0.57	84.16
Homogeneous-HCl	23.86	88.23
Homogeneous-H_2SO_4	4.09	71.01

Table 6 Properties of phenolated kraft lignins.

Lignins	Combined cresol(%)	Color	Solubility* MeOH	EtOH
Original KL	-	Brown	39	21
SS-KL	12.03	Gray	72	58
Homo-KL(HCl)	10.01	Black	7	8
Homo-KL(H_2SO_4)	21.25	Black	34	47

* Solubility of phenolated kraft lignin (20 mg) to 3 ml of solvent (% of lignin).

conjugated systems. On the other hand, Homo-KLs still had a broad peak around 350-400 nm. The conjugated systems in the side chains would influence the reactivity of phenyl nuclei and make the conversion of lignin to high performance products difficult.

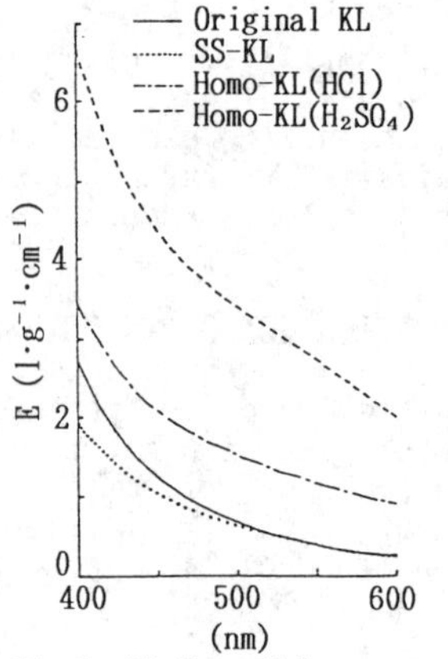

Fig.5 Visible light spectra of lignins.

CONCLUDING REMARKS

The swelling-solvation process is a new separation and modification system of phytomass constituents. The key point of this process is to destroy tissue structures of polysaccharides by concentrated acid under the solvation of lignin with cresol, that is, to allow lignin and carbohydrates to react in the different phases.

The characteristics of this process are as follows: 1. Rapid phytomass refining. 2. No special chemicals and equipment. 3. No selectivity for phytomass. 4. High reactivity of separated lignin. The separation of phytomass constituents proceeds rapidly at room temperature because the tissue structures are destroyed by the swelling and dissolution of carbohydrates by concentrated acid. There is no need for special equipment and high heat energy because the separation proceeds at room temperature under normal pressure. The chemicals required are

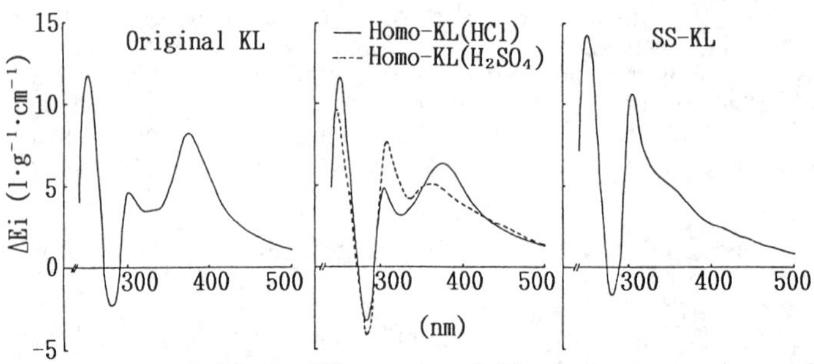

Fig.6 ΔEi spectra of lignins.

only sulfuric acid and cresol which are cheap. This process can be performed without large modification of conventional concentrated acid hydrolysis processes of which the technology has already been established. Furthermore, this process is applicable to any kind of phytomass to give perfect separation because of the decomposition of tissue structures during the treatment. The resulting lignin has various distinctive characteristics that conventional lignins do not have, such as no conjugated system, highly phenolic property, no self-condensation, white color etc. This lignin can be converted directly to various kinds of products without further modification.

REFERENCES

1. Funaoka, M.; Abe, I. *Tappi Journal* 72, 145-149 (1989)
2. Funaoka, M.; Abe, I. *Mokuzai Gakkaishi* 35, 1058-1062 (1989)
3. Funaoka, M. *6th ISWPC (Australia)* Vol.2, 397-403 (1991)

12

Anaerobic digestion of cotton and linseed plant stalks for the preparation of pulp and paper

A.J. Shaikh, Y.D. Pai, R.H. Balasubramanya and V.G. Khandeparkar - Central Institute for Research on Cotton Technology, (ICAR), Matunga, Bombay 400 019, India.

ABSTRACT

The preparation of good quality paper from cotton and linseed stalks by the conventional kraft pulping process is described. The possibilities of preparation of pulp and paper by anaerobic digestion have been explored. Evaluations of paper produced by a kraft process as well as anaerobic digestion showed that the papers were equivalent.

INTRODUCTION

In India, bamboo and softwood are the conventional raw materials for the manufacture of paper. The acute scarcity of these raw materials compelled the searching out of unconventional raw materials which could be readily available in sufficient quantity. Cotton plant stalk and linseed stalk are two such unconventional raw materials which could be exploited profitably by the paper industry. Apart from being agro-wastes, these are available in abundance, every year.

Cotton is mainly grown for fibre. India has about 8 million hectares of this crop under cultivation and produces around 2.1 million tonnes of lint per annum; in addition India generates around 15 million tonnes of cotton stalks each year. To a limited extent these stalks are used as fuel by the rural masses, but are mainly disposed off by burning in the fields after harvest of the cotton crop (1).

India is one of the major linseed growing countries, ranking third in the world. The total area of this crop under cultivation in India is around 2 million hectares with an annual seed production of about 50,000 tonnes. As per a rough estimate about 3.5 to 4.0 million tonnes of the stalks are generated as by-product annually mainly for the extraction of fibre popularly known as "flax" which has many specialised industrial applications (2). These stalks are not utilised for any useful purposes, though sometimes they are used as fuel by the rural people (3).

In order to find a better use for these renewable cellulosic agrowaste materials and to assess their potential as raw materials for the paper industry, attempts were made at this Institute to prepare pulp and paper using the conventional kraft pulping process. Results indicated that good quality pulp and paper could be produced from both these raw materials. However, the ehemical requirement during pulping was high and the cooking conditions involved were drastic (4,5).

Therefore, an attempt was made to reduce the chemical requirement and drastic cooking conditions by coupling a biological process of anaerobically digesting the materials before pulping (6).

EXPERIMENTAL

Dried cotton plant stalks belonging to G. hirsutum received from the Central Institute for Cotton Research, Nagpur (India) and dried linseed stalks of LCK.152 variety received from the Project Coordinator (Linseed), C.S. Azad University of Agricultural Technology, Kanpur (India) were used. All the chemicals used were of analytical reagent grade and analysis were made by TAPPI Standard Test Methods (7).

PRETREATMENT

Cotton stalks and linseed stalks were chopped to chips of about 1.5 to 2 cm and 7.5 to 12.5 cm size respectively and open boiled in different concentrations of sodium hydroxide (1% to 4% w/v) with a substrate to liquor ratio of 1.10. The materials were charged into anaerobic digesters after adjusting the C/N to 1:30 with urea. Digested slurry (10%) from a working batch digester of a biogas plant based on willow dust was used as inoculum. Batch digestions were carried out as before (8).
The digested materials were removed from the digesters at weekly intervals for the preparation of pulps. Open boiled stalks in different concentrations of alkali without anaerobic digestion served as controls.

PREPARATION OF PULP

The biodegraded cotton and linseed stalk samples along with their corresponding controls were converted into pulps by refining in a disc refiner in one pass at 15 thou (15/1000") clearance followed by defibrization in a valley beater to get the pulps of desired freeness, i.e. 250 CSF (Canadian Standard Freeness).

The cotton and linseed stalks were converted without pretreatment into pulps by cooking in a rotary digester with 16% kraft liquor ($NaOH + Na_2S$) on oven dry weight (o.d.) of the raw material, at a material to liquor ratio of 1:3, 160°C for 3 hrs, which included 1 hr required to raise the temperature of the digester from room temperature to the cooking temperature of 160°C. The cooked materials were washed thoroughly, followed by disintegration and then beating in a valley beater to get the pulps of desired freeness, i.e. 250 CSF.

BLEACHING OF PULP

Anaerobically digested samples and their controls were bleached with 1% sodium chlorite at 70°C for 2 hr. The pH during bleaching was maintained at 4.1.

The pulps prepared from cotton and linseed stalks by the kraft process were bleached using hypochlorite having 8% available chlorine, at 10% consistency, 40°C for two hr (1+1). The pH during bleaching was maintained at 8.5.

SHEET MAKING AND TESTING

Bleached paper sheets of 60 ± 1 g/m^2 were prepared from all the above pulps on a standard sheet making machine. The sheets were dried in air, conditioned at 65% R.H. and 27°C and then tested for various properties viz., breaking length, burst factor, tear factor, number of double folds, etc. using TAPPI Standard Test Methods.

RESULTS AND DISCUSSION

The results of the chemical analysis of cotton and linseed stalks indicated that both are rich in cellulose, 79% and 76% holocellulose respectively (5,9), and hence could serve as a potential source of raw material for the paper industry. The trials undertaken earlier on the preparation of pulps using different concentrations of alkali at 130°C indicated that the quality of pulp produced by using even 14% alkali was not satisfactory for making good quality paper (4,5). The reason might be the hardy nature of the stalks that inhibits the penetration of the chemicals inside the chips. This necessitates refining of the stalks, before cooking for opening up the lignocellulosic structure for better penetration of chemicals. The trials carried out after refining indicated that cooking the materials with 16% kraft liquor on o.d. weight of the raw materials at 160°C for 3 hr could give superior quality pulp. The pulps thus obtained were bleached in a two step hypo process and the paper sheets prepared from both the raw materials possessed all the desirable properties as given in Table 1.

TABLE 1: PROPERTIES OF PAPER MADE FROM KRAFT PULP

Properties	Cotton	Linseed
Freeness of Pulp (CSF)	250	250
Basis Weight (g/m^2)	60±1	60±1
Burst Factor	31.82	19.25
Tear Factor	154.8	112.2
Breaking Length (metres)	4346	3624
Double Folds (no.)	37	12

The data on various properties of the paper sheets prepared from the biologically softened cotton and linseed stalks are given in Table 2.

TABLE 2: PROPERTIES OF PAPER MADE FROM BIODEGRADED COTTON AND LINSEED STALKS

Canadian Standard Freeness - 250, G/M^2 - 60 ± 1

COTTON

Alkali Conc.(%)	Duration (Weeks)	Burst Factor	Tear Factor	Breaking Length(m)	Double Folds (no.)
1	Control	5.8	56	1323	2
1	1	6.9	79	1422	2
1	2	7.8	102	2170	19
1	2	9.3	107	2002	30
1	4	12.1	110	2323	20
2	Control	6.2	96	2068	10
2	1	14.1	98	2309	10
2	2	14.8	123	2389	41
2	3	15.6	128	2537	35
2	4	19.5	155	2634	63
3	Control	8.2	83	2259	14
3	1	14.6	87	2523	29
3	2	14.8	123	2389	41
3	3	19.5	155	2634	63
3	4	15.6	128	2537	35
4	Control	7.5	87	2645	24
4	1	28.2	102	3335	373
4	2	31.8	127	4346	951
4	3	26.5	102	3002	85
4	4	20.1	127	2463	34

LINSEED

Alkali Conc.(%)	Duration (Weeks)	Burst Factor	Tear Factor	Breaking Length(m)	Double Folds (no.)
1	Control	6.90	2435	127	1
1	1	9.30	1679	127	5
1	2	11.4	1967	205	20
1	3	12.5	2274	131	26
1	4	13.7	2515	105	40
2	Control	8.8	1668	128	6
2	1	9.1	1734	80	11
2	2	9.2	1967	80	11
2	3	13.2	2582	129	40
2	4	10.9	1976	99	17
3	Control	-	-	-	-
3	1	10.1	2001	121	9
3	2	14.2	2422	144	36
3	3	13.7	2001	139	25
3	4	10.8	2101	119	9
4	Control	10.7	2167	139	11
4	1	26.6	3190	139	45
4	2	20.4	2280	141	37
4	3	13.5	1937	119	21
4	4	11.2	1191	85	17

The results indicated that cotton stalks treated with 4% alkali and subjected to anaerobic treatment for two weeks produced good quality pulp as evidenced by the properties of the paper. The burst factor, tearfactor and breaking length recorded were on a par with that of the bleached paper made by the kraft process. However, the number of double folds was significantly higher in the case of the anaerobically digested sample. Similarly, the linseed stalks treated with 4% alkali and subjected to one week anaerobic digestion produced better quality pulp as indicated by the properties of the paper. In fact, the burst factor, tear factor and number of double folds were better than the bleached paper made by the kraft process. However, the breaking length was comparable. The probable reasons for the observed results could be that treatment with alkali the cellulose in the wet state results in an increase of strength as evidenced by the enzyme hydrolysis. When pretreated stalks were subjected to anaerobic digestion, the environment must have paved the way for the lignocellulose complex to attain its original biological volume. When such celluloses are treated with alkali, the microfibrils try to agglomerate. This has been observed in the case of never dried cotton cellulose (10). Another interesting observation is that increasing the concentration of alkali during pretreatment (1% to 4%) has an inverse relationship with the time of anaerobic digestion (Table 2). The study indicated that duration of digestion can be altered depending upon the quality of the pulp and paper required by changing the concentration of alkali.

CONCLUSIONS

This study, which encompasses a novel method for the production of pulp and paper from cotton and linseed stalks by anaerobic digestion has definite advantages over the conventional processes viz. low chemical requirement, mild cooking conditions, no toxic effluent problems, process economy,etc.

The quality of the paper produced from cotton and linseed stalks by anaerobic digestion was on a par with the quality of paper produced using 16% kraft liquor. The paper produced finds applications for writing and printing. This process is most ideally suited for adoption by the handmade paper industry in the rural areas.

ACKNOWLEDGEMENT

The authors thank Dr. N.B. Patil, Director, Central Institute for Research on Cotton Technology, for permitting to publish this article.

REFERENCES

1. Sundaram, V.; Balasubramanya, R.H.; Shaikh, A.J.; Bhatt, I.G.; Sitaram, M.S. J. Indian Soc. Cotton Imp. **1989**, 14(1), 94.

2. Shastri, S.B.N. The Wealth of India (Raw Materials), **1962**, 6, 122.

3. Anon. Proceedings of Annual Rabi Oilseeds Workshop, held at G.B. Pant University of Agriculture and Technology, Pantnagar, U.P., India, **1986**.

4. Pandey, S.N.; Shaikh, A.J. Indian Pulp & Paper **1985**, 40, 14.

5. Shaikh, A.J.; Varadarajan, P.V.; Sundaram, V. and Kulkarni, A.Y. Indian Pulp and Paper Ind. Tech. Asso., **1987**, 24 (2), 1.

6. Mehta, V.; Gupta, J.K. Res. & Ind. **1991**, 36, 161.

7. Tech. Asso. of the Pulp and Paper Ind. (TAPPI) **1980**, Standards of suggested methods, New York, T.200 - T.265.

8. Balasubramanya, R.H.; Khandeparkar, V.G.; Betrabet, S.M. and Sundaram, V. J. Text. Asso. **1981**, 42, 145.

9. Pandey, S.N.; Shaikh, A.J. Indian Pulp & Paper **1986**, 12(1), 10.

10. Bhatawdekar, S.P.; Balasubramanya, R.H.; Sreenivasan, S. and Paralikar, K.M. Textile Research Journal **1992**, 62(5), 290.

13

Possibilities of increasing dissolving pulp reactivity.

K. Fischer and I. Schmidt - Dresden University of Technology, Institute of Plant and Wood Chemistry, Pienner StraBe 23, O-8223 Tharandt, Germany.

ABSTRACT

According to our investigations there are two ways of increasing the reactivity of cellulose. It was shown that several levels of the cellulose structure were influenced by irradiation. By means of electron beam radiation the molecular weight distribution of the cellulose chains could be narrowed and the degree of polymerization decreased. The other way is to delignify the wood chips extended during the acid bisulphite process by using the lignosulphonates dissolved during an acid hydrolysis. In this way the distribution of the cellulose chains is also narrowed, and the viscose from these pulps has a good quality.

INTRODUCTION

The accompanying products of the pulp and the reactivity of cellulose, which is dependent upon the structure of the cellulose, have to be considered, especially when processing rayon fibers and filaments.

The first problem arising in connection with the accompanying products may be solved during the production of the pulps.

The second problem is not easy to solve as the problems become obvious only during the reaction between the cellulose and the chemicals in a heterogeneous system. The viscose fiber production is influenced by the reactivity of the hydroxyl groups which depends on the intramolecular and intermolecular interactions of the hydroxyl groups, the proportion of the crystalline and disordered regions within the fibrils of the

fibers and the conglomeration of the fibrils.
In this paper two possibilities have been investigated to improve the reactivity of pulp in order to obtain a better viscose quality.

THE EFFECT OF PRETREATING CELLULOSE WITH HIGH-ENERGY ELECTRONS

One possibility of influencing all the structural levels of a dissolving pulp is to treat the pulp with high-energy electrons. As a result of the high energy these beams have no problems when penetrating both the disordered and crystalline regions of the cellulose.
During irradiation free radicals are formed in the cellulose molecules. There are several possibilities of radical types which may be formed.
The radicals formed as a result of dehydration of the C_1 and C_4 position of the cellulose molecule play an essential rôle in the subsequent reactions, as the 1,4-glucosidic linkage is split off [1]. This means that changes at the macromolecular level occur. The so formed primary radicals may react subsequently, keeping their radical properties, whereby water is eliminated.
As a result carbonyl groups are formed and the radical moves to the neighbouring C-atom [2].
In the presence of oxygen the carbonyls may be oxidized. The subsequent radicalic reactions cause changes at the anhydroglucose unit which means changes at the molecular level occur.
The most obvious effect linked with the irradiation of cellulose is the decrease of the degree of polymerization. The number of chain splittings and the degree of scissions, respectively, grows directly in proportion with the irradiation doses (Fig.1) [3]. The uniformity of the molecular weight distribution increases at growing irradiation doses (Fig. 2). The determined molecular weight distributions of cellulose approach with growing doses the most probable distribution according to Kuhn and Flory [2]. That means

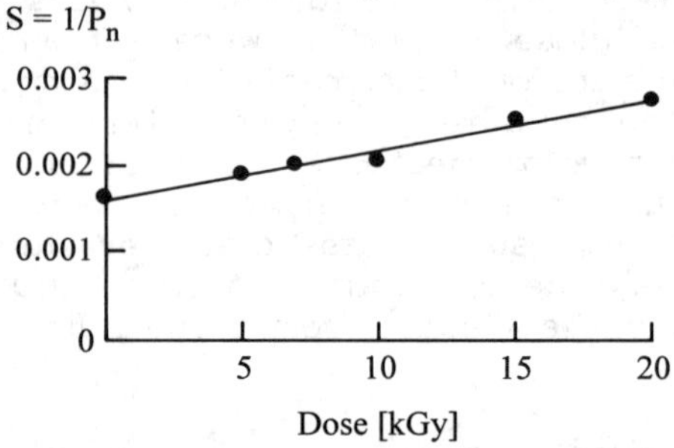

Figure 1. Scissions of cellulose chains

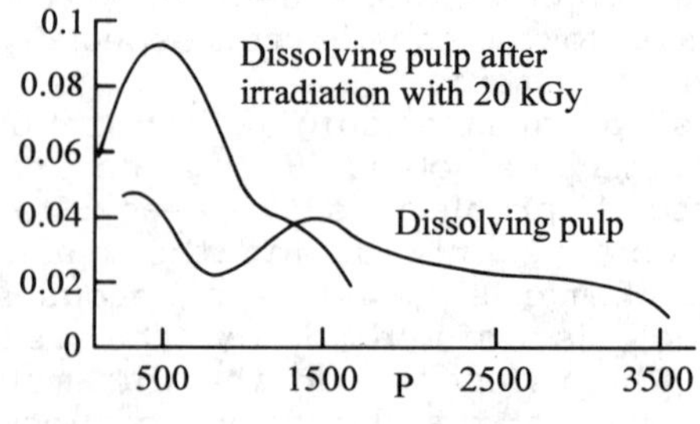

Figure 2. Molecular weight distribution

the cellulose chains are evenly cut according to statistic laws both in the disordered and crystalline regions [3].
It is not possible to describe changes in the supermolecular structure with one method only. Only the results of several methods allow a characterization of the system of hydrogen bonds and of the arrangements of the cellulose molecules to each other.
The ray-induced changes in the cellulose will be shown through the results from the determination of the water retention values (WRV) after alkaline treatment, and the distribution of the substitutes after xanthation.
The results from the determination of the water retention value subsequent to an alkaline treatment of the pulp in sodium hydroxide solutions within a concentration ranging from 2% to 20% show that the swelling of pulp in sodium hydroxide is very intensive in the range with values between 16 - 20%, whereas the irradiated pulp reaches the maximum of swelling at a lower concentration of alkali (Fig. 3). The earlier initiation of swelling, the higher value of swelling and the achievement of the swelling maximum at a lower sodium hydroxide concentration means that the accessibility for alkaline solution is improved, probably via the lowering of the interior order within the crystals.

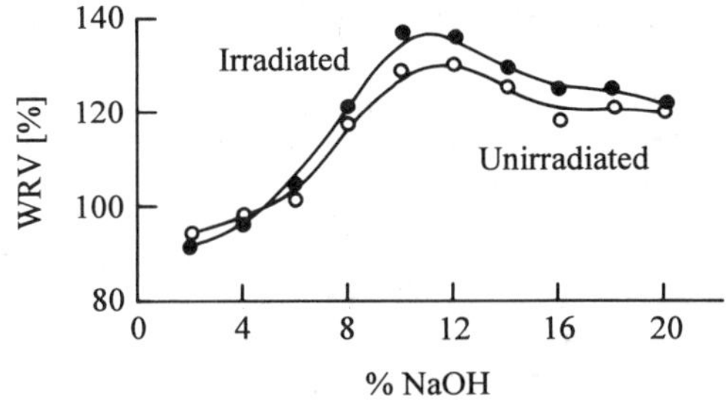

Figure 3. Water retention value after NaOH treatment

When assuming that the high-energy electrons cause defects within the crystalline regions of cellulose these changes are ineffective in intercrystalline reactions, because the disordered regions are surrounded by ordered crystals.
The centres of disorder within the crystals are only activated when using a sodium hydroxide concentration greater than 5%.
It was also possible to describe the changes in the supermolecular structure by determining the degree to substitute xanthogenate groups for hydroxyl groups. Obviously, by xanthation of the irradiated pulp the degree of substitution is increased due to a higher degree of substitution of the primary hydroxyl groups in the C_6 position (Fig.4). This means the

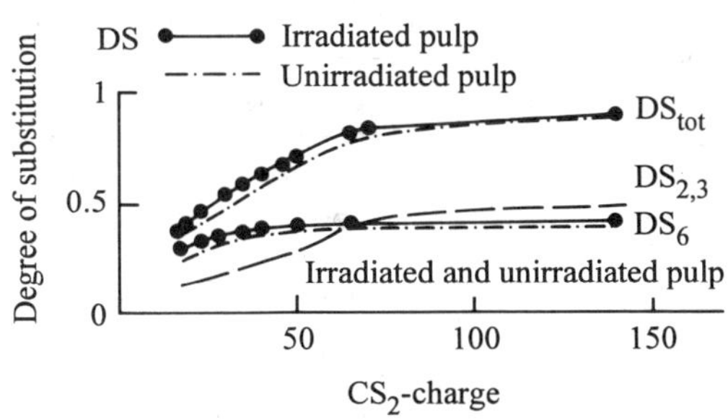

Figure 4. Distribution of xanthogenate groups at anhydroglucose units

primary hydroxyl groups became more accessible by irradiation which might have been caused by a splitting-off of intermolecular hydrogen bonds /O(6)H...O(3)"/ or, rather, the intramolecular bonds /O(6)H...O(2)'/, when applying the model of cellulose conceived by Blackwell [4]. The higher degree of substitution of the C_6 hydroxyl groups is also observed subsequent of the ripening process of the viscose.
The investigations as for the irradiation of cellulose show that the possibility to reduce the charge of carbon disulphide is given when the xanthogenate groups are uniformly distributed along the chain, the primary hydroxyl group is more accessible and the molecular weight distribution is smaller in connection with a degradation of high-molecular cellulose chains.
If these preconditions are met the xanthation may be done at a lower charge of carbon disulphide as is shown by the determination of the clogging values of the viscoses (Fig.5). The filterability of the viscoses is of very good quality up to a lower charge of the carbon disulphide of 6 - 8%.

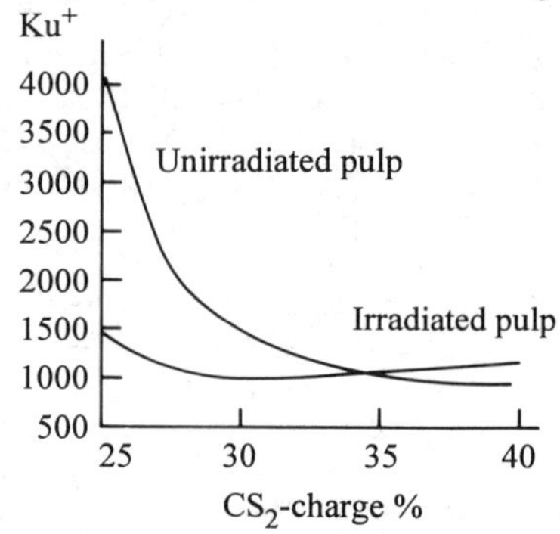

Figure 5. Influence of the CS_2-charge on the clogging value

THE EFFECT OF AN ACID HYDROLYSIS ON CELLULOSE

A second possibility to influence the several structure levels of the cellulose affects the acid bisulphite process.
During the acid bisulphite process for pulping dissolving pulp two major aims are pursued. Firstly, the lignin among and within the cells has to be dissolved. Therefore sufficient amounts of hydrogen and bisulphite ions are necessary.
Secondly, cellulosic substances have to be depolymerized, partially or completely. The depolymerization is caused by hydrogen ions.
Our experience shows that the pulp could be processed in a good way, which was determined by the clogging value of the processed viscoses if the lignin was extensively dissolved during the cooking process at a comparable degree of pulp viscosity (Table 1) [5].

permanganate number	16.1	44.0
clogging value, adjusted	1490	3010

Table 1. Influence of the permanganate number of sulphite pulp on the adjusted clogging values of the viscose

A further improvement in quality is only possible by a controlled hydrolysis of the cellulose during the cooking process with the aim to dissolve the low-molecular cellulosic components and depolymerizing the high-molecular cellulose molecules. A solution to the problem may be seen in a continuous digester. This digester has a possibility for a second hydrolysis to be effected within the hi-heat-washing zone. The pressure in this zone equals that of the cooking zone, the temperature, however, is lower and the water is fed from the washing station and is contaminated with a high concentration of lignosulphonates.

The temperature and the concentration of the lignosulphonates are adjustable in the hi-heat-zone. The conditions for the hi-heat-zone may be influenced as follows: The coefficient for the diffusion of the lignosulphonates from the chips into the liquor depends very strongly on the temperature [6]. At a higher temperature than usual in this zone the rate of dissolution of lignosulphonates is accelerated, thereby the acidity of the liquor is increased as the lignosulphonic acid is a strong one. It is expected that by a further removal of lignin new ranges at the cellulose matrix are accessible for hydrolysis.

pulp			ordinary	extended hydrolyzed
unbleached	kappa number		9.5	6.0
	viscosity	mPas	71.5	48.0
	α-cellulose	%	89.2	90.4
bleached	α-cellulose	%	90.5...91.5	92.0...93.5
	clogging value adj.		900...1100	400...600

Table 2. Influence of the hi-heat-washing on the properties of pulp from spruce

By variation of the concentration of protons and temperature a pulp could be produced with a lower kappa number and viscosity in comparison with a dissolving pulp produced under ordinary conditions in the hi-heat-washing zone.

The content of α-cellulose could be improved, too (Table 2).

A further investigation of such cooked pulps after bleaching with the bleaching sequence CEHH (the E was an extraction with hot alkali) showed that this pulp has a higher α-cellulose content and the clogging values of the test viscoses are essentially better in comparison with a pulp which is produced under ordinary conditions.

These great differences in the clogging values indicate serious changes in the reactivity of the pulp. A further indication for a more reactive pulp results form the determination of the clogging values at reduced charges of carbon disulphide. The results shown in Table 3 point to the clogging values of the pulps prepared by special procedures, to decrease in quality only at a charge of carbon disulphide lying below 28%. This implies with respect to this special treatment of the pulp in the digester the necessity of increasing the accessibility to the hydroxyl groups of the cellulose, so that the reaction with carbon disulphide may be carried out at a lower level in order to obtain viscoses that are very well filterable.

CS_2 charge %	34	32	30	28	24
clogging value, adjusted	443	440	552	558	1183

Table 3. Influence of the CS_2 charge on the adjusted clogging value of pulp prepared by special procedures

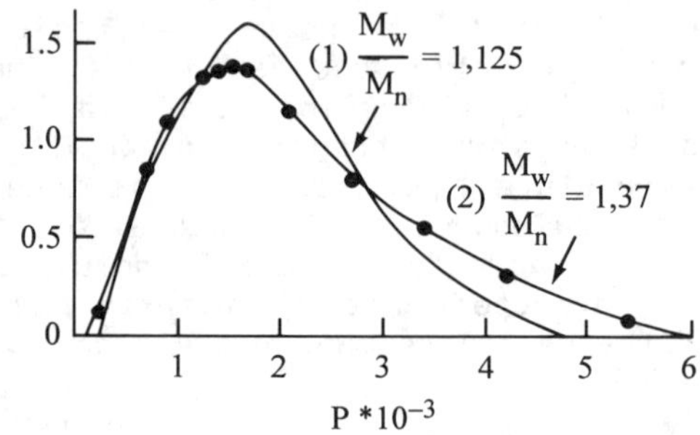

Figure 6. Molecular weight distribution

Having a further evaluation of the pulps in mind their molecular weight distribution curves were recorded (Fig. 6). It is clear to recognize that the molecular weight distribution of the special pulp (Fig. 6/1) has narrowed and that the factor for the non-uniformity (Mw/Mn) is lowered and approaches 1. This is a characteristic feature of a heterogeneous degradation.

REFERENCES

1 Fischer, K.; Goldberg, W.; Wilke, M. *Lenzinger Berichte* **1985**, *85*, 32-39
2 Goldberg, W.: Strahlenchemische Veränderungen der Cellulose. Doctoral thesis, Dresden University of Technology, 1989
3 Fischer, K.; Goldberg, W.; Schmidt, I.; Wilke, M. *Makromol. Chem., Macromol. Symp.* **1987**, *12*, 203-322
4 Gardner,K.H.; Blackwell,J. *Biopolymers* **1974**, *13*, 1975-2001
5 Fischer, K.; Rennert, S.; Wilke, M.; Goldberg,W. *Acta Polymerica* **1990**, *41*, 5, 279-284
6 Goring, D.A.J. *Proceeding XXI Eucepa Conference*, Torremolinos, Spain, **1984**, 63-72

14

Bleaching a spruce kraft pulp with light, air and semiconductor powders

Edward L. Springer - USDA Forest Products Laboratory, One Gifford Pinchot Drive, Madison, WI 53705-2398, USA.

ABSTRACT

Spruce kraft pulp has been bleached by exposing an aerated slurry of the pulp to sunlight or to near ultraviolet light. Addition of the polycrystalline semiconductor powders zinc oxide or titanium dioxide to the slurry greatly increases the final pulp brightness. With titanium dioxide in the slurry, both high and low pH result in even greater increases in brightness. When zinc oxide is present at high pH the highest brightness levels are attained. At low pH zinc oxide dissolves in the liquor and the brightness is the same as when no oxide is present.

INTRODUCTION

Chlorinated organic compounds produced in the chlorination, chlorine dioxide and hypochlorite stages of the bleach plant are the major source of pollution from pulpmills producing bleached kraft pulp. If pulp could be bleached without the use of chlorine containing compounds this source of pollution would be eliminated. Some work has recently been initiated at the Forest Products Laboratory on the use of light in bleaching kraft pulps.

It has long been known that near ultraviolet light will bleach cellulose and low lignin papers if the temperature is near ambient and if air and moisture are present (1). Recently Turner obtained a patent on a process for bleaching pulp wherein a dilute aqueous slurry of the pulp is irradiated with ultraviolet light while oxygen is bubbled through the mixture (2). Best results were obtained when the pH was 12, the temperature was near ambient and the pulps contained little lignin. No added photosensitizers or photocatalysts were present in the system. The lignin itself may have been acting as a photosensitizer. Addition of photosensitizers or photocatalysts might enhance the bleaching effects. Solid photocatalysts which could be separated from the pulp and the pulping liquor by screening would be of particular interest.

It has been known for many years that ordinary polycrystalline semiconductor powders, such as zinc oxide and titanium dioxide, can promote heterogeneous redox reactions under illumination (3). Nitrites, sulfites, and, cyanides in aqueous solutions have been oxidized by adding semiconductor powders and irradiating the resulting slurries with ultraviolet light (4-7). Phenols, alcohols, carboxylic acids and hydrocarbons have also been oxidized using such systems (8-18). It seemed reasonable to assume that, if a diffusible oxidant were produced in these systems, solid substrates such as wood pulp could also be oxidized (bleached). In some initial experiments a very finely divided zinc oxide powder was added to a dilute aqueous slurry of spruce kraft pulp, air was bubbled through the slurry to supply oxygen and keep the particles in suspension and the slurry was exposed to direct sunlight.

INITIAL EXPERIMENTS WITH SUNLIGHT

The initial two reactors employed were 500 ml soft glass gas washing bottles, 60mm in diameter,fitted with 8mm diameter open ended Pyrex glass tubes extending very close to the bottom. Distilled water, 420 ml, never dried spruce (*Picea glauca*) kraft pulp (Yield= 49.6%, Kappa No.= 31), 1.5g ovendry equivalent, and powdered ZnO, 1.2g, were added to each bottle and air was bubbled through using the Pyrex tube to keep the pulp and oxide in suspension. The oxide was omitted in some experiments. The filled reactors were exposed to sunlight for periods of from one half to twelve days. After reaction the pulp was separated from the oxide particles and the solution by screening the slurry using a 100 mesh screen held in a Buchner funnel. The oxide particles were small enough to pass through the screen. The pulp was then washed several times with distilled water and formed into a handsheet on No. 595 Schleicher and Schuell filter paper in a 100 mm Buchner funnel. After air drying, the handsheet brightness was determined using a Beckman Model B spectrophotomer with a reflectance attachment. Magnesium carbonate, the standard, was set to 100%. The initial pulp brightness was 33%.The results from these experiments are plotted in Figure 1.

When ZnO was present in the bottles, large brightness increases occurred. The presence of the semiconductor powder did indeed enhance the bleaching. At first the brightness increase was plotted as a function of the total days exposed regardless of whether the days were sunny or cloudy (see open circles and dashed lines). As can be seen the data were very poorly correlated. The brightness increase data were then plotted versus the sunny days exposed (see filled circles and solid lines). Here the correlation was much improved. Apparently the brightness increase is a function of the total light received and the light received on cloudy days is only a small fraction of that received on sunny days. The experiments where no ZnO was present (see triangles) gave much smaller brightness increases. Here the total day correlation appeared to be better. This is probably due to experimental error and the small amount of data taken.

In one experiment 1.2g of titanium dioxide powder (anatase form) was used in place of the ZnO. The result is given by the squares on Figure 1. Titanium dioxide was apparently somewhat less effective than ZnO in enhancing bleaching. This might have been due to differences in particle size or surface area between the semiconductor powders. Transmission electron microscope photomicrographs showed the particle sizes and shapes to be very similar being roughly spherical in shape with diameters of about 0.5 micrometer. Differences in size distribution of the particles and in surface roughness, although not readily apparent from the photomicrographs, could of course result in a considerable difference in surface areas. Differences in the range of wavelengths of the radiation absorbed by the oxides could also cause such a difference, however, in this case the band gap energies have been found to be identical for ZnO and anatase TiO2 at 3.2 eV. The reason for this difference in effectiveness is presently unknown.

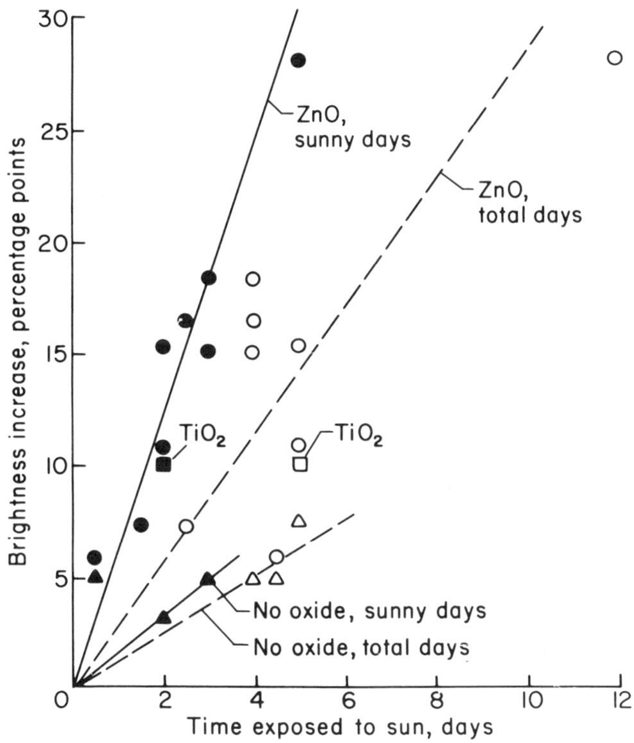

Figure 1-- Brightness increase of an aerated slurry of spruce kraft pulp exposed to the sun with and without zinc oxide present

Due to the difficulties involved in controlling the exposure when irradiating with the sun, and since near ultraviolet light was assumed to be the effective wavelengths, a Rayonet RPR-100 photochemical reactor was obtained and we began irradiating the pulp slurries with 350nM ultraviolet light. Two 660ml quartz reaction vessels, 54.3mm O.D., (Rayonet RQV-218) were used in the reactor. Open ended 5mm O.D. Pyrex glass tubes were centered in the vessels, with one end close to the bottom, to supply air for oxygenation and agitation. The vessels were placed 30mm from the nearest fluorescent tube.

EXPERIMENTS WITH 350nM ULTRAVIOLET LIGHT

In these experiments the quantities of distilled water, spruce kraft pulp and oxides used were identical to those used in the initial experiments. Exposure times of up to 96 hours were used. The brightness increases obtained when no oxide was present and when 1.2g of zinc oxide or 1.2g of titanium dioxide (anatase) was present are shown in Figure 2. At any given exposure time, the presence of the oxides greatly increased the brightness. For the conditions employed, the zinc oxide was again much more effective than the titanium dioxide.

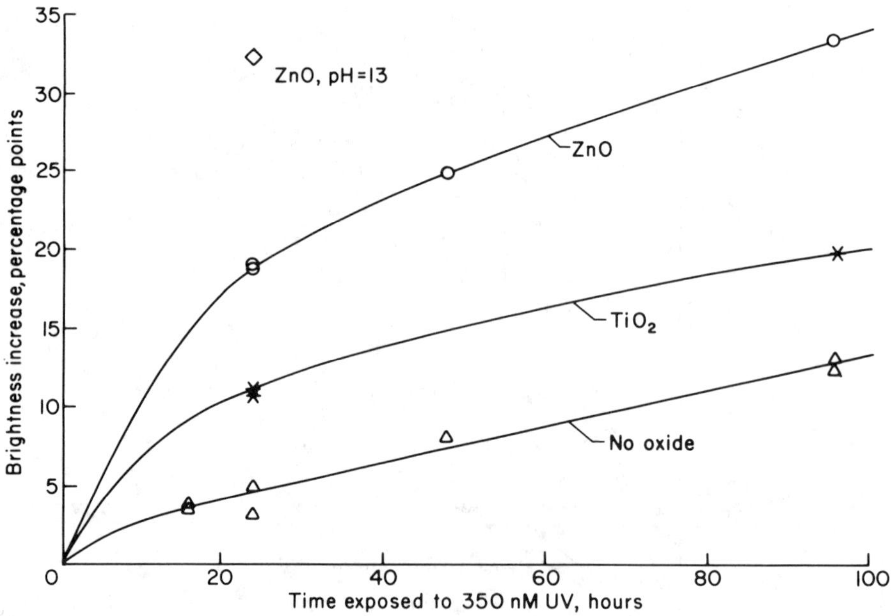

Figure 2 -- Brightness increase of an aerated slurry of spruce kraft pulp exposed to 350 nM UV light in a photochemical reactor with zinc oxide, titanium dioxide and no oxide present

Based on Turners findings, it seemed likely that pH might have a strong effect on the bleaching. In an initial experiment a slurry containing pulp and ZnO was adjusted to pH 13 by adding 2M NaOH and exposed to the 350nM UV light for 24 hours. The result is shown on Figure 2. High pH did indeed greatly increase the brightness.

Experiments on the effect of pH on brightness increase were then performed. All conditions employed were identical with those used previously except that the pH of the slurries was adjusted by adding either 2M NaOH or 0.7M sulfuric acid. The irradiation time for the pH experiments was 24 hours. To maintain constant pH at high pH levels, the carbon dioxide in the incoming air was removed by sparging the air through concentrated solutions of sodium hydroxide contained in two wash bottles connected in series. The air was then passed through a wash bottle containing water to remove any entrained sodium hydroxide solution. The effect of pH on brightness increase is shown in Figure 3.

High pH has a very strong effect in increasing the brightness in the presence of ZnO. The effect is less strong but still very significant when titanium dioxide (anatase) is present. Even when no oxide is present, high pH significantly increases the brightness,as was found by Turner. When no oxide is present, and

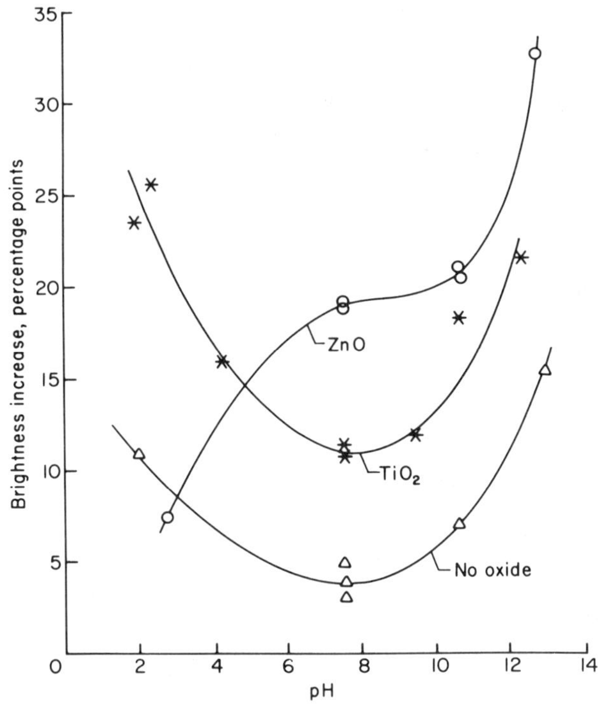

Figure 3 -- Brightness increase of an aerated slurry of spruce kraft pulp exposed to 350 nM UV light for 24 hours as a function of pH with zinc oxide, titanium dioxide and no oxide present

in the presence of titanium dioxide, low pH also increases the brightness. At low pH zinc oxide dissolves in the solution, therefore, no effect was found. In the presence of dissolved zinc oxide the brightness at pH 2.7 was essentially the same as that for no oxide. From the experiments performed to date it appears that the greatest brightness increases can be obtained using zinc oxide at high pH.

CONCLUSIONS

Addition of polycrystalline semiconductor powders such as zinc oxide or titanium dioxide to aerated spruce kraft pulp slurries being irradiated with sunlight or near ultraviolet light greatly increases the brightness of the pulps.

Both with and without the presence of semiconductors, the rate of brightness increase decreases with increasing irradiation time.

When titanium dioxide is present in the slurry, both high and low pH result in increased brightness. The same effect of pH is observed when no oxide is present, although at lower levels of brightness. When zinc oxide is present at high pH, the highest brightness levels are attained, however, at low pH the brightness level attained is essentially the same as that when no oxide is present. This is because ZnO dissolves at low pH.

The results obtained to date indicate that bleaching with light in the presence of semiconductors is an interesting phenomenon which warrants further study. Other variables such as pulp consistency, temperature, semiconductor to pulp ratio, semiconductor particle size and surface area, pulp type and so on should be investigated. To the present time, the strength of the bleached pulps has not been tested. This is another dependent variable which must be evaluated.

LITERATURE CITED

1. Singh, R.P.(Ed.), The Bleaching of Pulp, TAPPI Press, Atlanta, GA, 1979, p.362
2. Turner, R.H., U.S.Pat.4,294,654 (1981).
3. Bard, A.J., J. Photochem. 10:59 (1979).
4. Hori, Y.,Nakatsu, A.and Suzuki, S., Chem.Lett.1985(9):1429.
5. Kogo,K.,Yoneyama,H.and Tamura,H., J.Phys.Chem.84:1705 (1980).
6. Frank,S.N.and Bard,A.J., JACS 99:303(1977).
7. Frank,S.N.and Bard,A.J., J.Phys.Chem.81:1484(1977).
8. Markham,M.C.,Hannan,M.C.and Evans,S.W., JACS 76:820(1954).
9. Markham,M.C.,Hannan,M.C., Paternostro,R.M.and Rose,C.B., JACS 80:5394(1958).
10. Yesodharan,E.P.,Ramakrishnan,V.and Kuriacose,J.C., Ind.J.Chem. 15A:875(1977).
11. Yesodharan,S.,Ramakrishnan,V. and Kuriacose, J.C., Ind.J.Chem.2 lA:468(1982).
12. Kraeutler,B.and Bard,A.J., JACS 99:7729(1977).
13. Kraeutler,B.and Bard,A.J., JACS 100:2239(1978).
14. Kraeutler,B.and Bard,A.J., JACS 100:5985(1978).
15. Kraeutler,B.and Bard,A.J., Nouveau J.de Chimie 3:31(1979).
16. Miyake,M.,Yoneyama,H.and Tamura,H., Chem. Lett.1976:635.
17. Miyake,M.,Yoneyama,H.and Tamura,H., Denki Kagaku 45:411(1977).
18. Izumi,I.,Dunn,W.,Wilbourn,K.,Fan,F.and Bard,A., J.Phys.Chem. 84:3207(1980).

15

Effect of Cr(III) at the ClO_2 stage in the bleaching of kraft (sulfate) paper pulp

F. Taner and A. Olcay* - Ç.U. Faculty of Arts and Sciences Chem. Dept. 01330, Adana, Turkey. *A.U. Faculty of Sciences Chem. Eng. Dept. Ankara, Turkey.

ABSTRACT

The Kraft pulp produced at the Cellulose and Paper Factory, SEKA, Dalaman, Turkey, was bleached by the stages C (Chlorination), D (Chlorine dioxide bleaching), E (Extraction), D and Cr(III) was added at the last D (Chlorine Dioxide) stage at 20 °C in plastic bags. The bleached pulp was analyzed to determine the viscosity, Kappa number and brightness (MgO%), and the residual ClO_2 was determined. It was found that ClO_2 increases the brightness of the pulp, but the addition of the Cr(III) like other cations causes decrease in the brightness of the paper pulp. Cr(III) effects the viscosity of the pulp greatly. It can be concluded that cations like Cr(III) could not be present in the bleaching liquor.

INTRODUCTION

Bleaching of paper pulp is a process that uses various kinds and great amounts of chemicals that are oxidizing compounds. Delignification in connection with bleaching is often achieved by oxidation. Now chlorine dioxide is used to bleach both softwood and hardwood pulp. Chlorine dioxide and chlorite oxidation of lignin have been found to partly eliminate the absorption maximum at 280 µm, suggesting a destruction of aromatic systems [1,2]. ClO_2 in acid medium seems to be completely inactive towards carbohydrates [3-5]. Technical ClO_2, containing small amounts of chlorine, leads to limited carbonyl formation [4]. When applied in large excess and at high temperature, aliphatic compounds have been found to react with ClO_2, and then carbonyl compounds much more readily than alcohols [6].

Many researchers [7-11] have studied the delignification of lignocellulosics by using oxygen and they found that oxygen is an effective compound in delignification. Undesirable degradation of polysaccharides during oxygen bleaching is diminished by the presence of magnesium carbonate [12,13]. It has been shown that carbonyl groups formed during oxygen bleaching act as initiators for peroxide formation, and that peroxide concentration increases when Mg^{2+} ions are present [14]. The discovery that magnesium complexes in solution inhibit polysaccharide degradation was a major contribution to oxygen bleaching [15].

In bleaching it is necessary to prevent the cellulose from severe degradation. Bleaching of pulp may be achieved with oxidizing agents such as O (Oxygen), C (Chlorine), E (Alkaline Extraction), D (Chlorine Dioxide), E and D stages. The conventional bleaching process contains C E H (Hypochlorite bleaching) D E D stages. Hydrogen peroxide which is sensitive to heavy metal ions, has not been used effectively in the first stage of a kraft pulp bleaching sequence.

Some impurities, mainly iron, manganese and copper catalytically accelerate the air oxidation of unbleached pulp [16-18]. It has also been reported [19] that the presence of manganese decreases the brightness of the pulp.

In this study, we tried to bleach the Kraft pulp using the CDED process. Our main concern was to find the effect of Cr(III) at the last chlorine dioxide stage (D_2).

MATERIAL AND METHODS

The kraft pulp, produced by the Cellulose and Paper Factory, SEKA in Dalaman, Turkey, and bleached by conventional CEHDED stages was used as raw material. In this study the kraft pulp was produced from the black and red pine grown in the region of Denizli, Mugla and Antalya by digesting with 21-23% active alkaline at 170-173°C and 0.7-0.9 MPa pressure. After separation of the pulp from the black liquor, it was used as raw material. Some properties of the kraft pulp are given in Table I.

Table I. Some Properties of the Pulp

Brightness (MgO%)	: 20.00
Viscosity (0.5% N/ms)	: 17.64
Alkaline Solubility (18%)	: 8.10
Kappa Number	: 17

CHLORINE BLEACHING STAGE (C).

This stage was conducted in polyethylene plastic bags. The weighed amount of wet pulp with known moisture content was put into plastic bags and the bags were sealed at the mid and upper sections were filled with chlorine (Cl_2) solution that was produced by dissolution of chlorine gas in water (10°C) and stored in refrigerator. The upper section of the bag was closed, and the pulp and the Cl_2 solution were left to warm up to 20 °C. Then both of them were mixed together by loosening the plastic bags at the midpoint. The pulp has been treated for 1, 2, 3, 4, 5 and 24 hours while squeezing the plastic bags to mix the mixture from outside, manually. The Cl_2 was 0.2 g/g dry pulp and the initial pH was 2.2. Concentration of the pulp in the slurry was 3% by mass. After completion of the chlorine stage (C), the pulp was separated from the bleaching solution by filtering through a perforated porcelain funnel and then washed with water at 20°C. The bleaching solution was analyzed to determine the residual chlorine and pH after bleaching.

CHLORINE DIOXIDE BLEACHING STAGE (D_1)

This stage was carried out in the same way as the chlorine stage (C). This time, the pulp was treated with chlorine dioxide (ClO_2) at 20°C, and pH 2.1

for 180 min. At this stage (D_1), the amount of chlorine dioxide, used, was 1% by mass of the dry pulp. Concentration of the pulp in the slurry was 3% by mass. After the chlorine dioxide bleaching stage (D_1), the pulp was separated from the bleaching liquor by filtering through a perforated porcelain funnel and washed with water until the color of the washing filtrate disappeared. After bleaching with chlorine dioxide, the pH of the bleaching liquor was measured and the residual ClO_2 in the bleaching solution was determined. First the total amount of both Cl_2 and ClO_2 was determined by treating with KI and titrating the I_2 formed with $Na_2S_2O_3$ solution of known concentration in acidic media. Then the amount of Cl_2 was determined by titrating the bleaching liquor with $NaAsO_2$ in basic media, separately. Residual ClO_2 was calculated by subtracting the amount of Cl_2 from the total amount of Cl_2 and ClO_2. Then the pulp was extracted with NaOH.

EXTRACTION STAGE (E)

The pulp slurry (10% by mass) was mixed with NaOH (0.03 g NaOH/g of dry pulp) in a plastic bag in a water bath at 60°C. The pulp was extracted with NaOH for 45 min while squeezing the bags, manually. Then the extracted pulp was separated by filtering and the pulp was washed with water until the color of the washing filtrate disappeared. After extraction the pH's of the extractive were measured. The moisture content and amount of pulp were determined and then the pulp was processed by the second ClO_2 bleaching stage (D_2).

SECOND CHLORINE DIOXIDE BLEACHING STAGE (D_2)

This stage was carried out in the same way as the first (D_1) stage. This time, the pulp was bleached with chlorine dioxide (ClO_2) at 20°C, and at pH 2.0 for 60 min. At this stage (D_2), the amount of ClO_2 per gram of pulp was different. Concentration of the pulp in the slurry was 3% by mass. The experiments were carried out in two sets. One set of experiments were carried out by addition of Cr(III) chloride solution with the chlorine dioxide bleaching solution at the same ratio of chlorine dioxide to $CrCl_3$, 9.8 g ClO_2/g $CrCl_3$ by mass, while the other set of the experiments were carried out with chlorine dioxide solution, alone. After the chlorine dioxide bleaching stage (D_2),the pulp was separated from the bleaching solution and bleached pulp was washed with water until the color of the washing filtrate disappeared. The pH's and residual chlorine dioxide were determined.

After determining the moisture content and amount of pulp, the properties of the bleached pulp were determined. The viscosity of the pulp slurry (0.5% by mass) in Cu(II) ammonium solution [20], brightness [21], and Kappa number [22] were determined.

RESULTS AND DISCUSSION

The bleaching conditions of all the stages and the results of all the analyses are shown in Tables I-IV. The changing properties of the bleached pulp by addition of Cr(III) at the second chlorine dioxide stage (D_2) with the amounts of chlorine dioxide are shown in Table IV.

The pH's decrease when bleaching with chlorine, chlorine dioxide and extraction with NaOH. The drop in pH shows that there is some degradation in cellulose and residual lignin is left after Kraft cooking to form acidic compounds. It was concluded that some oxidative degradation occurs in all stages applied here.

Table II. Chlorination Stage (C), 20°C, pulp concentration of 3 % by mass, initial chlorine (0.2 g Cl_2/g pulp)

Sample	pH initial	pH final	Time(h)	Chlorine ($g\ Cl_2/g$ pulp) residual
1	2.2	2.1	1	2.70×10^{-2}
2	2.2	2.1	2	2.18×10^{-2}
3	2.2	2.1	3	1.75×10^{-2}
4	2.2	2.1	4	1.67×10^{-2}
5	2.2	2.1	5	1.60×10^{-2}
6	2.2	2.0	24	0.52×10^{-2}
7	2.2	2.1	1	2.65×10^{-2}
8	2.2	2.1	2	2.18×10^{-2}
9	2.2	2.1	3	1.70×10^{-2}
10	2.2	2.1	4	1.66×10^{-2}
11	2.2	2.1	5	1.65×10^{-2}
12	2.2	2.0	24	0.52×10^{-2}

As seen in Table IV, as the amount of ClO_2 increases, viscosity of the bleach pulp decreases. Decrease in viscosity is higher when Cr(III) was used in D_2 stage. This shows that when Cr(III) was used, extensive degradation of cellulose occurs. Reactions leading to decreased viscosity may be similar to the oxidative degradation which occurs in oxygen alkaline bleaching [23]. Similar reactions occur during aging of alkaline cellulose and also close similarities with degradative reactions which occur under various bleaching conditions [24]. Degradation of cellulose and other wood polysaccharides may be reflected in decreased pulp yield and viscosity. Peeling reactions are generally the most important source of yield loss in bleaching processes.

Table III. Chlorine dioxide stage (D_1): 20°C, 180 min, Pulp Concentration of 3 % by mass,
Extraction: NaOH (0.03 g/g pulp), 60°C, 45 min, pulp concentration of 10% by mass

Chlorine Dioxide Stage (D_1)					Extraction (E)	
	ClO_2 (g/g pulp)		pH		pH	
Sample	initial	residual	initial	final	initial	final
1	0.01	$5.02x10^{-3}$	2.1	1.7	12.2	11.6
2	0.01	$4.70x10^{-3}$	2.1	1.7	12.2	11.5
3	0.01	$5.24x10^{-3}$	2.1	1.7	12.2	11.5
4	0.01	$4.80x10^{-3}$	2.1	1.7	12.2	11.4
5	0.01	$4.80x10^{-3}$	2.1	1.7	12.2	11.4
6	0.01	$4.90x10^{-3}$	2.1	1.7	12.2	11.3
7	0.01	$5.02x10^{-3}$	2.1	1.7	12.2	11.5
8	0.01	$4.60x10^{-3}$	2.1	1.7	12.2	11.5
9	0.01	$5.20x10^{-3}$	2.1	1.7	12.2	11.4
10	0.01	$4.80x10^{-3}$	2.1	1.7	12.2	11.4
11	0.01	$4.80x10^{-3}$	2.1	1.7	12.2	11.3
12	0.01	$4.90x10^{-3}$	2.1	1.7	12.2	11.3

Kappa number of the bleached pulp decreases for both cases (Table IV). But Decrease in Kappa number of the bleached pulp, when Cr(III) was used, is so extensive as to reach to zero. This shows that the oxidation of the functional groups is completed in the presence of Cr(III) when the amount of oxidizing agent is enough. Cr(III) acts as a catalyst to complete the oxidation of the compounds.

Brightness of the pulps relating with the increasing amount of ClO_2 increases for both cases. Increase in brightness, when ClO_2 was used alone, is higher than the brightness when Cr(III) was used with ClO_2. This may be explained by the formation of some colored complexes with Cr(III) or by extensive degradation of the pulp to form some colored compounds.

Table IV. Chlorine dioxide Stage (D_2): 20°C, 60 min, pulp concentration of 3% by mass, 9.8 g ClO_2/g $CrCl_3$,

Chlorine Dioxide Stage (D_2)						
ClO_2 (g/g pulp)		pH		BLEACHED PULP		
initial	final (10^{-2})	initial	final	Kappa No.	Vis. (N/ms)	Brightness MgO%
0.095	0.65	2.0	1.8	5.00*	15.68*	58.00*
0.220	0.80	2.0	1.7	4.25*	13.52*	67.70*
0.436	1.40	2.0	1.9	2.50*	10.98*	72.80*
0.490	5.03	2.0	1.7	3.50*	12.15*	70.80*
0.522	6.94	2.0	1.8	1.25*	10.78*	69.42*
0.723	7.90	2.0	1.8	0.00*	10.76*	68.75*
0.095	0.65	2.0	1.8	5.00	15.68	58.20
0.220	1.05	2.0	1.8	3.85	15.09	69.50
0.436	2.22	2.0	1.8	3.05	14.95	73.80
0.490	6.58	2.0	1.9	2.95	14.70	73.00
0.522	7.25	2.0	1.8	2.75	13.33	71.70
0.723	9.20	2.0	1.9	2.50	13.72	68.90

* values when $CrCl_3$ was added, Vis.: viscosity

From the findings, it can be concluded that the presence of the Cr(III) in the chlorine dioxide stage has an adverse effect on the properties of the bleached pulp. The effect in viscosity which shows whether or not degradation of cellulose molecules occurs, seems to be very pronounced.

REFERENCES

1. Barton,J.S., Tappi, **1950**, 33, 496.
2. Purves,C.B., Pulp Paper Mag. Can. **1955**, 56, 5,117.
3. Rapson,W.H., Tappi, **1956**, 39, 286.
4. Samuelson,O., Svensk Papperstid. **1950**, 53, 155.
5. Schmidt,E., Ber., **1925**, 58, 1394.
6. Somsen, R.A., Tappi, **1960**, 43, 154, 157.
7. Sarkanen, M. and Palenius, I., Tappi, **1975**, 58, 7.
8. Palenius, I. and Hiisuirto, L., Pulp Paper Mag. Can., **1970**, 71 (22) T-465.
9. Palenius, I. and Saukkonen, M.,"Delig. with Oxygen and Soda", Paper Presented at the 4th Can. Wood

Chem. Symp., **1973**, July 4-6.
10. Chang, H. M., Mc Kean, W. J., and Gratzl, J. S., Tappi, **1973**, 56 (9), 116.
11. Charles, J. E., Angelier, R., Weisbuch, H., and Monzie, P., Revue Assoc. Tech. Ind. Panetiere, **1973**, 27 (4), 315.
12. Robert, A., Rerolle, P. Viallet, A. and Martin-Borret, C., A.T.I.P. Bull., **1964**, 18, 151.
13. Robert, A., Viallet, E., Rerolle, P. and Andreolety, J-P., A.T.I.P. Rev., **1966**, 20, 207.
14. Samuelson,O. and Stolpe,L., Svensk Papperstid., **1969**, 72, 662.
15. Noresus,O. and Samuelson,O., Swedish pat. **1970**, 314, 581.
16. Adler, E. and S.Haggroth, Svensk Papperstid. **1950**, 53,287,321.
17. Rollinson.S.M., Tappi, **1955**, 38, 625.
18. Wahlberg, H.E., Svensk Papperstid., **1924**, 27, 38.
19. Ingemar, C., J.Appl. Chem., **1964**, vol.14, p.240.
20. TAPPI Standard T 206 OS-63
21. TAPPI Standard T 217m, T452
22. Scan Standard Test Methods Scan-C1:59
23. Samuelson, O. and Kolmodin, H., Svensk papperstid. **1970**, 73, 93.
24. Theander, O., Tappi, **1965**, 48, 105.

16

Water soluble substances from bleached paper fibres

M. Oblak-Rainer*, D. Budin*, S. Černe*, B. Lipič** - *Pulp and Paper Institute, P.O. Box 366/VII SLO-61000 Ljubljana, Slovenia. **Pulp and Paper Mill Videm SLO-68270 Krško, Slovenia.

ABSTRACT

It has been known for a long time, especially for the high and ultra high yield pulps, that there is a loss of organic material during mechanical treatment. Carbohydrates and lignin are the main components of the water soluble material from these pulps. Many parameters such as wood species, pulp yield, pH, etc. which influence the amount of dissolved substances were investigated. In spite of these being only little information available on the release of carbohydrate and lignin substances during beating of bleached pulps, in our laboratory studies concerning losses of organic material of bleached pulps has been performed by HPLC. Some differences in lignin and carbohydrate contents were observed.

INTRODUCTION

When pulp is suspended and subjected to mechanical treatment different amounts of water soluble organic substances are released from pulp fibres. In paper mills it means loss in yield and undesirable accumulation of soluble and colloidal substances such as extractives, hemicelluloses, lignin and others in the white water system. These anionic substances interfere with the production process, reduce the performance of most chemical aditives, influence the production rates and finally impair the quality of the finished paper.

They also possess great adsorption ability and therefore can be involved together with other components of paper stock in forming the deposits (1 - 3).

Many studies concerning the effects of process variables on the leaching of lignin and carbohydrates from unbleached normal, high and ultrahigh yield pulps have been performed (4, 5). This investigation arose in connection with solving the problem of deposits formation during disintegration of bleached hardwood magnefite pulp in a high consistency pulper in one of our paper mills. With analysis of deposits composition we could not answer this phenomenon. As water soluble substances could be in direct connection with deposits and specks in paper we attempted to find how the release of organic substances varies with the pulp type.

EXPERIMENTAL

Two bleached paper grade pulp samples were investigated on laboratory scale: beech magnefite and mixed hardwood kraft pulp.

Each pulp sample was disintegrated using a standard wet disintegrator at 3 % consistency for 10 minutes. Pulp suspensions were filtered through filter paper and a fine sintered glass filter. One part of filtrate was taken for analysis directly or after concentration in a vacuum rotavapour to the appropriate volume. Fibres were redispersed in proper amounts of filtrate for the next beating step. Pulp beatings were performed in both a Jokro mill and Valley beater according to ISO 5264/3-1979 and ISO 5264/1-1979 at stock concentrations of 6 % and 1,6 % respectively. After 10 minutes refining the fiber suspensions were filtered as previously.

The total content of dissolved substances in the filtrates was determined by evaporation on a water bath and drying for 16 hours at 105°C. Soluble substances were analysed by ion exclusion chromatography on cationic exchange resin (Bio Rad AG 50 W-X4 Mg^{2+}) with water as eluant at a flow rate of 1,5 ml/min. Soluble substances in eluate were recorded using UV and refractive index (RI) detectors. For electrokinetic measurements of pulp filtrates a Model 500 Laser Zetameter and Streaming Current Detector Kolb SD 1000 were used.

THE COMPOSITION OF ORGANIC SOLUBLE SUBSTANCES IN FILTRATES

Fig. 1 shows the ion exclusion chromatogram obtained for standard mixtures of compounds such as lignin, cellobiose, mannose, acetic acid, furfural, which are usually found in spent liquors.

The separations of lignin, oligosaccharides and monosaccharides is good but almost the same retention times are exhibited by polysaccharides, uronic acids and aldonic acids respectively. Only lignin and furfural may be detected by UV absorbance.

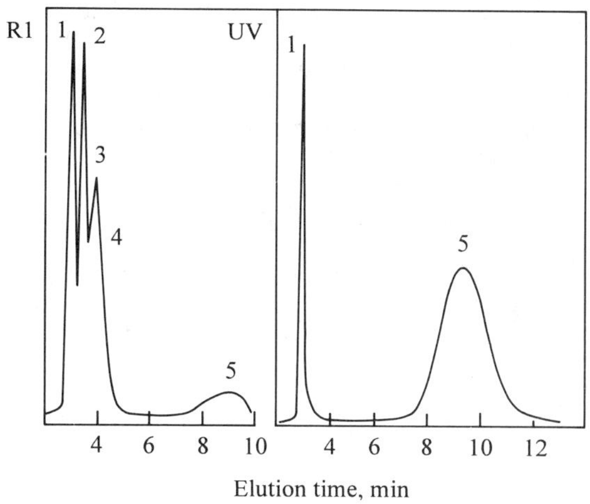

Figure 1. Ion-exclusion chromatogram of standard substances: 1 - lignosulfonate, 2 - cellobiose, 3 - mannose, 4 - acetic acid and 5 - furfural.

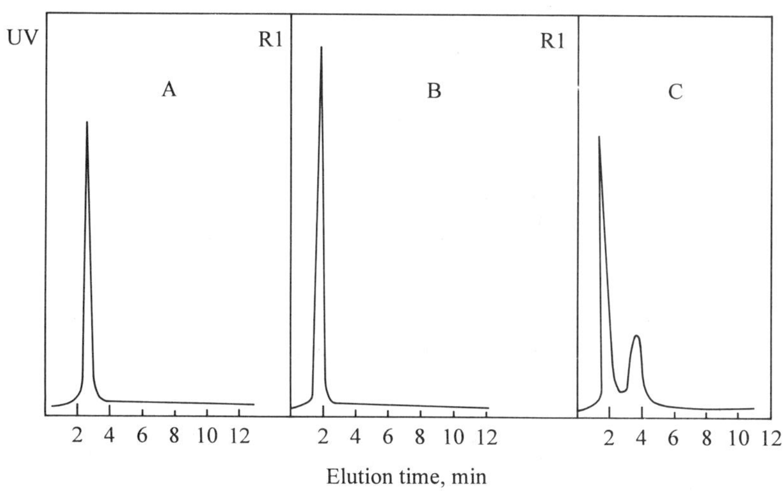

Figure 2. Ion-exclusion chromatogram of beech magnefite pulp filtrate before (A, B) and after hydrolysis (C).

Bleached pulps have been investigated only with regard to the solubilisation of carbohydrates during beating(6, 7). The ion-exclusion chromatogram of filtrate after preliminary beating of beech magnefite pulp shows one peak with retention time belonging to polysaccharides or lignin (Fig. 2 A and B). No other component is evident. The presence of polysaccharides was confirmed by appearance of the additional peak of monosaccharides after hydrolysis of substances in the filtrate (Fig. 2 C). From UV detection (Fig. 2 A) we concluded that lignin substance is also present. In spite of a predominant part of lignin being removed during the bleaching a certain amount of it can be trapped in fibre pores. UV absorption spectra of the filtrate after beating of bleached magnefite pulp are characteristic for lignin (Fig. 3).

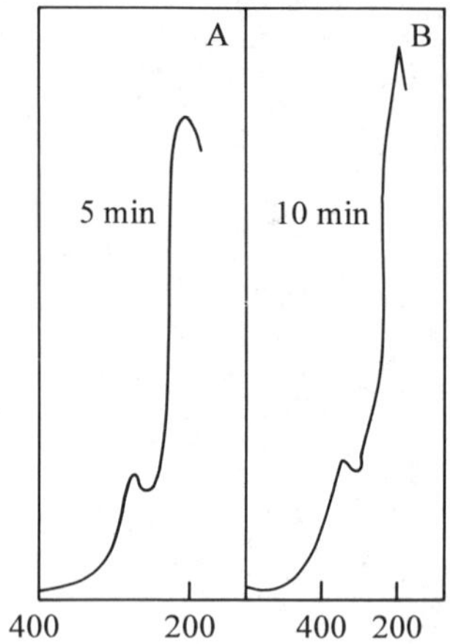

Figure 3. UV spectra of beech magnefite pulp filtrates, A-after disintegration, B-after beating.

For interpretation of the UV spectra it is necessary to consider enolate structures which absorb at the same wavelength as lignin (8). Since enol intermediates are unstable and pass into the tautomeric form the area determined by UV detection was assumed to be lignin. The area obtained by RI detection represents the total lignin and carbohydrate content and is in good agreement with the amount of total dissolved substances.

INFLUENCE OF BEATING ON THE DISSOLUTION OF ORGANIC SUBSTANCES

The amounts of soluble organic substances after disintegration and beating of both pulps are summarised in Table 1. The total amount was found to increase with beating. The tendency of this increase is largely expressed at the magnefite pulp stage. Substantially more carbohydrates than lignin are released in comparison to hardwood kraft pulp.

Beating of magnefite pulp in the Valley beater gives at the same time higher amounts of soluble substances compared with the Jokro mill (Table 2). With prolonged beating time there is pronounced increase in the release of total substances with higher carbohydrate proportions. Similar trends are observed for kraft pulp with the exception that the increase in total amount as well as in carbohydrates to lignin ratio is much lower.

Table 1: Composition of dissolved organic material after disintegration and beating bleached pulps in Jokro mill

Pulp	Beating time	Slowness SR	Dissolved material	Ratio Lignin:	Carboh.	Zeta potential	SCD consumption
	min.	No.	mg/g	%	%	mV	ueq/l
Beech magnefite	0	14	2,3	38	62	-34,3	11,7
	10	27	4,0	26	74	-33,7	22,5
Mixed hardwood kraft	0	15	2,6	55	45	-36,2	20,4
	10	24	3,6	48	52	-37,1	105,3

Table 2: Composition of dissolved organic material after beating bleached pulps in Valley beater

Pulp	Beating time	Slowness SR	Dissolved material	Ratio Lignin:	Carboh.	Zeta potential	SCD consumption
	min.	No.	mg/g	%	%	mV	ueq/l
Beech magnefite	5	18	3,6	55	45	-19,9	5,5
	10	23	6,1	16	84	-19,2	7,7
Mixed hardwood kraft	5	17	3,3	67	33	-20,1	10,7
	10	23	5,2	56	44	-20,2	12,6

The trends observed for leaching water soluble components during processing of magnefite and kraft pulps could be explained by differences in fine fibre structure. The most important is different swelling behaviour in water depending on pore size and content of acid groups. The higher pore size and acid group contents in the case of sulphite pulps means higher swelling and consequently higher leaching. Besides smaller pore size and carboxylic group content kraft pulp fibres are also stiffer.

Faster carbohydrate leaching in the case of sulphite pulps can be due to more extensive fibrillation and continuous fibre wall swelling during beating. Kraft pulps do not continue to swell after the first few minutes of beating (4, 9, 10).

Results for electrokinetic parameters do not correlate perfectly with these results but a general relationship can be observed. Increasing the proportion of lignin reflects in as increase of polyelectrolyte consumption for charge neutralisation (Tables 1, 2). This consumption is in all cases higher at kraft pulp beating.

CONCLUSIONS

During disintegration and beating of bleached pulps lignin and carbohydrates are leached at different ratios.

At higher beating intensity larger amounts of total organic substances are released from beech magnefite pulp than from mixed hardwood kraft pulp. Irrespective of the beating method hardwood kraft pulp gives more intensive leaching of lignin substance, whereas in the case of magnefite pulp, especially at higher beating intensity mainly carbohydrates are released.

REFERENCES

1. *Linhart F; Auhorn W.J.; Degen H.J.; Lorz R. Tappi J.* ***1987****, 70, 89.*
2. *Sidaway S. Tappi J.* ***1988****, 71 47.*
3. *Schmitt G.; Schempp W.; Krause Th. Das Papier* ***1990****, 44, V49.*
4. *Favis B.D.; Goring D.A.I. J. Pulp Paper Sci.* ***1984****, 10, 139.*
5. *Willis J.M.; Delafield C.; Goring D.A.I. J. Pulp Paper Sci.* ***1986****, 12, J10.*
6. *Sjöstrom E.; Haglund P. Sv.Papperstidn.* ***1963****, 66, 718*
7. *Sidaway S. PIRA report PBI (R),* ***1986****.*
8. *Böttger J.; Schultz M.L.; Krause Th.Das Papier* ***1978****, 32, V39.*
9. *Young R.A.; Rowell R.M., Cellulose Structure, Modification and Hydrolysis,* ***1986****, John Wiley & Sons, New York.*
10. *Mohlin V.B.; Alfredsson C., Proc. The World Pulp and Paper Week, 24th Eucepa Conf., Stockholm* ***1990****, 207.*

PART 2

17

Do fibre types make paper grades?

D.J. Ryder, H.W. Kropholler, C.P. Wilkins and E. Sheffield - Department paper sciences, UMIST, P.O. Box 88, Manchester M60 1QD, England, UK.

ABSTRACT

Using only a small quantity of pulp it is possible to characterise and assess the papermaking potential of fibres. It is known that a link exists between the paper grade and fibre characteristic, albeit tenuous.

In this paper, sheet properties such as strength, air permeability and optical characteristics are linked with the fibre traits: length, width, thickness, coarseness, shape and bonding potential.

INTRODUCTION

Paper grades are commonly marketed in seven categories: newsprint, printings and writings, packaging papers, packaging boards, corrugating case material, tissues and speciality grades.

TABLE 1 was constructed using the BPBIF classification of Grades[1] and the FACT FILE[2] for the relevant properties. Consistent trends concerning the qualities and technical aspects of production emerged.

From 1940 to the middle of the 1970's researchers tried to establish how fibre characteristics influenced paper properties. Full details of the research are given by

TABLE 1 QUALITIES AND ASPECTS OF CERTAIN PAPER GRADES

GRADE	NEEDS	TECHNICAL ASPECTS
NEWSPRINT	FAST CHEAP OPAQUE PRINTABLE	MECHANICAL AND WASTE PAPER USUALLY TWIN WIRE MACHINE NO ADDITIVES USUALLY
PRINTING AND WRITING	OPAQUE SMOOTH WHITE PRINTABLE	BLENDED, BLEACHED CHEMICAL PULPS SOME MECHANICAL PULPS MANY ADDITIVES, FILLERS, SIZE etc.
PACKAGING PAPERS AND BOARDS	CHEAP STRONG RIGID	WASTE AND MECHANICAL, SOME KRAFT SOFTWOOD FOR STRENGTH OFTEN STRENGTH ADDITIVES
TISSUES	FAST CHEAP SOFT	BLEACHED RECYCLED SOFTWOODS PROPERTIES OBTAINED FROM MACHINE TYPE
SPECIALITY	VARIABLE	LARGE CONTRIBUTION FROM MACHINE TYPE. SPECIAL ADDITIVES. FIBRE TYPE OFTEN VERY IMPORTANT

Howard[3]. Standard pulp evaluation procedures were used to ascertain the particular qualities of a pulp. These techniques have also been applied to assessing the potential of new pulp sources. However, this type of approach requires significant quantities of pulp.

Basic fibre properties were thought to be capable of predicting sheet qualities. The literature suggests that researchers are still undecided as to which fibre properties affect sheet characteristics. This indecision has undoubtedly arisen from the fact that few if any of the basic fibre properties are truly independent.

Basic fibre properties must contribute to and influence the behaviour of the sheet, and must therefore be an important consideration when designing a product. For example softwoods, hardwoods and mechanical pulps can be separated into their contributing factors as follows:

Softwood; higher strength (burst, tear, tensile) lower opacity and poorer formation.
Hardwood; good formation and opacity, lower strength properties.

Mechanical; very good bulk but subject to shade degradation, lower strength properties.

Other factors also contribute to a paper grade. The paper machine can be thought to control the sort of paper produced. Creped tissue cannot be produced on a newspaper machine, and so on. The total furnish which includes refining and additives in many cases actually makes the paper grade itself e.g. fillers, binders, dyes, starches etc, are added because of the product requirement. So the fibres contribute only 20-25 % to the paper type.

MATERIALS AND METHODS FOR BASIC FIBRE PROPERTIES

A known weight of fibre is deposited onto a filter paper on the wire of a British standard sheet former. These fibres are then transferred directly to glass slides by couching in a handsheet press. The substance of this sheet or network will be 1-2 g.The low grammage networks are then dried and stained to enhance optical visualization.

Fibre length and curl are determined using standard image analysis techniques, the proportion of area occupied by fibre is determined under both incident and transmitted illumination and the number of fibres per unit area are counted.

Less than 10g of fibre are required for the above. From the measurements it is possible to calculate the following basic fibre properties.

MEASURED BASIC FIBRE PROPERTIES

FIBRE LENGTH (mm)	500 measurements
FIBRE CURL INDEX	" "
AREA COARSENESS (gsm)	weight per unit area
DECIGREX (mg/100m)	weight per unit length
FIBRE WIDTH (um)	length/area
FIBRE THICKNESS (um)	gsm / cellulose density
CONTACT RATIO	incident / transmitted area

MEASURED HANDSHEET PROPERTIES

APPARENT DENSITY (gcm^{-3})
AIR PERMEABILITY ($ml\ min^{-1}$)
ZEROSPAN (Nmg^{-1})
TENSILE INDEX (Nmg^{-1})
BURST INDEX ($Nm^{-2}g$)
TEAR INDEX ($mN.m^{-2}.g$)

RESULTS and DISCUSSION

Tables 2 and 3 show the basic fibre properties and the sheet physical characteristics, for a variety of fibre types, respectively.

The length, width and thickness results (Table 2) can be used to describe fibre morphology, which in turn can be related to the sheet physical properties.
Fibre morphology not only affects the physical strength. It also affects opacity and air permeability (Table 4). For example air permeability is greatest with the rounded fibre cross-sections and least with the flattened. An even greater reduction in air permeability occurs when there are multiple fibre types present e.g. flax and straw added to longer fibres. The smaller cells of flax and straw fill between the other fibres producing a densely packed pad of low air permeability.

Opacity which is the light scattering ability of paper depends on the number of air fibre interfaces, the more interfaces the greater the opacity. Fibre morphology dictates the amount of air fibre interfaces. Fibres which can conform closely i.e. flattened or have lots of small cells tend to have a lower opacity because there are less air-fibre interfaces and more fibre-fibre interfaces. Conversely pulps which have more rounded fibre types which cannot associate quite as closely tend to have better opacities. Moreover those pulps which have a greater number of smaller less closely fitting fibres tend to have greater opacities because there are more fibre-air interfaces per unit area.

Fibre bonding is perhaps the most important fibre characteristic, without it paper would not really exist. Fibre morphology greatly influences the amount of fibre surface area available for bonding.

A flattened fibre morphology e.g. Softwood and Sunn Hemp, thickness 3.4 and 3.6 μm respectively, will provide more fibre surface for bonding and a stronger paper should result, a tensile index of 32 and 30 Nm/g.

Conversely a more rounded fibre morphology will bond less fully and thus form a weaker sheet e.g. Esparto and cotton, thickness 5.4 and 3.0 μm and tensile index 19 and 12 Nm/g. Obviously a longer wider fibre will have more area for bonding than will a short slim fibre and this must also be considered because it will also influence strength characteristics.

TABLE 2 BASIC FIBRE PROPERTIES

FIBRE TYPE	LENGTH (mm)	WIDTH (μm)	THICK -NESS (μm)	CONTACT RATIO	AREA COARSENESS (gsm)	DECIGREX mg/100m
SOFT-WOOD	3.0	35	3.4	0.45	5.1	9.9
HARD-WOOD	0.9	49	2.5	0.28	3.7	18.5
SISAL	2.5	9	5.4	0.57	8.2	7.6
SUNN HEMP	1.5	24	3.6	0.75	5.4	12.8
ABBACA	3.5	19	4.6	0.51	7.0	13.7
ESPARTO	1.3	11.5	4.0	0.30	6.0	7.0
FLAX	1.0	20	2.7	0.65	4.2	8.4
STRAW	0.9	38	2.1	0.80	3.2	12.1
COTTON				0.00		
COTTON + 6% STARCH	2.0	8.5	2.7	0.27	4.2	9.5
SYNTH-ETIC	5.2	20	15.0	0.37	22.2	30.65
CTMP	2.0	21	6.8	0.40	10.2	20.9

TABLE 3 PHYSICAL PROPERTIES OF HANDSHEETS (ISO)

FIBRE TYPE	APPARENT DENSITY (gcm-3)	AIR PERM ml\min	ZERO SPAN km	TENSILE INDEX Nm/g	TEAR INDEX mN m/g	BURST INDEX N/cm2/g
SOFT-	0.535	42	198	32	21	0.184
HARD-	0.530	48	159	24	6.8	0.116
SISAL	0.437	72	158	28	22	0.179
HEMP	0.526	14	150	30	21	0.221
ABBACA	0.430	75	144	26	30	0.143
ESPARTO	0.370	76	148	19	3.9	0.058
FLAX	0.481	1.0	134	42	8.6	0.253
STRAW	0.758	1.7	122	37	3.4	0.276
COTTON	0.377	72	127	12	15	0.065

A more complex situation develops when a pulp source has more than one fibre type. Flax and straw have many cell types contributing to their respective pulp types.

TABLE 4 EFFECT OF FIBRE MORPHOLOGY ON AIR PERMEABILITY

FIBRE TYPE	AIR PERMEABILITY
FLATTENED MORPHOLOGY SOFTWOOD AND SUNN HEMP	LOWISH AIR PERMEABILITY 42.6 AND 14.0 ml/min
MORE ROUNDED MORPHOLOGY SISAL AND ESPARTO	HIGH AIR PERMEABILITY 72.4 AND 76.4 ml/min
MULTIPLE CELL TYPES FLAX AND STRAW	V.LOW AIR PERMEABILITY 1.0 AND 1.7 ml/min

For the most part the flattened parenchyma and collenchyma act as bonding agents filling between the other cell types and increase the strength performance of these pulp types, thickness 2.7 and 2.1 µm and tensile index 41.7 and 37 Nm/g.

CONCLUSIONS

Basic fibre properties influence sheet properties. However the amount of influence is dependant upon the whole papermaking process. The use of additives, refining and the type of machinery used to make the paper can often mask the influences fibres are exhibiting within the paper. Despite this certain grades of paper will still rely heavily on the initial qualities of the pulp.

In short the assessment of basic fibre properties for future pulp sources is very useful. The information provided can be used when deciding the performance requirements of the particular grades of paper. However availability of the raw material and the economics of production will dictate what papers are produced from.

REFERENCES

1. British Paper & Board Industry Federation publications Papermakers House Swindon 1990.
2. Lines,C and Booth,G., Paper Industry Resource Pack, published by, paper publications Ltd 1990.
3. Howard,R.C., PhD Thesis, UMIST. England 1986.

18

Extract of short staple fibres from flax by an optimied steam explosion technique

R.W. Kessler, B. Goth, W. Kessler and M. Tubach - Institute für Angewandte Forschung, FH Reutlingen Alteburgstr. 150, D-7410 Reutlingen, Germany.

ABSTRACT

Linen short stable fibres from flax (linum usitatis-simum) with unique properties can be obtained by means of steam explosion technique. The "Reutlingen processing technique" offers the advantage of a fast chemical degumming process with a simultaneous mechanical separation of the fibre bundles. Variables to optimize the process are temperature, time, extrusion speed and additives to reduce the activation energy of the individual constituents. The morphology and the chemical composition (pectins, hemicellulose, lignin etc.) of the plant must be considered during processing since the properties of the individual fibres are influenced much more than in the case of cotton. The process can be designed according to the input quality and the final need of the fibres.

INTRODUCTION

An alternative to the surplus production of food may be the production of industrial crops. From the economic point of view, flax growing in Europe can be profitable as highly valuable fibres are produced, but also all other parts of the plant can be used to manufacture products of high value (1,2). Flax growing is more sensitive to the output of high quality material rather than to a maximum yield. This has also ecological advantages.

The traditional fibre extraction from the plant starts after harvest with a dew retting process which loosens the woody part of the plant and opens up the fibre bundles. After decortication, the material is hackled, scutched and successively chemically treated and wet spun. This processing leads to extraordinary valuable, but also high priced so called "continuous" filaments of linen.

The production and selling of these high grade fibres is limited. Thus new routes for a broader application of flax fibres must be developed. A possibility may be to blend short staple linen fibres with man made fibres, e.g. cotton and wool within the highly economic and productive staple fibre industry.

STEAM EXPLOSION TREATMENT OF DECORTICATED FLAX

A schematic diagram of the "Reutlingen processing" is shown in figure 1. The production of short staple linen can be economically feasible when the following criteria are fulfilled (2):

* substitution of the dew retting process on the field to achieve a constant quality without loss of harvest

* effective and fast degumming process of the fibre bundles with an integrated mechanical disintegration

* standard equipment for processing as used in paper industry to reduce production costs

The steam explosion treatment meets all these requirements. The "Reutlingen processing" starts with an alkali impregnation of the non retted or semiretted decorticated material. An effective degumming within 5 to 10 minutes takes place at high temperatures (8 - 15 bar saturated steam). The mechanical segmentation of the fibre bundles is integrated in the process when the material is suddenly released ("steam explosion"). After scouring (optional bleaching) and drying, the fibres may be further separated by carding as used in the cotton line. The short staple linen can be blended with other fibres and spun on OE- and Ring Spinning systems (3,4). An example of the cross section of fibre bundles before and the separated fibres after processing are shown in figure 2.

MECHANISM OF STEAM EXPLOSION TREATMENT

The major advantage of the process is, that the raw material can be treated according to the input quality and the final use of the fibre. The mechanism during steam treatment and explosion can described in four steps:

Step 1: penetration of steam into the material

Step 2: degradation of pectins, hemicelluloses etc.

Step 3: disintegration of the composite structure into short staple fibres by explosion

Step 4: degradation of the cellulose

The raw material parameters, temperature and processing time (severity factor) and extrusion speed influence the fibre quality. Therefore a highly lignified plant must be processed differently than a fibre of lower maturity. It is also possible to change the activation energy of the chemical reaction

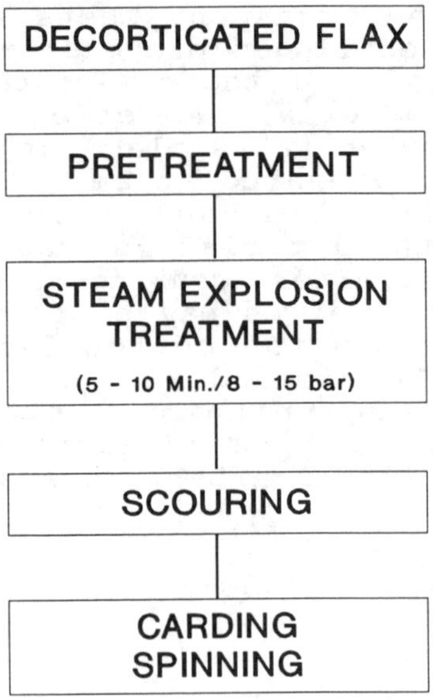

Figure 1: Schematic diagram of the "Reutlingen steam explosion processing" to extract short staple fibres from flax

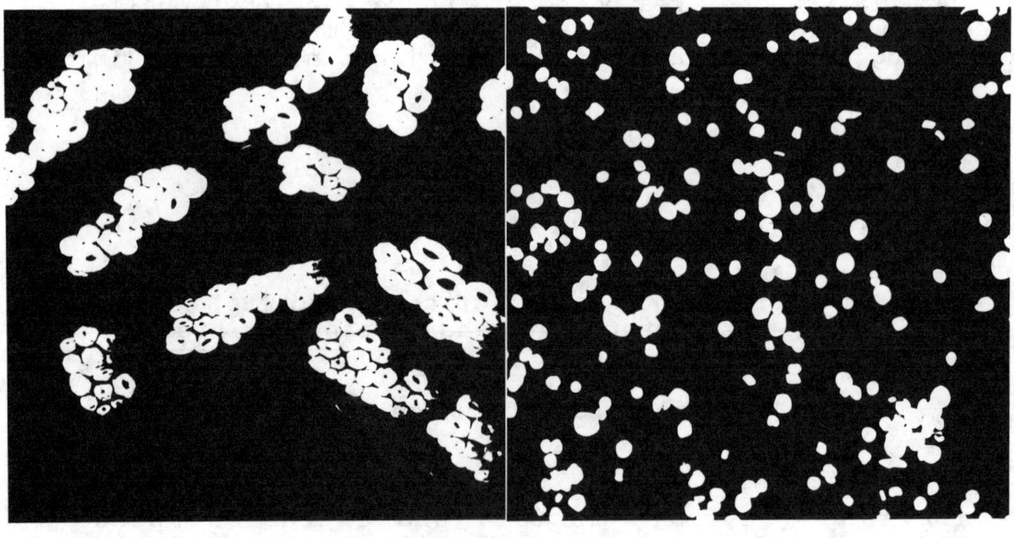

Figure 2: Cross section of a flax plant showing the fibre bundles (left) and cross section of disintegrated fibres after steam explosion

of the four steps by a different chemical pretreatment. It is thus possible to produce a constant quality short staple linen fibre. In any case it is necessary to determine the morphological and chemical structure of the plant.

DESIGN OF THE PROCESS

By means of experimental design and multivariate data analysis it is possible to determine the main factors for processing. On average the final fibre quality is predetermined by the raw material in the range of 50 - 60 %. The severity factor (temperature and processing time) accounts to 30 - 40%, the extrusion speed to 10 - 20% to the fibre properties like mean fibre length, CV, tenacity and fineness. The weighting differs according to the raw material and the final use of the fibre.

The main parameters for the raw material are:

* retted or non-retted
* immature or mature
* lignified or non-lignified

The lignification as well as the maturity determines the tenacity of the final fibre, whereas the fineness is mainly defined by chemical composition of the plant and the retting degree. Immature fibres need a treatment with low severity, mature fibres especially when they are highly lignified need a treatment of high severity.

ACKNOWLEDGEMENT

The authors wish to acknowledge the financial support of the Ministry of Agriculture of Germany.

LITERATURE

1) R. Kohler, R. W. Kessler and M. Wedler in: H. Paetsch (Editor), "Flachs - ein nachwachsender Rohstoff", Statusseminar des Bundesministeriums für Forschung und Technologie 2./3. May 1991, Bonn, Germany, p. 239

2) M. Tubach, R.W. Kessler, B. Goth and U. Dinkel in: H. Paetsch (Editor), "Flachs - ein nachwachsender Rohstoff", Statusseminar des Bundesministeriums für Forschung und Technologie 2./3. May 1991, Bonn, Germany, p. 131

3) S. Messmer and W. Topf, Melliand Textilberichte **10**, 1990, 734

4) R.W. Kessler, J. Wurster, U. Dinkel und M. Tubach in: B. Focher, A. Marzetti and V. Crescenzi (Editors), "Steam Explosion Techniques", Intern. Workshop on STEX, Milan, 20./21.10.1988, Italy, Gordon and Breach Sci. Publ., 1991, p. 233, ibid p. 243

19

Textile fibers from bagasse pulp

D. Paul, H.-J. Gensrich, O. Quintela* and E. Ramirez* - Institute of Polymer Chemistry, KantstraBe 55, O-1530 Teltow, Germany. *Academy of Science, La Habana, Cuba.

ABSTRACT

The rich resources of sugar cane in some countries makes this material interesting not only for sugar manufacture but also as a cellulose source. Bagasse, a by-product of sugar refining is a typical lignocellulosic waste containing about 40 % cellulose, and is useful for papermaking and other cellulose products. In this paper some electronmicroscopic and chemical investigations of alkalization and xanthation of bagasse pulp are presented. The manufacturing of textile fibers obtained from prehydrolysis sulfate pulp of bagasse by viscose process was studied monitoring the effect of the variation of some chemical and technical parameters on a pilot plant scale. The properties of the experimental fibers with 2 g/den strength and 24 % elongation are comparable to those of wood pulp fibers.

INTRODUCTION

Most of the world's rayon is made by the viscose process which is now more than 90 years old. Current plan investment, labor and operating and pollution

costs make the construction of any new viscose plants unlikely in many states. Several research centers and companies are exploring the possibility of modifying the overall viscose process. Shortage of pulp from wood in some countries and high production rates of cellulose in annual plants make bagasse from sugar cane of interest as a raw material for a modified viscose process.
In our work consideration is given to optimize the parameters for manufacturing staple fibers on a pilot plant scale in the viscose process.

ALKALIZATION

Due to the importance of the cellulose alkalization stage in the processing of viscose, certain alkalization parameters of bagasse pulp were changed and others were kept constant, in order to observe their influence on viscose properties (Fig. 1).
X-ray scattering indicates a rapid transition occurs from cellulose I - sodium cellulose - cellulose II for bagasse and beech pulp in the range 8 to 12 wt % NaOH /1/. The solubility of bagasse, NH_3-treated bagasse and beech pulp show a maximum at 8-10 % NaOH. Although the main D.P. decrease occurs during aging, it begins immediately upon exposure to NaOH on steeping and continues during shredding. Also D.P. loss of 75 - 125 monomer units occurs during xanthation.
The state of the alkali cellulose has an important influence at the filterability of the viscose (standard filtration, viscosity corrected, filter: paper 5896 Schleicher & Schüll). Realizing standard xanthation with 40 % CS_2, the best value of filterability K_r is reached at 17 - 18 wt % NaOH in the alkalization liquor. A high pressed-weight ratio decreases the filterability. The filterability of viscose from wood pulp is generally better than from bagasse pulp.

XANTHATION

Xanthation of alkali cellulose was realized with carbon disulfide in the gas-liquid state. During the course of the experiments made in order to find the optimal conditions of the viscose process a satisfactory value of filterability K_r was reached by application of > 34 % CS_2 related to cellulose.
Furthermore, alkali cellulose from bagasse pulp was converted into viscose using a surfactant distributed over the alkali cellulose.

pulp

α - cellulose	89.75 %
pentosan	4.41 %
ash	0.15 %
Fe	45 ppm
$D.P._{cuoxam}$	790

alkalization (steeping)

NaOH	18 %
cellulose	30.6 %
alkali	14.8 %
modifier	0.5 % Berol V 32 0.05 % Berol V 31

aging

	sample 417	418	427
temp/time (°C/h)	20/23	23/47	23/47
$D.P._{cuoxam}$	530	390	385

Xanthation

	sample 417	418	427
temp./time (°C/h)		20-28 / 2	
CS_2 (%)	40	40	34

dissolving

NaOH (%)	5.52
temp./time (°C/h)	10/3.5

filtration

pressure (kp/cm²)	12.5-18

ripening / deaeration

viscose

	417	418	427
cellulose	7.37	8.37	8.4
alkali	6.25	6.2	6.2
viscosity (rel.)	160	108	80
filtrability k_r	2400	1700	2600
xanth./100 AGU	57	58	52
ripeness (°H)		9-10	

spinneret

holes/d (μm)	1000/60

spinning	sample 417-1 (2)	418-1 (2)	427
- precipitation bath			
velocity (m/min)	12 (24)	12 (24)	24
spin-draw ratio	2 (0.97)	2 (0.97)	0.97
composition	120g H_2SO_4 /l,	197g Na_2SO_4 /l, 7g $ZnSO_4$ /l	
temperature		50-52°C	
- stretching			
velocity (m/min)	30	30	30
composition	60g H_2SO_4 /l,	100g Na_2SO_4 /l	
temperature (°C)	76	80	90
- washing			
velocity/temp.	30 m/min, 35°C		

Fig. 1 Viscose process, pilot plant scale

The number of particles with diameter greater than 9.7 μm (determined conductometrically with a TUR ZG-2 instrument) in viscose spinning solution reaches values around 800000 particles/g cellulose which decrease to 600000 after the addition of 0.25 % surfactant. In the same way the filterability of the viscose is improved. The value of filterability K_r of viscose is quickly reduced. Over 0.25 % surfactant K_r is practically constant.
Viscose γ-value does not show considerable variation, the values increase from 59 ± 0.8 to 65 ± 2, this range is approximately similar to that reported for spruce sulfate pulp, when microheterogeneity levelling occurs /2/.
According to these results it is possible to extend to cellulose from sugar cane annual plants the conclusion for wood cellulose /3/ showing that the best reaction conditions are reached when surface tension is diminished, the microheterogeneities in xanthation are levelled, and the influence of inhomogeneities in the raw material are decreased.

STAPLE FIBERS

The stage in the preparation of cellulose staple fibers by the viscose process on a pilot plant scale are the same for bagasse and wood pulp. The conversion of cellulose from bagasse pulp sheets to alkali cellulose is done by placing 1 - 6 kg sheets in a steeping press and filling the press with 18 wt % NaOH /4/. The initial exposure of cellulose to caustic solution is completed within 5 min. Alkalization for 1 h led to a hemicellulose concentration of 1.1 % in the steep caustic.
After steeping, the alkali cellulose pressed under high pressure to a pressed-weight ratio of 3.13 - 3.26 to give a material with 30.64 wt % cellulose and 14.64 - 14.93 wt % sodium hydroxide (Fig.1). By shredding we obtained a crumb-like material. Shredding gives a more uniform distribution of caustic and of sprayed surfactant in the alkali cellulose crumb and increases the temperature favorable for ageing. Ageing was carried out at different temperatures and times to decrease and control different cellulose D.P. (samples 417 and 418, 427). The cellulose xanthate was made by the dry-churn method operating with addition of CS_2 at reduced pressure into evacuated churns containing the alkali cellulose crumb. 34 wt % (sample 427) and 40 wt % (sample 417, 418) CS_2 were used, the temperature increase from 20 to 28°C in 2 hours.
A final high vacuum is applied to the churn to remove traces of CS_2 before the churn is opened and the xanthate crumb is dumped.

Xanthate is dissolved in a dissolver with moving blades propeller in a dilute caustic solution of 7.52 wt % NaOH at 10°C for 3.5 hours.
The filtration of the viscose (2 sheets of cotton fabrics) is followed by ripening. A redistribution of xanthate groups is the main function of the ripening stage. During this time the xanthate concentration drops significantly to 58 xanthate groups/100 AGU and is now ready for spinning (coagulation value appm. 10° Hottenroth).
The viscose samples differ in cellulose concentration (417: 7.37 wt %; 418, 427: 8.38 wt %), viscosity (417 > 418 > 427), filterability K_r (427 > 417 > 418) and D.P. (417 > 418, 428). The poor filtration (sample 427) is caused by a low CS_2-concentration in the xanthation, producing too many undissolved fibers and gels.
The chief goal in the spinning of viscose in the pilot plant process (Fig. 1) from bagasse pulp was manufacturing textile staple fibers dried without tension.
Finally, the fibers are washed with water at 35 °C, finished in a bath with 3 g Elvausan/l, dried without tension and cut. Cross-sectional views of fibers produced by the pilot plant experiment presents a relatively dense morphology. Fiber properties are strongly dependent on the spinning conditions.
The three samples described had a tenacity between 1.6 and 2.5 p/den (Fig. 2). The tenacity increases by drying the fibers with tension. A high D.P. of cellulose (sample 417) is related to a higher tenacity. The difference between conditioned and wet elongation is relatively high. The physical properties of staple fibers from cuban bagasse pulp are comparable with other rayons /4/. Naturally, high wet modulus-types /5/ had a higher tenacity.

Sample	417-1	417-2	418-1	418-2	427
fineness (den)	1.6	3.5	1.3	2.8	2.2
strength (p/den)					
cond.	2.5	2.1	2.0	1.8	1.6
wet	2.2	1.4	1.3	1.6	0.9
elongation (%)					
cond.	14.7	19.2	24.8	20.5	22.8
wet	31.0	32.0	29.4	34.5	33.5
breaking length (km)	23.1	19.2	17.5	16.8	14.3

Fig. 2 Properties of textile fibers; staple, tensionless dried

In the conditioned stress-strain curves of rayons, experimental bagasse fibers are in the middle class range /6/.
A high demand for fibers with rayon-like properties will exist in the future. Bagasse pulp may be an alternative raw material.

References

/1/ Quintela, O.; Paul, D.; Schleicher, H.; Philipp, B.: ATAC (Rev. de la Association de Tecnicos Azucareros de Cuba), 44(1985), 23 - 28

Ramirez, E.; Paul, D.; Paul, C.; Tormo, B.: Cuba Azucar Jan/März 1984, 17 - 23

/2/ Schleicher, H.: Faserforschg. u. Textiltechnik 18(1968), 350 - 354

/3/ Quintela, O.; Paul, D.; Paul, C.; Rodriguez, G.: Cellulose Chem. Technol. 18(1984), 401 - 403

/4/ Paul, D.; Quintela, O.; Gensrich, H.J.: Reporte de Investigation del Instituto de Quimica y Biologia Experimental, No. 4(1983), 2 - 19

/5/ Kao, G.; Li Wei, R.; Zheng Han, M.: Cellulose Chem. Technol. 14(1983), 441 - 455

/6/ Klare, H.: Acta Polymerica 33(1982), 93 - 108

20

A novel solvent system for the production of regenerated cellulose fibers

A.A. Sioumis - CISRO Division of Forest Products, Private Bag 10, Clayton, Victoria, 3168, Australia.

ABSTRACT

The dissolution and regeneration of cellulose, employing the solvent system formic acid/dichloromethane/sulphuric acid, have been studied with a view to obtaining fibers suitable for textiles and nonwovens.

Cotton linters and a wide range of wood pulps were found to dissolve and were spun to produce fibers with properties akin to those of viscose rayon.

Compared to the viscose process, the solvent system is simpler to operate, requires fewer chemicals and is much quicker. Bleaching of the product fiber is not required.

INTRODUCTION

It has been reported [1-4] that cellulose formate can be obtained by the action of formic acid on cellulose with or without catalysts and reaction times from hours to months have been noted. The extent of formylation depends upon the conditions of the reaction and up to three formyl groups may be introduced. The ease of formylation of the cellulose hydroxyl groups is C-6>C-2>C-3 whereas in the formation of xanthates the hydroxyl group at the

C-3 carbon is the one that is mainly substituted [5]. The formyl groups can be removed by hot steam [2,6] and hence it is possible to avoid the use of alkali for the hydrolysis to regenerate the cellulose if so desired.

A new approach to the formation of cellulose formate is proposed in this paper by using dichloromethane. It is proposed that dichloromethane breaks the intermolecular hydrogen bonds between cellulose molecules and thus facilitates formylation. Furthermore, dichloromethane may slow down the degradation of cellulose that takes place in the presence of acids and offer the possibility of better control of the degree of polymerization desired for particular applications.

RESULTS AND DISCUSSION

Dissolution of Cellulose

The dissolution of cellulose in the dichloromethane/formic acid solvent system is easily carried out at ambient temperature and factors that affect the rate of dissolution include the amount of sulphuric acid used, the temperature of the reaction and the ratio of solvent to cellulose.

The formylation reaction, and hence the dissolution of cellulose, may be carried out in the temperature range from about 9°C to 30°C.

It is convenient to make up a stock solution of dichloromethane and formic acid (10:7 v/v) which can be added to the cellulose sample concurrently with sulphuric acid or prior to the addition of the sulphuric acid. This avoids significant reaction between the formic acid and sulphuric acid which results in the production of carbon monoxide. The dissolution of cellulose takes place in 15-30 minutes. Most of the dichloromethane separates from the clear viscous solution and is easily recovered by decantation. The dope is birefringent indicating the presence of liquid crystals. The spun fibers are regenerated in the water bath and the temperature range may be from 12-30°C. Additives to the water bath alter the properties of the fibers.

Stretching is carried out either in air or in a water bath. The formyl groups are easily removed by hydrolysis with dilute sodium hydroxide or other bases at ambient temperatures.

Pulps

A distinct advantage of this solvent system is that regenerated cellulose fibers with good mechanical properties can be obtained from chemical papermaking pulps, which have not been converted to dissolving pulps. Dissolving pulps,

TABLE I. Pulps Dissolved, Spun and Regenerated

Sample	α-Cellulose	Viscosity[a] (mPa.s)	Pentosan[b] (%)
Cotton Linters	99.4	63.3	0
Nitration Cell-Q (ITT-Rayonier) Bleached Softwood Sulphite	92.5	14.9	2.4
Acetakraft (IPC) Hardwood Prehydr. Sulphate	98.7	11.2	1.2
Viscokraft (IPC)	95.6	4.6	1.9
PB-1 (Apcel) Unbleached, Bisulphite *P. radiata*	-	17.6	6.0
PB-30 PB-1 CEH-bleached	87.8	13.7	6.1
PB-78 Peroxide bleached	88.9	20.9	6.2
PB-79 Peroxide bleached, alkali extracted	91.6	19.9	3.7
EK-1 (APM) Eucalypt kraft, bleached	89.5	22.0	12.4

[a] **TAPPI official method T230 om-82**
[b] **TAPPI historical method T223 hm-84**

pulps prepared for papermaking and pulps prepared in our laboratory from *Pinus radiata* wood (Table I) were dissolved and spun to give fibers with properties as shown in Table II. The pentosan content of the pulps that were dissolved ranged from 0% for cotton linters to 12.4% for eucalypt kraft.

Furthermore the pentosan content of the regenerated cellulose fibers is negligible which indicates that the hemicellulose component is removed during the process thus enhancing the tensile strength of the fibers.

TABLE II. Properties of Regenerated Cellulose Fibers

Pulp	Sample No.	Denier	Tenacity (gpd)	Elongation (%)	Modulus (gpd)
Cotton Linters	497 Godet 497 Bulk	2.60 2.53	3.77 3.01	6.5 14.2	91.0 45.4
Bleached Softwood Sulphite	502G 502B	2.21 2.76	2.78 2.60	8.8 15.1	45.6 37.8
PB-1, Apcel Unbleached, Bisulphate, *P. radiata*	503	2.76	2.23	16.0	42.9
PB-32, Apcel PB-1 CEH-bleached	509	3.17	2.22	18.2	45.1
PB-78 Peroxide Bleached	507	3.70	2.33	20.5	42.3
PB-79, PB-78 Alkali Extracted	504	3.61	2.44	17.3	55.4
EK-1 (APM) Eucalypt Kraft, Bleached	508	3.46	2.16	22.7	52.7
Acetakraft	506	2.18	2.17	13.4	40.9

Fiber Properties

The fibers were coagulated as cellulose formate and were hydrolysed with 3% sodium hydroxide solution at room temperature. After washing with water they were dried either at room temperature (~ 20°C) or in the oven at approximately 60°C.

There is strong adhesion amongst the fibers and when dried in the oven they tend to form a mat. This increased bondability may be of use in nonwovens applications. Fibers with tenacity values up to 3.8 gpd were obtained as shown in Table II.

The tensile strength of regular rayon is of the order of 2 gpd. It is envisaged that as our experimental techniques improve fibers with even higher tensile strengths will be produced. The elongation ranges from 6% to 30%.

The modulus values ranged from 10 to 120 gpd. In Table II only values for the specific experiment quoted in the Table are given. Over a broad range of experimental techniques the denier values ranged from 1 to 7.

ACKNOWLEDGEMENTS

The author is indebted to Ms Vanessa Dusting for competent technical assistance and to Dr A.F.A. Wallis for providing some of the pulps.

REFERENCES

1. T. Fugimoto, S.I. Takahashi, M. Tsuji and T. Miyamoto. J. Polymer Science: Part C: Polymer Letters, **1986**, 24, 495-501.

2. E. Heuser. The Chemistry of Cellulose, New York, Wiley, **1944** p.284-6.

3. Y. Uyeda and T. Shimada. Cellulose Industry, (The Journal of the Cellulose Institute, Tokyo, Japan) **1939**, 15, 212-17. (Abstracts [in English] 49-50 [1939].)

4. Y. Uyeda, S. Nakamura and T. Shimada. Cellulose Industry, **1939**, 15. Abstracts pp. 49-50.

5. S.I. Takahashi, T. Fujimoto, B.M. Barua, T. Miyamoto and H. Inagaki. J. Polymer Science: Part A: Polymer Chemistry Edition **1986**, 24, 2981-93.

6. A.F. Turbak, R.B. Hammer, R.E. Davies and H.L. Hergert; Chemtech, **1980**, 51-58.

21

Using biotechnology to improve cotton fiber quality: progess and perspectives

B.A. Triplett - USDA, ARS, Fiber Physics and Biochemistry, Southern Regional Research Center, New Orleans, Louisiana, USA.

ABSTRACT

In thousands of years of use as a textile fiber, cotton has been improved empirically without direct knowledge of the biological basis for fiber quality. Biotechnology offers a new set of tools capable of unraveling the complexities of plant cell walls and relating the biology of the fiber to important physical characteristics. Already, biotechnological approaches have permitted the introduction of novel genes into cotton plants to improve insect and herbicide resistance. These same approaches have been successfully used to characterize genes coding for cellulose biosynthetic enzymes from bacteria. The tools of bio-technology will be invaluable for determining how cellulose microfibrils and other cell wall components are organized and interact. Detailed mapping of the cotton genome with restriction fragment length polymorphisms (RFLP's) will identify key sets of genes involved in determining the architecture of cotton fiber cell walls. Together these approaches will allow an unparalleled ability to manipulate cotton quality.

INTRODUCTION

Cotton is an important commodity for American commerce with the value of the raw crop estimated to be in excess of $ 5 billion annually. The value of finished goods manufactured with cotton adds an additional $ 50 billion to the US economy. Cotton is a versatile, comfortable, and adaptable textile fiber with end-uses as diverse as flannel, denim, and corduroy fabrics as well as finely woven products. As a major component of home textiles, blue jeans, carpeting, industrial products, and fine apparel, cotton is a multi-purpose fiber.

While the quality of US cotton is quite high at present, future needs of the world-wide textile industry demand that continued attention be focused on improving cotton fiber quality. In the last decade, faster and more efficient spinning, weaving, knitting, and processing technologies have reinforced the need for improvements in fiber physical properties. Foreseeable technological improvements in cotton processing machinery in the next several decades will require cotton of greater strength, fineness, and maturity [1]. As a minimum time frame, introduction of a new variety with improved fiber characteristics takes 7 years merely to increase the seed supply to commercial levels. Clearly, basic research work conducted today is focused on identifying factors necessary for cotton fiber quality improvement designed for varieties to be grown in the early part of the next century.

COTTON FIBER DEVELOPMENT

Fiber quality must be considered within the context of the unique features of how the cotton fiber grows and matures. Cotton fibers are trichomes, or plant hairs, that grow from the epidermal surface of cotton seeds. Approximately 1 out of every 4 seed epidermal cells develops into a fiber [2]. Each fiber is a single cell which elongates for a period of approximately 3 weeks to a final length of 1 - 6 cm depending on the variety. Fiber cell diameter is typically 12-20 μm.

During the elongation stage of fiber development, the primary cell wall must be both rigid and sufficiently plastic enough to accommodate the incorporation of new cell wall material. This contradiction is not unique to cotton fiber cells. All growing plant cells must satisfy this apparent incongruity, yet very little is known biochemically about how this process is accomplished (for a recent review see [3]). In elongating cotton fiber cells, accumulation of solutes such as K^+, malate, and low molecular weight sugars has been demonstrated [4]. Somehow the primary cell wall must "loosen" to permit the deposition of new cell wall polymers. Expansion of the cell is believed to be driven by the influx of water to maintain cellular turgor pressure. In a process similar to most higher plant cells, a large central vacuole develops during fiber growth.

During the latter part of the elongation stage, secondary cell wall synthesis begins. Cotton fiber secondary walls are nearly pure cellulose, > 90% of the dry weight. Cellulose is deposited in organized helical arrays in the secondary wall with successive layers of microfibrils having slightly different orientations with respect to the long axis of the fiber cell. Periodically, the orientation of cellulose microfibril deposition within a layer changes gyre. The points where cellulose microfibrils switch orientation are called reversals. Secondary wall synthesis continues until the wall is 2.5 - 6 µm thick. At 45 to 60 days after flowering, the boll (carpel wall) dehisces, exposing the mature fiber cells to air and sunlight. The cell's cytoplasm dries against the inner layer of secondary wall leaving a lumen where, previously, the central vacuole was located. With drying, the fiber cell collapses into a flattened, twisted ribbon.

FIBER QUALITY

Fiber quality is determined by several measured physical and mechanical properties of mature cotton fiber. Fundamentally, the physical and mechanical properties are determined by fiber structure and biochemistry. Unlike many commodities where producers receive compensation for yield and perhaps one quality measurement, seven quality measurements are taken for every bale of cotton produced under the US loan program. Fiber strength, length, fineness, non-fiber content, and length uniformity are measured by an electronic high volume instrument (HVI). Color and grade are visually evaluated by trained specialists and measured by HVI colormeter.

Each physical property is determined by complex and, for the most part, undetermined relationships between several biologically determined factors. Fiber strength, for example, is correlated with the molecular weight distribution of cellulose polymers [5], the orientation of cellulose microfibrils [6,7], the frequency and distribution of reversals [8,9], and fiber surface properties [10,11]. As yet, systematic measurement of US cotton germplasm for these four strength-related parameters has not been performed. Multivariate analysis of these parameters for a group of cotton varieties should reveal the general contribution of each factor to strength and identify other potential determinants.

Of more fundamental concern is the lack of basic knowledge of how these proposed strength-related parameters are biologically regulated. We know very little about how cellulose biosynthesis initiates in plant cells, however, even less is known about how or why cellulose biosynthesis stops to give polymeric products with a fairly narrow molecular weight distribution. This issue may be a function of a) some inherent property of the cellulose synthase complex, b) the mechanism for cellulose crystallization, c) a restricted supply of cellular cofactors, activators or inhibitors, and/or d) simply a limitation on substrate availability. None of these potential sites for regulating cellulose synthesis are likely to be single gene products.

The ultimate strength of a fiber is a composite of many contributing genes. While details about how cellulose is synthesized in plant cells may help improve fiber quality, other processes involving other genes also play important roles. Cellulose microfibril orientation is related to the orientation of the fiber cytoskeleton [12], however, the cellular components which regulate the organization of microtubules, microfilaments, and related proteins remain elusive for all plant cell types. Fiber breakage occurs near or at reversal points [8,9], yet why reversals occur and if there is an optimum number in fiber cells are unknown questions. Fiber strength is typically measured in bundles of fibers, so inter-fiber interactions are an important consideration. Curiously, there has never been a systematic study of the chemical composition of fiber cuticles to identify a relationship between fatty acid chain length or degree of saturation or modification with fiber friction potential and fiber strength.

Like fiber strength, cotton fiber length is very closely associated with the genetic background of a variety. Turgor pressure-driven cell expansion could be regulated by a number of possible gene products including vacuolar proton ATPases [13]. Wall-associated hydrolases and peroxidases are still other potential enzymes which could help determine fiber length [3,14].

On average, longer cotton fibers tend to be finer fibers [7]. Longer fibers can result from either an extended period of elongation during development or an increased rate of elongation. The regulatory triggers for cessation of cell elongation and initiation of secondary wall deposition are unknown at present.

BIOTECHNOLOGY

Biotechnology is a set of experimental tools that can be used to identify, characterize, mobilize and manipulate genetic information. For modification of higher plants, three principle activities are necessary: 1) genes must be identified which will confer desirable traits, 2) a system for the stable introduction of new genes must be optimized, and 3) recovery of transgenic plants via regeneration or meristem culture must be possible.

In cotton, progress in identifying new genes for fiber improvement lags behind the other two activities. First success came from stable regeneration of cotton from embryogenic suspension cultures [15]. Next, *Agrobacterium*-mediated transformation of cotton suspension cultures with marker genes and recovery of mature plants was perfected [16]. Since these early reports, several groups have introduced *Bacillus thuringensis* toxin genes for pest resistance [17] and various genes for herbicide resistance [18,19]. Cotton has also been transformed via particle bombardment [20]. While these accomplishments are noteworthy, introducing resistance genes may have both positive and negative effects on fiber yield and quality. These effects may be due to somaclonal variation arising from the processes of transformation and regeneration. It is also possible that the inserted gene or the location of the insertion may lead to unknown pleiotrophic effects on fiber quality.

Identification of new genes that will improve cotton fiber quality can take two complimentary approaches. For the first approach, genes suspected of being involved in some aspect of fiber growth can be isolated and characterized. This targeted approach would include genes for cellulose synthase, the fiber cytoskeleton, wall-associated proteins, or cuticle lipid synthesis, for example. Quantitative comparisons of mRNA and/or protein product levels in cotton varieties differing in a physical property would narrow down the field of potential genes to examine in greater detail. The finding that antibodies and photo-affinity labels to *Acetobacter* cellulose synthase recognize specific proteins in cotton fiber cells offers the hope of using heterologous probes for identifying cellulose synthase genes in cotton [21].

Efforts to identify fiber-specific or stage-specific genes from fiber suffer from the same problems as identifying genes by the targeted approach. Namely, substantial efforts to isolate, characterize, quantify, and localize the gene transcripts and products may prove fruitless if the genes do not have a direct impact on cotton fiber quality. Additionally, the methods used to characterize genes in a targeted approach rely on the genes being abundantly expressed. Perhaps the most important determinants of fiber quality will be regulatory molecules and would be missed entirely by such an approach.

The second approach for identifying new fiber quality-related genes involves the use of genome mapping techniques. This strategy starts with genotypes which differ in fiber physical properties. Reference markers in the genome are identified and differences are correlated with the phenotypic variation. These markers can be identified either by restriction fragment length polymorphism (RFLP) analysis [22] or by PCR methods [23]. Although time consuming to generate the number of markers needed to adequately cover the entire cotton genome, the advantage to the method is that loci directly involved with the quantitative trait (fiber quality) will be identified regardless of whether they are abundantly expressed or are regulatory.

Biotechnology offers a new opportunity to examine the development of cotton fiber at a level never before possible. Integration of the targeted approach of gene identification with genome mapping capabilities extends the possibilities.

ACKNOWLEDGEMENTS

The author gratefully thanks W.R. Meredith, Jr., R.H. Kloth, J.D. Timpa, M.A. Rousselle, and N.L. Trolinder for critically reading the manuscript.

REFERENCES

1. Duessen, H. *Proc. Beltwide Cotton Prod. Res. Conf.* **1987**, 47.
2. Stewart, J. McD. *Amer. J. Bot.* **1975**, *62*, 723.
3. Fry, S. *Ann. Rev. Plant Physiol.* **1986**, *37*, 165.
4. Dhindsa, R.S., Beasley, C.A. and Ting, I.P. *Plant Physiol.* **1975**, *56*, 394.
5. Timpa, J.D. and Ramey, H. *Text. Res. J.* **1989**. *59*, 661.
6. Meredith, R. *Text. Prog.* **1975**, *7*, 4.
7. Gupta, V.B., Moharir, A.V. and Panda, B.C. in Harrison, P.W. (Ed.) *Cotton in a Competitive World*, The Textile Institute, Manchester, 1979, 83.
8. Hearle, J.W.S. and Sparrow, J.T. , *Text. Res. J.* **1971**, *41*, 736.
9. Wakeham, H. and Spicer, N. *Text. Res. J.* **1955**, *21*, 197.
10. Nelson, M.L., Andrews, F.R., and Grant, J.N. *Text. Res J.* **1959**, *29*, 260.
11. Tripp, V. Moore, and Rollins, M.L. *Text. Res. J.* **1957**, *27*, 419.
12. Seagull, R. *Can. J. Bot.* **1986**, *64*, 1373.
13. Wilkins, T. *Proc. Beltwide Cotton Prod. Res. Conf. 1991,* 1028.
14. Triplett, B.A. *Proc. 5th Cell Wall Meeting*, Scottish Cell Wall Group, Edinburgh, U.K. **1989**
15. Trolinder, N. and Goodin, J.R. *Plant Cell, Tissue Organ Cult.* **1988**, *12*, 43.
16. Umbeck, P., Johnson, G., Barton, K. and Swain, W. *Bio/Technology*, **1987**, *5*, 263.
17. Deaton, R. *Proc. Beltwide Cotton Prod. Res. Conf.* **1991**, *2*, 576.
18. Collins, H.B. *Proc. Beltwide Cotton Prod. Res. Conf.* **1990**, 7.
19. Trolinder, N., Quisenberry, J., Bayley, C., Ray, C., and Ow, D. *Proc. Beltwide Cotton Prod. Res. Conf. 1991,* 840.
20. Finer, J.J. and McMullen, M.D. *Plant Cell Rep.* **1990**, *8*, 586.
21. Amor, Y., Mayer, R., Benziman, M. and Delmer, D. *The Plant Cell* **1991**, *3*, 989.
22. Paterson, A.W., Lander, E.S., Hewitt, J.D., Peterson, S., Lincoln, S.E., and Tanksley, S.D. *Nature* **1988**, *335*, 721.
23. Williams, J.G.K., Kubelik, A.R., Livak, K.J., Rafalski, J.A., and Tingey, S.V. *Nucleic Acid. Res.* **1990**, *18*, 6531.

22

Spectroscopic evaluation of plant fibers and fibre products

I.M. Morrison and D. Stewart - Fibres Group, Scottish Crop Research Institute, Invergowrie, Dundee DD2 5DA, UK.

ABSTRACT

Fibres can be obtained from most plants but the usefulness and economic value of the fibre depends on its composition and properties. The primary composition can be obtained with considerable accuracy from chemical analyses but the secondary and tertiary structures, which influence the fibre properties, require more sophisticated methods. Plant fibres from oat straw, flax and forage rape stems were modified by chemical treatments to remove specific components. The products were analysed by Near Infra-red Reflectance Spectroscopy (NIRS), Fourier Transform Infra-red Spectroscopy (FTIR) and solid state and solution Nuclear Magnetic Resonance Spectroscopy (NMR). NIRS did not reveal any significant differences between treatments but FTIR showed subtle and significant spectral variations, especially in lignin absorbancies. CP MAS ^{13}C NMR did show loss of lignin but, under the conditions used, appeared most useful in predicting the crystallinity of the cellulosic component.

INTRODUCTION

Plant fibres are derived from and can be considered synonymous with "cell wall". However, the properties and composition of a fibre will depend on the type of cell from which it is derived. As plants contain a number of different cell types, whose proportions will vary during the growth cycle of the plant, the usefulness and economic value of the fibre will change during growth. The major constituents of a fibre (cellulose, non-cellulosic polysaccharides

and lignin) can be determined with considerable accuracy [1] while minor constituents, such as acetyl and phenolic acid esters, can also be determined. Although quantitatively minor, such constituents may have fundamental rôles in cell structure. The properties of a fibre cannot be predicted from the primary composition. An understanding of the secondary and tertiary structures are required to assess physical parameters.

EXPERIMENTAL

The fibre fraction was obtained from oat (*Avena sativa*) straw, flax (*Linum usitatissimum*) and forage rape (*Brassica napus*) stems. They were modified by a wide range of chemical treatments which were designed to remove specific components of the fibre. Most of these have been described elsewhere and, since a reduction in the lignin content is desired for most applications of fibres, particular emphasis was placed on this aspect. Cellulose was determined as the difference between the D-glucose content in a 72% sulphuric acid hydrolysate and that in a 2M trifluoroacetic acid (TFA) hydrolysate [2,3]. The non-cellulosic polysaccharide content was determined as the total neutral monosaccharide content of the 2M TFA hydrolysate supplemented, where appropriate, by a determination of uronic acid by the *m*-hydroxydiphenyl method [4]. Lignin was determined by the acetyl bromide method [5].

Near infra-red reflectance spectroscopy (NIRS) was carried out on a Neotec 6350 scanning instrument. Fourier transform infra-red spectra (FTIR) were run on a Bruker instrument. CP-MAS ^{13}C nuclear magnetic resonance spectra (NMR) were recorded on a Bruker AM300WB instrument.

CHEMICAL ANALYSES

The composition of the straw and treated straws is shown in Table 1. The only treatment which had any effect on the cellulose content of the oat straw was that involving the carbohydrate-degrading

TREATMENT	CELLULOSE	NCP*	LIGNIN
Water	49.6	33.9	11.8
Hypochlorite	49.5	34.6	6.9
Driselase	43.7	41.4	12.5
Permanganate	54.1	36.2	3.3
Chlorite (70°C)	55.2	34.9	3.8
Acetic/nitric	91.8	3.2	3.2
Peroxide	56.1	29.3	9.9
3.75M NaOH (60°C)	79.4	6.2	6.8
4.5M NaOH (20°C)	78.3	8.3	9.9
Chlorite + NaOH	89.9	2.6	3.3

* Non-cellulosic polysaccharide

Table 1. Composition of intact and treated oat straw fibre.

enzyme mixture, Driselase. Almost 30% of the cellulose was removed but it remained the major component of the residue. The non-cellulosic polysaccharide (NCP) contents were considerably reduced by a number of the treatments especially those involving alkali of high molarity. The acetic/nitric treatment was also very effective but the alkaline peroxide treatment, which uses a low molarity, only removed a small proportion. The lowest NCP was found in the chlorite + NaOH-treated residue. A number of treatments removed high proportions of the lignin and the most efficient were the strong oxidative treatments such as chlorite and permanganate (>80%). However, the acetic/nitric treatment was also very efficient while alkaline solutions of high molarity, especially when hot, were as efficient as weak oxidising agents such as peroxide or hypochlorite.

Flax has a much lower lignin content than cereal straw (*ca.* 4%) which was effectively removed by chlorite treatment. Peroxide removed about 40% but mild extractants such as sodium methoxide and borohydride had negligible effects. The alkaline treatments only had a slight effect on the flax NCP which confirms other results that this fraction is largely mannose-based and requires alkaline borate for removal. Forage rape lignin appears to have a different structure to the cereal lignins since the strongly oxidative treatments only removed 10-15% of the fraction. Only a further treatment with strong alkali removed *ca.* 90% of the lignin.

NIRS

The NIRS spectra of a selection of the treated oat straw and treated forage rape residues are shown in Figure 1.

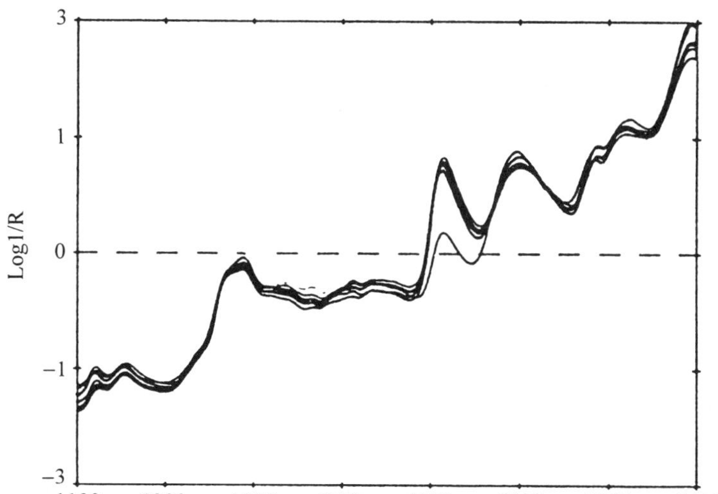

Figure 1. NIRS spectra of treated oat straw samples.

It was not expected that any significant differences would be found in the carbohydrate fraction since cellulose and NCP both absorb in the same regions of the spectrum. The principle peaks for carbohydrate are around 1500 and 2100nm and, indeed, the spectra in these regions are virtually indistinguishable. It was expected, however, that major differences would be seen around 2300nm since this is the region where lignin is reported to absorb and the residues obtained from the oat straw after the strong oxidising agents had lost about 80% of the lignin. The reduction in this absorbance was not observed. Indeed, the only difference between the treated oat straw samples was the low absorbance at 1950nm from the acetic/nitric residue but this absorbance is reported to be due to water.

FTIR

The FTIR results (see Figure 2 for examples) appeared more promising.

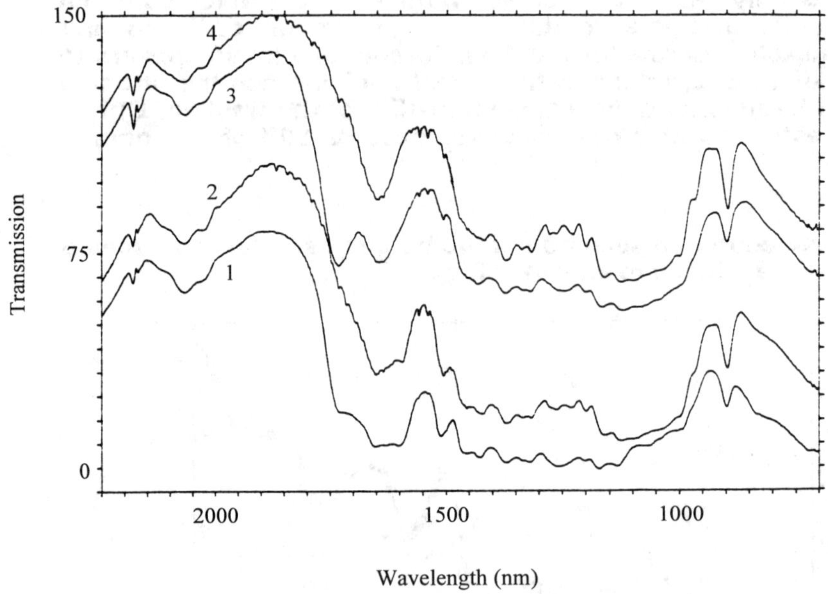

Figure 2. FTIR spectra of treated oat straw samples.
1 - Untreated: 2 - NaOH: 3 - Chlorite: 4 - Chlorite + NaOH

At first glance, an absorbance around 1740nm in the oat straw samples appeared to be due to NCP since it is present in the intact straw and the chlorite-delignified sample but is absent from the alkali-treated residue with or without prior delignification. However, it is probably due to ester bonds between the NCP and acetyl and/or phenolic acid groups. Another, though less well defined, absorbance at *ca.* 1250nm may also be related to NCP since it is reduced in intensity after alkaline treatment but remains strong after delignification. An absorbance *ca.* 1600nm is presumably due to lignin since it is present in the untreated and alkali-treated samples but absent from the permanganate and chlorite delignified samples. In agreement with the chemical analysis, hypochlorite, a weak oxidising agent, reduces the intensity of absorbance but does not remove it. There is one surprising difference between the permanganate- and chlorite-treated samples. The absorbance at 1750nm has almost disappeared from the permanganate-treated sample. This suggests that the permanganate treatment has altered the ester bonds in this material. Other differences were observed in the spectra but they are less well defined and require further investigation.

The flax samples show similar results especially in the absorbance at 1750nm. There is negligible absorbance at 1600nm which agrees with the low lignin content found in the flax.

NMR

In the experiments reported here, the CP-MAS ^{13}C NMR conditions were selected for optimisation of the carbohydrate signals. The lignin signals are observed but are not well defined. Selected spectra from flax and oat straw samples are shown in Figure 3.

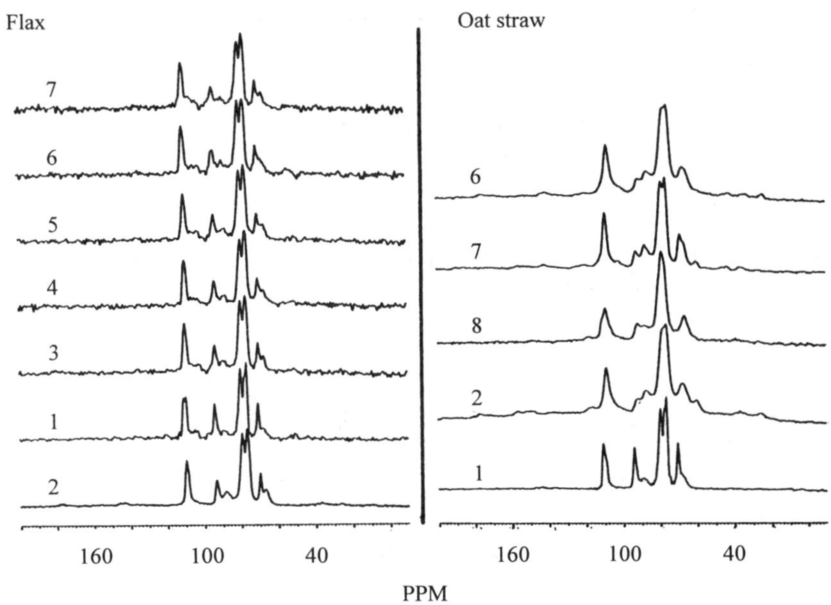

Figure 3. CP-MAS ^{13}C NMR spectra of treated flax and oat straw samples. 1 - Cellulose standard: 2 - Untreated: 3 - NaOH: 4 - NaOMe: 5 - $NaBH_4$: 6 - Chlorite: 7 - H_2O_2: 8 - Chlorite + NaOH.

For comparisons, a long fibre product from cotton cellulose (Whatman CF1) was included in each set. The flax results are easier to interpret since flax fibre has a higher cellulose and lower lignin and NCP content. The doublet from the cotton cellulose for C-1 at 110.7 and 109.4ppm is replaced by a skew signal at 109.8 to 109.4ppm for the flax samples. Since the skewness is on the downfield side, the chemical shift has been reduced by *ca.* 1ppm. The chemical shifts for C-2, C-3 and C-5 are virtually indistinguishable at 79.8-79.3 and 76.9-76.6ppm. The major differences are in the signals for C-4 and C-6. Compared to the cotton cellulose where the major signals for these atoms are at 93.9 and 70.4ppm, the values for the flax samples ranged from 93.1-93.6 and 69.7-70.0ppm. The other significant difference, especially in the treated flax samples, is the presence of another signal in each of these regions which is less sharp and indicative of the non-crystalline regions in the sample. The signals are found in the untreated flax at 88.2 and 67.0ppm; in the treated samples, they are more diffuse and accompanied by a reduction in the signal for the crystalline segment.

The spectra of the oat straw samples are far more difficult to interpret. The signal for C-1 is at 108.7-108.1ppm which is a further 1ppm lower than for cotton cellulose. The C-2, C-3 and C-5 signals are not separated (except for the peroxide-treated residue) which may be due to the higher concentrations of NCP in the oat straw. A further confusion is in the C-4 and C-6 signals for crystalline and non-crystalline segments. The C-4 signals suggest a lower percentage crystallinity but the C-6 signals suggest the opposite.

CONCLUSIONS

The spectroscopic methods are able to give a far greater insight into the usefulness of plant fibres for industrial applications.

ACKNOWLEDGEMENT

This work was financially supported by funds from the Scottish Office Agricultural and Fisheries Department.

REFERENCES

1. Selvendran R.R. *Amer.J.Clin.Chem.* 1984, *39*, 320-337.
2. Morrison I.M. *J.Sci.Food Agric.* 1988, *42*, 295-304.
3. Morrison I.M. *J.Sci.Food Agric.* 1991, *54*, 521-533.
4. Blumenkranz N.; Asboe-Hansen G. *Analyt.Biochem.* 1973, *54*, 484-489.
5. Morrison I.M. *J.Sci.Food Agric.* 1972, *23*, 1463-1469.
6. Hirai A.; Horii F.; Kitamura R. *Cellul.Chem.Technol.* 1990, *24*, 703-711.

23

Application of the image analyzer to the analysis of cellulose fibers in newsprint and cotton/bast fiber blends

Le Tiet Hecking - U.S. Customs Service, Room 7113, 1301 Constitution Ave., N.W., Washington, DC 20229, USA.

ABSTRACT

The U.S. Customs laboratories analyze various imported commodities for proper classification under the Harmonized Tariff Schedule of the United States (HTSUS). Wet chemistry, instrumental analysis, and physical tests are applied as appropriate. Microscopical methods are used for analyzing products which contain cellulose fibers such as papers and textile garments [especially cotton/bast (ramie and linen) fiber blends]. In both cases, the procedures are tedious, time-consuming, and cause eyestrain. This paper discusses the use of the image analyzer (IA) to determine the percent composition of chemical and mechanical pulps in newsprint, and the content of cotton and bast fibers in cotton/bast fiber blend garments. The IA, with computerized microscopical accessories which automate certain mechanical manipulations and related calculations, proves to be fast, accurate, and causes less eyestrain.

INTRODUCTION

The U.S. Customs laboratories analyze imported goods to provide technical support to Customs mission of protecting the revenue, enforcing trade laws and regulations, and protecting the consumers.

Both paper and textile products are analyzed to ensure correct tariff classification; in addition, textile articles are analyzed to ensure compliance with quota/visa and labeling regulations.

Microscopical methods are often tedious, time-consuming, and cause eyestrain - all factors which drive the search for improved analytical techniques. This paper discusses the application of the image analyzer (IA) to determine the percent composition by weight in newsprint and cotton/bast fiber blend garments (1, 2).

INSTRUMENTATION

Figure 1

The IA used, as shown in Fig. 1, was the Bioquant System IV which was composed of:

- ✦ polarized light microscope (Leitz model Laborlux 12 Pol)
- ✦ software: the Bioquant Meg IV *(R&M Biometrics, Inc.)*
- ✦ hardware:
  - personal computer Everex 286/12 *(Everex, Inc.)*
  - color video camera-RGB/ YC/NTSC *(Micromage Video Systems)*
  - color video monitor *(Sony model CPD-1302, Sony Corp.)*
  - b/w monitor (MTI HR 2000, *DAGE-MTI, Inc.*)
  - digitizing pad - SummaSketch II with a 4-button cursor or stylus *(Summagraphics)*, and
  - printer (Epson LX-800, *Epson America, Inc.*)

GENERAL ANALYTICAL PROCEDURES

✦ Create a routine for each type of application, following the software instructions, and store it in the computer.

✦ Place a cross mark, which is used as a pointer for the fiber counting, in the center of the video monitor.

✦ Follow the specific method to prepare sample and slides.

✦ Mount the slide on the microscope stage after applying the appropriate mounting medium. Adjust the focusing, using the 20X objective lens, until sharp images appear on the video monitor screen. Readjust when out of focus.

✦ Identify and count the fibers by looking at the video monitor screen, and by moving the slide horizontally for paper, or both vertically and horizontally for textile samples. Click on the cursor button and switch from one fiber to another, as they pass through the cross mark, to record the fiber counting.

✦ Execute the calculations when all data are collected.

✦ Analyze the sample in triplicate (each analysis consists of three microscope slides), and take the average.

I. NEWSPRINT

Newsprint is "uncoated paper used for the printing of newspaper" (3). It must contain at least "65% by weight of woodpulp obtained by a mechanical process" (3), and have other physical characteristics which are not discussed in this paper.

Method of analysis: *The Strelis Method* (4)

This is a stain and count method which is used to obtain the percent composition by weight of chemical and mechanical pulps. Cotton pulp *(0.05 g of high quality filter paper either Whatman or Schleicher and Schuell),* used as an internal standard, was added to 0.2 g of sample. Sample and slides were prepared as specified in the method. The Herzberg or

Graff's C stain (2-3 drops) was applied. Mechanical pulp fibers turned yellow, cotton became red, and chemical pulp fibers turned blue (Fig. 2). Cotton and chemical pulp fibers were counted. The percent composition was calculated using the fiber counts and relative weight factor for each type of fiber (4).

Data obtained from the 20 samples analyzed on the IA are presented in Table 1 (1).

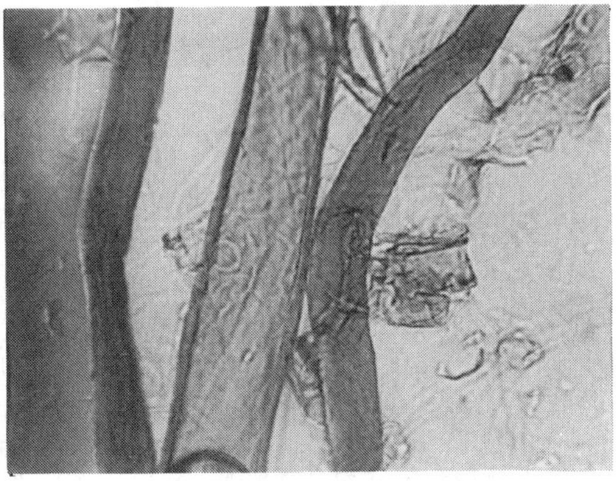

Figure 2

Table 1

	Percent Composition				
	Image Analysis Results			Stated Composition*	
Sample #	Chem. pulp	Mech. pulp	RSD	Chem. pulp	Mech. pulp
1	15.20	84.80	1.07	15.00	85.00
2	13.23	86.77	1.48	14.00	86.00
3	17.36	82.64	0.28	20.00	80.00
4	4.59	95.41	0.37	7.00	93.00
5	10.80	89.20	0.84	10.00	90.00
6	24.13	75.87	1.12	23.50	76.50
7	22.35	77.65	1.17	22.00	78.00
8	32.93	67.07	0.66	32.40	67.60
9	21.60	78.40	0.80		
10	6.56	93.44	1.41		
11	0.00	100.00	0.00		
12	28.05	71.95	2.84		
13	18.69	81.31	1.53		
14	13.73	86.27	0.52		
15	12.19	87.81	0.90		
16	33.55	66.45	4.43	34.00	66.00
17	43.89	56.11	3.12	43.00	57.00
18	19.67	80.33	0.77	20.00	80.00
19	20.60	79.40	1.65	20.00	80.00
20	12.52	87.48	0.12	12.00	88.00

*Information furnished by the manufacturers

II. COTTON/BAST FIBER BLEND GARMENTS

All cotton garments imported into the United States are subject to quota/visa regulations. Headnote 2, Section XI of the HTSUS states: "Goods made of a mixture of two or more textile materials are to be classified as if consisting of wholly of that one textile material which predominates by weight over any other single textile material." (3)

Method of analysis: *AATCC Test Method 20A-1981, Sections 13 and 14* (5)

Figure 3

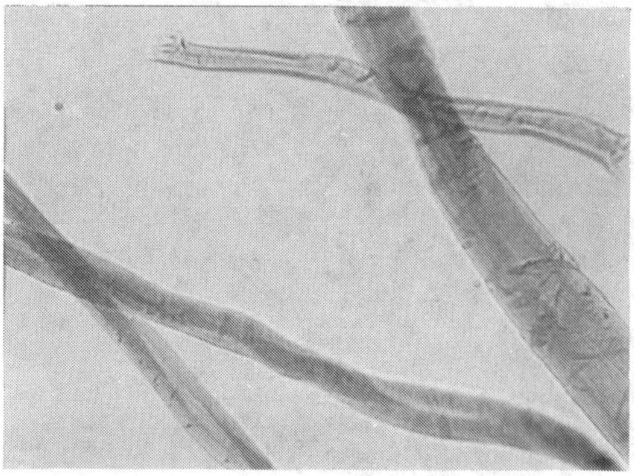

This method is used to analyze fiber blends which cannot be chemically or mechanically separated.

a. **Fiber counting**

Sample preparation was done as instructed. The fibers were identified and counted in 2-3 drops of Glycerin/Water (50/50 v/v) at magnification 200X (Fig. 3). The total fiber count must be at least 1000 for each slide.

b. **Cross sections**

The yarns were embedded in Collodion *(Fisher Scientific)*. Cross sections were made using a Hardy microtome, and were permanently mounted in a mixture of Permount *(Fisher Scientific)* and toluene (1:1 v/v) so that they could be used to trace 100 areas for each type of fiber, on both the IA (Fig. 4) and microprojector (Fig. 5).

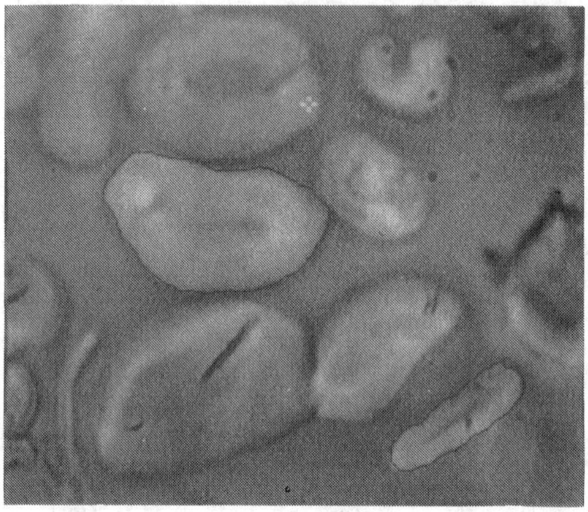

Figure 4

Figure 5

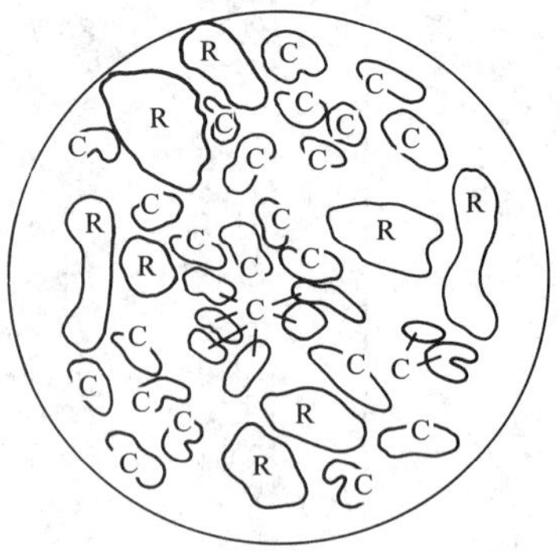

The percent composition by weight was calculated using the fiber counts, specific gravities, and averages of the cross section areas (5).

Table 2 presents the data obtained from six samples analyzed by three analysts, using the IA, the microprojector, and published diameters (2).

Table 2

Percent Composition

		Image analyzer		Microprojector		Publ. diam.	
Sample#	Fiber	Operator1	Operator2	Operator1	Operator 2	Operator 3	Stated on label
1	Cotton	45	46	46	43	42	45
	Ramie	55	54	54	57	58	55
2	Cotton	43	41	45	45	40	45
	Ramie	57	59	55	55	60	55
3	Cotton	45	44	47	43	48	45
	Ramie	55	56	53	57	52	55

Table continued on page 5

Table 2 continued

Percent Composition

		Image analyzer		Microprojector		Publ.diam.	
Sample#	Fiber	Operator1	Operator2	Operator1	Operator 2	Operator 3	Stated on label
4	Cotton	46	45	47	47	46	45
	Ramie	54	55	53	53	54	55
5	Cotton	46	48	46	49	48	45
	Ramie	54	52	54	51	52	55
6	Cotton	45	48	43	48	47	45
	Ramie	55	52	57	52	53	55

The average standard deviations for the IA analyses were1.01 - 1.39.

CONCLUSION

The data obtained clearly indicate that the IA can be used, in place of the microscope and microprojector, for the determination of fiber contents in newsprint and in cotton/bast fiber blends. It is not only fast and accurate, but also causes less eyestrain; therefore, it reduces analysis time and greatly improves the productivity.

REFERENCES

1. "Application of the Image Analyzer to Newsprint Analysis", Le Tiet Hecking—Customs Laboratory Bulletin, Volume 2, No. 1, 1990

2. "Quantitative Analysis of Cotton/Bast Fiber Blends Using an Image Analyzer and a Microprojector", Le Tiet Hecking, John L. Sullivan, Melanie Louise Mealy—Customs Laboratory Bulletin, Volume 3, No. 1, 1991

3. Harmonized Tariff Schedule of the United States (HTSUS), USITC (U.S. International Trade Commission) Publication 2333, 1991

4. "Improved Method for Fibre Analysis of Newsprint", I. Strelis–Pulp and Paper Reports, PPR No.4, Pulp and Paper Research Institute of Canada, 1968

5. Technical Manual, Volume 63 – American Association of Textile Chemists and Colorists (AATCC), An American National Standard,1988

Note: *Reference to any products does not constitute Customs endorsement.*

24

X-Ray diffractometric measurement of cellulose microcrystallite size

D.W. Foreman*, K.A. Jakes** - *Oral Biology, **Textiles and Clothing, The Ohio State University, Columbus, OH 43210-1295, USA.

ABSTRACT

A procedure was designed by which Full-Width-at-Half-Maximum X-ray line breadth is determined by regression analysis of the angular-intensity pairs against the Gaussian function and corrected for instrumental broadening to calculate crystallite size by the Scherrer technique. The design provides "goodness of fit" assessment for the regression and experimental error determination for FWHM, the Gaussian parameters, and the crystallite size. The microcrystalline sizes of linen textiles recovered from a 133 year old shipwreck as well as of test fabrics submerged for 3 months at the same deep ocean site are discussed.

INTRODUCTION

The crystal structure of cellulosic fibers is a fundamental characteristic of those fibers; one method by which crystallinity is studied is X-ray diffraction. While X-ray diffractometric studies of fibers have been conducted for years, the work reported herein sought to develop a data processing method to assess experimental errors through regression analysis and propagation of error techniques applied to the Scherrer method of crystallite size evaluation.
As a test of the method, a group of textiles recovered from a deep ocean environment were subjected to diffractometric analysis. One group of marine textiles was recovered from the interior of a trunk which sank to the ocean floor in 1857 and was raised in 1990. A second group of textiles examined

in this work was composed of modern textiles submerged at the same site for a period of three months and retrieved in October 1991.

REVIEW OF LITERATURE

X-ray diffractometric examination of the crystal structure of fibers has been applied to the cellulosic fibers in particular. Knowledge of crystal size and of crystalline and amorphous character is essential in describing the fiber's fine structure (e.g. 1, 2) and influential in determining the behavior of these materials (e.g. 2, 3, 4). Measurement of crystalline spatial dimension is described in Tripp and Conrad (2). Various approaches to the correction of instrumental line broadening are noted, including the necessity of computerized calculations. Crystallinity and crystal structure are often cited in degradation studies of cellulosic polymers as well (e.g. 5,6).

The Scherrer technique for determining crystallite size has been employed successfully for decades in many fields of investigation. Rau (7) summarized and demonstrated use of the general method for determination of the sizes of crystallites ranging from 20 to 5,000 Å. Limitations in use of the method involve variations in the Scherrer-compensated diffraction angle, but also internal crystal strain, mosaic structure, crystal imperfection, and stacking faults which are not adjusted by means of the method. The original Scherrer proportionality constant, for example, is known to vary considerably under the circumstances of experimentation and sample type. Hence, it was decided to conduct a calibration process in advance of this study to enhance utilization of the method.

In general, the Cellulose Type I fiber structure contains highly ordered crystalline domains interposed among amorphous zones and, in varying degree, ordered along the fiber axis. Gjonnes and co-workers (8) suggested that a significant portion of the presumed amorphous character is likely due to imperfections within the crystalline component itself, rather than arising from a preponderant amorphous component. Preparation and examination of specimens in this study were conducted in such a manner as to enhance characterization of the crystalline component of the structure and to provide a reliable means of determining precision of the crystallite sizes so that sample group distinctions could be discerned with confidence.

In order to give an example of the usefulness of the technique developed in this work, it was applied to the determination of microcrystallite size of textile materials recovered from a marine environment. These materials were chosen because textiles of this type are rarely recovered and little is known concerning the degradation processes which they have undergone and the consequent effects of those processes on the chemical and physical structure of the textiles.

The chemical and physical structure of degraded marine textiles is of particular interest in the case of materials recovered from historic or archaeological sites. So little is known concerning the degradation of textiles in marine environments that conservation of the few historic examples which have been found has taken an experimental approach; no comparable fabric models were treated to judge the suitability of the treatment applied to the historic items. Characterization of the chemical and physical structures of marine textiles would not only contribute to the understanding of the degradation processes which these materials have undergone but also would aid the determination of the appropriate methods for conservation treatment of marine textiles of historic significance.

OBJECTIVES

The objectives of work reported herein include those dealing with the design of the X-ray diffractometric technique and the application of that technique to textiles recovered from the marine environment. Specifically: 1. to develop a method using standard X-ray diffractometers for the calculation of cellulose fiber microcrystallite sizes from their principal reflection profiles; 2. for ultimate specimen comparison, to derive a method of analysis including all the principal profile data points in determination of experimental parameters and their precision; 3. to apply the method to a group of marine textiles; and, 4. to contribute to the description of the physical macromolecular structure of degraded linen textiles from the marine environment.

MATERIALS AND METHODS

Sampling. Samples for research were obtained from the many textiles retrieved from the site of the *S.S. Central America*. The ship, which sank off the coast of the Carolinas in 1857 is the subject of continued study by the Columbus-America Discovery Group (For further information see 9, 10). A trunk was retrieved in 1990 and the textiles within were examined, cataloged, and frozen until identification and characterization of the fibers could be conducted and methods for their conservation could be developed. Jakes and Mitchell (11) describe the recovery operation as well as experiments in the appropriate methods of drying waterlogged materials. Marine textile samples taken for this work included: three samples of lining from the trunk; three from handkerchief (#14f); three from handkerchief (#60); one, each, from inconspicuous areas of men's long underwear (#33) and men's shirts (#22 and #27). Individual fibers removed from each of these samples were examined with the techniques of optical microscopy to identify their composition. In an effort to minimize damage to the historic marine textiles, the samples removed were small, < .5cm square. A linen fabric with neither finishes nor dyes (*Testfabrics* L-53) was employed in this work as a paradigm of modern linen.

Another group of samples were taken from modern linen fabrics which had been placed on the ocean floor near the site of the shipwreck for a period of three months. Polyethylene jars containing the closely packed textiles were constructed with tubing vents at each end to allow ocean water to enter during the period of the test. Upon retrieval, however, pinch clamps on the tubing were closed to seal the water inside the immersion jars. Individual jars contained cotton and linen fabrics alone (group D) and in combination with copper (E) or iron (F); similar jars with wool and silk also were immersed. One set of jars was retrieved in October 1991; others remain on the ocean floor for prolonged exposure and future retrieval. Three linen textile samples, intended for X-ray diffractometric analysis and approximately 2 cm square, were removed from three randomly chosen fabric pieces contained within each of the three jars.

X-ray diffraction. Bundles of yarns were aligned manually and lacerated with a scalpel oriented at slightly less than 40 degrees to the bundle axis thereby providing short segments of fibers which were subsequently placed into a sliding disk mill. The mill comprised a means of randomly mixing and distributing the small fiber lengths in various orientations at conditions of surface pressure far below those which would induce crystallite fracture. The sample mass was then placed into a standard holder within the stage of the diffractometer. A Philips Electronics PW 1316/90 wide range goniometer with XRG 3100 X-ray generator, DMS-41 measuring system, theta-compensating slit, graphite monochrometer and copper target were operated at 35 kV, 15 mA to obtain spectral data pairs, (°2θ,intensity), at increments of 0.05 °2θ in the range $15° \leq 2\theta \leq 55°$.

Data treatment. By using the expressions below, where the summations ran from i=1 to i=50, each reflection center was determined from the mean value $2\bar{\theta}$ of the distribution of intensity *I* over diffraction angle; the precision of the determination, from the standard deviation of the mean: $2\bar{\theta} = \sum[I_i(2\theta)_i]/\sum I_i$; $s_{2\theta} = \sum[I_i\,(2\theta_i - 2\bar{\theta})^2\,]\,/\,\sum I_i$. Assuming a Gaussian distribution $I = a\exp[-b(2\theta_i - 2\bar{\theta})^2]$, regression of Ln(I) onto $-(2\theta_i - 2\bar{\theta})^2$ provided slope *b* which can be related to the spectral line breadth, specified as "FULL-WIDTH-AT-HALF-MAXIMUM" (FWHM) intensity: $FWHM = 2[\,(Ln2)\,/\,b\,]^{1/2}$. Scherrer (12) derived the relationship between wavelength λ of incident X-radiation, line-broadening due to instrumentation B_i and total line breadth *B* as: $B = K[\,\lambda/\sigma\cos\theta\,] + B_i$. Warren and Briscoe (13) subsequently demonstrated a more precise relationship which takes the following form when the crystallite dimension σ is explicit: $\sigma = K\lambda/\,[(B^2 - B_i^2)^{1/2}\cos\theta]$

Under identical conditions of operation, the parameters of the expression for σ were calibrated for subsequent study of a large number of specimens by a series of crystallites of sizes determined previously from scanning electron microscopic observation. A regression analysis of *B* onto $(\sigma\cos\theta)^{-1}$ provided B_i by extrapolation to the B-axis and provided, from the slope of the regression, a parameter which was a composite of the coefficient of proportionality and the wavelength of radiation, which appears in the following calibration expression: $\sigma = 5.81/[\cos\theta(B^2 - B_i^2)^{1/2}\,]$, where $B_i = 0.00161$ radians, and the units are Å for σ. The calibration included determination of sizes for four sets of crystallites over a broad range of $(\sigma\cos\theta)^{-1}$ values; the coefficient of correlation was 0.9.

Experimental precision was assessed by a variety of techniques: error in establishing the center of the (002) reflection derives from determination of the sample distribution mean and standard deviation; error s_b in the slope of the Gaussian fit of data pairs, from regression sums; error s_B in the derived FWHM and error s_σ in the crystallite size, from propagation of error in the Gaussian slope. Specifically, reduced to use of the familiar regression sums in propagation of error analysis, these statements of experimental precision are:

$$s_B = [kLn2/(k-2)]^{1/2}\,[k\Sigma x_i y_i - \Sigma x_i \Sigma y_i]^{-3/2}[k\Sigma x^2_i - (\Sigma x_i)^2]^{-2}[\Sigma y^2_i - Q],$$

where $Q = \{ (\Sigma x_i)^2(\Sigma y_i)^2 - 2\Sigma x_i y_i \Sigma x_i \Sigma y_i + k(\Sigma x_i y_i)^2 \} / \{ k\Sigma x^2_i - (\Sigma x_i)^2 \}$;

and $s_\sigma = 5.81 \sec\theta (B^2 - B_i^2)^{-1/2} [(Bs_B)^2 (B^2 - B_i^2)^{-2} + (s_\theta \mathrm{TAN}\theta)^2]^{1/2}$

These precision measurements are standard deviations because the values from which they were propagated were standard deviations. (See, for example, 14.)

RESULTS

The historic linen fabrics examined show variation in staining and physical condition. The trunk lining and handkerchief #60, in particular, show evidence of severe biodeterioration and degradation while fibers from handkerchief #14F revealed no biodeterioration but were very blackened with deposits and stains. X-ray microanalysis of samples of this piece revealed significant quantities of iron and sulfur (11).

The standard linen fabrics immersed alone with no metals were white-looking upon visual examination, but showed some black deposits on the surface when the fibers were examined microscopically. Samples exposed to copper in the immersion experiment looked green in some areas and the fiber nodes protruded more than in an untreated fiber. The textiles immersed with iron showed some black staining and their fibers displayed more fibrillation than fibers from either of the other two textile groups. The differences in degradation which occurred with each of these textile groups is a subject for another manuscript; change in microcrystallite size reported herein is one indicator of the changes which the textiles have undergone.

X-ray diffractometric spectra of each of the samples exhibit the structure typical of cellulose Type I: (101), 14.73° ; (10$\bar{1}$), 16.56° ; (002), 22.55°; (040), 34.74°. Tripp and Conrad (2) summarized the interplanar spacings $d_{(002)}$ of Cellulose Type I as ranging from 3.90 to 3.96 Å with an average interplanar spacing of 3.94 Å. The 24 specimens listed in Table I have mean interplanar spacing, $d_{(002)}$,3.93 ± .03 Å. The microcrystallite size of each of the samples is reported in Table I.

Applying the t-test of means (Table 2), the microcrystallite size of the historic marine samples is significantly larger than that of the modern samples ($p <$ .01). The crystallite size of linen which had been submerged for a three month period is smaller than that of untreated linen ($p < .02$); the crystallite size of linen which had been submerged for a three month period with copper

TABLE 1

GROUP MEAN CRYSTALLITE SIZES, DIFFRACTION ANGLES, AND INTERPLANAR SPACINGS

Specimen	n	$\bar{\sigma}$(Å)	$\overline{2\theta}$	$\bar{d}_{002}$(Å)
clothing	3	214± 9	22.76± .17	3.90± .03
trunk lining	3	171± 5	22.70± .10	3.91± .01
handkerchief	6	190±17	22.66± .16	3.92± .03
modern linen	3	150± 2	22.43± .12	3.96± .02
submerged specimens				
linen	3	144± 5	22.54± .08	3.94± .01
linen + Cu	3	126± 5	22.51± .01	3.95± .006
linen + Fe	3	140± 8	22.51± .08	3.94± .01

For all regressions, mean coefficient of correlation, r= 0.97; $p^* < .001$.

TABLE 2

STATISTICAL EVALUATION

t-TEST OF MEAN σ	t	p
"pooled" historical linen vs. modern linen	3.44	<.01
modern linen vs. submerged linen	4.46	<.02
modern linen vs. submerged linen+Cu	7.30	<.01
modern linen vs. submerged linen+Fe	2.21	ns
trunk lining vs. handkerchiefs	1.97	ns

on the ocean floor also is smaller than that of untreated linen ($p < .01$). No significant difference was found between the microcrystallite size of group F (immersed with iron) and the untreated modern linen.

DISCUSSION

While the number of samples examined from each material was small, the results of the measurement of microcrystallite size may be indicative of real differences between the groups of materials. The size of the crystallites of linen textiles recovered from a marine environment is increased relative to that of the modern linen. It is unlikely that the larger crystallite size of the marine linen textiles examined in this work is attributable to differing initial microcrystallite size of the historic materials relative to that of the modern linen. Rather, the increased microcrystallite size is more likely attributable to the conditions of long term storage in the marine environment. Due to the accessibility of the amorphous areas of fibers relative to that of the crystalline domains, amorphous areas are more readily swollen with water and subsequently attacked by degrading forces. As less ordered regions are swollen with water, they can rearrange to form more orderly crystalline patterns. As amorphous areas are damaged, rearrangement of associated chains can occur, resulting in an increased quantity of crystallinity (3). As degradation ensues, the polymer chains decrease in length, and their mobility increases, once again encouraging the formation of more ordered crystallite regions (6).

In addition to the increased crystal size, the range of variation of size is more narrow in the historic samples than in the modern ones indicative of a more homogeneous microstructure than that of the modern linen materials. An equilibrium between the forces which reduce crystallite size and those which induce increased crystallinity was reached.

The larger microcrystallite size of the degraded historic fibers holds implications for the textile conservator. The increase in crystal structure relative to modern linen materials is a factor in the stiffer, more brittle character of these materials and should be considered in determination of methods for their conservation treatment. As Atalla (6) suggests, a desirable treatment would be one which modifies the crystal domains in such a way that fibers can respond to deformation more elastically.

Of the immersed linens, fibers from the linen fabric (group D) submerged for three months showed a decreased crystallite size relative to those obtained from untreated linen ($p<.02$); fibers of linen fabric submerged with copper (E) showed a significant decrease in size ($p<.01$). Those submerged with iron (F) showed no significant change in crystallite size. Further exploration of the conditions within the immersion jars as well measurement of chemical, physical, and performance changes of the submerged fabrics relative to the untreated fabric are being conducted and are the subject of a future publication.

CONCLUSION

The X-ray diffractometric method presented in this paper provides the analyst with a method by which a standard X-ray diffractometer can be used to determine microcrystallite size. Concomitantly, the technique provides a means by which experimental errors can be assessed. This exploratory work also indicates changes in crystallite size over time in the marine environment. Further work will compile all of the effects of the textile degradation in the deep ocean, including changes in crystallite size, in order to better understand marine textiles and to better preserve those of historic interest.

ACKNOWLEDGEMENTS

The services provided by Wenwei Wang, Katherine Fritsch, Department of Textiles and Clothing, and Franklin Jones, Department of Agronomy, are greatly appreciated. The authors also express their appreciation to Dr. Bigham for the use of the X-ray diffractometer located in the Department of Agronomy, The Ohio State University. The support of the Columbus-America Discovery Group in the conduct of this research is gratefully acknowledged. In addition, salaries and research were supported in part by state and federal funds appropriated to the Ohio Agricultural Research and Development Center, The Ohio State University. (manuscript no. 112-92)

REFERENCES CITED

1.Shenouda, S.G.; in Happey, F. (Ed.) *Applied Fibre Science,*Vol.3; Academic Press Inc., New York, 1979, pp.275-309.

2.Tripp, V.W.; Conrad, C.M. in O'Connor, R.T. (Ed.) *Instrumental Analysis of Cotton Cellulose and Modified Cotton Cellulose*, Marcel Dekker, Inc, New York, 1972, pp.339-379.

3.Scallan, A.M. *Tex. Res. J.* 1971,41,647-653.

4.Hindeleh, A.M. *Tex. Res. J.* 1980,20,667-674.

5.Varma, I.K.; Anantha Krishnan, S.R.; Krishnamoorthy, S. *Tex. Res. J.* 1989,59,368-370.

6.Atalla, R.H. in Williams, J.C., (Ed.) *Advances in Chemistry Series, No.193*, American Chemical Society, Washington, D.C., 1981, pp.169-176.

7.Rau, R.C. *Advances in X-ray Analysis V*, Plenum Press, University of Denver, 1962, pp.104-116.

8.Gjonnes, J.; Norman, N.; and Viervoll, H. *Acta Chem. Scand. 1958, 12*, pp.489

9.Conrad, J. *Am. Hist. Illus.*, 1991,72,March/April, pp.58-65.

10.Herdendorf, C.E.; *Am. Hist. Illus.*, 1991, 72,March/April, pp.66-71.

11.Jakes, K.A.; Mitchell,J. *J. Am. Ins. Cons.*, In press.

12.Scherrer P. *Nachrichten des Gesellschaft für Wissenschaften*, Göttingen, 1918,2,98-100.

13.Warren B.E.; Briscoe J. *J. Am. Ceram. Soc.*, 1938,21,49-54.

14.Beers, Y. *Introduction to the Theory of Error,* Addison-Wesley, New York, 1958.

25

Inorganic composition of cotton processing dust - should its role in byssinosis be re-examined?

J.K.S. Warrier, P.K. Chidambareswaran and V. Sundaram - Central Institute for Research on Cotton Technology, (CIRCOT), Post Box 16640, Bombay 400 019, India.

ABSTRACT

X-Ray Fluorescence Spectrometry (XRFS) investigations of the inorganic elemental compositions in processing wastes of Indian cottons have revealed the presence of over twentyfive elements, many of which have known histories of respiratory and dermatological toxicities. Some notable differences are found between the results of the present study and those reported in the literature on elemental composition, possibly due to the diferences in agricultural practices as well as mechanical processing methods employed. While the role of gram negative bacteria and other agents on the incidence of byssinosis cannot be ignored, the present data suggest that the potential for inorganic elements playing a direct role has to be re-examined.

INTRODUCTION

A large volume of work has been reported, mainly from USA, on the elemental composition of cotton processing dusts which are known to cause the occupational hazard, byssinosis, in cotton textile mill workers. Although gram negative bacteria and endotoxins found in the dusts are believed to be the causative agents of this disease, the direct role of toxic inorganic elements present in Respirable Dusts (RD) cannot be ruled out in this context [1]. The elemental composition of dusts can highly vary depending upon cotton growing practices and processing conditions. No information is available on dust compositions related to cottons grown in India. Besides, information on the standardisation of XRFS analytical technique is also scanty.

EXPERIMENTAL

Plant samples of the varieties T7 (G.hirsutum), P4 (G.barbadense), Jayadhar (G.herbaceum) and G27(G.arboreum) raised in the same location and season with identical agricultural practices were investigated. Each plant was divided into Top, Middle, Bottom and Root regions, and leaf, bract, stem, etc., from each region was collected.

The plant materials were dried in an oven at 80 °C and ground for 5 min in a domestic grinder subsequently. The material was further pulverised using pestle and mortar to 15 μm particle size. 3 g powder, mixed with equal amount of boric acid, was filled into Al sample cups and subjected to 10 ton pressure for 2 min in a die. Pellets with excellent surface finish and long term stability could be thus prepared.

Wastes from different processing stages were sieved to isolate RD(<15 μm) fraction. The RD and Non-Respirable Dusts (NRD) were analysed separately. Pellets of NRD were made as described; no grinding was necessary for RD. All pellets contained boric acid, 50% by weight.

We used Philips PW 1404 Sequential XRFS facility provided with X 40 software. The de Jongh model [2,3] was found to be quite satisfactory for correcting matrix influences. Sc was the excitation X-ray source, and K_{α} was the analytical line for the 26 inorganic elements detected.

BDH Analytical Grade chemicals (mainly oxides, nitrates and sulphates) were employed for preparing Calibration Standards, which also contained 50% boric acid. The standards were divided into 3 groups, to represent High, Medium and Low Concentration Elements (HCE, MCE and LCE). This strategy, which simplifies calibration, also yielded acceptable matrix corrections.

RESULTS AND DISCUSSION

Tables 1 and 2 summarise the findings, and constitute the Elemental Map of an average Indian cotton plant. Table 1 gives the ranking of different plant parts for 9 elements which occur as HCE or MCE. In Table 2, the HCE and MCE in different plant parts are identified. Besides, the distributions of these elements in the top, middle and bottom regions of the plant are also presented. Using the Elemental Map, information can be obtained on aspects such as the likely concentration of a particular element in the plant, the part of the plant in which the concentration of the element is expected to be highest or least, and the region of the plant in which the identified plant part is likely to have the maximum and minimum concentration of the element.

This type of information will be useful in studies on nutrient requirements of different plant parts as well as nutrient translocations. Efforts are under way to genetically manipulate the plant physiology with a view to achieve maximum utilisation of pertinent nutrients for the production of commercially important parts such as fiber and seed and to reduce nutrient diversions to less important biomass.

Ratios of concentrations of elements such as Ca, K and S have been found to be useful for plant part identification. Ca/S was noted to be distinctly different for fiber and seed as compared to other plant parts. Ca/K was indicated for individually distinguishing leaf, bract fiber and seed. K/S was particularly found useful for differentiating seed from bract. Fibers and seed coat fragments were noted to possess extremely high values of K/S. These ratios, therefore, can be used for identifying plant part components in respirable processing dusts, thereby opening up possibilities of reducing toxic elements which might be located in selected plant parts.

TABLE 1

RELATIVE STATUS OF IMPORTANT HCE AND MCE IN DIFFERENT PLANT PARTS

HCE : High Concentration Element (> 1%)
MCE : Medium Concentration Element (0.1-1.0%)
LCE : Low Concentration Element (< 0.1%)

Lf=Leaf; Bt=Bract; Sm=Stem; Rt=Root; Sd=Seed

Na	:	Lf	>	Sm	>	Rt	>	Bt	>	Sd
Mg	:	Lf	>	Rt	>	Sd	>	Sm	>	Bt
Al	:	Rt	>	Lf	>	Bt	>	Sm	>	Sd
Si	:	Rt	>	Lf	>	Sd	>	Bt	>	Sm
P	:	Sd	>	Lf	>	Bt	>	Sm	>	Rt
S	:	Lf	>	Bt	>	Sd	>	Sm	>	Rt
Cl	:	Sd	>	Lf	>	Bt	>	Sm	>	Rt
K	:	Bt	>	Lf	>	Sm	>	Sd	>	Rt
Ca	:	Lf	>	Bt	>	Sm	>	Rt	>	Sd

LCE : Ti, Cr, Mn, Fe, Co, Ni, Cu, Zn, Ge, Br, Sr, Y, Zr, Mo, Pd, Cd, Ba.

TABLE 2

REGION-WISE RANKING OF IMPORTANT HCE AND MCE IN DIFFERENT PARTS OF THE COTTON PLANT

@ : maximum, $: intermediate, x : minimum, | -> TOP
<> : nearly same, E : Element, | -> MIDDLE
Integers : Primed (') : HCE, Unprimed : MCE; * : LCE | -> BOTTOM
Concentration decreases progressively from 1 to 7

E -> Part	Na	Mg	Al	Si	P	S	Cl	K	Ca
Lf	3 | x | $ | @	4 | $ | @ | x	5 | $ | @ | x	2 | $ | @ | x	1 | x | $ | @	3' | x | $ | @	4' | x | $ | @	1' | x | $ | @	2' | x | $ | @
Bt	3 | x | x | @	5 | @ | x | x	6 | x | @ | x	4 | x | @ | x	2 | x | x | @	2' | x | x | @	1 | x | x | @	1' | x | x | @	3' | x | x | @
Sm	3 | <> | <> | <>	5 | @ | x | $	7 | @ | x | $	6 | @ | $ | x	4 | <> | <> | <>	1 | <> | <> | <>	2 | <> | <> | <>	1' | <> | <> | <>	2 | <> | <> | <>
Sd	*	3 | x | $ | @	*	2 | @ | $ | x	1 | x | $ | @	3' | x | $ | @	2' | $ | x | @	1' | $ | x | @	4 | x | $ | @
Rt	5	6	4	1	7	3	2	1'	2'

Certain inorganic elements are known to be toxic to respiratory organs even at concentrations as low as 1 ppm [4]. The elements which fit into such classification, and their detected levels in the RD fractions of cotton processing wastes were :

Bromine	:	340	to	390	ppm
Nickel	:	236	to	250	ppm
Cadmium	:	0.5	to	90	ppm
Barium	:	0.5	to	45	ppm

In addition, elements which cause respiratory toxicity at concentration ranges of 50 to 500 ppm were found to be present at the following levels in the RD fractions :

Silicon	:	610	to	2490	ppm
Cobalt	:	377	to	400	ppm
Zirconium	:	240	to	340	ppm
Manganese	:	3	to	40	ppm

Thus, Br, Ni, Cd and Ba were judged to be the most dangerous elements. In the second group, Si, Co and Zr merit attention. Although Mn levels were below the threshold toxic limit, it is possible that prolonged deposition of Mn could create problems. Incidentally, Co has been also implicated in dermatological disorders.

The toxic elements listed above were found in cotton plant parts at comparable concentration levels, thereby suggesting means of their control by reducing the incidence of such parts in harvested cotton.

CONCLUSIONS

1. The XRFS technique was standardised for quantitatively analysing 26 inorganic elements found in cotton plant parts and processing dusts.
2. An Elemental Map of the average Indian cotton plant was constructed using the elemental composition data.
3. The findings suggested the need for a re-examination of the direct role of toxic elements present in respirable cotton dusts in the context of byssinosis as well as dermatological disorders.
4. Reduction of cotton plant parts in harvested cotton lint is disclosed to have the potential for decreasing toxic element levels in respirable processing dusts.

ACKNOWLEDGEMENT

The authors thank Dr. N.B. Patil, Director, CIRCOT, for his keen interest and encouragement.

REFERENCES

1. Rooke, G.B., Textile Res. J., 1981, 51, 168.
2. de Jongh, W.K., X-Ray Spectrometry, 1973, 2, 151.
3. Operation Manual of PW 1404 Automatic Sequential Spectrometer, 1986, N.V. Philips Gloeilampenfabrieken, Eindhoven, The Netherlands.
4. Kirk-Othmer Encyclopaedia of Chemical Technology, 1978, (Third Edn.), 15, 580.

26

Lateral order in microfibrils of native and regenerated cellulose

H.-P. Fink, D. Hofmann and H.J. Purz - Institute of Polymer Chemistry, O-1530 Teltow-Seehof, Germany.

Introduction

Microfibrils are accepted as the fundamental morphological subunits of both native and regenerated cellulose fibres. However, different opinions do exist with regard to the supermolecular structure or the lateral order of the chains inside the microfibrils /1-10/. However, the lateral crystallite size seems to be a very important structural factor e.g. in alkalization /11/.

We investigated a variety of cellulosic materials by wide angle X-ray scattering (WAXS) using a new method /12,13/ for the separate determination of crystallite sizes and lattice distortions. For native cellulose samples we compared lateral crystallite sizes determined by WAXS with lateral dimensions of microfibrils obtained from electron micrographs to select the supermolecular model next to reality. Furthermore, lateral crystallite sizes and lattice distortions of regenerated fibres of the viscose-type have been determined by our microstrain based WAXS-evaluation of samples similar to those discussed earlier by the Hosemann-group /14,15/.

Experimental

Scattering curves were measured and processed as described in ref. /13,16/ . Neglecting lattice distortions, weight average crystallite sizes D were determined for the (101), $(10\bar{1})$, and (002) peaks of cellulose I and the (101) peak of cellulose II samples (Meyer and Misch-notation) using the Scherrer-Equation.

Completely corrected number average crystallite sizes and root mean square lattice strain parameters were determined using a single line Fourier-method /12,13/ based on the microstrain-concept and developed particularly for the investigation of polymers.
Micrographs were taken from suspended fibrils studied in a transmission electron microscope (60 kV) . For this preparation, a dilute aqueous cellulose suspension was distributed on a carbon film covered gold grid using an ultrasonic technique preferentially. Subsequently the samples were dried and shadowed with Pt-Ir/C for contrast enhancement.
The investigated samples are listed in Tables 1-3.

Results and Discussion

The morphological features of the native fibres obtained by TEM (cf. Figure 1) are summarized in Table 1.
From WAXS measurements we got lateral crystallite dimensions of native samples as presented in Table 2. The weight average D usually determined represents a minimum value because of neglected lattice distortions. For valonia and ramie celluloses we additionally calculated fully corrected number average crystallite sizes L and r.m.s. lattice strain parameters $\langle e_l^2 \rangle^{1/2}$ according to the single line Fourier method. (Note: The WAXS peak resolution was not sufficient for the other native cellulose materials.)

Lateral crystallite dimensions of different regenerated fibres were calculated from the well resolved (101)-peak using both, the Scherrer equation and the Fourier method. Because of the Gaussian

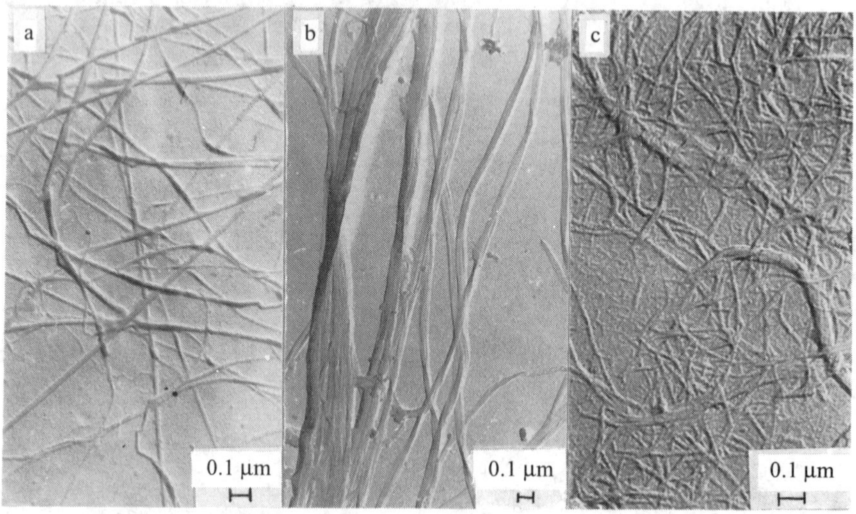

Figure 1 Electron micrographs of microfibrils of (a) valonia cellulose, (b) cotton linters, and (c) spruce sulfite sulfite dissolving pulp

Table 1 Lateral dimensions and features of native cellulose morphology

cellulose type	morphological features of microfibrils	lateral dimensions [nm]	
		microfibril	fibrillar strand
valonia	rather uniform dimensions, rigid, kink-like bending	10...35	——
bacterial	bimodal distribution, flexible structure, branching	4...7	35...95
cotton linters	slightly bimodal distribution, highly flexible structure	7...9	50...80
ramie	non-uniform dimensions flexible structure, branching	10...15	40...80
dissolving pulps	non-uniform features	10...30	up to 100

Table 2 Lateral crystallite dimensions of native cellulose

cellulose type	(hkl)	crystallite size [nm]		lattice distortions
		D	L	$\langle e_1^2 \rangle^{1/2}$
valonia	(101) $(10\bar{1})$ (002)	10.1 9.7 8.9	9.2 - 7.9	0.022 - 0.01
bacterial	(101) $(10\bar{1})$ (002)	5.3 6.5 5.7		
cotton linters	(101) $(10\bar{1})$ (002)	4.7 5.4 6.0		
ramie	(101) (002)	4.6 5.0	4.2 4.4	0.017 0.012
dissolving pulps	(002)	4.1...4.7		

like shape of the (101)-reflection we have been able to obtain also the true weight average values D^*. Table 3 contains the respective results. For a comparison with results /14,15/ based on paracrystal theory $\alpha^* = [\langle e_l^2\rangle L/d]^{1/2}$ was calculated assuming that $\langle e_l^2\rangle^{1/2}$ corresponds to Hosemann's g-Parameter.

Table 3 Lateral crystallite dimensions and lattice distortions of regenerated fibres from the WAXS (101)-reflection

fibre sample	crystallite size [nm]			lattice distortions $\langle e_l^2\rangle^{1/2}$	α^*
	weight average		true number average L		
	minimum D	true D^*			
mercerized linters	5.1	5.9	5.8	0.027	0.078
High wet modulus fibres A	4.6	5.1	4.8	0.025	0.066
B	4.6	5.0	4.7	0.026	0.067
C	3.9	4.4	3.8	0.030	0.075
D	3.9	4.2	3.9	0.031	0.074
rayon filament	3.8	4.2	3.7	0.032	0.077
rayon staple	3.6	3.9	3.3	0.034	0.078

Generally the fibrillar morphology of native cellulose has been confirmed by our TEM-investigations (cf. Figure 1 and Table 1). But morphological features and lateral dimensions of the microfibrils as defined by TEM strongly depend on the source of cellulose with lateral dimensions of single microfibrils ranging from about 4 nm to 100 nm.

Lateral crystallite sizes obtained by WAXS also depend on the native cellulose source, these dimensions vary in the range of 10 nm for valonia down to 4 nm for dissolving pulps (cf. Table 2). Lattice distortions from Fourier analysed WAXS of valonia and ramie celluloses are between 1 % and 2 %.

For model discussions the suggestions of a) microfibrils being single crystals /8-10/, b) microfibrils are composed of straight /1-3/ or zig-zag type /4/ elementary fibrils with a cross section of 3.5 nm x 3.5 nm, and c) a non-uniform two phases fringed fibrillar model /5-7/, drawn schematically for comparison in Figure 2 should be considered.

Comparing our WAXS and TEM results we have to reject both the suggestion of a single crystal microfibril and the 3.5nm x 3.5nm elementary fibril as the general supermolecular architecture of native cellulose. Taking into account the more or less broad distribution of the crystallite size, the lateral microfibrillar TEM dimensions usually being considerably higher than the corresponding lateral crystallite sizes L, as well as a degree of crystallinity usually clearly below 100 % /17/, only the fringed fibrillar model seems to approach the reality at this structural level.

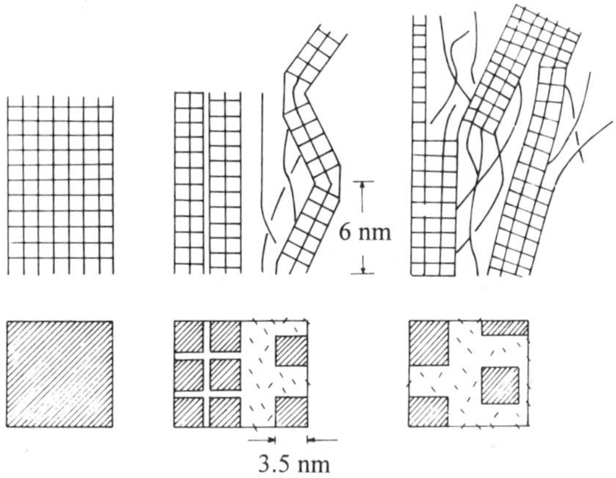

(a) (b) (c)

Figure 2 Models of native cellulose supermolecular structure
a) single crystal microfibril
b) microfibril composed of elementary fibrils
c) fringed fibrillar model

The morphology of regenerated fibres looks like a network with fibrillar units much shorter than the non-branched elements of native cellulose microfibrils. From TEM investigations of a variety of rayon fibres we found lateral microfibrillar dimensions in the range of about 5 nm to 20 nm which are usually clearly higher than lateral crystallite sizes from WAXS measurements (cf. Table 3). Lateral crystallite sizes vary between about 6 nm for slack mercerized linters and about 3 nm for conventional rayon staple fibres. High-wet-modulus fibres range between 4 nm and 5 nm at a somewhat higer level than conventional rayon filament and staple.
A comparison of the true weight and number average crystallite sizes in general indicates a more narrow size distribution than that of native cellulose. Especially the crystallite size distribution of linters has been sharpened by the mercerization thus confirming earlier results with regard to the mechanism of this process /11/. As we showed in that paper, by alkalization small crystallites are dissolved and bigger ones are split into smaller parts. The lattice distortion parameter of the regen-

erated samples vary between 2.5 % and 3.4 % with a tendency of increasing lattice distortions with decreasing crystallite sizes. This behaviour is in accordance with the suggestion of Hosemann that the growth of crystals is limited by lattice distortions. Indeed, the α^* parameter is about constant as demanded for materials of the same class and the figure of 0.08 (cf. Table 3) does agree quite well with that published by the Hosemann group for similar fibres /14,15/. Despite the somewhat greater difficulties in getting instructive electron micrographs of man made fibres, the fringed fibrillar model seems to be the most suitable suggestion also for the cellulose II samples.

Acknowledgement

The authors thank Prof. H. Chanzy from Centre de Recherches sur les Macromolecules Vegetales of the University Grenoble for providing the valonia cellulose and Prof. B. Philipp from the Institute of Polymer Chemistry Teltow for supporting this work.

References

/1/ Frey-Wyssling,A. and Mühlethaler,K.: Makromol. Chem. 62 (1963) 25
/2/ Heyn,A.N.J.: J. Appl. Phys. 36 (1965) 2088
/3/ Blackwell,J. and Kolpak,F.J.: Appl. Polymer Symp. 28 (1976) 751
/4/ Tsuji,M., Frank,J., and Manley,R.St.J.: Colloid & Polymer Sci. 264 (1986) 89-96
/5/ Hearle,J.W.S.: J. Polymer Sci. 28 (1958) 432
/6/ Krässig,H.: Das Papier 38 (1984) 571
/7/ Fink,H.-P. and Philipp,B.: J. Appl. Polymer Sci. 30 (1985) 3779
/8/ Revol, J.-F.: Carbohydrate Polymers 2 (1982) 123
/9/ Sugiyama,J., Harada,H., Fujiyoshi,Y., and Uyeda,N.: Planta 166 (1983) 161
/10/ Chanzy,H., Heurissat,B., Vuong,R., and Revol,J.-F.: Holzforschung 40 (1986) Supp. 25-30
/11/ Fink,H.-P., Fanter, D., and Philipp,B.: Acta Polymerica 36 (1985) 1
/12/ Hofmann,D. and Walenta,E.: Polymer 28 (1987) 1271
/13/ Hofmann,D., Fink,H.-P., and Philipp,B.:Polymer 30 (1989) 237
/14/ Haase,J., Hosemann,R., and Renwanz,B.: Kolloid-Z. u. Z. Polymere 251 (1973) 871
/15/ Haase,J., Hosemann,R., and Renwanz,B.: Cellulose Chem. Technol. 9 (1975) 513
/16/ Fink,H.-P., Hofmann,D., and Purz,H.J.: Acta Polymerica 41 (1990) 131
/17/ Fink,H.-P. and Philipp,B.: Proceedings of the conference "Progress in Cellulosic Man-made Fibres", Göteborg, Sweden, 9.-12. September 1991

27

New findings on the lateral order distribution in cotton fibers

J.K.S. Warrier and P.K. Chidambareswaren - Central Institute for Research on Cotton Technology, (CIRCOT), Post Box 16640, Bombay 400 019, India.

ABSTRACT

Recent work done at CIRCOT has shown that four distinct regions of order are exhibited in the Lateral Order Distribution (LOD) curves of the cotton fiber, obtained using aqueous NaOH solutions as structural probes and Lattice Conversion Ratio (LCR) as the measured property. The investigations have been extended to LiOH, KOH, RbOH and CsOH, and it has been found that these reagents give information on the regions having different extents of order in the fiber. It must be stressed that the LOD curves obtained in the present work, carried out using a well-aligned x-ray diffractometer, exhibit features which were probably lost in similar studies done earlier, mainly with the photographic x-ray techniques. The LOD results obtained using the aforesaid alkali metal hydroxides have been examined on the basis of the genetic background and cell wall development.

INTRODUCTION

The existence of a spectrum of order in cotton fiber has been recognised for a long time [1-4]. The Lateral Order Distribution (LOD) can be conveniently determined by evaluating the crystal lattice conversion from cellulose I to cellulose II induced by alkaline swelling agents of appropriate concentrations. The very early attempts to describe the LOD in cotton fibers[1] were handicapped by the use of photographic x-ray techniques which were incapable of yielding highly accurate experimental data. Besides, an assessment of the potential of LOD aspects in explaining physico-chemical properties of cotton fibers which are not amenable to a complete understanding on the basis of the conventional and oversimplified two-phase model appeared to be topical and warranted.

EXPERIMENTAL

Since reproducibility of results is extremely important, the conditions of treatment of cotton fibers with aqueous solutions of LiOH, NaOH, KOH, RbOH and CsOH, were standardized through repeated experimentation and confirmation. Giza7 cotton was used in the experiments. Treatments with different concentrations of alkali were carried out at 29 ± 0.1°C for 30 min in a thermostatted water-bath. Initially, wider intervals of normality were used, and the ranges in which maximum changes in LCR [5] occured were ascertained. Subsequently, each interval of interest was narrowed down to the extent of 0.01 N. Alkali-treated fibers were washed with flowing tap water and subsequently with distilled water; thereafter, they were dried at the same temperature as washing. The washing and drying conditions were kept strictly constant for all the treatments. The LCR values were reproducible within the error limits ± 0.01 when the above procedures were adhered to.

The x-ray diffraction pattern of the treated sample was obtained from pellets of Wiley Mill cut fibers passed through a 40 mesh screen. Only a nominal pressure of 5kg was applied on the powder filled in a standard sample holder during pellet preparation. The pellets were mechanically stable for much longer periods than the time for which they were exposed to x-rays. Ni-filtered Cu K_{α} radiation from a highly stabilised PW 1130 Philips X-ray Generator fitted with a high-resolution diffractometer, scanning/ counting arrangements and a recording system was employed. The divergence and anti-scatter slits were 0.5° and the receiving slit was 0.1mm. All intensities were measured above the background level at $2\theta=30^{\circ}$.

RESULTS AND DISCUSSION

The LOD curves obtained using LiOH, NaOH, KOH, RbOH and CsOH are given in Fig.1. The LOD curve generated using NaOH has been discussed by us earlier in a short communication [6]. It is obvious from the LOD data that LiOH is the most appropriate structural probe insofar as regions with highest orders are concerned. On the other hand, regions of poorest orders are best brought out in the LOD curve obtained using CsOH. This finding is in accordance with the concepts of fiber swelling in relation to ionic size [7]. In the series of alkali metal hydroxides used, Li has the smallest ionic size and hence possesses maximum power to penetrate highly ordered regions in the fiber. Regions with progressively decreasing order are sampled as the ionic size increases, from Li to Cs in the series. The differences in the LOD curves for various alkali metal hydroxides arise due to this reason. The present work thus shows that the order spectrum in cotton fibers can be investigated using different alkali solutions. It is interesting that the LOD curve of NaOH presently obtained, reveals more details than those reported earlier [1], due to the differences in the X-ray techniques used. It is also apparent that NaOH would be the best structural probe if a choice of LiOH and CsOH is not possible.

Efforts are being made to quantify the relative proportions of regions of different order in cotton fibers and examine the influence of this distribution on their textile properties.

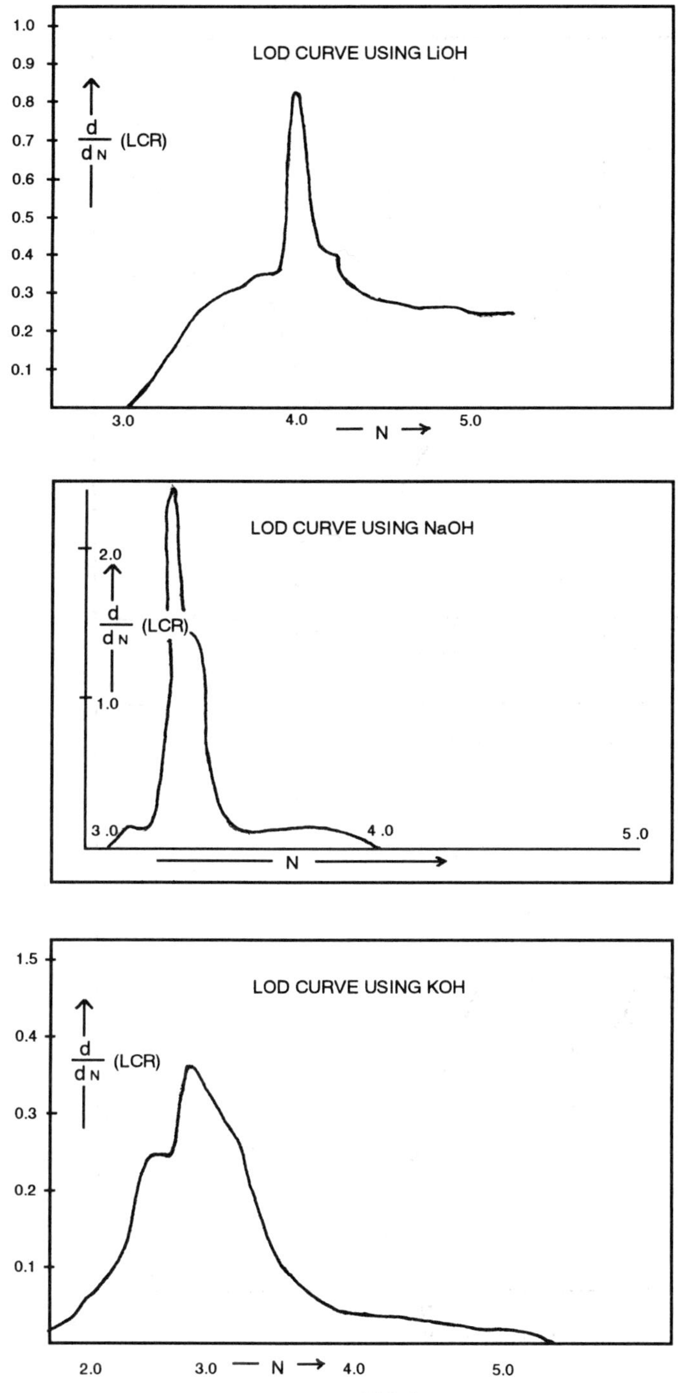
LOD CURVE USING LiOH
1.0
0.9
0.8
0.7
0.6
0.5
0.4
0.3
0.2
0.1
d/dN (LCR)
3.0
4.0
— N →
5.0
LOD CURVE USING NaOH
2.0
d/dN (LCR)
1.0
3.0
4.0
5.0
N
LOD CURVE USING KOH
1.5
0.4
0.3
0.2
0.1
d/dN (LCR)
2.0
3.0
— N →
4.0
5.0

FIG.1

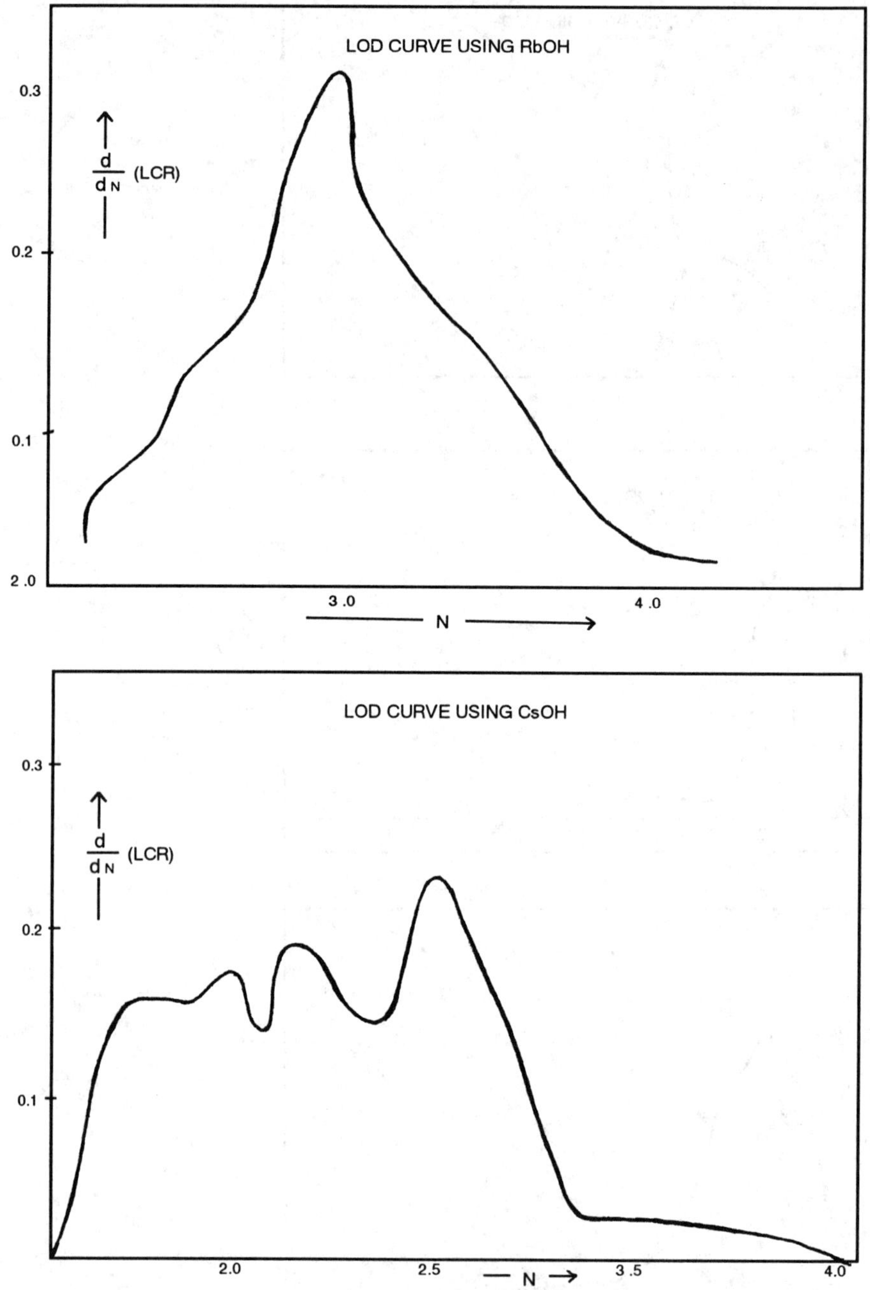

FIG.1(CONTINUED)

ACKNOWLEDGEMENT

The authors thank Dr. N.B. Patil, Director, CIRCOT, and Dr. V. Sundaram, Emeritus Scientist and former Director, CIRCOT, for their interest.

REFERENCES

1. Howsmon, J.A., and Sisson, W.A. in Cellulose and Cellulose Derivatives, Part I, Ott, E., Spurlin, H.M., and Grafflin, M.W., Eds., Interscience, New York, 1954, 251.
2. Chidambareswaran, P.K., Patil, N.B., and Sundaram, V., J. Appl. Polym. Sci., 1976, 20, 2297.
3. Rowland, S.P., and Roberts, E.J., J. Polym. Sci., 1972, A-1, 10, 2447.
4. Chidambareswaran, P.K., Patil, N.B., Pawar, A.D., Shukla, S.R., and Lokhande, H.T., J. Appl. Polym. Sci., 1984, 29, 837.
5. Vigo, T.L., Wade, R.H., Mitcham, D., and Welch, C.M., Textile Res. J., 1969, 39, 305.
6. Warrier, J.K.S., and Chidambareswaran, P.K., J. Appl. Polym Sci., 1990, 40, 1059.
7. Warwicker, J.O. in A Reveiw of Literature on the Effect of Caustic Soda and Other Swelling Agents on the Fine Structure of Cotton, 1966, Pamphlet 93, Shirley Inst., Manchester.

28

New insight into orientation - property associations in cotton fibers

J.K.S. Warrier and P.K. Chidambareswaren - Central Institute for Research on Cotton Technology, (CIRCOT), Post Box 16640, Bombay 400 019, India.

ABSTRACT

Crystallite orientation and tensile properties determined on 80 cotton varieties have revealed that a single average index of orientation is inadequate to predict cotton fiber tensile properties. The results indicate that fibrils with different inclinations to the fiber axis have different extents of influence on a specific fiber property. Fibrils with relatively larger inclinations (or, poorly oriented fibrils) have been found to be more important in deciding tenacity, while fibrils with steeper angles (or, highly oriented fibrils) play a more dominant role in determining elongation at break and toughness. The findings are discussed in terms of mechanisms of fiber tensile deformation and rupture.

INTRODUCTION

The Crystallite Orientation Distribution Profile (CODP) of cotton fibers describes the relative numbers of crystallites oriented at different angles to the fiber axis. An average crystallite orientation index, such as the Hermans Orientation Factor f(x) [1], is derived using the entire CODP and has been considered to be the most representative average measure of orientation characterising a cotton variety in many earlier studies. Accordingly, correlations have been sought between f(x) and those fiber properties (breaking strength, elongation, etc.) expected to be closely associated with fibrillar orientation and f(x).

Preliminary studies made by us earlier [2] had led us to suspect that the influences of fibrils having different inclinations to the fiber axis on a particular fiber property are not identical.This led us to a more detailed investigation, results of which have been partly published [3-5]. In this presentation, the new findings are discussed to provide a clearer understanding of the mechanisms involved in the fractography of cotton fibers.

EXPERIMENTAL

The CODP, fiber breaking tenacity (T) and elongation (E) of 80 cottons were determined. The cottons were selected to encompass the four major cultivated species, viz. G.(=Gossypium) hirsutum, G. barbadense, G. herbaceum and G. arboreum. Within each species, the varieties were chosen to represent as wide a range in fiber properties as possible. The CODP was obtained as described by us in an earlier publication [3]. The strongest reflection, (020 - fiber axis being 'c'), was used for CODP evaluations. At least two fiber bundle specimens were analysed for each variety. T and E were determined on the stelometer adopting standard procedures [6].

RESULTS AND DISCUSSION

The correlation coefficients obtained between fiber properties and orientation parameters are given in Table 1. In addition to f(x), the 20% and 75% X-ray angles [4] were also included in the analysis. From the Table it can be concluded that, in general, 20% X-ray angle shows the best correlation with T, while the 75% X-ray angle exhibits the highest correlation with E. In fact it was noticed that correlation for T progressively decreases from the 20% X-ray angle to the 75% X-ray angle when the 40% and 50% X-ray angles were also considered. On the other hand, an opposite trend was noticed as regards E.

Table 1. Table of correlation coefficients

Parameter --> Species ↓	Fiber Property	20% X-ray angle	75% X-ray angle	f(x)
G. hirsutum	T	- 0.86 *	- 0.75 *	+ 0.74 *
	E	+ 0.70 *	+ 0.86 *	- 0.67 *
G. barbadense	T	- 0.84 *	- 0.75 *	+ 0.83 *
	E	+ 0.65 $	+ 0.74 *	- 0.64 $
G. herbaceum	T	- 0.60 $	- 0.49 @	+ 0.73 *
	E	+ 0.69 *	+ 0.73 *	- 0.55 @
G. arboreum	T	- 0.83 *	- 0.81 *	+ 0.75 *
	E	+ 0.62 $	+ 0.70 *	- 0.57 $

Significance Levels : * 0.1%, $ 1%, @ 5%

The findings on 20% X-ray angle being the best orientation parameter related to T appeared to be upset for G-herbaceum cottons. This is not surprising in view of the structural peculiarities, notably as regards microfibrillar dimensions, known to characterise the species [7,8]. For the purposes of discussion of the findings, the exceptional trend of G. herbaceum cottons need not be considered as a serious drawback. Besides, these cottons do not constitute the predominant bulk of raw material for yarn and fabric manufacture.

Fiber bundle toughness, evaluated as 0.5 TxE [9] exhibited the following correlations with 20% X-ray angle, 75% X-ray angle and f(x): + 0.40, + 0.80 and - 0.30 respectively. The first two coefficients were significant at 0.1% level, while the last was significant only at 1% level. Thus toughness behaved similar to elongation.

The present data point out that fibrils or crystallites with relatively larger inclination to the fiber axis (represented by the 20% X-ray angle) are more important in the context of fiber breaking strength. Elongation, however, is principally determined by fibrils aligned closer to the fiber axis (denoted by 75% X-ray angle). These conclusions are understandable on the basis of the model given in Fig 1. As the fiber is strained axially, fibrils making larger angles to the fiber axis have a greater possibility of being separated laterally due to the comparatively weak interfibril bonds which must bear the main component of the applied load. The outer regions of the fiber could thus easily become flaw initiation regions, the flaws propagating inward as the fiber test proceeds. Just as the weaker strands in a lea of yarn have a bearing on the average lea breaking strength, the weak bonds in the peripheral region in a fiber can significantly influence the average fiber breaking strength. As the fiber is progressively extended, the inner fibrils take up increasing loads; however, they need not rupture due to the fact that the major component of the applied force tends to be in the direction of the cellulose chain molecule. Ultimate breaking elongation is decided by the limiting highest orientation which can be attained by the inner fibrils, and this explains the predominant role of these fibrils in the context of elongation. The similarity of results between elongation and toughness indicates that fiber toughness is determined largely by fiber elongation.

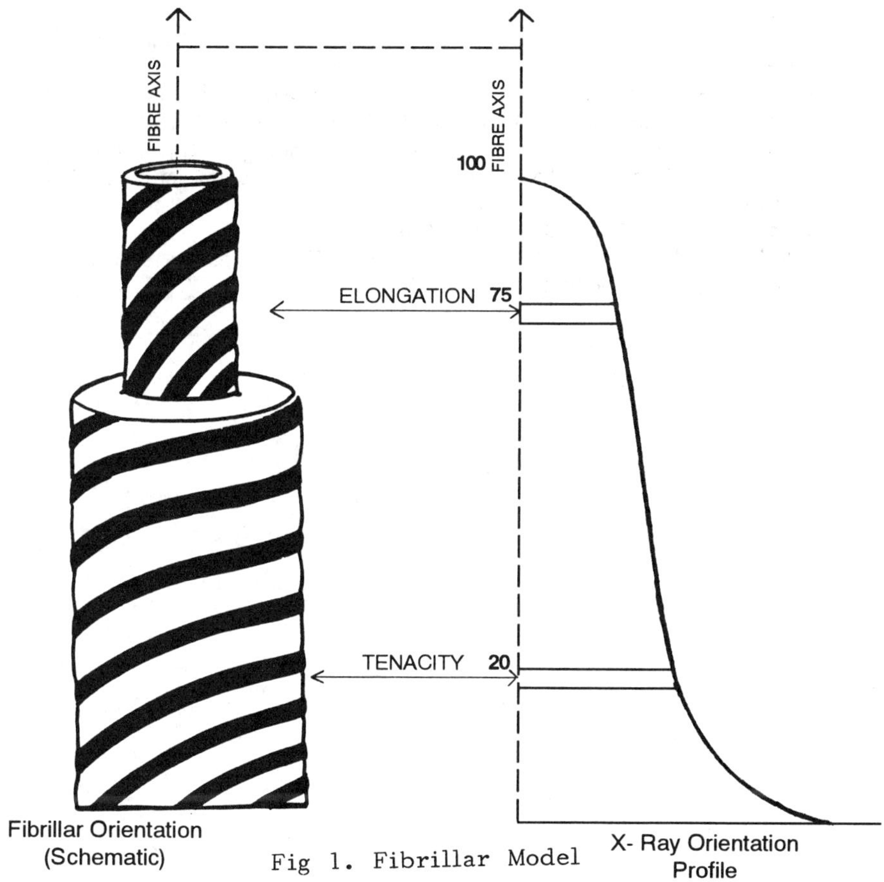

Fig 1. Fibrillar Model

CONCLUSIONS

1. An average measure of crystallite orientation distribution is not the appropriate parameter for structure-property investigations in cotton fibers.
2. Fibrils with small and large inclinations respectively control elongation and strength of fibers at break.
3. The outer cellulose layers are indicated to be the zones of flaw initiation during tensile loading leading to fiber fracture.

ACKNOWLEDGEMENTS

The authors thank Dr. N.B. Patil, Director, CIRCOT, for encouragement and permission to publish this work. Acknowledgments are also due to Dr. V. Sundaram, Emeritus Scientist and former Director, CIRCOT, for his keen interest.

REFERENCES

1. Hermans, J. J., Hermans, P. H., Vermaas, D., and Weidinger, A., Rec. Trav. Chim. Pays-bas, 1946, 65, 427.
2. Warrier, J. K. S., Munshi, V. G., and Chidambareswaran, P. K., Textile Res. J., 1979, 49, 617.
3. Chidambreswaran, P. K., Warrier, J. K. S., and Sundaram, V., J. Indian Soc. Cotton Improve., 1988, 13, 129.
4. Warrier, J. K. S., Munshi, V. G., and Chidambareswaran, P. K., ibid., 1988, 13, 140.
5. Warrier J. K. S., Munshi, V. G., and Chidambareswaran, P. K., ibid., 1090, 14, 80.
6. Sundaram, V., Iyer, K. R. K., Munshi, V. G., Parthasarathy, M. S., and Ukidve, A. V., Handbook of Methods of Tests for Cotton fibers, Yarns and Fabrics, 1979, CTRL, Bombay.
7. Betrabet, S. M., J. Sci. Industr. Res., 1967, 26, 201.
8. Betrabet, S. M., Datar, K. L., and Sundaram, V., Textile Res. J., 1973, 43, 718.
9. Meredith, R., J. Textile Inst., 1952, 43, 785.

29

Structural response to tensile deformation in cotton fibers

P.K. Chidambareswaren and J.K.S. Warrier - Central Institute for Research on Cotton Technology, (CIRCOT), Post Box 16640, Bombay 400 019, India.

ABSTRACT

X-ray diffraction methods have been applied to the study of the response of fibrillar orientation to tensile loading of cotton fibers. Well-parallelised bundles of fibers were kept extended to different levels by static loading and the fibrillar orientation evaluated. The Extension-Orientation Curve (EOC) differed for various cotton varieties. In general, the curves could be classified into four types, mainly on the basis of orientation-response to very small and very large extensions. Cottons with smaller convolution angles registered noticeable orientation changes at small extensions, while highly convoluted cottons did not show this behaviour. At very high extensions, diploid varieties exhibited considerable disorientation, unlike tetraploid cottons. Orientation in mature cottons was found to respond better to extension than that in immature ones. At average extensions all varieties had a linear orientation response to stretch; however, the rate of change of orientation with extension depended upon the normal orientation, convolution angle and maturity of the fibers.

INTRODUCTION

Although numerous publications are available on the differences in Crystallite Orientation Distribution Profile (CODP) as well as quantitative estimates of orientation between cottons and the relationship of orientation with fiber tensile properties [1,2], there are very few reports on the change in orientation in a fiber brought about by tensile loading. This may be because of the relatively smaller changes in orientation that could be expected from a fiber possessing smaller extensibility compared to the more easily deformable man-made fibers. Large changes in orientation can be brought about in cotton fibers only by stretching them after treatment with suitable swelling agents and because of the improvement in certain properties obtained by such treatments, most of the attention has been centered upon fibers which are swollen and then stretched. However, a few studies have indicated that when raw cotton fibers are held under different loads, their X-ray orientation is measurably different and there can be differences in the response of orientation to external loading. Manjunath and Peacock [3] provided data on two cottons which when analysed indicates that the rate of change of the 40% X-ray angle with extension was different for the cottons. Kalyanaraman [4-6] reported differences in the improvement of orientation on loading different cottons and concluded that stress orientation relationship deviates from a regular Hookean trend and interpreted the results on the basis of a crystalline-amorphous model. However, his work was based upon cottons all drawn from the American Upland types. There has been no comprehensive study on either the effect of fiber morphological parameters such as convolutions, maturity, etc., or the influence of species on the response of orientation to stress. In the absence of such information an identification or understanding of the factors responsible for these differences has not been possible. The present study was motivated by these considerations.

EXPERIMENTAL

In order to cover samples with as wide a range of morphological details as possible a few cottons for the study were chosen from the diploid species (G.arboreum and G. herbaceum) besides the commercially more prevailing tetraploids (G. hirsutum and G. barbadense). It is well known that in general, the diploids have smaller convolution angles, 50% X-ray angle (φ), higher degree of cell-wall thickness and circularity of cross section in comparison with the tetraploids. However, these parameters do not significantly vary within the diploids.

Well-parallelised flat bundles of cotton fibers were prepared and mounted in the jaws of the fiber bundle specimen mounting gadget used with an X-ray flat plate camera. The bundle was checked for parallelism of fibers after mounting the gadget on the stage of a projection microscope. Next, tension was applied using a screw head in the gadget till the fibers were just taut, as seen on projection screen. The normal length of the fiber bundle was measured at this stage as the distance between the edges of the jaws holding the bundle, to an accuracy of 0.02 mm. Flat film X-ray photograph of the specimen was taken using a 0.5 mm collimator and Cu K_α radiation, with 30 mm specimen to film distance. The highly stabilized generator was operated at 45 kV and 35 mA; the exposure time was 3 hr.

The exposed film was microdensitometered for obtaining the azimuthal intensity distribution of the strongest reflection, viz., 020 ('c' axis being the fiber axis). The usual corrections were applied for film background and non-linearity of film characteristics as well as detecting system response. The optical density data collected over all the four quadrants were averaged out to give the intensity profile in one quadrant, from 0° to 90°. The equator of the film where the maximum intensity occurs when the specimen axis is vertical was assigned the value 0° for the azimuthal angle. The profile was normalized to yield an intensity value of 100 units on the equator. This represented the CODP of the bundle specimen maintained at normal length.

The CODP of the specimen was subsequently obtained after stretching the bundle by small increments (1 - 2% of normal length) using the screw head in the gadget. For each level of stretch the diffraction pattern was obtained after allowing sufficient time for stress relaxation effects to occur. The exact length after giving any particular stretch was accurately measured using the projection microscope. At least two bundles were studied for each variety and the results averaged out. The maximum extension that could be given to a bundle was determined by visible signs of breakage of fibers. The 50% X-ray angle was evaluated from the CODP as its width at an intensity of 50 units. The reciprocal of ψ which increases with orientation was used as the X-ray orientation index (ψ-1).

RESULTS AND DISCUSSION

The plot of ψ-1 against % elongation gave the Extension Orientation Curve (EOC) of a cotton. The EOC showed noticeable differences between cottons. In general, the EOC of any cotton could be fitted into one of the four typical curves given in Fig. 1. The classification into these four types could be done mainly on the basis of the response of ψ -1 to very small and very large extensions. The former show smaller disorientation at large extensions compared to the latter. Cottons from tetraploids with small convolution angles were found to have EOC shown in Fig. 1 (a). This type is characterised by increase in ψ -1 even at very small E values. After a linear increase at medium elongation levels, the curve shows a tendency to decrease near breaking E values.

Highly convoluted tetraploids behave as shown in Fig. 1 (b). It may be noted that at small E values, ψ -1 does not increase markedly, indicating that extension is not primarily associated with fine structural deformation. After this initial sluggish region, a sharp increase in observed. This linear region has a higher slope compared to tetraploids with lower convolution angles. Besides, the disorientation at extensions near break is also not so pronounced in the highly convoluted variety.

The diploids, as mentioned, show sharper disorientation at high extensions (Fig. 1, c & d). In comparison to the tetraploids, the drop in orientation in the high extension region is more abrupt in the case of the diploids. The more mature of these varieties exhibit changes in even at low extensions (Fig. 1, c). Thus, in this respect, they are similar to less convoluted tetraploids. Immaturity tends to retard the orientation response (Fig. 1, d), although not to the same extent as higher convolution angle does in the case of tetraploids.

In short, convolution angle and maturity, respectively, affect orientation behaviour of the tetraploids and diploids, respectively, at very small and very large extensions. All the varieties have a linear region in the EOC at medium elongation levels. It may be seen from Fig. 1. that the slope of this region is comparatively higher in highly convoluted tetraploids and immature diploids.

The differences in the EOC between tetraploids and diploids at extensions near break indicates that the failure of the structure is more abrupt in coarse short staple cottons which are normally diploids. In view of earlier evidence that the microfibrillar dimensions are likely to be larger in such fibers, it appears that the role of interfibrillar slippage is less in their fractography [7].

Within the tetraploids, a cotton which is highly convoluted will tend to deconvolute more as the fiber starts extending initially as the load is applied. Apparently, the initial sluggish region of the EOC arises due to extension of the fiber on account of deconvolution. This can happen without significant alteration in orientation, to bring about which larger forces are required. The immature fibers in the diploids normally have a slightly higher level of convolution as compared to fibers from highly mature types. Besides, since the diploids in general are less convoluted than the tetraploids, the effect of maturity or cell wall thickness is probably revealed more noticeably in diploids. Thus, both convolution angle and maturity influence the orientation response of a cotton to external load.

There was a positive association between ψ-1 and the slope of the linear part of the EOC.

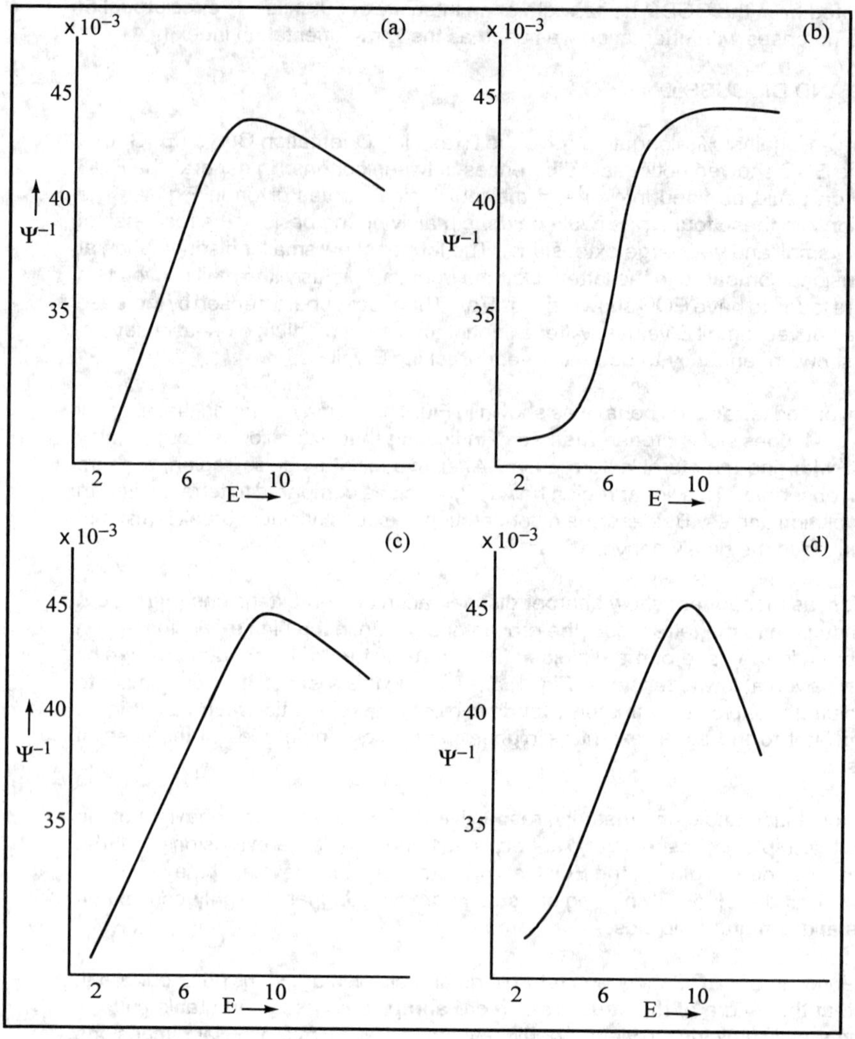

FIG 1. EOC CURVES

Assuming that a smaller spiral angle, or higher orientation, means fewer number of turns of the molecular helix per unit length along the fiber, the change in the spiral angle will be more when the spiral angle is smaller, because the helix length available to accommodate fiber extension is smaller. A cotton with a large spiral angle has a longer helix length to contribute to extension by virtue of the larger number of turns per unit length. Consequently, the detected change in the spiral angle itself will be small. Possibly, deconvolution effects may also partly contribute to extension even at medium elongation levels of less oriented cottons with larger convolution angles. Although deconvolution is expected to lead to the observation of an increase in the X-ray orientation index, this increase should be much less than that expected by a straightening of the spiral structure itself.

CONCLUSIONS

The EOC of different cottons differs mainly with regard to its shape at very low extensions as well as stretch levels approaching breaking elongation. While convolution angle is related to the behaviour of the curve at low extensions in the case of tetraploids, maturity has a predominant influence in diploid cottons which are relatively less convoluted. Highly oriented cottons were noted to undergo extension with relatively larger changes in orientation.

ACKNOWLEDGEMENT

We acknowledge the encouragement received for this work from Dr. N.B. Patil, Director, and Dr. V. Sundaram, Emeritus Scientist & former Director, CIRCOT.

REFERENCES

1. Warwicker, J.O., A Review of the Literature on the Effect of Caustic Soda and Other Swelling Agents on the Fine Structure of Cotton, 1966, Pamphlet No. 93, Shirley Inst., Manchester.
2. Chidambareswaran, P.K., X-Ray Diffraction Studies on Crystallite Orientation in Cotton Fibers, 1982, Ph.D. Thesis, Bombay University.
3. Manjunath, B.R., and Peacock, N., Proc. 11th Technol. Conf., 1969, II, 49.
4. Kalyanaraman, A.R., Textile Res. J., 1978, 48, 366.
5. Kalyanaraman, A.R., ibid., 1978, 48, 582.
6. Kalyanaraman, A.R., J. Appl. Polym. Sci., 1985, 30, 3801.
7. Betrabet, S.M., J. Sci. Industr. Res., 1967, 26, 201.

30

Rate control method for enzyme treatment of cellulosic fiber

M. Ueda*, Y. Yoshimura*, K. Kumita and M. Morita**** - *Osaka Municipal Technical Research Institute, Morinomiya 1-6-50, Joto-ku, Osaka 536, Japan. **Nagase and Co., Ltd., 4jo-dori, Nishinotoin-nishi, Shimokyo-ku, Kyoto 600, Japan.

ABSTRACT

Utilization of cellulase for the treatment of cellulosic fiber has been widely developed to improve its softness and hygroscopicity. Cellulase is known to hydrolyze cellulose and to produce many kinds of polysaccharide. Reaction products are too complicated to chemically analyze. Alternative methods, therefore, have been desired to monitor the cellulase reaction. A novel technique to monitor the rate of cellulase reaction is proposed. Mass of the reaction products can be detected by total organic carbon (TOC) measurement. Since the increase of TOC in the treating bath was found to be proportional to the loss of the cellulose substrate, the enzyme reaction can be monitored with the aid of TOC measurement of the bath. Appropriate end point of the treatment which has been difficult to predict can be determined well with this method [1].

INTRODUCTION

The treatment of cellulosic fiber with cellulase has been practically applied in order to improve its handle properties such as softness and water absorbability [2]. Cellulase hydrolyzes cellulose fiber producing various kinds of polysaccharide. The partial hydrolysis of cellulose chains of the cellulosic fiber by the treatment provides appreciably softer and more water absorbable characteristics.

The enzyme reaction proceeds preferentially with less crystalline regions of the substrate, with primary wall in the case of cotton fiber, for example, rather than with the secondary wall. The depression of the tensile strength of the fiber by the treatment, therefore, is less than one may expect from the weight loss of the fiber. Nevertheless, the treatment must not exceed 10% in weight loss of the fiber. The tensile properties of over-treated fiber would be too poor to be textiles.

Generally, in the practical process of the treatment, the rate of the treatment has been checked by the weight loss of the fiber substrate and the suitable end point of the treatment has been roughly expected from the enzyme reaction time. However, the reaction is remarkably influenced by woven fabric structure, nature of cellulose substrate, temperature and pH of treating bath, etc. The end point expected by the enzyme reaction time would not be reliable to obtain good reproducibility in this treatment. An alternative method to monitor the process, therefore, has been desired. This work gives a novel technique to monitor the rate of cellulase reaction detecting the concentration of total organic carbon (TOC) in the treating bath with a TOC analyzer which is commercially available.

EXPERIMENTAL

Cellulase (from Trichoderma viride, 1600FDUN/mg) was obtained from Nagase Biochemicals Ltd., Kyoto, Japan, and diluted to the concentration of 4 g/l as a treating solution. Anionic surfactant (1 g/l) was added in the solution as a penetrating agent. The treating solution was buffered with 0.6 g/l acetic acid and 1.2 g/l sodium acetate at pH 5.

Mercerized cotton knit, cotton broad cloth and linen were treated with the bath-ratio 1:20 at 50°C for 3-4 hours. After the treatment, the fabrics were immersed in boiling water to stop the enzyme reaction for 10 min. and dried.

The weight loss of the substrate and TOC of the treating bath were measured at one hour intervals gravimetrically and with a Shimadzu TOC analyzer (type 10A), respectively.

WEIGHT LOSS OF FIBER BY CELLULASE TREATMENT

In Figure 1, the weight loss of mercerized cotton knit and cotton broad cloth were plotted against the cellulase treating time. The figure shows that the weight loss of the substrates is not proportional to the reaction time and depends remarkably on the nature of the substrate even treated under the same conditions. Figure 2 shows the temperature and pH dependences of the cellulase activity determined with filter paper as a substrate [3]. As can be seen, the activity of the enzyme is remarkably influenced by temperature and pH. From this point of view, if the process is controlled only with reaction time, good reproducibility of the treatment would not be obtained.

Alternative simple techniques are, therefore, desired to detect the progress of the enzyme reaction direct and to control the process.

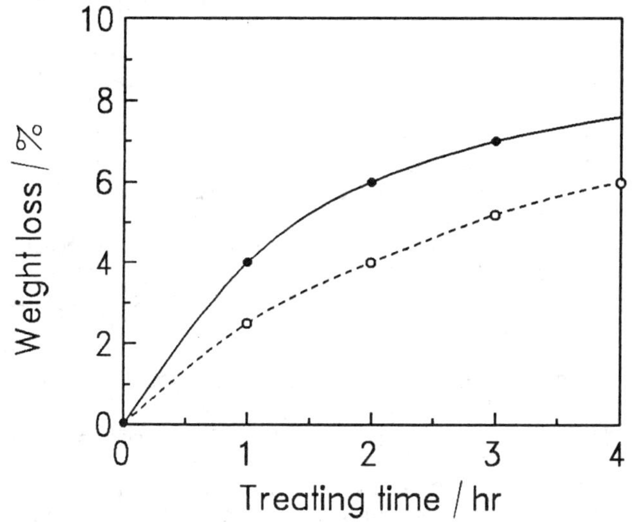

Figure 1. Weight loss of mercerized cotton knit (filled symbols) and cotton broad cloth (open) by cellulase treatment.

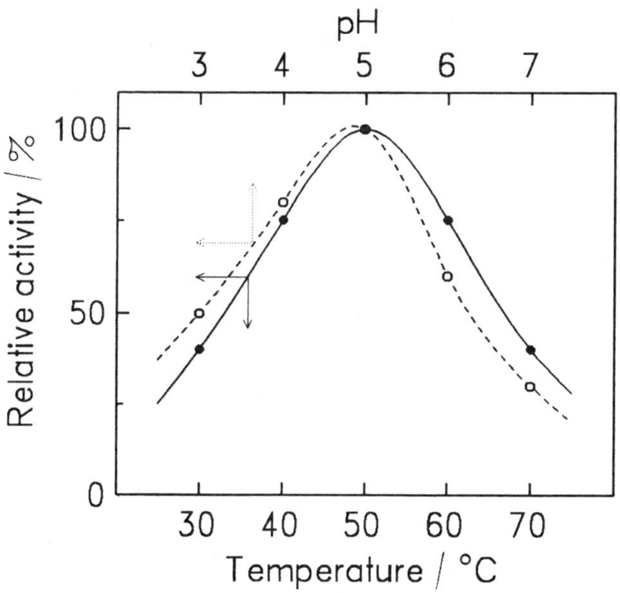

Figure 2. Temperature (filled symbols) and pH (open) dependences of the cellulase activity.

As mentioned above, cellulase produces various kinds of polysaccharides. The chemical analysis for the products can not be applied to monitoring the enzyme reaction. Nor the mass of the substrate (weight loss) can be detected direct during the treatment.

TOTAL ORGANIC CARBON IN TREATING BATH

The cellulose substrate is hydrolyzed and loses its weight with the progress of the enzyme reaction, whereas the relative amount (in the sense of the number of carbon atoms) of products should be produced and exist in the treating bath as a mixture of oligosaccharides. Therefore, it seems to be possible to know the progress of the enzyme reaction by means of the simultaneous measurement of the increase of total organic carbon (TOC) in the bath during the treatment.

The relationship between TOC in the bath and actual weight loss of the cotton fiber by the treatment is shown in Figure 3. This shows that TOC in the bath proportionally increases with increasing the weight loss of the fiber. This means that all the hydrolysate of cellulose remains stoichiometrically in the treating bath as water soluble polysaccharides, and has been detected accurately by the TOC measurement. The TOC value at weight loss being zero corresponds to the initial TOC in the treating bath due to the existence of enzyme, surfactant and buffer reagents. The correlation coefficients of the linear regression between weight loss of the cellulose and TOC in the bath were more than 0.99 in all cases measured in this work.

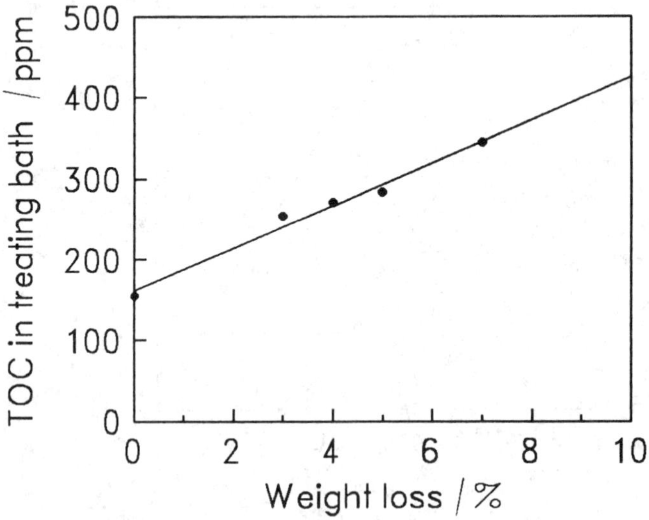

Figure 3. Relationship between TOC in the bath and actual weight loss of the cotton broad cloth by the treatment.

TOC analysis is based on the catalytic oxidation of the organic carbon compounds in the test solution to carbon dioxide gas, followed by quantitative analysis by means of infrared absorption. The analysis takes only 30-60 sec. from the start of the introduction of the test solution to the completion of obtaining the data. This time lag is short enough to neglect compared to the total treatment time (3-4 hr.), and it can be regarded that the TOC analysis would be a good quick-response sensor for the cellulase reaction.

RATE CONTROL SYSTEM FOR CELLULASE TREATMENT

As mentioned above, the cellulase reaction is influenced remarkably by temperature and the pH of the bath. Taking simultaneous measurement of TOC in the bath during the treatment, one can always monitor the reaction from the increase of TOC value even if the reaction condition, temperature or pH, may change. Appropriate end point of the treatment, which is dependent on the nature of fiber to be treated, activity of cellulase, and treating system used, can be accurately determined. With this system, best reproducibility of the treatment can be provided and excessive treatment due to unexpected change of the treating conditions can be checked.

The possible system for the rate control of cellulase treatment of cellulosic fiber are illustrated in Figure 4. This system gives another economical advantage that one can repeatedly use the same cellulase solution several times for the treatment even if the activity of it may be reduced.

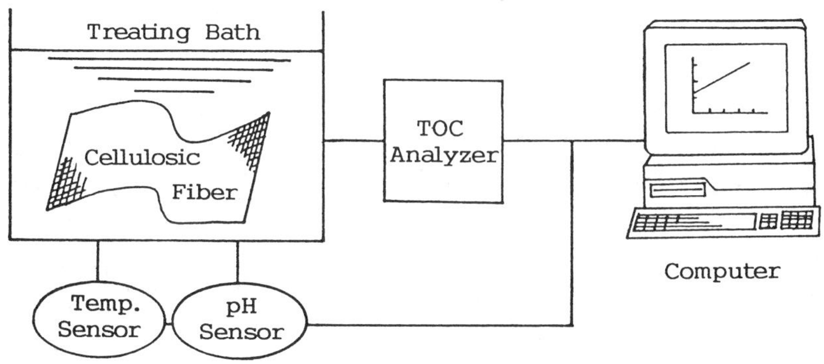

Figure 4. Rate control system for cellulase treatment.

REFERENCES

1. M. Ueda, Y. Tagawa, and K. Kumita, Japanese Patent pending, 1990.
2. O. Tanida, Dyeing Industry (Senshoku-kogyo), 37, 122 (1989).
3. Technical data of Nagase Biochemicals Ltd.

31

Water diffusion in wood pulp fibers

T.Q. Li*, U. Henriksson* and L. Ödberg** - *Department of Physical Chemistry, Royal Institute of Technology, S-100 44 Stockholm, Sweden. **Swedish Pulp and Paper Research Institute, P.O. Box 5604, S-114 86 Stockholm, Sweden.

ABSTRACT

The self-diffusion of water in pulp fibers was studied using the pulsed gradient spin-echo (PGSE) NMR method. The spin echo can be decomposed into contributions from bulk water between cellulose fibers and water trapped in pores inside fibers. The self-diffusion coefficient of bulk water is independent of the diffusion time but dependent on the concentration of the fibers in the sample, which can be explained by the obstruction and hydration effects of the fibers. The apparent self-diffusion coefficient of water in pores is approximately one third of bulk water when the diffusion time is $12ms$ and is reduced further by a factor of 3 when the diffusion time is increased to $80ms$. The apparent self-diffusion coefficient of the confined water was found to be anisotropic but independent of the fiber concentration. These observations indicate that the pores in cellulose fibers are elongated along the fiber axis having a length of the order of tens of μm. From the relative amplitude of the two components, which is nearly constant at different diffusion times, the amount of the restricted water is calculated to $1.6 \pm 0.3g/g$ of cellulose fibers. This value confirms results obtained by the solute-exclusion technique.

INTRODUCTION

The aim of this study was to investigate the mobility of water molecules in the pores of wood pulp cellulose fibers, and from these results gain some insight as to the geometric arrangement of these pores [1]. For this purpose we have used the pulsed field gradient spin-echo method (PGSE). PGSE is an non-invasive method and can be used to accurately determine self-diffusion coefficients in various bulk liquids [2,3]. Since the diffusion time in a PGSE diffusion experiment is well-defined, it can also be applied to study heterogeneous systems where the apparent diffusion coefficient usually depends upon the diffusion time (transient diffusion). Information on the structure of these systems can then be extracted from the dependence of the diffusional mobility on the diffusion time. The PGSE method has been applied to study a variety of heterogeneous systems, such as, plant and yeast cells [3,4], emulsion droplets [3,5,6], and some other systems [7]. The diffusion length that can be detected in a PGSE diffusion experiment is in the range $0.1\text{-}100\mu m$.

EXPERIMENTAL

A never-dried bleached softwood kraft pulp (Imperial Anchor, Iggesund AB, Sweden) and an unbleached softwood linerboard kraft pulp with $\kappa = 103$ (SCA Nordliner Munksund, Sweden) were investigated. The pulps were treated and purified by procedures described in reference [8]. A pulp sample with 18.5% fiber content was prepared from the bleached kraft pulp (beaten to 5000 PFI) by slightly pressing the washed pulp. The same pulp was also used to make 20 g/m^2 sheets. The wet sheets were cut into circles of 8 mm in diameter and packed into 10 mm NMR tubes with the sheet plane perpendicular to the tube axis. In order to obtain an appropriate NMR signal to noise ratio the water content in the sheet sample was adjusted to 88.0%. Furthermore, fiber suspensions of various concentrations (0-6 weight %) were made from both the bleached and unbleached pulps by adding distilled water with 50% D_2O. The fiber concentration of the suspensions is accurate to within ±10%.

The PGSE measurements of the pulp sample and the sheet sample were carried out on a Bruker MSL-200 spectrometer equipped with a Bruker microimaging probe. The measurements of the fiber suspension were performed on a Jeol FX-100 NMR spectrometer equipped with a home-made pulsed magnetic field gradient unit. As shown in Figure 1, in a PGSE experiment, two magnetic field gradient pulses of amplitude G and duration δ are applied symmetrically before and after the 180^0 pulse of the spin-echo pulse sequence with a time interval Δ between the gradient pulses. The diffusion time in a PGSE experiment is $\Delta - 1/3\delta$. In this investigation, water diffusion in the cellulose samples was studied as a function of the diffusion time and of the direction of the applied gradient.

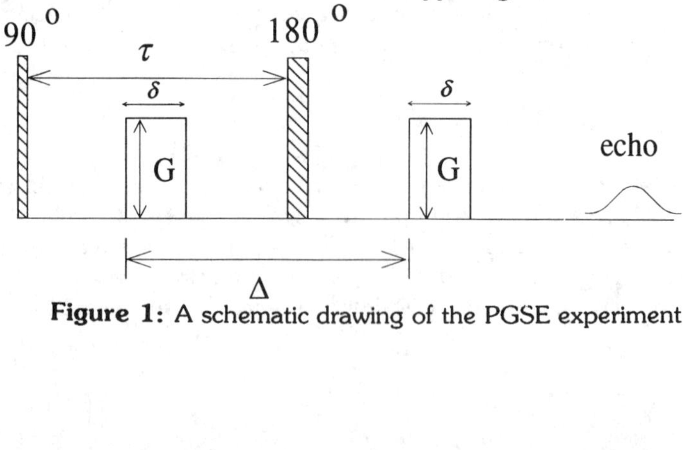

Figure 1: A schematic drawing of the PGSE experiment

RESULTS AND DISCUSSION

Figure 2 shows the observed echo attenuation profiles for water in the wood pulp sample at various diffusion times. The semi-logarithmic plot of the echo attenuation versus the variable $k = \gamma^2 G^2 \delta^2 (\Delta - \delta/3)$ (where γ is the gyromagnetic ratio of the monitored nuclear spin) does not give a straight line as is normally the case for bulk liquid samples. From Figure 2, it is immediately clear that diffusional behaviour of water in cellulose fibers differ from bulk water. All the cellulose samples used in the present study contains both bulk water and water confined in pores inside fibers. The water molecules inside the cellulose fibers are confined by the walls of the pores and should have transient characteristics. Measurements of spin-spin relaxation time, T_2, for the pulp sample by Carr-Purcell method [9] have shown that the two types of water with different diffusional characteristics have the same apparent T_2. Furthermore, in our samples the amount of water inside fibers is relatively low compared with bulk water between fibers and the diffusion time used in our PGSE experiment is short ($\Delta \approx 100ms$). Under these conditions, the observed echo attenuation R, at a fixed Δ, can be approximated by a simple population-weighted sum of the bulk and confined water *i.e*:

$$R(k) = P\exp(-D_1 \cdot k) + (1-P)\exp(-D_2(\Delta) \cdot k) \quad (1)$$

where D_1 represents the self-diffusion coefficient for the bulk water between fibers, $D_2(\Delta)$ is the apparent diffusion coefficient for the fiber water and P the fraction of bulk water.

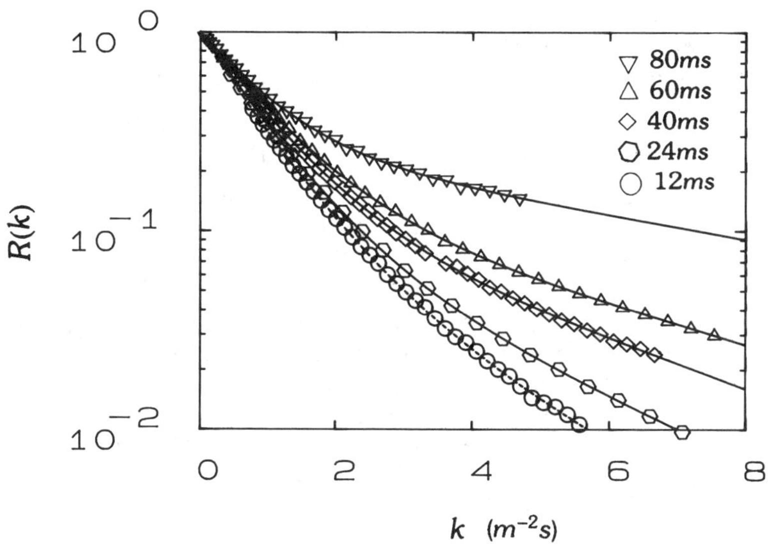

Figure 2: PGSE experimental results for the pulp sample observed at 5 different diffusion times and the numerical fittings of equation (1) to the experimental data.

As shown by the solid lines in Figure 2, the fit of equation (1) to all the experimental data at different diffusion times is quite good. Similar results were also obtained for the sheet sample as well as for the fiber suspensions. The self-diffusion coefficients and the fractions obtained from the nonlinear curve fittings are given in Table 1 for the pulp and sheet samples and in Figure 3 for the suspensions. As shown in Table 1 and Figure 3, the self-diffusion coefficient D_1 and the fraction parameter P for the bulk water between fibers are relatively constant for different diffusion times but they are dependent on the concentration of the fibers in the suspension samples. As expected, the apparent diffusion coefficient $D_2(\Delta)$ of the restricted water in pores inside fibers is concentration independent and is only a function of the diffusion time used.

The concentration dependence of the self-diffusion coefficients of the bulk water between fibers is due to two well-known phenomena: obstruction and hydration [10]. As shown in Figure 3, our theoretical calculations indicate that the obstruction is the dominant effect whereas the hydration is of minor importance. Furthermore, the calculations also indicate reasonable amounts of hydrated water [1,8]. The fraction parameter P for the suspensions is linearly correlated with the fiber content in the samples. From this linear relationship and the known overall water content in the samples, the amount of the water trapped in pores inside the fibers is estimated to $1.6 \pm 0.3 g/g$ of cellulose fiber. This independently confirms the results obtained from the solute-exclusion technique [11].

Table 1: D_1, $D_2(\Delta)$ and P obtained from the PGSE experimental data for the pulp and the sheet sample at various diffusion times.

Samples	Δ (ms)	P (%)	D_1 ($10^{-9}m^2s^{-1}$)	D_2 ($10^{-9}m^2s^{-1}$)
	12	84.3 ± 0.4	1.31 ± 0.03	0.49 ± 0.03
The pulp	24	85.3 ± 0.4	1.29 ± 0.01	0.40 ± 0.01
sample with	40	85.8 ± 0.5	1.15 ± 0.01	0.28 ± 0.01
18.5% fibers	60	81.9 ± 1.0	1.19 ± 0.04	0.25 ± 0.03
	80	69.9 ± 1.5	1.12 ± 0.04	0.16 ± 0.01
The sheet sample	15	85.1 ± 0.7	1.35 ± 0.02	0.29 ± 0.01
oriented $\perp G$	20	85.1 ± 0.7	1.35 ± 0.02	0.23 ± 0.01
	30	81.5 ± 0.4	1.34 ± 0.02	0.14 ± 0.01
The sheet sample	15	85.5 ± 0.6	1.47 ± 0.04	0.41 ± 0.02
oriented $\parallel G$	20	84.8 ± 0.5	1.37 ± 0.04	0.35 ± 0.01
	30	84.0 ± 0.9	1.33 ± 0.06	0.27 ± 0.02

As presented in Table 1, at $\Delta = 12ms$, $D_2 = 0.49 \times 10^{-9} m^2 s^{-1}$, which is roughly one third of D_1. A straightforward interpretation is that the water molecules in pores inside fibers are allowed to diffuse more or less freely in one direction at short diffusion times but are narrowly confined in the other two directions. This agrees with the picture of the pores being extended in the direction of the fiber axis but otherwise rather narrow. When the diffusion time is increased from 12 to 80 ms, the diffusion in the extended direction of the pores is also restricted and the observed apparent diffusion coefficient of the confined water is further reduced by a factor of 3. This indicates that the size of the pores in this direction is comparable with the free diffusional displacement of the water molecules during the same time, which is the order of tens of μm. This is also supported by the measurements on the sheet sample. As shown in Table 1, the apparent diffusion coefficient $D_2(\Delta)$ obtained by applying the field gradient perpendicular to the sheet is always smaller than what is observed with the field gradient parallel to the sheet. *i.e.*, the water diffusion inside the fibers perpendicular to the fiber direction is much more restricted than that in the direction along the fiber. It should be pointed out that the pore size that is measured by solute exclusion is of the order of 20-200 Å[11]. This, however, represents the smallest dimension of the pores which determines if a probe molecule can enter the pore or not.

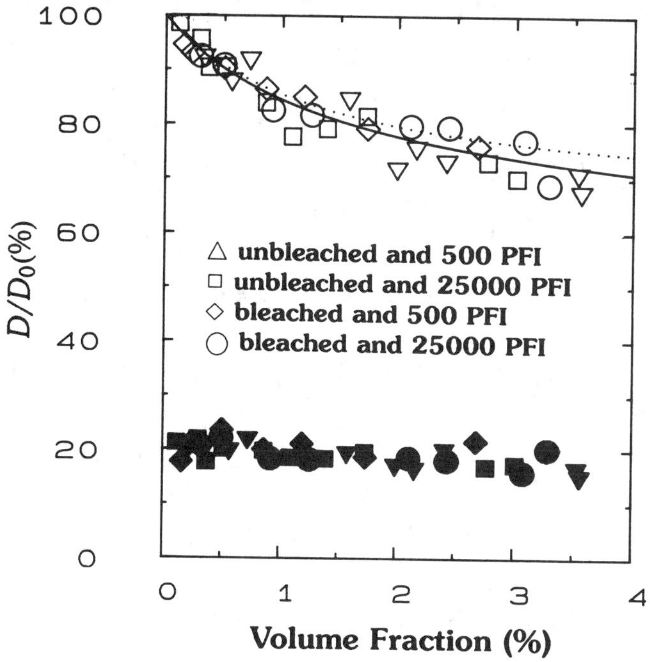

Figure 3: Diffusion coefficients obtained from PGSE experiments on the suspension samples as functions of the volume fraction of the fibers. The unfilled and filled symbols represent D_1 and D_2, respectively. The dotted line and solid lines refer to the calculated data for the suspension of large oblate particles (axial ratio = 120) with hydration equal to 0.08 and 0.8 $g\ H_2O/g$ fiber respectively.

CONCLUSIONS

Investigations of water diffusion with the PGSE method can distinguish water between the cellulose fibers and water confined inside the fibers. These studies give a new way of estimating pore water. The figure that is obtained (1.6*g* of H_2O/g fiber) confirms the results from solute exclusion studies. The PGSE results also show that the pores are anisotropic being extended along the fiber axis. The length is estimated to be a couple of tens of μm.

ACKNOWLEDGEMENTS

A scholarship from the Swedish Paper and Pulp Research Institute to one of the authors, Tie-Qiang Li, is gratefully acknowledged. We are also grateful to Mrs. Gunnel Risinger and Ms. Margareta Lind for experimental assistance.

References

[1] Li, T. Q., Ph.D. Thesis, Royal Inst. of Tech., Stockholm, Sweden (1991).

[2] Stilbs, P., *Prog. in NMR Spectr.* **19**, 1 (1987).

[3] Callaghan, P. T., *Aust. J. Phys.* **37**, 359 (1984).

[4] Tanner, J. E. and Stejskal, E. O., *J. Chem. Phys.* **49**, 1768 (1968).

[5] Cory, D. G. and Garroway, A. N., *Magn. Reson. in Medicine* **14**, 435 (1990).

[6] Lönnqvist, I., Khan, A. and Söderman, O., *J. Colloid and Interface Sci.* **144**, 401 (1991).

[7] Chachaty, C., Korb, J.-P., van der Maarel, J. R. C., Bras, W. and Quinn, P., *Physical Review B* **44**, 4778 (1991).

[8] Li, T. Q., Henriksson, U., Eriksson, J. C. and Ödberg, L., *Langmuir*, in press.

[9] Heink, W., Kärger, J. and Pfeifer, H., *J. Chem. Soc., Chem. Commun.*, **1990**, 1454.

[10] Jönsson, B., Wennerström, H., Nilson, P. G., and Linse, P., *Colloid and Polym. Sci.* **264**, 77 (1986).

[11] Scallan, A. M., *Cell. Chem. Tech.* **21**, 215 (1987).

32

Potassium hydroxide: a better mercerising agent as revealed by physico-mechanical and structural studies

S. Sreenivasan, P. Bhama Iyer and G.S. Patel - Central Institute for Research on Cotton Technology, Matunga, Bombay 400 019, India.

ABSTRACT

This paper establishes the superiority of aqueous potassium hydroxide (KOH) over sodium hydroxide (NaOH) in producing desirable swelling effects in cotton fibres. Irrespective of concentration and temperature of swelling, KOH swollen fibres show higher tenacity at all gauge lengths. These fibres also possess comparable or higher extensibility. In spite of marginally lower swelling, KOH treated fibres have higher accessibility and better fibrillar orientation, besides more uniform swelling as evidenced by electron microscopy. Extension of the studies to 17 cottons belonging to different species also revealed that KOH maintains its superiority, irrespective of the species and the initial strength uniformity of the variety. In the case of cottons belonging to **G. arboreum**, the change in tensile properties could be largely attributed to the change in fibrillar orientation. However, with **G. barbadense** cottons, this seems to result from reduced crystallite dimensions and closer packing of the structural elements in the secondary cell wall of the fibre.

INTRODUCTION

It is well known that mercerisation of cotton enhances and imparts certain desirable physical and structural characteristics. Our earlier studies [1,2] as well as those by other researchers (references cited in 1, 2), have shown that the fine structure of the swollen fibres is highly influenced by the size of the cation present in the swelling agent. Swelling in potassium hydroxide (KOH) was found to help in greater retention of tensile properties commensurate with better fine-structural characteristics such as higher accessibility, higher crystallite orientation and more flexible and shorter crystallites. These studies have been now extended to cover a wide range of temperatures of swelling in order to understand the behaviour of this reagent under extremely different temperatures. Fibres swollen in KOH at 0°C showed better retention of tensile properties due to conducive changes in the structural parameters resulting from a more uniform swelling. The present paper summarises some of the pertinent findings about the influence of temperature on swelling of cotton fibres in KOH vis a vis that in NaOH.

Furthermore, swelling of cotton has been known to be influenced by the fineness of the fibres in addition to the weak links as characterised by the strength uniformity of a variety. In the present study, by choosing 17 different varieties spread over the three major botanical species, the influence of strength uniformity on swelling has been analysed.

EXPERIMENTAL

Kier-boiled cotton fibres belonging to the variety Suvin **(G.barbadense)** were swollen at temperatures 0°C, 15°C, 30°C, and 45°C in KOH and NaOH of different concentrations. In addition fibres from different varieties having a wide range of properties were swollen at 30°C for studying the effect of strength uniformity. The physical, mechanical and structural evaluations of the treated fibres have been carried out as per standard procedures described elsewhere [1,2].

RESULTS

Bundle Tenacity and Elongation : Figures 1 and 2 illustrate the variation in the bundle tenacity at 3.2 mm gauge ($T_{3.2}$) and breaking elongation (E%) respectively of the fibres swollen in KOH and NaOH at the four different temperatures. While swelling in NaOH brings in deterioration in tenacity with increased swelling, irrespective of temperature, fibres swollen in KOH improve and maintain higher tenacities. The elongation of the fibres swollen at temperatures 15°C and above in both KOH and NaOH is comparable. The excessive swelling in NaOH particularly at 0°C weakens the interlamellar linkages resulting in reduced elongation, whereas KOH swollen fibres retain higher elongation as the swelling remains moderate even at 0°C.

Water Retention Values: The water retention values (WRV%) measured on swollen washed and undried fibres were used to get an idea about the extent of swelling. The extent of swelling in KOH was lower at all temperatures.

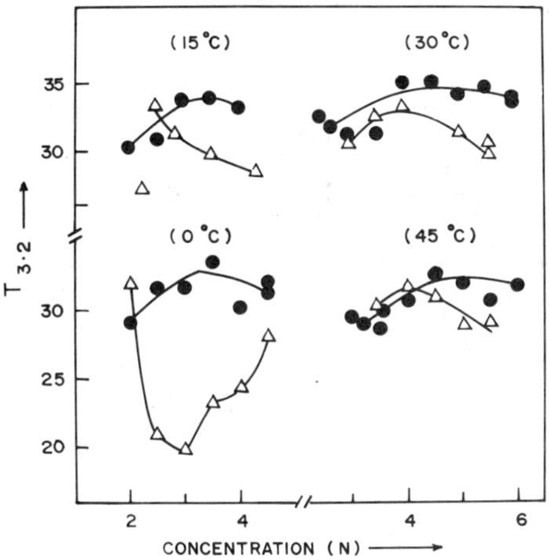

Figure 1. Variation in tenacity with concentration.
●——● KOH △——△ NaOH

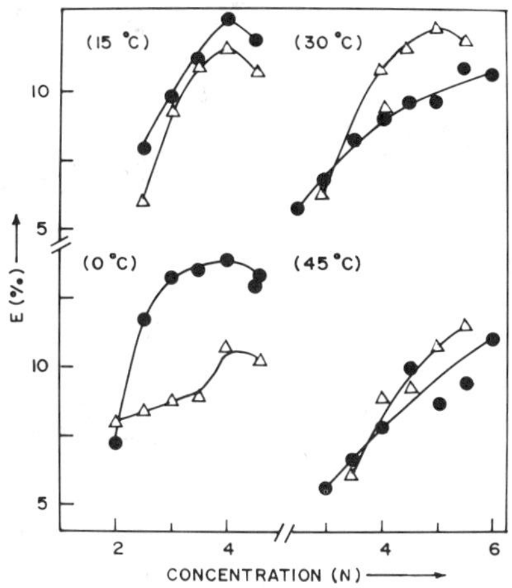

Figure 2. Variation in elongation with concentration.
●——● KOH △——△ NaOH

Amorphous Content and Crystallite Orientation : The amorphous contents for fibres swollen in KOH were found to be higher. Higher amorphous content despite the lower swelling suggests that the higher cationic size of KOH imparts a more uniform swelling of the entire structure of the fibre. As already noted [2] the reduction in crystallite length, measured as half width at half maximum (HWHM) of (004), after swelling in KOH is higher both at 0°C and at 30°C. The higher amorphous content and lower crystallite dimensions give the crystallites sufficient mobility for realignment during tensile loading.

Figure 3 highlights the changes in fibrillar orientation, measured as Hermans' orientation factor (f_x), for the swollen fibres. Irrespective of the temperature and concentration of swelling, KOH treated fibres retain higher crystallite orientation. Higher tenacities found in KOH swollen fibres might be mostly due to this. The excessive swelling of cotton fibres in NaOH at 0°C, particularly at 3N, leaves them with a very high disorientation.

The TEM micrographs obtained on the layered cross sections from fibres swollen at 30°C in both NaOH and KOH revealed that the bilateral structure present in native fibres does not disappear completely even after NaOH swelling. However swelling in KOH is more uniform as noted from cross sectional morphology. Uniform swelling all along the cross section, in spite of lower bulk swelling corroborates higher amorphous contents obtained in x-ray studies.

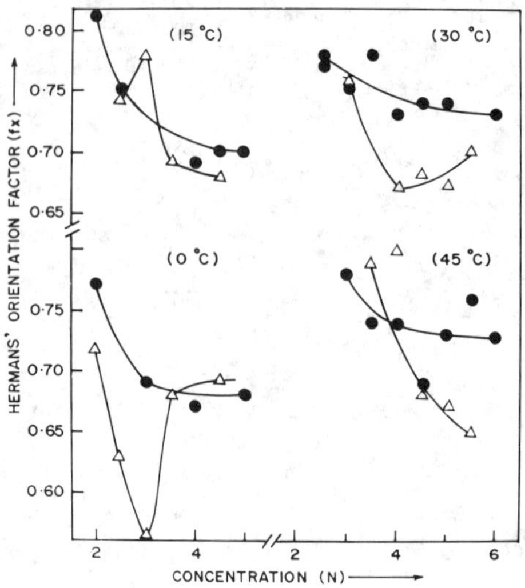

Figure 3. Hermans' orientation factior (f_x) with change in concentration. ●—● KOH △—△ NaOH

Studies conducted on swelling in KOH at different temperatures discussed above reveal that **KOH swollen fibres perform better at all temperatures.** As noted in the introduction, swelling of cotton is highly influenced by the fineness as well as by the weak links in the fibres. However the superior performance of KOH is maintained irrespective of fineness and strength uniformity. This is substantiated by the results given below.

Influence of Strength Uniformity on Tenacity : Figure 4 illustrates the relationship between the change in tenacity and the initial strength uniformity of cottons belonging to both **G. arboreum** and **G. barbadense** species. Irrespective of the species, more uniform cottons deteriorated after NaOH swelling. The current results for highly uniform cottons of **G. arboreum** also go in line with the findings [3, 5] that more uniform cottons, irrespective of species, decrease in tenacity at 3.2 mm gauge length after swelling in NaOH. With KOH, all cottons improve during swelling, although the improvement was less for more uniform cottons.

Tenacity-Structure Relations : Cotton fibres belonging to **G.arboreum** decreased or increased in tenacity in accordance with the changes in fibrillar orientation produced by swelling. A high correlation of 0.93 between $T_{3.2}$ and f_x is observed in these fibres. However, this correlation is very poor for **G.barbadense** cottons. Hence the tenacity of swollen fibres in this case seems to be influenced by factors other than orientation.

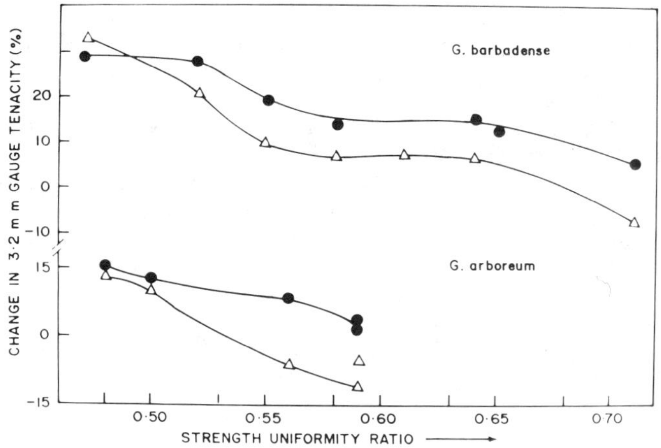

Figure 4. Change in tenacity with strength uniformity ratio
●——● KOH △——△ NaOH

As noted earlier the reduction in crystallite dimensions, particularly length, influences the changes in tenacity. This is evident from the results presented in Table I where HWHM values for the swollen fibres are reproduced. The changes in HWHM are also maximum for KOH swollen fibres belonging to **G.barbadense.** This suggests the predominant influence of reduced crystallite length on tenacity.

Table I
Data on Crystallite Length and Wall Thickening

Sample	Treatment	HWHM ($\beta_{\frac{1}{2}}$) in degrees		Degree of wall thickening (D)	
		Actual	% Increase	Actual	% Increase
G. arboreum	Control	0.33		0.75	
	NaOH	0.34	3.0	0.93	24.0
	KOH	0.36	9.1	0.88	17.3
G.barbadense	Control	0.29		0.65	
	NaOH	0.34	17.2	0.87	33.8
	KOH	0.40	37.9	0.75	15.4

Another parameter of interest in explaining the tenacity changes in cottons of **G.barbadense**, is the change in packing density of the fibrillar elements in the cell wall of the fibres. From measurements of area of wall and lumen of the fibre cross-sections as well as the cross-sectional perimeter, the degree of wall thickening (D) [6] which is inversely related to packing density has been worked out and are given in Table I . The apparent increase in the degree of thickening effected by swelling is higher in NaOH. Further the difference between NaOH and KOH is more marked in **G.barbadense** cottons than in **G.arboreum**. In other words packing decreases substantially in **G.barbadense** cottons after NaOH swelling. However, with the uniform swelling produced by KOH, the packing does not decrease drastically and concomitant strengthening of weak places in the fibres, helps in realising higher tenacity.

From the results presented above, it can be stated that, **for finer and more uniform cottons, swelling in KOH imparts better mechanical and structural characteristics.** The higher accessibility noticed in KOH treated fibres can be expected to help in the finishing treatments. Finer yarns spun from finer and more uniform cottons on swelling in KOH (results are not reported here) also revealed the superiority of the treatment.

A commercial exploitation of the KOH treatment at present, would be hampered by its high cost as compared to NaOH. Potassium is as abundant in nature as sodium. Development of suitable technologies for economic production of KOH would go a long way in its commercial acceptance as a mercerising agent. The present studies are expected to give sufficient impetus to realise the above mentioned objective.

ACKNOWLEDGEMENT

The authors thank Dr. K.R. Krishna Iyer and Dr. N.B. Patil for the kind encouragement.

REFERENCES

1. Sreenivasan, S., Bhama Iyer, P., Patel, G.S., Chidambareswaran,P.K.; *J. Appl. Polym. Sci.* **1989**, 37, 2191.
2. Bhama Iyer, P., Sreenivasan, S., Patel, G.S., Chidambareswaran, P.K., Patil, N. B.; *J. Appl. Polym. Sci.*, **1989**, 37, 1739.
3. Rajagopalan, A., Venkatesh, G.M., Dweltz, N.E.; *Textile Res. J.*, **1975**, 45, 4409.
4. Venkatesh, G.M.; *Textile Res.J.* **1975**, 45, 813.
5. Lawson, R., Hertel, K.L.; *Textile Res. J.* **1974**, 44, 975.
6. Booth, J.E. in *"Principles of Textile Testing"*, National Trade Press Ltd., London, **1961**, 172

33

Comparison of unique staple-core and conventional cotton-blend fabrics with durable-press finish

L.B. Kimmel, A.P.S. Sawhney, G.F. Ruppenicker, and R.J. Harper - USDA, ARS, Southern Regional Research Center[1], P.O. Box 19687, New Orleans, Louisiana 70179 USA.

ABSTRACT

This study compares unique staple-core and conventional intimate-blend fabrics before and after durable-press finishing. The yarn for both fabrics contains the same proportions of the same cotton and polyester fibers. However, the intimate-blend yarns are a homogeneous mix of these fibers, while the staple-core yarns consolidate the high tenacity polyester in the center of the yarn, with cotton on the outside. All yarn was spun on the same ring-spinning machine, with minor modifications that are described. The staple-core fabrics exhibit better retention of physical properties and appearance ratings than the intimate blend or 100% cotton control fabrics after durable-press finishing. The comparatively better performance of the staple-core fabrics is attributed to their unique yarn structure.

INTRODUCTION

Cotton is often blended with polyester to combine the strength, uniformity, and easy-care properties of synthetic fibers with the superior comfort and aesthetics of natural cotton. The fibers are usually combined early in processing to produce intimate-blend (IB) yarns and fabrics. However, cotton-rich IB fabrics lack the aesthetics of cotton, are usually weaker and less durable than pure cotton or synthetics, are considerably weakened by durable-press (DP) treatment, and may pill excessively. Patented staple-core (SC) spinning [1] produces cotton-rich fabrics that look and feel like cotton but are superior to cotton and IB fabrics.

This study compares SC and IB yarns and fabrics of identical composition, before and after DP finishing. The composite SC, IB and cotton control yarns were spun on modified or unmodified spindles of the same spinning frame. Others have used other methods to produce somewhat similar yarns, but generally failed to commercialize them. We continue to improve the technology from that described by previous publications [2,3,4,5].

MATERIALS AND METHODS

The SC and IB yarns contain the selected fibers in 67% cotton and 33% polyester proportions. The polyester is Hoeschst Celanese[2], 1-denier, 1.5-inch staple, high-tenacity Trevira #121[2], with a 60.3 g/tex tenacity and 23% elongation. The cotton is a 1.125-inch staple, strict low middling, rain-grown Delta cotton, with 2.5% and 50% span lengths of 1.15 and 0.51 inches respectively, a 24.7 g/tex tenacity, and 7.5% elongation. The cotton is carded at 60 lb/hr to 55 gr/yd, twice drawn into 55 gr/yd sliver, and processed into 2-hank roving for the SC yarn, and 1-hank roving for the cotton. The polyester is carded at 25 lb/hr on a granular card, drawn into 55 gr/yd sliver, and into 2-hank roving for the SC yarn. A cotton/polyester blend lap is similarly processed into 1-hank roving for the IB yarn.

The SC spinning method incorporates four additional devices on a conventional ring-spinning frame, depicted in Figure 1. The roving condensers and spacers control and separate the rovings in the drafting zone. A grooved bar mounts below the top front roller and incorporates adjustable twist-controllers to restrain yarn twist from the nip area. The bar is a metal rod with carefully machined grooves to stabilize the yarn at each spindle.

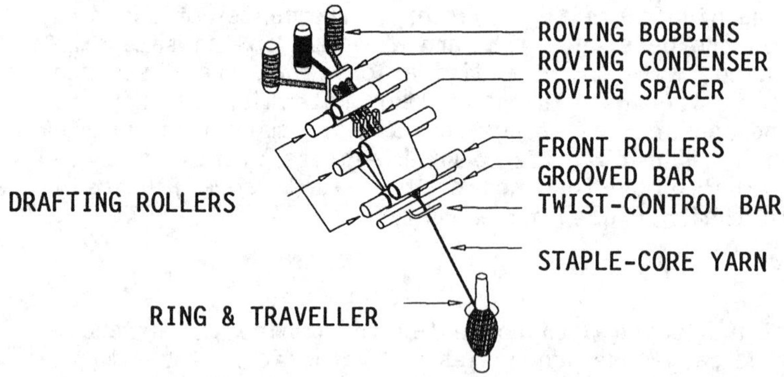

Figure 1. Staple-core spinning for producing core/wrap yarn [1].

During SC spinning, the core fibers are confined in the groove, while the sheath fibers migrate and interlock beside it. The fibers converge near the mid-point of the groove and simultaneous twist insertion transforms the fibrous sandwich into a SC yarn. The nominal 29.5-tex (Ne 20/1) SC yarns are spun as follows on a Roberts Arrow[2] spinning frame with 5-cm rings:

Cotton Rovings (two)	2.0 hank each (nominal)
Polyester Roving (one)	2.0 hank (nominal)
Combined Rovings (three)	0.67 hank (total)
Approximate Draft	30
Spindle Speed (rpm)	9,000
Twist Multiple	4.0
Traveller	#6

The yarns are wound, warped, and sized with a 6% PVA solution. Identical fabrics are plain woven on a 24-inch Leesona X-2[2] loom, using the subject yarns in warp and filling directions. The fabrics are sequentially subjected to boil off, a 0.7% caustic scour, another boil off, peroxide bleaching, and a 0.1% acetic acid scour. Finally, the fabrics are padded and dried in a tenter frame for 3 minutes at 195°F.

The SC, IB, and cotton fabrics are treated with two kinds of DP agents, using existing dimethyloldihydroxyethyleneurea (DMDHEU) and butanetetracarboxylic acid (BTCA) methods [6]. The DMDHEU treatment includes 5% DMDHEU, 2% magnesium chloride hexahydrate (catalyst), 0.5% polyethylene (fabric softener), and 0.1% non-ionic wetting agent in a 4.25 pH bath. The fabrics are padded twice to an average 72% wet pick up, dried for 3 minutes at 190°F, and cured for 3 minutes at 320°F. The BTCA method uses 6% BTCA, 3% sodium hypophosphite (catalyst), 0.5% polyethylene, and 0.1% non-ionic wetting agent, in a 2.24 pH bath. The fabrics are padded twice to an average 75% wet pick up, dried for 3 minutes at 190°F, and cured for 90 seconds at 356°F. All fabrics are backwashed twice in water and redried.

RESULTS AND DISCUSSION

Random cross sections show good segregation of the cotton and polyester in the SC yarns and good dispersion of the fibers in the blends (Figure 2). The 16 g/tex SC yarn strength is 19% stronger than the cotton yarn but 14% weaker than the IB yarn. Although unfinished SC fabric strength, elongation, and tear resistance remain lower than the IB fabrics, there is less differentiation between these fabrics than their constituent yarns. Note that the bleached and scoured (B&S) SC fabric strength approaches that of the IB to within 5% (Table 1). Yarn and fabric testing is in accordance with standard ASTM and AATCC procedures [7,8].

Figure 2. Photographs of Hardy cross sections [9] of the yarns.

Table I. Properties of bleached & scoured fabrics.

	Weight (oz/yd^2)	Count (WxF)	Break load (lb)	Elongation (%)	Tear resist. (g)
100% cotton	4.6	72 x 51	83 x 61	8 x 40	1250 x 993
Intimate-blend	4.7	72 x 51	106 x 66	12 x 46	1733 x 1300
Staple-core	4.5	71 x 52	96 x 64	10 x 43	1567 x 1120

The cotton fabric exhibits the greatest strength losses after DP finishing, remaining the weakest and least extensible fabric. Whereas the cotton fabric loses up to 55% of its strength after DP treatment, the SC and IB fabrics only lose between 30 and 40% (Figure 3a). It is well known that the cotton fibers are weakened by the associated crosslinking, and that the polyester is inert to these chemicals. The relative rankings of SC and IB fabric strength are the same after both DP treatments, but strength losses are slightly greater for the DMDHEU process.

The elongation of the SC fabric is lower than the IB before finishing, but higher after either DP treatment (Figure 3b). Whereas the DP finished cotton fabrics retain only 54-58% of their warp elongation, the IBs retain 65-74%, while the SC fabrics retain up to 91%. Woven fabrics are typically less extensible in the warp direction, but the unbalanced thread counts cause greater differences than usual in these fabrics. The deviation is the result of a combination of excess finishing tension and a faulty pick counter. The higher extensibility of the SC fabrics after resin treatment can be attributed to the consolidation of the higher elongation polyester in the yarn axis.

The unfinished SC fabrics are intermediate in tear resistance to the cotton and IB fabrics, but are superior to both after either DP treatment (Figure 4). Whereas the pure cotton fabrics retain about 80% of their warp tear resistance after resin treatment, the IB fabrics retain 80-87%, and the SC fabrics retain 95-100%.

Appearance ratings for the SC and IB fabrics are substantially better than the pure cotton fabrics before DP finishing, and remain higher to a lesser degree after finishing (Figure 5). B&S cotton ratings improve from 1.0 to 1.7 after 5 washes, while the IB and SC fabric ratings range from 2.3 to 2.9. Durable-press treatments bring the cotton fabrics to between 3.1 and 3.7, the IB fabrics to between 3.2 and 3.6, and the SC fabrics to between 3.3 to 3.7. Ratings tend to drop in DP fabrics after repeated laundering. Minor defects in the SC fabrics detracted from their ratings.

The SC fabrics are substantially superior to the IBs in abrasion and pilling resistance, especially after DP finishing (Figure 6). The crosslinking of the cotton in both DP processes reduces the abrasion resistance of the fabrics. Pilling is a significant problem in IB fabrics, whose average DP ratings of 1.0 to 1.7 are

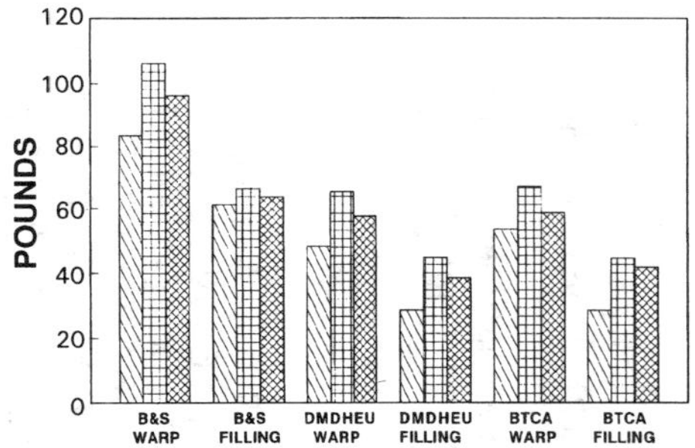

Figure 3a. Fabric breaking load.

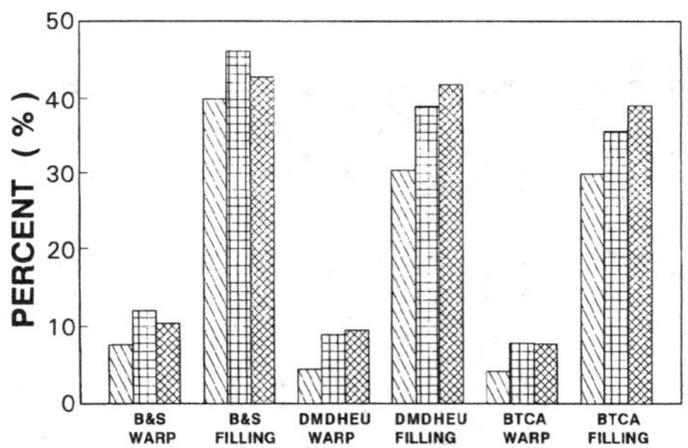

Figure 3b. Fabric elongation.

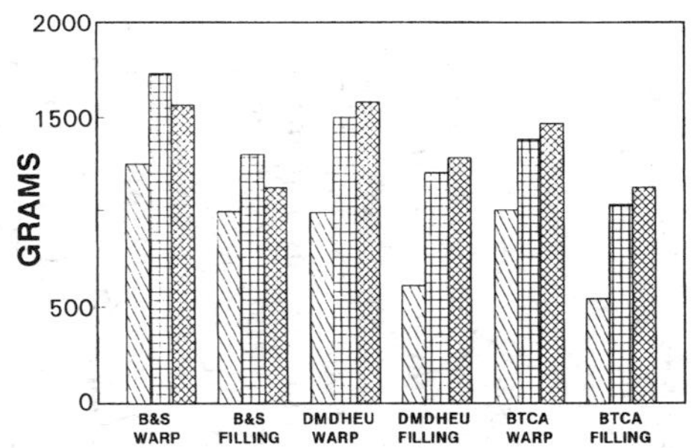

Figure 4. Tear resistance.

	100 % COTTON	INTIMATE BLEND	STAPLE CORE
BLEACHED & SCOURED	1 WASH: 1.0 5 WASH: 1.7	1 WASH: 2.6 5 WASH: 2.9	1 WASH: 2.3 5 WASH: 2.6
DMDHEU FINISH	1 WASH: 3.7 5 WASH: 3.1	1 WASH: 3.6 5 WASH: 3.2	1 WASH: 3.7 5 WASH: 3.6
BTCA FINISH	1 WASH: 3.4 5 WASH: 3.3	1 WASH: 3.6 5 WASH: 3.6	1 WASH: 3.3 5 WASH: 3.6

Figure 5. Fabric DP ratings.

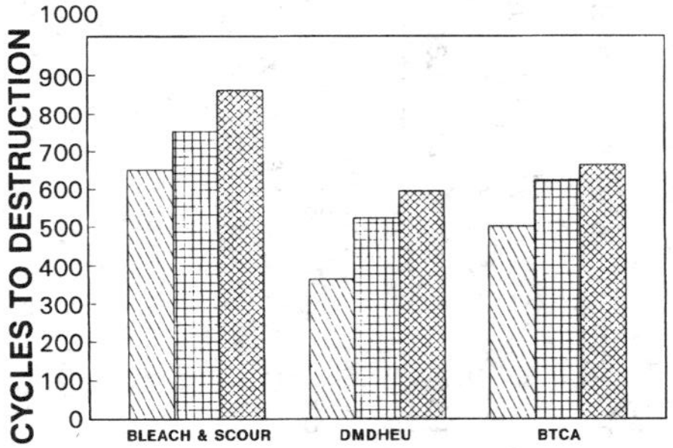

Figure 6a. Abrasion resistance.

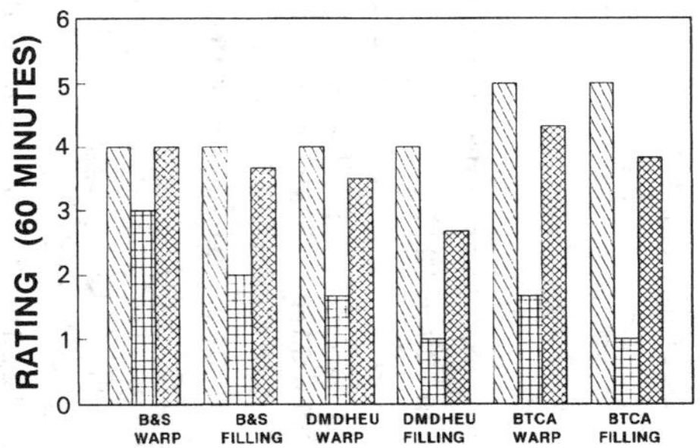

Figure 6b. Pilling resistance.

100% COTTON INTIMATE BLEND STAPLE CORE

extremely poor. The containment of the polyester in the SC yarn core reduces its deformation associated with contact abrasion.

CONCLUSIONS

Staple-core spinning enables the production of unique composite yarns and fabrics which exhibit the desirable characteristics of core and wrap fibers without the functional drawbacks of comparable blends. The patented yarns are easily spun from triple-creeled roving, on simply-modified, commercial ring-spinning frames. The SC yarns exhibit excellent core coverage, are highly strip resistant, and are easily woven into fabric.

In general, the high-tenacity polyester core/cotton sheath SC fabrics look and feel like cotton, but have properties superior to pure cotton, and selectively better than blends of the same fibers. The strategic consolidation of the stronger, more extensible, and chemically inert polyester in the core of the SC yarn is responsible for the superior abrasion resistance and greater retention of other fabric properties after DP treatment.

The cost of SC spinning is similar to conventional ring spinning, but producers may need additional roving or creeling capacity. Minor SC fabric defects have detracted from reported properties and appearance ratings, but refinements continue to improve yarn quality and uniformity. The successful commercialization of the technology would favor the use of SC fabrics for industrial or military applications which require a combination of fabric strength, durability, comfort, and appearance.

ACKNOWLEDGMENT

We gratefully acknowledge the assistance of our co-workers in the Fiber Quality and Textile Finishing Research Units.

ENDNOTES

1. One of the facilities of the Agricultural Research Service, U.S. Department of Agriculture.
2. Names of companies or commercial products are mentioned solely to provide specific information and their mention does not imply U.S.D.A. recommendation or endorsement.

REFERENCES

1. Sawhney, A.P.S., Folk, C.L., and Robert, K.Q., System for Producing Core/Wrap Yarn, U.S. Patent 4,976,096 (12-11-1990).
2. Louis, G.L., Salaun, H.L., and Kimmel, L.B., Ring-Spun All-Staple Core/Wrap Yarn, Tex. Res. J. 59(4), 244-245 (1989).
3. Sawhney, A.P.S., Harper, R.J., Ruppenicker, G.F., and Robert, K.Q., Comparison of Fabrics Made with Cotton Covered Polyester Staple-Core Yarn and 100% Cotton Yarn, Tex. Res. J. 61(2), 71-74 (1991).

4. Sawhney, A.P.S., Robert, K.Q., and Ruppenicker, G.F., Device for Producing Staple-Core/Cotton-Wrap Ring-Spun Yarns, Tex. Res. J. 59(9), 519-524 (1989).
5. Sawhney, A.P.S., Ruppenicker, G.F., and Kimmel, L.B., New Types of Ring-Spun Yarns, Tex. Tech. Int. 143-144 (1989).
6. Welch, C., Durable-Press Finishing Without Formaldehyde, Tex. Chem. and Col. 22(5), 13-16 (1990).
7. American Association of Textile Chemists & Colorists, AATCC Technical Manual, volume 62, Research Triangle Park, NC, 1987.
8. American Society for Testing & Materials, Committee D-13, Annual Book of ASTM Stds, vol. 7.01-02, Phil., PA, 1991.
9. Hardy, J.I., A Practical Laboratory Method for Making Thin Cross Sections of Fibers & Yarns, U.S.D.A. circ. 378, (1935).

34

Core-yarn fabrics with special finishes for military uses

G.F. Ruppenicker, R.J. Harper, A.P.S. Sawhney and K.Q. Robert - Southern Regional Research Center[1], New Orleans, Louisiana 70179, USA.

ABSTRACT

Although cotton has many natural advantages, its use for military fabrics has declined, largely because high strength requirements cause the cotton fabrics to be excessively heavy. Many cotton fabrics treated with modern flame-retardant and easy-care finishes cannot meet the high performance standards required by the military. Stronger, more durable fabrics were produced from cotton yarns reinforced with high-tenacity filament cores. Filament reinforced yarns containing nylon, polyester, and glass cores were spun on a modified ring-spinning frame. Fabrics were treated with flame-retardant and durable-press finishes. Because the yarns were more than 60% cotton, softness, absorbency, breathability and other desirable properties of all cotton were preserved. Core-yarn fabrics designed for both uniform and tentage uses are discussed.

INTRODUCTION

Cotton fabrics have many properties that make them ideally suited to military uses, including high wet strength, natural water resistance, and good breathability. The breathability property is especially important in tarpaulin and tentage uses where condensation on the under-side of the fabric is undersirable. The cotton fabrics have the necessary porosity to allow the passage of moisture or water vapor when dry, and then, because of the unique swelling property of the cotton fiber, when the fabrics are wet they become highly water resistant.

Despite its many advantages, the use of cotton in these products has declined, largely because to meet high strength requirements the cotton fabrics must be excessively heavy. In addition, flame retardant (FR) finishes for cotton usually require a substantial weight add-on, which can cause fabric stiffening and a subsequent loss in tearing strength.

One approach to improving the strength properties of the cotton fabrics is through the use of core yarns. Core yarns are structures consisting of two component fibers, one of which forms the center axis or core of the yarn, and the other the covering. Generally, the core is a continuous multifilament yarn, while staple fibers are employed for the outer covering or sheath. Core yarns can be used to improve strength, durability, aesthetic, and functional properties of fabrics.

1/ One of the facilities of the Agricultural Research Service, U. S. Department of Agriculture.

Earlier work has demonstrated that core yarns could be used to produce fabrics that were significantly stronger than those made from comparable intimate blends (1). The core-yarn fabrics also had better abrasion resistance, and maintained their superior strength and durability after various finishing treatments. Other research has also shown that a conventional cotton treatment could be successfully used as a flame retardant finish for fabrics containing low levels of man-made fiber combined with cotton (2).

PROCESS

The core-yarn process is readily adaptable to conventional ring spinning frames with only minor modifications. In the arrangement shown in Figure 1, a frame with a typical double-apron drafting system was modified to

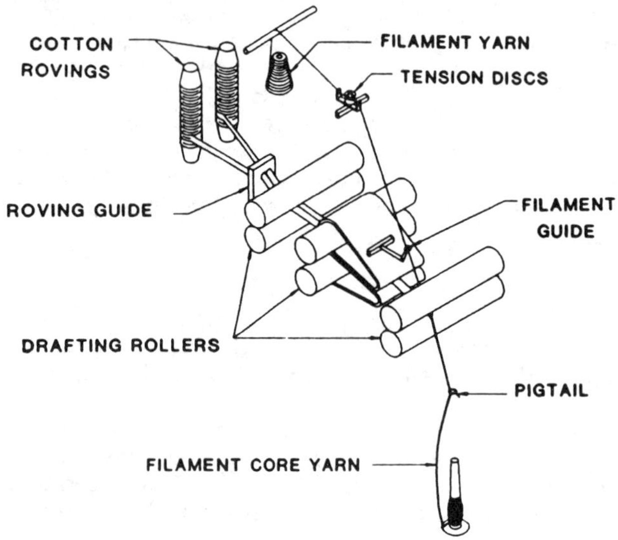

Figure 1. Diagram of core-yarn process.

accommodate the packages of filament yarn. Either single-or double-creeled roving can be used for the cotton supply. If double-creeled roving is used it is best to use a split guide to keep the individual strands of roving from crossing. The filament yarn was fed overend from the package, through a tension control disc, and then inserted behind the top front roll directly in the center of the strand of drafted cotton fibers via a porcelain guide. The traverse of the filament guide was synchronized with that of the roving guide at the back of the drafting system. A front roll speed of approximately 300 rpm was used to spin the yarns.

COTTON/POLYESTER CORE-YARN TENT FABRIC

As already mentioned, cotton is an ideal fabric for tentage uses. For many years the U. S. Army used a 100% cotton sateen fabric for modular tents.

These are tent structures that can be put together in sections for a variety of uses requiring a large continuous space. A problem arose when Army specifications were changed to require that the fabric be made fire resistant. The flame-retardant treatment reduced the tearing strength of the fabric to such an extent that it could no longer meet specifications. In order to obtain a suitable fabric with maximum cotton content, a cotton/polyester filament core-yarn canvas was designed. The fabric was woven from 36/2 yarns (cotton count), and each component of the ply contained a 55 denier, 27 filament, flat, high-tenacity polyester multifilament. This resulted in a fiber ratio of about 63% cotton and 37% polyester. The cotton content was high enough to enable the fabric to be properly treated with a conventional flame retardant normally used for cellulosic fabrics.

The yarns were woven as warp and filling into a 3/1 twill with a thread count of 86 ends and 81 picks per inch, and weight of slightly over 7 oz/sq yd. Finishing included scouring and treatment with a nitrogen-phosphorous flame retardant. The fabrics were first padded to a 60 to 65% wet pickup with a 30% acquous solution of tetrakis(hydroxymethyl) phosphonium chloride-urea precondensate (THPC-urea), dried, cured in an ammoniator, and then oxidized with 0.8% hydrogen peroxide solution. Because this finish has no effect on the polyester, the treatment level was slightly higher than that normally used for cotton to insure satisfactory flammability resistance.

Table I. Properties of Canvas Tent Fabrics.

Fabric Property[1/]	Military Specification	100% Cotton	Cotton/ Polyester Core	Percent Change
Char Length (in)	5.0	2.2	3.9	77
Breaking Strength (lbs)	175	124	190	53
Tearing Strength (lbs)	6.0	2.9	6.3	117
Flex Abrasion (cycles)	-	1095	>5000[2/]	-
Flat Abrasion (cycles)	-	2371	6431	171

[1/] Warpwise.
[2/] Test terminated at 5000 cycles.

Properties of the experimental core-yarn canvas are compared with a similar 100% cotton fabric and the military specifications in Table I. Both fabrics passed the Army minimum requirements for flammablity resistance as measured by the vertical burn test. The experimental core-yarn fabric easily exceeded the military strength specifications, whereas the all-cotton fabric failed in both breaking and tearing strength. The breaking strength of the core-yarn fabric was over 50% greater than that of the all-cotton fabric. The core-yarn fabric also had about twice the tearing strength and much better flex and flat abrasion resistance than the all cotton fabric. Although abrasion tests could be a good indicator of wear-life for a tentage fabric, no standards were available.

COTTON/NYLON CORE-YARN UNIFORM FABRICS

The Army currently uses a wind resistant poplin made from 100% combed cotton for hot-weather battle-dress uniforms. To increase tearing strength, the fabric is designed with reinforcing ribs in both the warp and filling directions which form a uniform checked effect commonly referred to as a rip-stop pattern. The reinforcement ribs appear as one quarter-inch squares in the fabric. Normal finishing includes dyeing and then overprinting with a camouflage design. In this state, the fabric performs reasonably well from a strength and durability standpoint. However, there is some interest in imparting flame-retardant and durable-press properties to this garment. Since these finishes can adversely affect the strength and durability of all-cotton fabrics, research was initiated to develop a stronger predominantly cotton fabric.

A cotton/nylon core-yarn fabric with a nylon content of only 30% was designed for this purpose. The fabric was woven from 40/2 yarns, and each component of the ply contained a 40 denier, 13 filament, flat, high-tenacity nylon multifilament. A plain woven rip-stop design with a greige thread count of 96 ends and 56 picks per inch was used, which was similar in construction to the original all-cotton fabric. Flame retardant finishing included treatment with a 30% aqueous solution of THPC-urea as already described. The fabrics were then treated with a 10% aqueous solution of dimethylol-dihydroxyethyleneurea (DMDHEU) to impart durable press (DP) properties. The DP finish was applied by a standard pad-dry-cure procedure.

Table II. Properties of Rip-Stop Uniform Fabrics.

Fabric Property[1/]	Military Specification	100% Cotton		Cotton/Nylon Core	
		FR	FR-DP	FR	FR-DP
Char Length (in)	5.0	1.5	1.4	2.4	1.6
Breaking Strength (lbs)	100	103	77	120	107
Tearing Strength (lbs)	4.0	3.7	2.1	6.2	5.4
Flex Abrasion (cycles)	-	652	1058	1510	5240
Flat Abrasion (cycles)	-	810	411	1858	818

[1/] Warpwise.

All the fabrics had excellent flammability resistance, and exceeded Army minimum requirements (Table II). The fabrics with the combination FR-DP finish tended to perform slightly better than those with only the FR finish. Breaking strength of the cotton nylon core-yarn fabric was only about 17% stronger than that of the 100% cotton fabric after FR finishing. However, the 100% cotton fabric lost much more strength from the DP finish which reacts with the cotton, and then did not meet minimum strength specification. The core-yarn fabric retained much of its initial strength because most of this strength is derived from the filament core. Therefore, when the cotton component is weakened by reaction with the DP finish, overall strength loss is minimal.

Tearing strength generally followed the same trends as breaking strength. The treated core-yarn fabrics had adequate tearing strength, whereas neither of the 100% cotton fabrics met minimum specifications. The core-yarn fabrics also had much better abrasion resistance than the all-cotton fabrics.

GLASS REINFORCED COTTON CANVAS

Glass has a wide range of properties that make it suitable for military applications including exceptionally high tensile strength, excellent resistance to heat, mildew, moisture, and sunlight, and good dimensional stability. However, it has extremely poor flex and abrasion resistance. In a core yarn structure, the glass filament core provides increased strength while the cotton wrapping helps to protect the glass from damage caused by flexing and abrasion. The glass reinforced fabrics have excellent sewability and produce strong non-slip seams.

For possible tent applications, a glass reinforced cotton canvas weighing approximately 9.00 oz/sq yd with a thread count of 60 ends and 47 picks per inch was produced. The fabric was woven from 17.5/2 yarns, and each component of the plied yarn contained a 100 denier (ECD 450) glass core. The yarn was approximately two-thirds cotton and one-third glass by weight. Each glass core was comprised of approximately 200 individual strands or filaments measuring 0.00025 inches and having a linear density of about 0.5 denier. By contrast, the cotton fiber averages about 1.8 denier. Electrical grade glass filament was used for the core because of its high strength and availability in a wide range of yarn numbers.

Because glass does not burn, a lower treatment level can be used on the cotton component to obtain acceptable flammability resistance. Accordingly, the fabric was treated with a 20% acqueous solution of the THPC-urea formulation. Properties of the treated fabric compared with a similar 100% cotton canvas are given in Table III.

Both fabrics had very good flammability resistance but the level of performance was slightly better for the glass reinforced fabric. The glass reinforced fabric was generally stronger than the all-cotton fabric.

Table III. Properties of Cotton/Glass Fabric Compared with 100% Cotton Fabric.

Fabric Property[1/]	100% Cotton	Cotton/ Glass Core	Percent Change
Char Length (in)	1.8	1.6	(11)
Breaking Strength (lbs)	156	169	8
Tearing Strength (lbs)	5.5	8.7	58
Flex Abrasion (cycles)	6082	1134	(81)
Flat Abrasion (cycles)	1371	1558	14

[1/] Warpwise.

Differences were particularly evident in tearing strength, in which the glass reinforced fabric was almost 60% stronger. Flat abrasion resistance of the two fabrics were comparable. As would be expected, the all-cotton fabric was superior in the flex abrasion test. However, a minimum abrasion of 1000 cycles is usually considered adequate for this type of fabric, and is much greater than that normally obtained with 100% glass.

SUMMARY

The filament core-yarn technique provides an excellent method of producing mostly cotton fabrics with exceptional strength and durability, particularly for military fabrics with high performance requirements. The fabrics still retain their superior performance after treatment with flame-retardant and easy-care finishes. Because of the unique core-yarn structure, fabric hand, appearance, breathability, and other desirable inherent properties of all-cotton are preserved.

ACKNOWLEDGEMENTS

The authors wish to express their appreciation to co-workers of the Fiber Quality and Textile Finishing Chemistry Research Units for their contributions in processing, finishing, and testing the yarns and fabrics; and to J. H. Carra for preparing the yarn cross sections.

REFERENCES

1. Ruppenicker, G. F., Harper, R. J., Sawhney, A. P. S., and Robert, K. Q. Comparison of Cotton/Polyester Core and Staple Blend Yarns and Fabrics, Textile Res. J. 59, 12-17 (1989).
2. Moreau, J. P., Beninate, J. V., and Drake, G. L., Jr., THPOH-NH_3 Flame Retardant: Its Effect on Cotton Blends, J. Fire Retard. Chem. 4, 7-17 (1977).

35

Construction and properties of cotton-rich yarn and fabric, using ARS-patented core-spinning technology

A.P.S. Sawheny, G.F. Ruppenicker and L.B. Kimmel - Southern Regional Research Center, ARS, USDA, P.O. Box 19687, New Orleans, Louisiana, 70179.

ABSTRACT

A super strong, staple-core yarn of predominantly cotton content and almost 100% cotton surface has been produced by using a high molecular-weight, man-made staple fiber in the core and utilizing ARS-patented, core-spinning technology developed at the Southern Regional Research Center. A core yarn containing 2/3 cotton (as sheath) and 1/3 man-made fiber (as core) has more than twice the tensile strength of an equivalent 100% cotton yarn. The construction and mechanical properties of the remarkable yarn and corresponding fabric are described.

INTRODUCTION

To achieve improved yarn and fabric properties, different textile fibers that compliment each other have been blended for a long time. Although conventional blends are generally superior in overall performance to their respective 100% fiber fabrics, the almost randomized distribution of their constituent fibers in the yarn cross-section is not as effective and useful as the total segregation of these fibers in the core-sheath yarns (Figure 1) developed and patented by the U.S. Department of Agriculture [1,2,3].

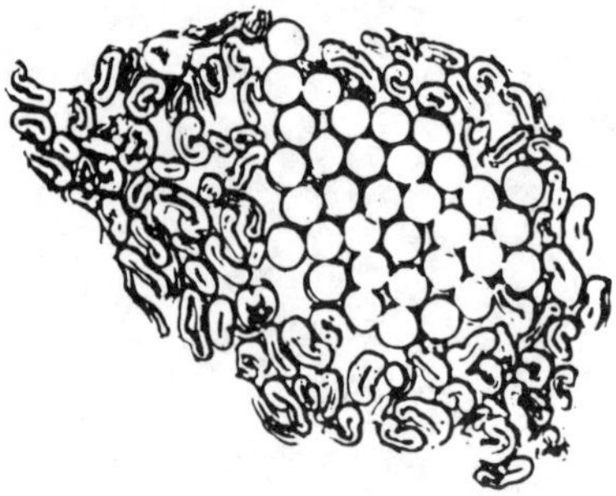

Figure 1. Cross section of a polyester staple-core/cotton-wrap yarn.

The main advantages of the fabrics made with the core-sheath yarns include:

- High strength and durability, attributed to the consolidation of high tenacity man-made fibers in the yarn core.

- Elimination of the fiber pilling problem, since the pilling-prone high tenacity fibers are buried under a sheath of relatively weak, pill-resistant cotton fibers.

- Improved wear comfort due to the presence of mostly natural fibers on the surface of the true core-yarn fabric.

- Improved dimensional stability achieved by heat setting the man-made (core) component of the yarn.

- Improved dyeing and special finishing options, because of the 100% cotton surface of the core yarn, which serves as an excellent substrate.

EXPERIMENTAL

a) Yarn production method:
Figure 2 shows the ARS-patented method of spinning a staple-core/cotton-wrap yarn [4].

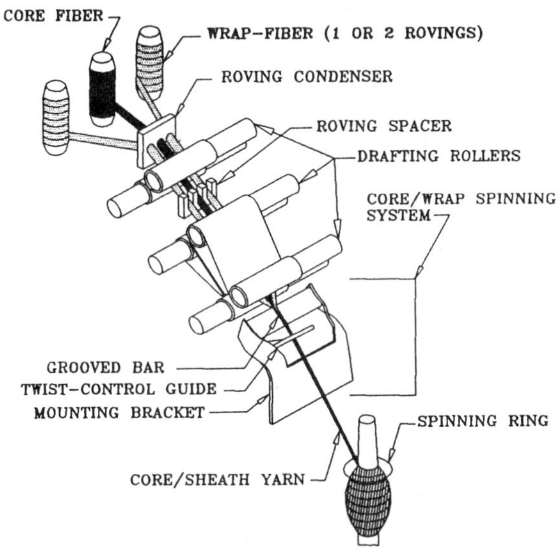

Figure 2. All-staple, core-wrap spinning system.

As shown in the figure, three normally prepared rovings of cotton and polyethylene staple fiber are positioned per spindle on the ring spinning frame. The drafted strands of the materials are merged onto a specially grooved bar placed in front of the front drafting rollers. The fiber assembly takes place on the bar instead of under the nip of the front rollers. The twist control guide, as shown in the figure, prevents the twist from flowing past the bar. The staple-core/cotton-wrap yarn is wound onto a ring bobbin in conventional manner. It may be noted that, during the yarn formation, the core fiber always stays in the groove of the bar.

b) Yarn construction:

- Cotton roving size (tex)	295
- Polyethylene roving (tex)	295
- Total draft	17.5
- Spindle speed (rpm)	10,000
- Twist Multiplier	4.0
- Traveler No.	8
- Temp. ^{0}C/R.H. %	20/48
- Yarn size (tex)	49
- Cotton content (%)	67
- Polyethylene content (%)	33
- Twist level (tpi)	16

c) Yarn and fabric testing:
Yarn tensile strength and fabric tensile and bursting

strengths were performed according to standard ASTM methods and procedures [5].

RESULTS

Figure 3 shows tenacity of the staple-core yarn and comparable 100% cotton yarn. The core yarn has more than twice the strength of an equivalent 100% cotton yarn, even though the core yarn has 67% cotton content and almost 100% cotton surface.

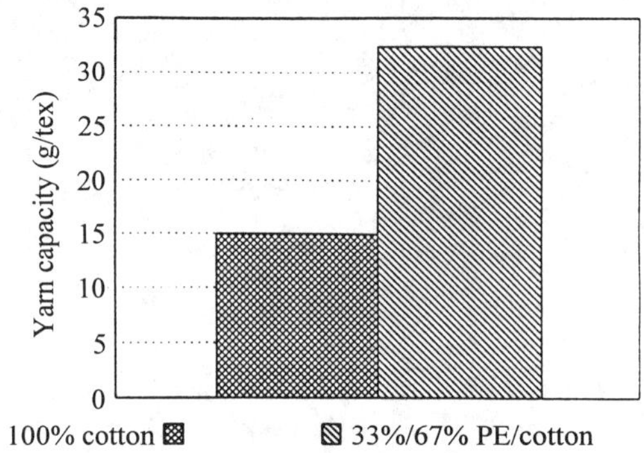

Figure 3. Single-strand tensile strength of the core and control yarns.

Figures 4 and 5, respectively, show the tensile strength and bursting strength of the plain-jersey knit fabrics made with the core yarn and the 100% cotton yarn. The tensile strength of the core-yarn fabric is 63% higher than that of the comparable 100% cotton, control fabric. The ball bursting strength of the core-yarn fabric is 48% higher than that of the control, 100% cotton fabric. Significant improvements in the fabric tensile and bursting strengths obviously reflect a significant increase in fabric durability.

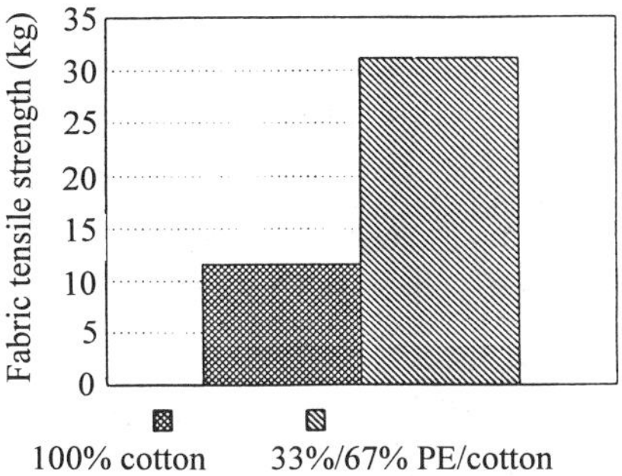

Figure 4. Grab tensile strength of the fabrics made with the core and control yarns.

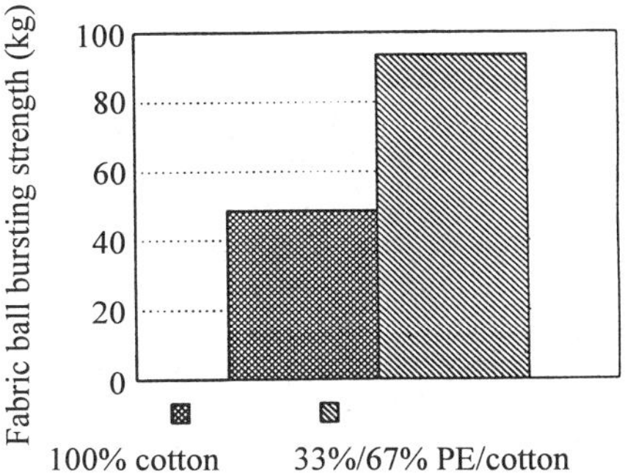

Figure 5. Ball-bursting strength of the fabrics made with the core and control yarns.

CONCLUSION

Totally new, predominantly cotton yarns of almost 100% cotton surface can be spun using the core-wrap spinning technology developed and patented by the Agricultural Research Service of the U.S. Department of Agriculture.

Yarns can be custom engineered to exhibit unique fabric features, such as extremely high strength and durability, excellent flame retardancy, and any desirable appearance, handle, and feel, to meet specific requirements. For example, the super strength of the 100% cotton-covered yarn referenced in this paper may be most useful for bullet-proof vests, furnishings, civilian and military tents, and other such applications, where good breathability and high tensile strength, durability, impact or bursting strength, and tear resistance of the fabric are rather critical.

ACKNOWLEDGMENT

The authors take this opportunity to thank Craig Folk and Mike Hamlin for their significant input in the production of yarns and the preparation of the manuscript.

REFERENCES

1. Sawhney, A.P.S., Robert, K.Q., Ruppenicker, G.F., Harper, R.J., and Kimmel, L.B., 'Engineering and evaluation of a unique cotton-covered, polyester staple-core yarn', Proc. 1990, Beltwide Cotton Production Research Conferences, January 9-13, 1990.
2. Sawhney, A.P.S., Robert, K.Q., and Ruppenicker, G.F., 'Device for producing staple-core, cotton-wrap ring-spun yarn', Textile Research Journal, 59 (9), 519-524 (1989).
3. Sawhney, A.P.S., et-al, 'Improved method of producing cotton-covered, polyester staple-core yarn on a ring spinning frame', Textile Research Journal (scheduled for publication in the December 1991, issue)
4. Sawhney, A. P. S., Folk, C., and Robert, K.Q., 'System for producing core-wrap yarns', U.S. patent No. 4,976,096 (December 11, 1990).
5. American Society for Testing Materials - ASTM Designations D-1424-83; D-3787-80a; D-3386-80.

36

Kawabata evaluation of cotton/pet blended fabrics after various chemical treatments

Dong-Hwa Shin, Linda B. Kimmel and Robert J. Harper - USDA, ARS, Southern Regional Research Center, P.O. Box 19687, New Orleans, LA. 79179, USA.

ABSTRACT

Fabrics produced from PET/cotton intimate blend and PET staple core/cotton wrapped yarns were examined by the Kawabata Evaluation System (KES) after each chemical treatment. The fabrics were tested 1) untreated; 2) after scouring and bleaching 3) after durable press finishing (DMDHEU, BTCA); and 4) after 25/50 wash cycles. Tensile, shear, bending, and compressional deformations as well as frictional properties of these two fabrics were compared with those of 100% cotton control fabric. In addition, fabric comfort assessments were made to study the effects of the two different DP finishes. The deformation data were used to evaluate the advantages/ disadvantages of the respective fabrics and the effect of each processing step.

INTRODUCTION

During durable-press finishing of cotton and cotton/synthetic blended fabrics, mechanical and handle properties undergo noticeable changes, particularly compared with the properties of the greigh-state fabric. There have been various attempts, in the past, to follow changes in physical properties after different stages of finishing. These studies (1,2), however, did not include quantification of sensitive handle properties, such as can be obtained with the Kawabata Evaluation System (KES). Modifications of the wrinkle resistance and smoothness appearance of fabrics are the major factors driving investigations of chemical finishing processes for cotton. At the same time, the fabric must be acceptable to the garment manufacturer, not only in terms of handle, but also with regard to mechanical strength and tailorability. Fabric objective-evaluation systems, such as KES, based on measurement of low-stress mechanical and surface properties, are now being utilized extensively. This is particularly true in fabric design and new product development by both industrial companies and academic research organizations (3,4).

In this study, three different fabrics (100% cotton, cotton/PET intimate and strategic blend) were examined by using KES at the various finishing stages in order to compare the effects of chemical treatments on handle-related properties.

EXPERIMENTAL

Fabric Preparation

Three different fabrics were prepared from three yarns as outlined below. The method of formation of these fabrics and their characteristics were presented previously by Kimmel (5).

Fibers : Cotton - Delta pine
Polyester - Trevira 121 (high tenacity)

Yarns : 100% Cotton
Cotton/PET = 67/33 Intimate Blend
Cotton/PET = 67/33 Strategic Blend
(cotton/PET = sheath/core)

Fabrics : Plain Weave (all three fabrics)
Weight - Medium weight shirting fabrics
(150-155 g/m^2)

Fabric Finishing

KES measurements were made on fabrics after the following operations.

* **Scour and bleach** (0.7% caustic scour, 2% H_2O_2)

* **Durable-Press Finish** :
DMDHEU (Dimethyloldihydroxyethyleneurea)
5% DMDHEU, 2% $MgCl_2$, 0.5% polyethylene,
0.1% Triton X-100, pH=4.
BTCA (Butanetetracarboxylic acid)
6% BTCA, 3% NaH_2PO_2, 0.5% polyethylene,
0.1% Triton X-100, pH=2.4.
Above chemical formulations were chosen in order to achieve a durable press rating of 3.5 (6), a typical target value, as suggested by other studies (1,2).

* **Washing** : 25 and 50 times for three fabric samples treated with two types of durable-press agents.

Measurements of Kawabata Evaluation System (KES).

Objective handle properties were measured on the KES-F instruments (Kato Tekko). The fabric sample size was 20x20 cm, and three replications were made in the warp and filling directions. All samples were conditioned at 20°C and 65% RH before the measurements. Details of the procedures used in these measurements were described by Kawabata (7).

RESULTS AND DISCUSSION

Application of KES can provide a sensitive and systematic method of measuring mechanical and surface properties of textile materials. Measurements are usually expressed in the form of 16 parameters shown in Table I.

Table I. Mechanical parameters measured by the KES instruments.

Parameter	Abbreviation	Unit
Tensile		
Tensile linearity	LT	-----
Tensile energy/unit area	WT	gf.cm/cm^2
Resilience	RT	%
Extension at specified load	EMT	%
Shear		
Shear stiffness	G	gf/cm degree
Hysteresis at 0.5°	2HG	gf/cm
Hysteresis at 5°	2HG5	gf/cm
Bending		
Bending stiffness	B	gf.cm^2/cm
Hysteresis width	2HB	gf.cm/cm
Compressional		
Compressional linearity	LC	-----
Energy of compression	WC	gf.cm/cm^2
Resilience	RC	%
Thickness at 0.5 g/m^2	To	mm
Surface		
Coefficient of friction	MIU	-----
Mean deviation of MIU	MMD	-----
Geometrical roughness	SMD	micron

In this study, measurements were made on the fabric samples after each of the finishing stages. Representative results are displayed in Table II. Each value in this table is the mean of three test replications.

During the scouring and bleaching processes, most waxy substances on the surface of the cotton are removed. This causes a dramatic reduction in tensile energy (Figure 1), but a small increase in tensile resilience. As expected, most of the mechanical properties were adversely affected by after durable-press finish treatments, as shown in Table II and Figure 1. For the cotton/PET

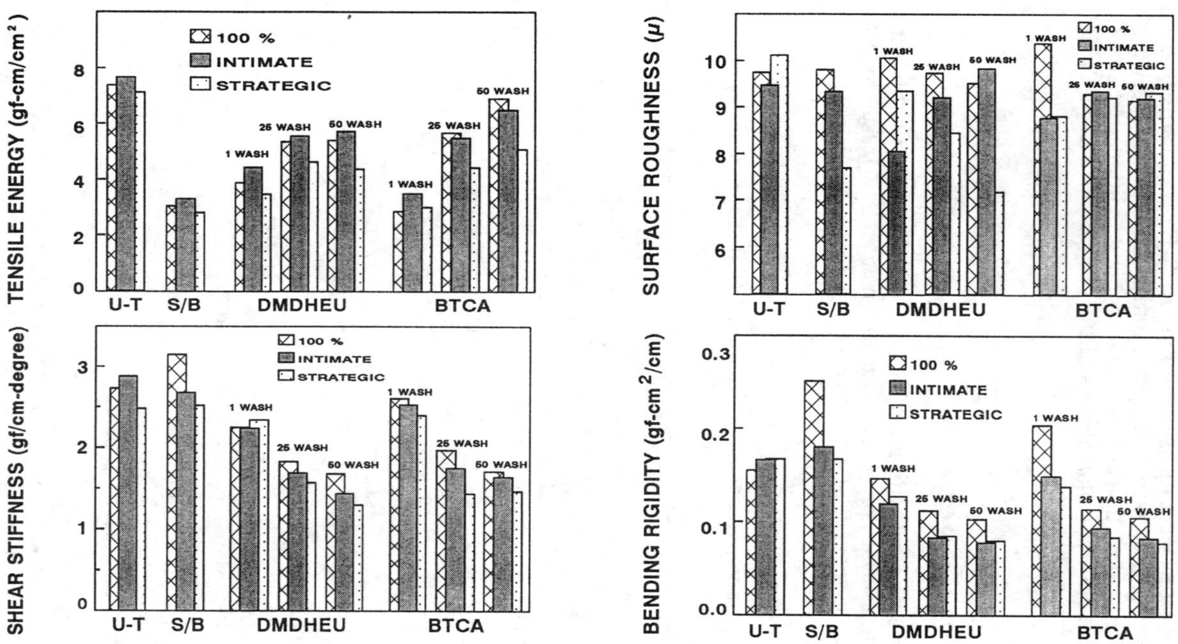

* U-T : Untreated cotton, S/B : Scoured and Bleached

<u>Figure 1.</u> Comparison of some KES parameters measured in the warp direction after various fabric finishing stages.

intimate-blend fabric, the two DP agents produced similar mechanical properties. However, for the 100% cotton and cotton/PET strategic-blend fabrics, BTCA performed better than DMDHEU in tensile, shear, and bending resilience. BTCA-treated fabrics had smoother surfaces but were slightly stiffer than DMDHEU-treated fabrics. After 50 washings, DMDHEU-treated, intimate-blend fabrics showed better tensile resilience. With dimensional stability, the 100% cotton and strategic-blend fabrics generated greater hysteresis in bending and shear than did intimate-blend fabric. The strategic-blend fabric with cotton on its surface gave higher frictional coefficients but lower mechanical property values compared with intimate-blend fabrics.

Table II. Comparison of selected mechanical properties of three sample fabrics after various chemical treatments.

	#	U-T	B/S	DMDHEU			BTCA		
				1 wash	25 wash	50 wash	1 wash	25 wash	50 wash
2HB	1	0.246	0.364	0.188	0.199	0.176	0.264	0.194	0.176
	2	0.301	0.301	0.185	0.144	0.122	0.238	0.136	0.126
	3	0.276	0.323	0.254	0.147	0.134	0.224	0.126	0.128
2HG	1	8.070	11.540	7.640	7.870	7.410	7.760	8.000	7.530
	2	8.250	8.630	7.080	7.780	6.640	8.270	6.880	6.460
	3	7.720	9.510	7.520	6.740	6.420	8.010	6.500	6.410
RT	1	56.340	55.540	52.590	48.920	49.380	60.580	47.360	44.070
	2	57.010	60.400	56.050	54.060	54.230	59.370	53.200	51.410
	3	56.360	63.550	60.880	56.720	59.320	63.840	57.140	54.430
RC	1	41.100	42.350	47.490	35.890	34.400	48.940	34.680	31.400
	2	40.610	38.500	51.290	38.980	35.510	51.880	37.680	35.920
	3	41.840	45.510	52.060	37.000	35.240	50.940	34.320	34.530
MIU	1	0.204	0.225	0.194	0.292	0.397	0.232	0.306	0.406
	2	0.195	0.217	0.207	0.247	0.310	0.249	0.280	0.293
	3	0.213	0.213	0.196	0.255	0.333	0.262	0.298	0.308
SMD	1	9.740	9.800	10.060	9.730	9.510	10.370	9.280	9.140
	2	9.460	9.320	8.050	9.190	9.830	5.840	9.340	9.180
	3	10.110	7.690	9.340	8.460	7.170	8.800	9.200	9.310
To	1	0.762	0.630	0.543	0.710	0.656	0.558	0.658	0.651
	2	0.740	0.605	0.536	0.675	0.675	0.548	0.663	0.661
	3	0.755	0.606	0.548	0.690	0.643	0.555	0.658	0.663

1 : 100% cotton,
2 : 67% cotton/33% PET Intimate Blend,
3 : 67% cotton/33% PET Strategic Blend.

U-T : Untreated
B/S : Bleached & Scoured

CONCLUSIONS

The remarkable influence of various finishing steps on mechanical and surface properties of fabrics can be specified quantitatively by the data obtained using the KES instruments.

During this study, it was demonstrated that 100% cotton and strategic-blend fabrics when treated with BTCA gave better mechanical properties than the intimate-blend fabric. After 50 washings, DMDHEU-treated fabrics retain better tensile resiliency. Intimate-blend fabric has slightly higher tensile properties and better dimensional stability than does strategic-blend fabric. Strategic-blend fabric has cotton dominated surfaces and better comfort properties than the intimate-blend fabric. These conclusions, as well as many others, are readily apparent from consideration of the wealth of KES data.

It is obvious from this study that KES is an excellent method for quantitative evaluation of the mechanical properties of chemically treated cotton and cotton-blended fabrics. Based upon such valuable data, it is possible to design appropriate fabrics for specific end-uses.

ACKNOWLEDGEMENT

Our sincere appreciation goes to Dr. Billie Collier for her cooperation in permitting use of the KES instruments at Louisiana State University.

REFERENCES

1. Harper, R.J. and Ruppenicker, G.F., Woven fabrics prepared from high tenacity cotton/polyester core yarn, Text. Res. J., 57 147-154 (1987).
2. Welch, C., Durable press finishing without formaldehyde, Text. Chem. Color., 22(5), 13-16 (1990).
3. Carnaby, G., The utilization of New Zealand wools in tropical fabrics in "Objective measurement: Applications to product design and process control," TMSJ, Osaka, 75-82 (1986).
4. Harwood, J., Weedell, P., and Carr, C., The use of the KES for product development and quality control, J. Soc. Dyers & Col., 106(2), 64-68 (1990).
5. Kimmel, L., Sawhney, A.P.S., Ruppenicker, G., and Harper, R., Comparative physical properties of conventional and staple-core cotton-blend fabrics with durable-press finish, this book.
6. American Association of Textile Chemists and Colorists, AATCC Technical Manual, 124-1982, 61-1980, Research Triangle Park, N.C., 1989.
7. Kawabata, S., The standardization and Analysis of Hand Evaluation, 2nd Ed., Text. Mech. Soc. Japan, Osaka (1980).

37

Chemical modifications of cotton fibers for improved water absorbency

C. Mizutain, Y. Onogi, H. Inagaki and T. Miyamoto* - Department of Home Economics, Mukogawa Women's University, Nishinomiya, Hyogo 633, Japan.
**Institute for Chemical Research, Kyoto University, Uji, Kyoto 611, Japan.

INTRODUCTION

A number of methods have been developed for preparing cotton fibers with a higher degree of water absorbency. One method for preparing such fibers is to chemically modify cotton fibers. However, the handling characteristics and textile properties of chemically modified cotton fibers prepared so far have been considerably different from those of natural cotton fibers because of relatively high degree of substitution (DS).

The objective of this work is to prepare cotton fibers which have high water absorbency and yet maintain their natural handling characteristics as well as other textile properties. These cotton fibers hold promise as new materials for sports underwear. To this end, the interior of cotton fibers was lightly modified with monochloroacetic acid and 2,3–epoxypropanol and their textile properties were compared with those of intact cotton fibers.

EXPERIMENTAL

Sample preparation

Carboxymethylation and dihydroxypropylation of cotton fibers were carried out according to the following schemes.

$$\text{Cell–OH} \xrightarrow[\text{aqu. NaOH in acetone}]{ClCH_2COOH} \text{Cell–OCH}_2\text{COONa} \qquad (1)$$

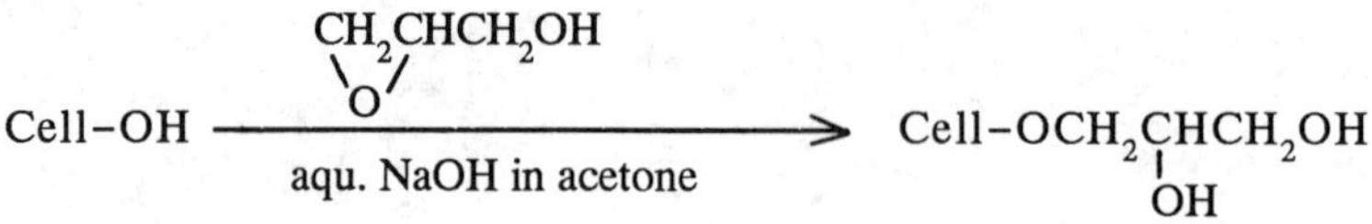

(2)

CARBOXYMETHYLATION: The sample fibers were immersed in an acetone solution containing 33% monochloroacetic acid for 2 h at room temperature. Following a thorough washing with acetone to remove monochloroacetic acid attached to the fiber surface, the fibers were carboxymethylated. Sample fibers with different DS values were prepared by changing the concentration of NaOH in the reaction medium. The concentration of NaOH was controlled by changing the mixing ratio of acetone and 20% aqueous NaOH solution.

DIHYDROXYPROPYLATION: Dihydroxypropylation was carried out in a manner similar to that for carboxymethylation. The fibers(5g of cotton) were reacted in a mixture of 480ml of acetone and 20 ml of a 2% aqueous solution of NaOH at 50 °C for 1 h. The sample fibers having different DS values were prepared by repeating the procedure described above.

Measurements

Determination of DS: The DS values of carboxymethylated samples were determined by back–titration, while those of dihydroxypropylated samples(DHPC) were estimated from ^{1}H–NMR spectra of fully acetylated DHPC according to the method proposed by Y.Tezuka et al. [1].

Water Content: Water content (%) was estimated by the water–of–imbibition method ; samples(0.5g) were immersed in water for 24h, then centrifuged (4000rpm; 2000g in our centrifuge) for 4min, weighed, and dried.

Moisture Regain:Sample fibers were conditioned at 20°C, 65% RH for more than 24h and weighed. Values of moisture regain were calculated by the following formula:

$$\text{Moisture regain (\%)} = (W_c - W_o)/W_o \times 100$$

where, W_c and W_o are weights of the fibers before and after drying (g), respectively.

Tensile Properties: Tensile strength and elongation at break of fibers were determined using an Instron type tensile tester, Tensilon UTM–II–20.

Light and Scanning Electron Microscopy: The sample fibers were dyed in 0.06% aqueous solution of methylene blue (100:1 liquor to sample ratio) at 70°C for 1h. The dyed fibers were embedded in Epon 812 in the usual way, and ultrathin–sectioned using a Sorvall MT–1 Porter –Blum with a glass knife. The cross sections were then examined in a light microscope(Nikon Model Optiphoto–pol, Japan). For scanning electron microscopy, the sample specimens were coated with a thin layer(100–150A) of gold using an Eiko 1B–3 ion coater and then observed with a Hitachi S–410 scanning electron microscope.

X–ray Diffraction: The X–ray diffraction spectra were taken by a Rigakudenki model RU–3h X–ray diffractometer with flatplate camera, using a Ni–filtered Cukα radiation.

DSC Measurement: A Rigakudenki differential scanning calorimeter DSC–10A equipped with a cooling cell was used to measure the content of water bound onto cellulose molecules, thus showing no phase transition at 0°C(referred to as the bound water). DSC curves were obtained by cooling from room temperature to −80°C at a scanning rate 8°C/minute and then heated to room temperature at the same rate as the cooling.

RESULTS AND DISCUSSION

Tables 1 and 2 show the water content of lightly carboxymethylated and dihydroxypropylated cotton fibers prepared, respectively. The number attached to each sample code indicates the water content of the sample. Figure 1 demonstrates the relationship between DS value and water content of the samples. From the handling tests for cotton fibers and fabrics modified differently, it was found that (1) the optimal DS value for obtaining suitable fibers was in the region of 0.01 to 0.02 for carboxymethylation and 0.1 to 0.2 for dihydroxypropylation, respectively, indicating that carboxymethylation is more effective than dihydroxypropylation in promoting water absorbability and (2) the cotton fibers thus obtained absorbed about 150 ~ 160% as much water as the untreated fibers.

Table 1. Water Content and Moisture Regain of Carboxymethylated Cotton Fibers

Sample code	Conc.of NaOH in medium(g/l)	Water content(%)	Moisture regain(%)
Untreated	–	48	7.1
I–55	1.5	54	7.2
I–60	3.0	59	7.3
I–70	5.0	70	7.7
I–80	6.0	79	7.9
I–100	7.0	105	8.6

Table 2. Water Content and Moisture Regain of Dihydroxypropylated Cotton Fibers

Sample code	Repeated number of treatment	Water content(%)	Moisture regain(%)
Untreated	0	48	7.1
N–60	1	59	7.4
N–70	2	70	7.7
N–90	3	88	8.1
N–100	4	103	8.5

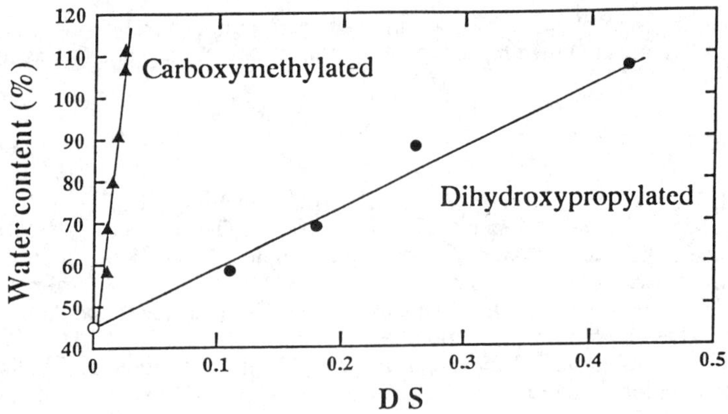

Fig. 1. DS – dependence of water content of chemically modified cotton fibers.

The cotton fibers modified in this way were found to show only small changes in dyeability for a cationic dye, methylene blue (Fig.2) and tensile strength (Table 3) when the water content of the sample is less than 100 %. Figure 3 shows typical scanning electron micrographs of the cross sections of chemically modified fibers, as compared with that of intact cotton fibers. The results indicate that the fine structure of lightly modified samples changed little up to DS values of about 0.02 for carboxymethylation and about 0.2 for dihydroxypropylation. From a comparison of X-ray diffraction spectra, it was also found that the crystallinity of the sample fibers thus obtained was little changed, as compared with that of intact cotton fibers.

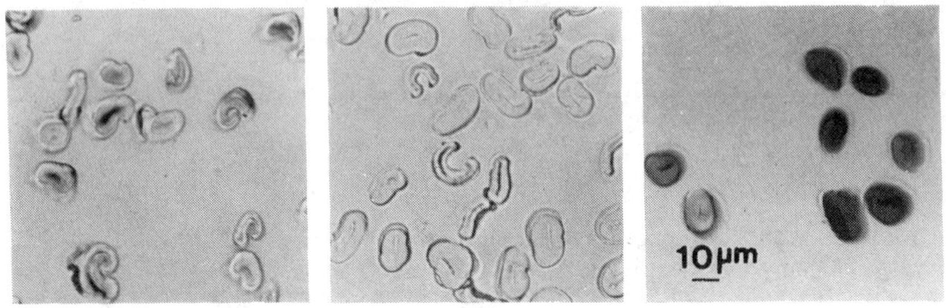

Untreated (48 %) I–70 (70 %) I–120 (120 %)

Fig. 2. Light micrographs of the cross sections of untreated and carboxymethylated cotton fibers dyed with methylene blue. The percentage in parentheses indicates the water content of the sample.

Table 3 Tensile Strength and Elongation of Chemically Modified Cotton Fibers

Sample code	Tensile strength(g/d)	Elongation (%)
N -70	3.3	1.0
N -100	3.7	2.8
I -70	3.4	2.4
I -100	3.4	1.7
Untreated	3.2	2.0

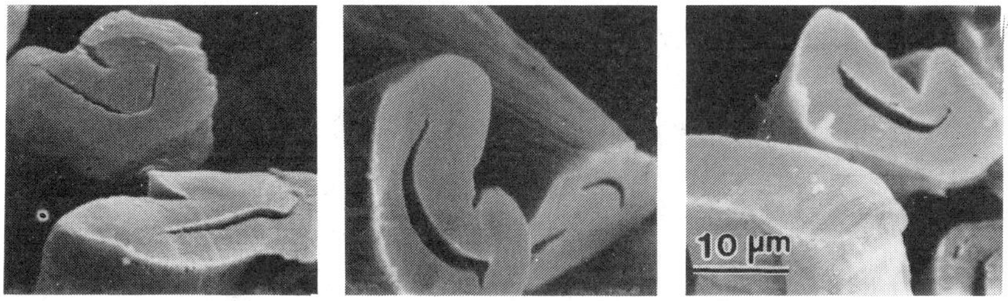

(a) Untreated (b) I–70 (c) N–70

Fig. 3. Scanning electron micrographs of untreated (a), lightly carboxymethylated (b) and dihydroxypropylated cotton fibers (c). (Magnification, 1200x)

The carboxyl groups attached to the fiber were completely transformed into their Na–salt form by alkali–treatment with aqueous Na_2CO_3 solution, and their water absorbency was compared with that of the acid form. The results are shown in Figure 4.

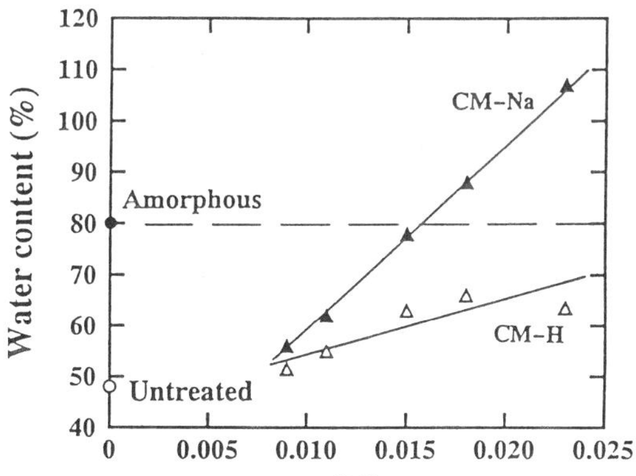

Fig. 4. Relationship between DS value and water content of lightly carboxymethylated cotton fibers.

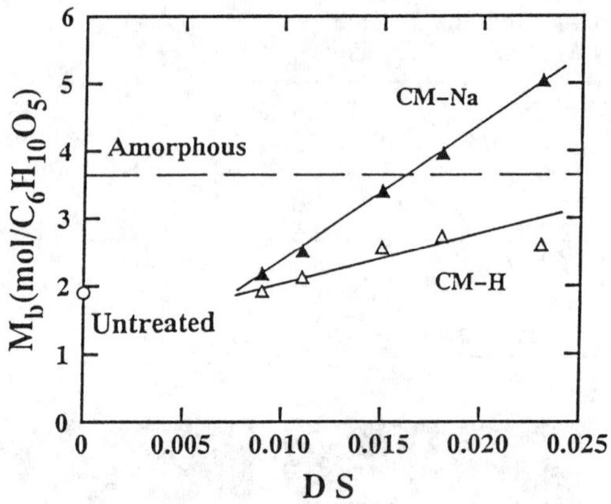

Fig. 5. Relationship between DS value and bound water absorbed per glucose unit of cellulosic samples.

On the other hand, the amount of bound water in the sample fibers was estimated quantitatively from DSC measurements[2]. Figure 5 shows plots between M_b and DS values for lightly modified cotton samples, where M_b is the moles of bound water adsorbed per glucose unit. An M_b value for amorphous cellulose was obtained from the DSC measurements for amorphous cellulose prepared according to the method of Isogai et al.[3]. It can be seen that (1) in the case of the acid form, the modified fibers absorbed about 120–130% as much water as the untreated fibers, (2) the percentage increase in the water absorbency of the fibers was further enhanced to twice that of the acid form by transforming carboxyl groups into their Na–salt form, and (3) the water absorbency and M_b value of carboxymethylated cotton fibers having a DS value of 0.02 were much higher than that for amorphous cellulose, indicating that the effect of the cleavage of hydrogen bonds by incorporating substituents, i.e. carboxymethyl groups, on the water absorbency of modified cotton fibers is negligible.

REFERENCES

1. Y. Tezuka, K. Imai, M. Oshima, T.Chiba, *Polymer*, **30**, 2288 (1989).
2. H. Sakabe, H. Ito, T. Miyamoto, H. Inagaki, *Text. Res. J.*, **57**, 66 (1987).
3. A.Isogai, H. Attalla, *J. Polym. Sci*, **29**, 113 (1991).

38

Fibrets: cellulose acetate microfibers for filtration applications

J.E. Smith - Hoechst Celanese Corporation, Fibers and Film Group, P.O. Box 32414, Charlotte, North Carolina, USA.

INTRODUCTION

Cellulose acetate microfibers are valuable tools for the development of specialty filtration applications. Their highly fibrillated physical structure provides high surface area and reduced pore size. This structure allows entrapment of small particles, mechanical binding with other fibers, and improved opacity and cover.

Because of the cellulose acetate polymer, these fibers are hydrophilic and thermoplastic. They are also biodegradable and non-toxic. They have selective absorption properties and are suitable for processing by wet-laid, fiber-coating, and molding techniques to form an array of products. This combination of characteristics allows cellulose acetate microfibers, such as fibrets, to meet requirements for specialty applications in liquid and air filtration in unique ways.

The development of this product was stimulated by the higher demands now being placed on filtration and separation systems to meet the 1990s' requirements for a cleaner environment and for energy conservation and control.

Other current business conditions also make fibrets useful. Workplace air quality concerns have gained new prominence. Global competition and a tougher marketplace have led to ever-greater demands for product quality through tighter raw material and process controls. The transfer of new technologies—biotechnology, for example—from the laboratory or pilot plant stage to viable commercial processes has heightened demands placed on filtration and separation systems.

PHYSICAL AND CHEMICAL PROPERTIES

Research indicates that cellulose acetate microfibers are qualified to address these increasing demands on both air and liquid filtra-tion systems. A look at the chemical and physical properties of fibrets shows why.

To begin with, fibrets are produced from cellulose acetate polymer, without the addition of topical finishes. They have the same chemical properties as acetate fibers and films.

Cellulose acetate is hydrophilic and has good surface wettability. Unlike other cellulosic fibers, it is also thermoplastic. Therefore, acetate products can be thermally bonded, ultrasonically welded and thermally embossed. Cellulose acetate has a melting point of 260 degrees C and begins to soften in the range of 190 to 200 degrees C. Acetate products are not brittle as is shown by their performance in shatter-resistance face shields and safety glasses and in the flexibility and drape of fabrics.

Fibrets are suitable for disposable applications. Acetate is biodegradable and will disintegrate when buried in landfills. It can be incinerated without toxic by-products. Through dissolving with solvents, it can be recycled without polymer degradation.

Cellulose acetate products have a long history of safe use and have been used in many food contact and sensitive medical applications. Acetate can be sterilized by common methods, and the polymer is "generally regarded as safe" by the FDA for food contact applications, including filters. Interestingly, in the light of current health concerns, cellulose acetate has very low ex-tractables in aqueous-based liquids, especially for heavy metal content.

Acetate can selectively absorb certain organic compounds. For example, it can remove limonin, which causes bitterness in citrus juice. Experiments have shown that a fibrets paper can selectively remove this bitter component from navel orange juice without affecting flavor, color, or significant levels of vitamin C. Other applications being investigated are the removal of trace components that can affect appearance and taste in beverages and absorption of volatile organic compounds from air.

The safe pH range of acetate for extended exposure and normal conditions is 4 to 8. For shorter exposures or less severe conditions, this range is wider—pH 3 to 9. Acetate is attacked by strong acids and moderate and strong bases. It is soluble in several common solvents, such as acetone and other low-molecular-weight ketones, phenols, and selected aldehydes. It is not soluble in gasoline, kerosene or turpentine, and it is highly resistant to mineral, vegetable, and animal-based oils.

In terms of physical properties, cellulose acetate fibrets have a highly fibrillated, pulp-like form, as shown in Figure 1. Their high fibrillation provides a surface area of 20 square meters per gram, which is 60 to 75 times the area of standard textile fibers. The high fibrillation enables fibrets to act as a binder in synthetic papers by entangling the staple fibers. It allows fibrets to mechanically entrap small particles in a structure, to increase cover, and to reduce pore size. The structure also enhances fibrets' performance as related to chemical properties. Combined with cellulose acetate's ability to remove organic com-pounds, the high surface area can increase a filter's absorption rate. This is valuable where there is a need to remove low levels of chemicals from a gas or liquid.

Fibrets are 20 to 200 microns in length and 0.5 to 5 microns in width. The fine fiber diameters give the material a strong advantage for use as a matrix to increase the efficiency of filtration.

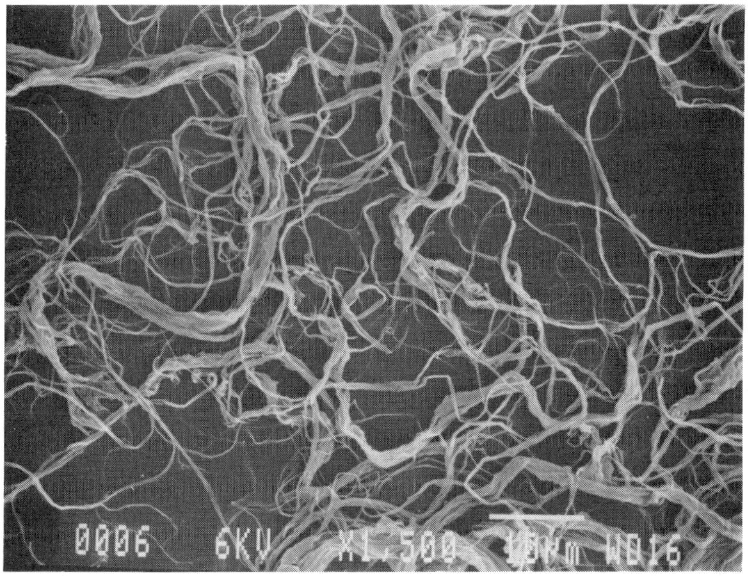

Figure 1. The highly fibrillated structure of cellulose acetate fibrets.

MANUFACTURING PROCESS

The fibrillated structure of fibrets is formed as the fibers are extruded rather than in a later refining stage. The cellulose acetate polymer is dissolved in acetone to form cellulose acetate dope, which is extruded into a precipitation bath of water under conditions of high shear. Process conditions such as dope solids, temperature, and conditions of shear can be changed to modify the fibrets' surface area and dimensions. The fibrets undergo further treatment to set the structure by expanding the voids and expelling residual acetone.

PROCESSING INTO FILTER MEDIA

Fibrets can be used in both wet-laid processes and coating applications. Coatings allow microfibers to be used in combination with dry-laid nonwovens and other existing structures. In both paper and coatings, fibrets have been found to be very effective even when included at a low percentage of the total structure—such as 2 to 25 percent.

In wet-laid processes, fibrets are used in filter media for tighter pore size, resulting in small particle retention. They can be successfully processed in both inclined wire and conventional flat paper-making processes and can be blended with natural and/or synthetic fibers.

Nonwovens, formed through the various dry-laid and bonding processes, are important for filtration because of their strength and dirt-holding capacity. Fibrets can be incorporated in these structures in coating applications, such as by spray coating methods. Their use allows the inclusion of a microfiber in these nonwovens, to enhance filtration efficiency.

LIQUID AND AIR FILTRATION APPLICATIONS

Because fibrets are a unique product by nature of both their physical and chemical structures, they are exceptionally well suited for the development of new structures for specialty applications. Certain filtration applications can benefit from both the fibrillated structure and the cellulose acetate polymer properties. As with any new fiber structure, media optimization will be required to develop the right balance of performance properties.

First, in terms of air filtration, the demand for high-efficiency particulate air filters continues for use in semiconductor and electronics applications, pharmaceutical processes, and precision assembly areas. In these applications, microglass—from 0.2 to 5 microns in diameter—is used to achieve high efficiences. Now many participants in the industry are looking for a synthetic microfiber to address the issues associated with glass brittleness. Brittleness can cause fiber breakage in use, which contaminates and causes blemishes. It also causes worker handling problems, both when the filters are assembled and when they are replaced.

Fibrets offer an alternative to microglass for high-efficiency removal. Their fine fiber dimensions, resulting from high fibrillation, allow ultrafine particle retention. Since they are not brittle, fibrets can reduce handling problems and fiber breakage. However, they cannot be considered a direct replacement for microglass and will require the development of new filter media designs to meet HEPA requirements.

In conventional heating and air conditioning applications as well, there is increased emphasis on improving the efficiency of filtration systems. There is growing concern about the indoor air quality in office buildings and homes because of the adverse health effect of respirable particles, biological agents, and gases that build up in well insulated and energy efficient structures. Fibrets can be used in pleated filter modules or pocket filters to address the need for good retention in these medium- to high-efficiency systems.

The selective absorption properties of cellulose acetate microfibers may also be beneficial in improving air quality. As a component in an air cleaning system, microfibers could address odors or volatile organic compounds that can accumulate in tightly closed buildings. Their use could also be applied to air filtration systems in airplanes and other public transportation vehicles.

In terms of liquid filtration, fibrets present special advantages for industrial manufacturing processes that usually use cartridge or module filters. These are typically disposable filters. Most are for water- or aqueous-based applications, while others are for petroleum derivatives. Because of the diversity of applications served by industrial process cartridge filters, suppliers tailor the filter to the specific liquid and contaminant to be filtered.

Filtration trends also influence this product area. For example, there is a continuing demand for finer filtration allowing better contamination control. Other trends are (1) an increased use of disposable products because of concerns for worker safety in handling toxic or noxious materials, (2) growing awareness that leachables from the filter media can affect the final product, and (3) the use of sonic or heat welding in filter assembly can affect the final product.

Fibrets' properties provide a means to engineer paper or nonwoven filter media to meet the demands of these product trends. High fibrillation allows higher retention and reduction of pore size. Fibrets meet concerns about leachables in food and water applications because of the no/low toxicity of the cellulose acetate polymer and lack of metallic extractables.

Additionally, the polymer provides a disposable product that can be thermally or ultrasonically bonded, resulting in reduced leakage in a filter assembly and the absence of extractables. Based on selective absorption properties, fibrets can remove low levels of specific organic compounds, such as phenol, from water and other liquids. Their rate of absorption is enhanced by their high surface area. Of course, fibrets do not fit all industrial filtration applications. For example, they cannot be considered for certain industrial chemical processes because of their solvent sensitivity.

SUMMARY

Cellulose acetate fibrets can enhance the performance of specialty filtration products by contributing both the fibrillar structure and cellulose acetate polymer properties. The high fibrillation provides a means to improve surface area, to act as a binder in synthetic papers, and to reduce pore size. The cellulose acetate polymer offers a fiber that is hydrophilic, thermally bondable, biodegradable and non-toxic, with selective absorption properties. Fibrets are suitable for wet-laid processing and coating applications, encouraging the development of new filter media to meet today's increasingly demanding requirements.

39

Polarized raman spectra of oriented cellulose triacetate fibers

Fran Adar, Diane Armellino and Herman D. Noether* - Instruments S.A., Inc., 6 Olsen Ave., Edison, N.J. 08820. *Tri-Princeton, P.O. Box 625, Princeton, New Jersey 08542, USA.

In the late 1950's, Sprague, Riley and Noether[1] published the x-ray diffraction pattern of cellulose triacetates I and II and indicated their correlations with celluloses I and II as well as a method of regenerating cellulose I from CTA I. CTA II could be prepared by acetylation of cellulose I or II and then spun into fibers from methylene chloride - methanol. There was no way to convert a cellulose triacetate solution to cellulose triacetate I or to cellulose I.

Recently, research at Dupont[2] showed that a solution of cellulose triacetate in trifluoro acetic acid (a liquid crystalline solution) could provide fibers or films which on annealing would give a crystalline x-ray diffraction pattern of CTA I. Additionally it showed that saponification of this material gave a mixture of cellulose II and IV (on careful hydrolyis at 80°C).

We had fiber samples of Arnel (Celanese triacetate) heat treated at 230°C, Ramie triacetates, heterogeneously prepared by H. Steinman at Celanese during their triacetate research (also heat treated at 230°C). It had the oriented CTA I x-ray diffraction pattern and could be hydrolyzed to cellulose I. E.J. Roche (at Dupont at the time of the trifluoroacetic acid spinning) graciously sent us small samples of oriented film annealed at 230°C. These materials were used for our Raman spectroscopic investigation with the microscope Raman equipment (Mole) of Instruments S.A.

Raman spectroscopy like I.R. can provide molecular structural information and thus should indicate differences and similarities of the three triacetates. The shape of the triacetate chains in CTA I and II as seen from x-ray diffraction investigations[3,4] are practically identical. Differences in the spectra should be based solely on parallel or anti-parallel arrangements of the chains, since the strong inter-chain hydrogen bonding of the celluloses has been eliminated by the acetylation. The Ramie and Roche materials have similar degrees of orientation (based on their x-ray diffraction pattern), the Arnel CTA II fibers are somewhat less oriented. The unit cell of Ramie CTA I has two parallel chains, the unit cell of CTA II has four chains. These chains are arranged in pairs of parallel chains but each pair is antiparallel with respect to the adjacent pair. If Raman were to see only intra-chain motions and conformations, all spectra should be identical; however the influence of parallel or anti-parallel nearest neighbors might produce differences.

I.R. spectra depend on changes of dipole moments. Raman bands require a change of polarizability to be active. Additionally a polarized laser beam allows several measurement options: (1) the laser and the emitted Raman radiation polarized parallel to the fiber direction (ZZ), (2) both polarized perpendicular to the fiber direction (RR) and (3) one of the two being parallel the other perpendicular to the fiber direction (ZR or RZ). These measurements provide differentiation between polarized and depolarized Raman bands; also the degree of orientation of the fibers or films can be correlated with differences in band intensities for the three options of polarization.

Raman spectra of the three materials were obtained for the (ZZ), (RR), and (ZR) options, resp., for the range from 150 to 1800 cm^{-1}. As expected, the overall patterns look very similar, but there are subtle differences in individual band intensities and positions. Only one previous pertinent publication on Raman CTA was found: Firsov and Zhbankov[6] analysed

I.R. and Raman spectra of CTA II in film and fiber form for the range from 300-1800 cm^{-1}. Our spectrum of CTA II agrees in general with their data; it includes additional data at 200cm^{-1} and the C-H stretch region(at 3000 cm^{-1}). We used and confirmed their data concerning polarized and depolarized Raman bands in all our spectra.

In the (ZZ) spectra major intensity differences exist at 1735 cm^{-1} (carbonyl band) and at about 200 cm^{-1} (both very strong for CTA II), the intensities of the 511 and 656 cm^{-1} bands are 511 > 656 for Ramie and Roche and 511 < 656 for CTA II. Also the shape and intensity of the five bands between 1040 and 1160 cm^{-1} are at variance. In all these, the Ramie and Roche pattern are very similar; the Arnel CTA II intensities are different. The peak near 200 cm^{-1} for the CTA I materials is at 206, for the CTA II at 195 cm^{-1}. All the major bands in these spectra with the exception of 1374 and 1430 cm^{-1} are polarized.

In the (RR) data the three strongest bands for all three samples are at 200, 906, and 1735 cm^{-1}. For CTA II the band at 200 appears double and there is a broad band at 377, (this CTA I band is sharp), the bands at 656 cm^{-1} vary in intensity of satellites from Ramie to Roche to CTA II and the latter is missing the band at 695. Again the shape of the 1040 to 1160 cm^{-1} bands is very similar for the two CTA I materials and is different in shape and intensity for CTA II. Except for the depolarized bands at 1374 and 1430 the major bands are polarized.

In the (ZR) option we encounter the increased intensity of the depolarized bands which are not or very little affected by orientational factors. The two bands at 1374 and 1430 are the strongest and also their shape and intensity as well as those in the 830 to 900 and 1040 to 1160 ranges are very similar. At 200 cm^{-1} both CTA I materials have peaks at 202 with a shoulder at 225 while CTA II has multiple bands between 200 and 226 cm^{-1}. For all three materials the strong peak at 511 in weak or absent and a peak at 491 or 506 has appeared. The two CTA I have a weak shoulder at 692 which is absent in CTA II.

In the C-H stretching range for (ZZ) and (RR) only the C-H symmetrical stretch at 2940 cm^{-1} is strong with very weak satellites at 2896 and 3020 cm^{-1}, in (ZR) pattern the depolarized bands at 2992 and 3020 cm^{-1} are equal or only slightly lower in intensity than the 2940 band.

Polarized bands change in intensity for the three polarization options; the depolarized bands seem to change little if at all. Thus the (ZR) spectrum can identify the depolarized bands. Table I lists these bands for the three samples:

Table I: Depolarized CTA Raman Bands*.

Ramie: 506, 598, 836, 1121, 1260, 1374, 1434 cm^{-1}
Roche: 506, 596, 836, 1116, 1264, 1372, 1429 cm^{-1}
CTAII: 491, 603. 839, 1121, 1266, 1377, 1432 cm^{-1}

*The depolarized C-H stretch bands were listed in the text above.

A comparison of the relative intensities of the polarized bands in the (ZZ), (RR) and (ZR) spectra can be made by assuming the strongest depolarized band at about 1370 cm^{-1} to have constant intensity for all three polarization options. These data are collected in Tables II-IV.

Table II RamieCTA I

cm^{-1}	intensity	pol.
206	RR>>ZZ ZR	p.
512	RR >ZZ >ZR	p.
658	RR >ZZ >ZR	p.
907	RR >ZZ >ZR	p.
971	ZZ >RR ZR?	p.
1070	ZZ >RR >ZR	p.
1125	ZZ >RR ZR	p?
1157-60	ZZ >RR >ZR	p.
1372	ZZ =RR =ZR	dp.
1432	ZZ RR ZR	dp.
1735	RR>>ZZ >ZR	p.

Table III RocheCTA I

cm^{-1}	intensity	pol.
206	RR>>ZZ >ZR	p.
511	RR >ZZ >ZR	p.
656	RR >ZZ >ZR	p.
905-12	RR >ZZ >ZR	p.
907	RR >ZZ ?	
972	ZZ >RR ?	p.
1045	ZZ >RR ?	p.
1080	ZZ >RR >ZR	p.
1126	ZZ >RR ZR	p?
1157	ZZ >RR >ZR	p.
1375	ZZ =RR =ZR	dp.
1432-35	ZZ RR >ZR	dp.
1734-38	RR >ZZ >ZR	p.

Table IV Arnel CTA II

cm^{-1}	intensity	pol.	cm^{-1}	intensity	pol.
194(204)	RR >ZZ >ZR	p.	1124	ZZ >RR ZR	p?
511-13	ZZ >RR ZR	p.	1162	ZZ >RR ?	p.
658	ZZ RR >ZR	p.	1372-79	ZZ =RR =ZR	dp.
909-14	ZZ RR >ZR	p.	1432-36	ZZ RR ZR	dp.
972-82	ZZ >RR ?	p,	1737-42	RR ZZ >ZR	p.
1075	ZZ >RR >ZR	p.			

The spectral data allow a distinction between the two types of CTA[7]. The fact that the Roche CTA I is hydrolysed to Cell II and IV, while Ramie CTA I can be hydrolysed to Cell I seems to be due to the option that under the spinning conditions (liquid crystal solutions or effect of the solvent) a CTA I structure can form whether the chains are arranged parallel or anti-parallel. This would be equivalent to the crystal structures of Cell IV[5], which can only be distinguished by spectroscopy or by reverting the Cell IV's to Cell I or II via their respective triacetates. While the present work finds the two CTA I Raman pattern more similar to each other than to CTA II, some differences might be found by further spectroscopic work. Also according to Roche et al.[2] this CTA I seems less stable than the Ramie CTA I; it converts to Cell IV by hydrolysis at low temperatures and also changes to CTA II starting around 260°C while Ramie CTA I changes in the melt at about 300°C.

References.

(1) Sprague, B.S., Riley, J.L. and Noether, H.D. Textil Research J., 28, 275 (1958)

(2) Roche, E.J., O'Brien, J.P. and Allen, S.R. Polymer Communictions, 27, 138 (1986)

(3) Stipanovic, A.J. and Sarko, A. Polymer, 19, 3 (1978)

(4) Roche, E., Chanzy, H., Boudoulle, M., Marchessault, R.H. and Sundrarajan, P. Macromolecules, 11, 86 (1978)

(5) Gardiner, E.S. and Sarko, A.E., J. Appl. Polym. Scie., Appl. Polym. Symp., 37, 303 (1983)

(6) Firsov, S.P. and Zhbankov, R.G. Translated from Zhurnal Prikladnoi Spektroskopii, 37, (2), 276-284 (1982) Original paper submitted 3/9/1981.

(7) Detailed spectra will be published elsewhere.

40

Resist finishing and dyeing of cationic fabrics

R.J. Harper, Jr. and A.H. Lambert - Southern Regional Research Center, ARS, USDA P.O. Box 19687, New Orleans, LA 70179, USA.

ABSTRACT

Cotton fabrics, which exhibit differential color uptake on dyeing, are of potential interest to the garment dyeing industry and would be of utility for just-in-time marketing procedures. Differential dyeability arises due to differential finishing in the case of fabrics that are partly cationic and partly non-cationic. Selective dyeing of cationic areas can be accomplished while non-cationic areas remain dye resist. Approaches for differential finishing and dyeing of cationic cotton garments are demonstrated.

INTRODUCTION

Garment dyeing has become a growing segment of the textile dyeing and finishing market in recent years. The majority of these cotton garments do not have smooth dry performance. Treatments which crosslink cotton to be smooth drying make fabric dye resistant. The development of a one-step process to make cotton crosslinked and cationic simultaneously not only provides a route to dyeable smooth dry cottons, but makes differentially dyeable cottons feasible.

CATIONIC COTTON

Cationic cotton fabrics are cottons that have been chemically modified to have a positive charge. Cationic cottons are dyeable with anionic (negatively charged) dyes such as reactive or acid dyes under acidic conditions. Under the same conditions, untreated cotton and crosslinked-only cotton are dye resist. Because of this, cationic cottons broaden the range of fabrics available to the garment dyer to achieve differential or resist effects on piece dyeing. A facile route to cationic cotton is that based upon the use of reactive additive, choline chloride, together with a crosslinking agent and an acid catalyst (1-2). As shown in the accompanying structure, this compound contains a primary alcohol group which permits bonding to the cellulose moiety and a quaternary group that confers cationic character to the finished cotton.

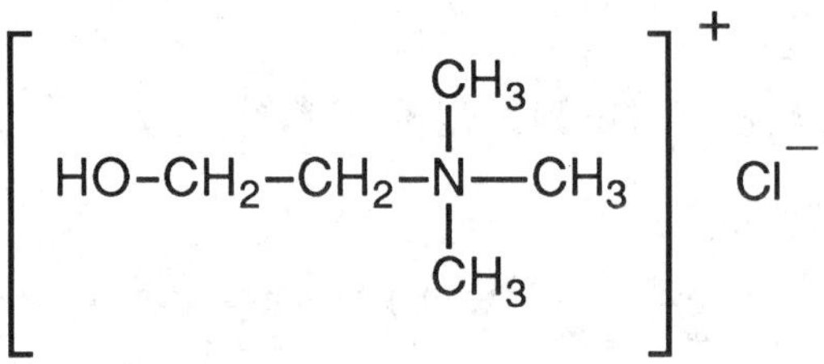

Choline Chloride

DIFFERENTIAL FINISHING FOR PRODUCING PARTLY CATIONIC FABRICS

Differential finishing of fabrics can be done in a number of ways. Examples of techniques employed include printing, spraying, yarn or slashing treatments, rope finishing, transfer techniques, stone finishing, and differential applications using the curved blade applicator. The fabric can be 1) either completely crosslinked, but only partly cationic, or 2) partly crosslinked and cationic and partly unmodified cotton. A piece dyed DP skirt, in which uneven dyeing arises from differential application with a cationic using a modified curve blade applicator (3), is shown in Figure 1.

APPROACHES FOR RESIST TREATMENTS

In terms of cationic fabrics, there are two general approaches that can be utilized to achieve resist effects on fabrics or garments using a cationic grafting system. One approach might be termed an enhancement technique. In this case, a cationic finishing system is applied differentially to a fabric or garment. The areas of the fabric that become cationic on curing show enhanced dye absorption while the remainder of the fabric is dye resist. A second approach is to sensitize the entire fabric (or face only) with the crosslinker, acid catalyst and quaternary additive. Then, apply catalyst poison (alkali) differentially to the fabric or garment to prevent crosslinking and grafting. By this means, the areas treated with the alkali are not cationic and as such are dye resist when dyed with anionic dyes under acidic conditions.

ENHANCEMENT TECHNIQUES FOR MAKING PARTLY CATIONIC FABRICS

There are two general approaches. The entire fabric can be sensitized with crosslinker and acid catalyst, then the cationic agent can be applied differentially. The result is a smooth-dry fabric that is partly cationic. On piece dyeing, only the cationic areas dye. The differentially dyed skirt shown in Figure 2 is an example of this process. The cationic agent was added to the garment via a stone treatment (4). In the second approach, the cationic system containing crosslinker, acid catalyst and cationic is differentially applied to the fabric. The resulting fabric is partly crosslinked and cationic, while the remainder is unmodified cotton. This fabric can then be differentially piece dyed.

CATALYST POISONS WITH CATIONIC FABRICS

Fabrics were treated with a formulation containing crosslinker, acid catalyst and cationic additive. Methods of application included conventional padding, or finishing of the face of the fabric, with a foam machine or a curved blade applicator. The fabrics were then dried. The sensitized fabric was then given a differential treatment with an alkaline poison, such as sodium carbonate. Methods of application of catalyst poison include spraying with open width fabric or stone or sponge treatments with garments. The fabrics were then cured and dyed under acidic conditions. Areas that are treated with alkali are dye resistant. An example of this approach is the skirt shown in Figure 3.

TWO COLOR PIECE DYEING WITH CATIONIC FABRICS

Fabrics that are part-DP cationic and part untreated, can be double dyed with one color under acid conditions and one color reactive under alkaline conditions. Tonal two-color effects can be achieved by dyeing with direct dyes under acidic conditions. The red and yellow skirt shown in Figure 4 is an example of double dyeing.

Figure 1. Piece dyed CBA finished cationic skirt.

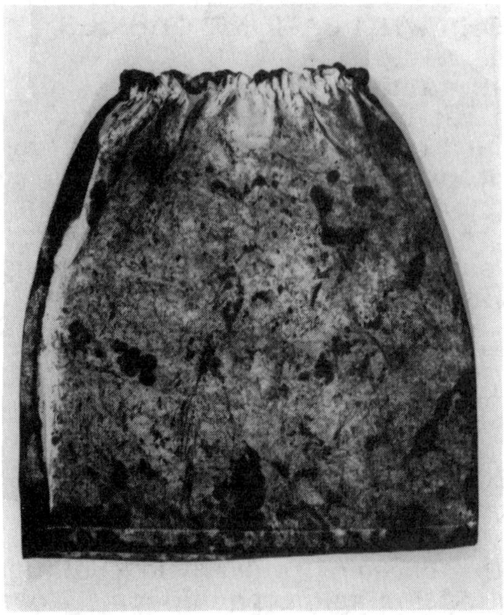

Figure 2. Piece dyed blue and white DP skirt, stone finished with cationic.

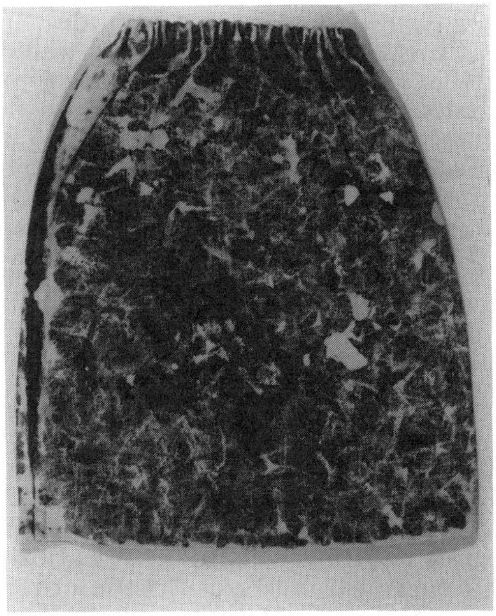

Figure 3. Piece dyed brown and white skirt, foam finished with alkaline resist.

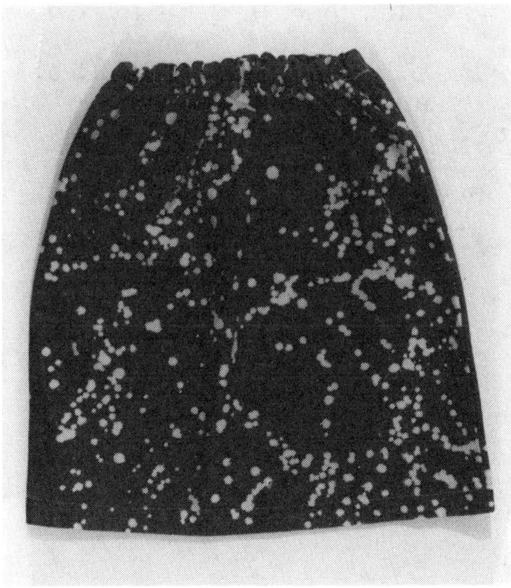

Figure 4. Double piece dyed red and yellow cationic skirt, spray finished with alkali.

The alkaline resist treatment can also include a reactive dye, so that alkali can function as dye catalyst as well as poison for the crosslink system. Then, the cationic areas of the fabric can be dyed to achieve a two - color effect.

Several cationic systems can be used in finishing. In the most diverse case, a grafting cationic system (choline based) and a cationic polyacrylate can be employed. On dyeing, areas with cationic agent grafted to cotton absorb anionic dye and areas treated with cationic polyacrylate absorb pigment dyes.

CONCLUSION

Chemical finishing can be performed to produce fabrics or garments, which exhibit unique color effects on piece dyeing. The chemical approach utilized is based upon differential finishing to make part of the fabric cationic, while the remainder is non - cationic. Two different approaches are possible. In one case, the crosslinker - cationic system is applied directly to part of the fabric. In the other, differential application of a catalyst poison is employed on fabric sensitized with the crosslinker - cationic system.

REFERENCES

1. Harper, R. J., Jr. and Stone, R. L. *Textile Chem. and Color.* **1986**, *No. 11*, 32.
2. Harper, R. J. Jr., U.S. Patent 4,629,470, Dec. 16, 1986.
3. Harper, R. J. Jr. and Lambert A. H. *Amer. Dyestuff Reptr.* **1988**, *77, No. 2*, 13.
4. Harper, R. J. Jr. and Lambert, A. H., 1991 AATCC Conference, Charlotte NC, Oct. 1991, Book of Papers, 278.

41

Photofading characteristics of dyed easy-care finished cotton cellulosic substrate

F.M. Tera*, A. Hebeish, N.A. Baset*** and M.N. Michael*** - *National Institute for Standards, Dokki, Giza, Egypt. **Academy of Scientific Research, Cairo, Egypt. ***National Research Center, Dokki, Giza, Egypt.

ABSTRACT

Fading studies were performed for dyed easy-care finished cotton samples produced by the conventional two-stages and the combined one-stage process for dyeing and finishing using polyfunctional N-methylol compound (DMA) as a resin with latent acid catalysts. The study comprised the fading kinetic order at the different exposure times, fading constant (K), fading rate constant (R_f), colour difference (ΔE), lightfastness characteristic, also colour strength and the change in the extinction ratio(Y/X) at the different dye bath concentrations. The obtained results revealed that fading of the one-stage processed samples is lower than that of the two-stage for both catalysts used.

INTRODUCTION

When dyed easy-care finished cotton samples are exposed to light, their colour change depending to a considerable extent on a large variety of factors[1], among which are the finishing agent and catalyst as well as type of bond between the dye and the substrate.

The proposed mechanism of photosensitisation[2] of the resin treated cellulose can be represented as follows:

i) $R + \{O_2\} \text{-----}> R\{O_2\}$, ii) $D \text{------}> D^*$,

iii) $D^* + R\{O_2\} \text{-----}> D\{O_2\} + R'$

D = dye, D^* = dye in the excited state,

R = resin and R' = a change form of the resin.

The differences in colour strength and light fading of the dyed fabric samples treated with various catalytic systems were associated with variation of many factors[4-6]:
a) Type and nature of the catalyst. b) Rate of dissociation and decomposition of the catalysts. c) The nature of the interactions between reagents, catalyst and the substrate. d) The mode of interaction i.e. either intermolecular or intramolecular.
e) The length and number of crosslinks.

The rate of light fading can be represented by the equation[6]: $\log C/C_o = -Kt/2.303$, hence K may be calculated.

EXPERIMENTAL WORK

The following materials were used:
Mill scoured, bleached and mercerized cotton fabric (lino). Polyfunctional heterocyclic N-methylol compound (DMA) as a finishing agent. The latent acid catalysts, namely ammonium dihydrogen phosphate ($NH_4H_2PO_4$) and ammonium nitrate (NH_4NO_3). The reactive dye Levafix Brilliant Red E-4BA (C.I 158) Bayer.

Dyeing and finishing were performed according to the conventional method, and by the combined process[7]. Reflectance spectra for all samples in the visible region(300-700 nm) were measured before and after the different exposure periods at the different dye bath concentrations (5,10,15,20 gm/L.) by a Shimadzu Double Beam UV 300-Spectrophotometer. The colour strength values (K/S) based on the reflection measurement was evaluated using the Kubelka-Munk equation.

Irradiation of the dyed-finished samples and the Blue Scale was conducted using the Tera Light Fastness Tester[8] for different exposure periods (20,60,100,160 and 220 hours).

RESULTS AND DISCUSSION

Table (1) shows that colour strength (K/S) increases by increasing the dye concentration[9], due to immobility of the cellulose hydroxyls. Also, the combined process brings about samples of colour strength values higher than those of the conventional one in accordance with:[10]. a) reaction of the reactive dye with the methylol group of the finishing agent, b) decomposition of N-methylol finishing agent leading to amide groups with higher reactivity to the reactive dye, c) possible trapping of the dye molecules in the polymeric products of the finishing agent in the cotton matrix, and d) the dye molecules are perhaps enmeshed in crosslinked cotton cellulose and/or embodied in crosslinked cotton containing polymeric product of DMA. Table (1) also indicates that (NH_4NO_3) acquires higher colour strength than that catalyzed by ($NH_4H_2PO_4$) due to the formation of the stronger (HNO_3) acid which is highly crosslinked with the cotton cellulose containing DMA.

The results of the Extinction (Y/X) Ratio with the (K/S)values of table(1) showed that increase of (K/S) is always accompanied by an increase (Y/X) value indicating the presence of aggregated particles of larger size[10]. It is also clear that at nearly equal colour strength the (Y/X) ratios for the samples processed by the combined one-stage are higher than those of the conventional two-stage. This supports the presence of higher covalent bonding among dye molecules cellulose hydroxyls and DMA. Moreover, samples catalysed by (NH_4NO_3) showed a higher (Y/X) value than by ($NH_4H_2PO_4$).

Table (1)

Catalyst	Dye Conc. g/L	The conventional process k/s	y/x	L.F.	The combined process k/s	y/x	L.F.
$NH_4H_2PO_4$	5	0.47	1.8	3	0.66	2.4	4
	10	0.54	2.2	3-4	1.29	2.5	4-5
	15	1.06	2.3	3-4	1.37	2.8	4-5
	20	1.39	2.6	3-4	1.54	3.2	4-5
NH_4NO_3	5	0.41	2.2	3	0.68	2.9	4
	10	0.56	2.4	3	1.20	3	4-5
	15	1.16	2.7	3	1.33	3.1	4-5
	20	1.41	3.0	3-4	1.68	3.3	4-5

(DMA), 100 g/L, (catalyst), 5g/L, curing time 3 min., curing temperature 140°C.

Table (1) also shows that the light fastness of the samples processed by the combined method are higher than those of the conventional ones. This was attributed to the formation of highly covalent bonds and aggregated dye particles in the cross-linked cellulose structure leading to considerable resistance of fading factors[11,12].

The characteristic fastness grade (C.F.G.) curves of figure (1) showed higher positions for samples of the combined one-stage than the conventional two-stage. Also, the (C.F.G) curves acquired lower slopes regardless of the finishing process or the catalysts used indicating uniform particle size distribution[13].

Evaluation of Fading Rate Constants (K, R_f)[11,12]: The fading kinetic curves of figure (2)(a,b) clarify that the fading rate of the reactive dye on cotton samples processed according to the two processes in question can be represented by first order kinetics[14] depending upon the initial colour strength. Also, the magnitude of fading according to the combined process are less than those processed by the conventional one, which means that the dye molecules in the first process are strongly covalently bonded with the cellulose and of larger aggregated particles than those processed according to the other method.

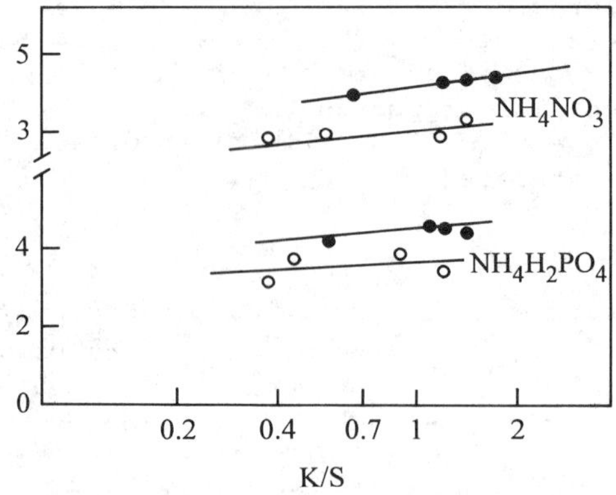

Figure 1. (C.F.G) curves for samples processed according to the (o) conventional and the (•) combined processes.

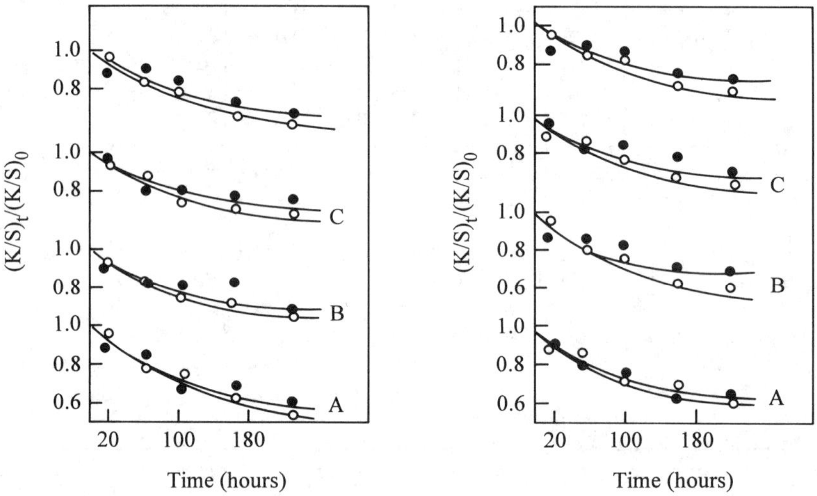

Figure 2. Fading curves of samples catalyzed by: {"a" $NH_4H_2PO_4$, "b" NH_4NO_3} and DMA. (o) Conventional (•) combined (A,B,C,D) are dye concentrations (5,10,15,20 g/L).

When log $\{(K/s)_t/(K/s)_o\}$ is plotted against the exposure time figure (3) (a,b) a straight line is obtained the slope of which gives (K) the fading constant. As is evident (K) decreases with increase dye concentration[12], resulting more in aggregation, and an increase in the covalent bond strength, and in turn the fading rate decrease.

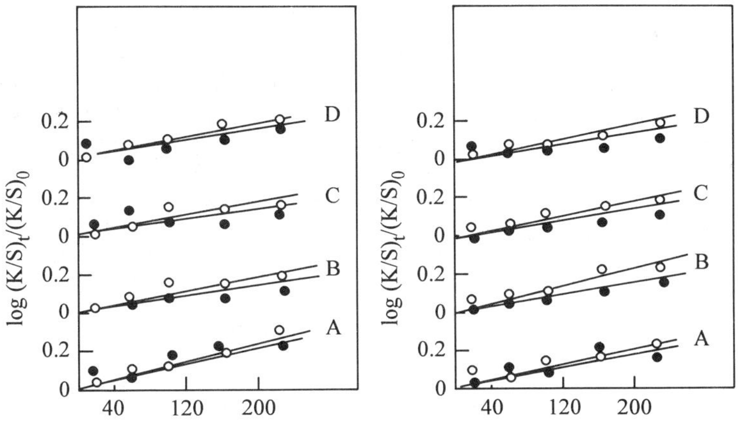

Figure 3. Log$(K/s)_t/(K/s)_o$ against the exposure time for samples catalyzed by:{"a" $NH_4H_2PO_4$,"b" NH_4NO_3} ande DMA.(o) Conventional process; (•) combined process (A,B,C,D) are dye concentrations (5,10,15,20 g/L).

The fading rate constant(R_f) - which is concentration dependent increases with the decrease in (K) and both decrease with rise in dye concentration[12] as indicated in table(2). Also, both (K) and (R_f) of the conventional two-stage are higher than those of the combined one-stage. The results also indicate that (R_f) values are related to the light fastness, i.e. the higher the (R_f) value the greater the magnitude of fading and the lower the light fastness.

Table (2)

Catalyst	Dye Conc. g/L	The conventional two-stage process			The combined one-stage process		
		$K(10)^{-3}$	$R_f x10^{-3}$	L.F.	$K(10)^{-3}$	$R_f x10^{-3}$	L.F.
$NH_4H_2PO_4$	5	1.25	0.59	3	1.10	0.45	4
	10	1.00	0.35	3-4	0.98	0.32	4-5
(5 g/L)	15	0.95	0.31	3-4	0.90	0.27	4-5
	20	0.80	0.30	3-4	0.83	0.30	4-5
NH_4NO_3	10	1.20	0.50	3	1.00	0.32	4-5
	15	0.94	0.36	3	0.89	0.30	4-5
(5 g/L)	50	0.90	0.33	3-4	0.80	0.28	4-5

Curing temp. 40°C, curing time 3 min.,[(K) hr^{-1}][(R_f) $g/L^{-1}hr^{-1}$]

The results revealed that within the range of the dye concentrations studied only two values of light fastness were recorded, while definite value for (R_f) was evaluated for each dye concentration assessed. This finding assured the fact that applying (R_f) values for assessing the light fastness is much more precise, and is very sensitive for the dye concentration change. The colour difference values (ΔE)[7,15] were taken at the highest dye concentration (20g/L). It is seen from figure (4) that (ΔE) increases by prolonging the exposure time and $d(\Delta E)/dt$ is higher for the conventional than the combined process. This finding reflects differences in molecular particle size distribution of the dye which may be more aggregated in the combined process. In addition, samples catalyzed by (NH_4NO_3) exhibit less change in (ΔE) than those catalyzed by ($NH_4H_2PO_4$) showing the highest rate of change.

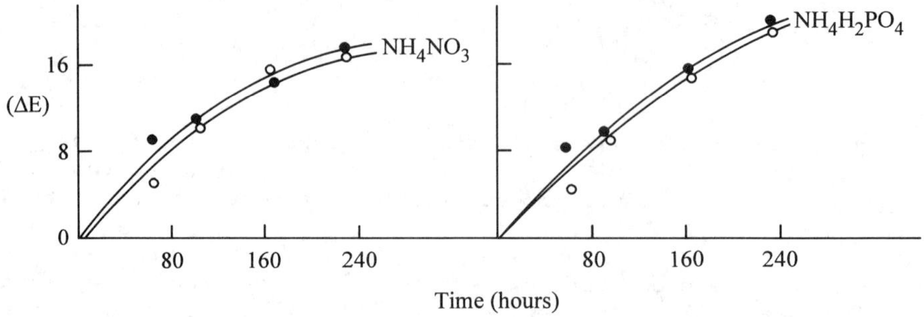

Figure 4. (ΔE) against the exposure time for sample catalyzed by two catalysts and DMA. (o) Conventional, (•) combined process.

REFERENCES

1. Giles,C.H. and Mukay,R.B., Review Tex.Res. J.,(1963) 33,527.
2. Schenk, G.O., Angew-Chemie, (1952), 64, 12.
3. Reinhardt,R.M. and Cashen,N.A., Tex. Res.J., (1972), 42,335.
4. Ibrahim, N.A., El Kashouti, M.A. and El Khardaly, E.A., Cell Chem., Tech., (1985), 12, 69.
5. Rowland, S.R., Blouin, F.A., Tex. Res. J., (1978) 48, 73.
6. Wan Shikha, M.Sc. Chul. Joo Lee, Melliand Tex. Berichte, (1986), 10, 724-727.
7. M.N. Micheal, Ph.D. Thesis, Cairo University, Egypt, (1990).
8. Tera Light fastness Tester, Egypt. Pat. (1981), 15182.
9. Kamel, M.M., Kharadly, E.A. and Youssef, B.M., Amer. dyestuff Rep., (1983), 33, 422.
10. Hebeich, A., Kamel, M.M., El-Rofie, M.H., El-Alfy,E.A., Cell Chem. Tech. (1978)12, 317.
11. Mitsuo, U. and Tsaneaki, H., Melliand Tex., (1986), 42, T 584.
12. Vanshik Ha., Melliand Tex. (1986), 42, 724.
13. Giles,C.H., Haslam, R.Tex. Res.J. (1977), 47, 347.
14. Grindberg, O.Y., Dubinsii, A.A., Gombrete, Y., Krichevskii, G.E., Kolor-Ertesito (1977), 19,Nr. 1,14.
15. Judd, D.B. and Wyszechi, G., "Colour in Business, Science and Industry", John Wiley and Sons (1975), 3rd. Ed.

42

Differences between the properties of cotton fibers treated with urea and carboxylic acid derivatives

S.H. Zeronian*, N.R. Bertoniere, K.W. Alger*, L.D. Fisher*, G. Buschle-Diller*, S.A. Holmes* and P.A. Slaven*** - *Division of Textiles and Clothing, University of California, Davis, CA 95616. **U.S.D.A., ARS, Southern Regional Research Center, New Orleans, LA 70179.

ABSTRACT

Differences were found between the mechanical properties of BTCA and DMDHEU-treated fibers when cotton was treated with the finishes with or without a preswelling treatment at low reagent contents. However at the higher reagent contents required for durable press properties, many of their mechanical properties were similar or insignificantly different. The properties of BTCA-treated mercerized fibers were unaffected by whether the BTCA had been applied to dry mercerized or never dried mercerized products.

INTRODUCTION

In the search for formaldehyde-free durable press finishing agents for cotton textiles, considerable attention is being paid to polycarboxylic acids such as 1,2,3,4-butanetetracarboxylic acid (BTCA) [1,2]. However the single fiber mechanical properties of BTCA-treated cotton remain to be investigated. Here, we compare such properties of BTCA crosslinked fibers with those reacted with dimethyloldihydroxyethyleneurea (DMDHEU), a commonly applied durable press finish. It is well known that normally the strength of cotton is reduced by crosslinking [3]. Thus if BTCA is to be considered as a replacement for such reagents as DMDHEU the mechanical properties of BTCA- and DMDHEU-treated cotton need to be compared.

Since interest first rose in crosslinking cotton much of the reported mechanical properties data has been obtained on fabrics. However for basic studies on the effects of crosslinking on the fine structure and physical properties of cotton, measurements of fabric properties do not reflect accurately the changes occurring to the fibers. Thus in our work on the fundamental aspects of crosslinking with DMDHEU we focussed on single fiber properties [4-6] and have continued to do so in the present study.

EXPERIMENTAL

The cotton variety studied was Pima S-5 (G. barbadense series). It was scoured and prepared for the application of BTCA in the manner described previously [4]. Four sets of samples were processed. In three sets the cotton was preswollen before BTCA application and in the fourth, the cotton was treated without prior swelling (NPT series). Of the preswollen series, two were mercerized as reported before [4]. However in one of these sets after the samples had been washed with water to remove the caustic soda the products were not dried before BTCA application (ND MERC series). In the other set they were dried first (MERC series). The third array was treated with liquid ammonia in the manner described earlier [6] before BTCA treatment (LAT series).

The conditions for BTCA application were based on the paper by Welch [1]. The catalyst used was $NaH_2PO_4H_2O$. The mole ratio of catalyst to BTCA was 1.13. The fiber assembly was immersed in the aqueous solution of BTCA and catalyst, passed through a pad and then wetted out again and repassed through the pad to obtain a pickup of about 120%. The assembly was placed flat in an oven at 85°C for 5 min. and then cured at 180°C for 90 sec. followed by rinsing several times in hot water before drying at 85°C for 5 min.

The procedures for characterizing tensile properties, breaking twist angle (BTA) and water retention value (WRV) have been described before [4-6]. BTA is a measure of fiber brittleness. As defined a high BTA value corresponds to high brittleness. With the exception of the LAT series, the BTCA content of the products was obtained by FTIR by N. M. Morris (USDA, SRRC). In the case of the LAT series the reagent content was determined from the difference in weight of the conditioned sample before and after BTCA application.

The application of DMDHEU to various samples is described elsewhere [4-6].

RESULTS AND DISCUSSION

The BTCA and DMDHEU contents of samples were determined for all treatments. However, the number of crosslinks actually induced by these reagents is unknown. Thus WRVs were used to compare the effects of BTCA with those of DMDHEU on the swelling properties of the fibers, since swelling is affected by crosslinking. Initially in all cases, WRV fell sharply with increasing reagent content and then decreased more slowly (Figure 1). For NPT cotton, at equivalent BTCA and DMDHEU contents, the BTCA-treated samples had higher WRVs. The difference may be due to the location of these reagents in the fibers. For LAT and MERC samples, the WRVs of the BTCA and the counterpart DMDHEU-treated products were roughly the same at similar reagent contents. Thus differences in the effects of BTCA and DMDHEU on the swelling of cotton in water are masked once the cotton receives a pretreatment with caustic soda or liquid ammonia. It should be noted that the WRV of the BTCA-treated products did not appear to be affected by the mercerization mode (cf MERC and ND MERC samples). We had hypothesized that if the sample was not dried before crosslinking, the BTCA would penetrate the fiber more uniformly and the product would be more accessible to water.

The extensibility and strength of NPT BTCA cotton fell and brittleness increased as WRV decreased, or as BTCA content increased. When a plot was made of strength versus percent of reagent on the fiber, the data for the BTCA and DMDHEU NPT samples fell close to the same curve (Figure 2). Thus the stress-bearing components of the BTCA NPT and DMDHEU NPT fibers appear to be similar. The data can be separated if the plot was made versus WRV (Figure 3). Thus fiber/crosslinking agent interactions may differ for DMDHEU and BTCA. The same observations were made when the BTA of these samples was plotted against percent reagent and WRV. At higher WRVs, for a given WRV, the brittleness of the NPT BTCA samples was greater than that of their DMDHEU counterparts and the strength of the former was lower. This observation may only be of academic interest since at the higher BTCA contents required for durable press properties [1], the WRVs of the NPT BTCA and NPT DMDHEU samples of the same reagent content have similar low values and their strength and brittleness values are close.

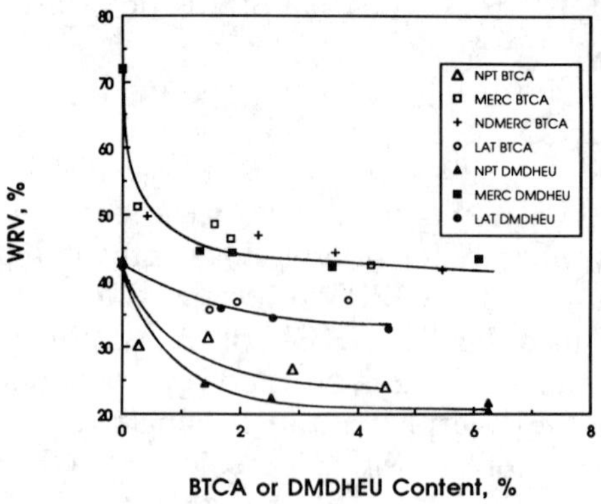

Fig. 1. Effect of reagent content on WRV.

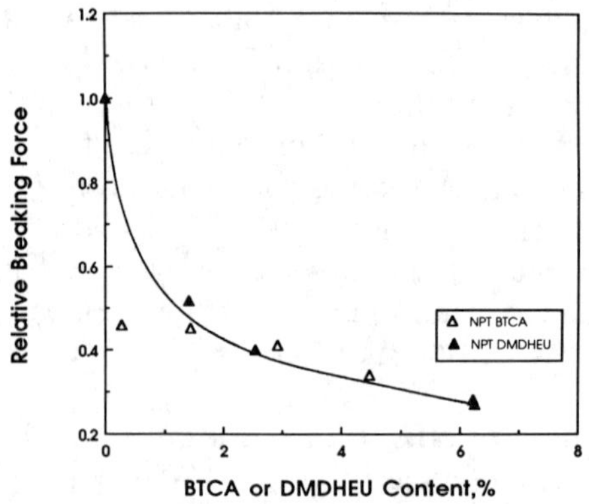

Fig. 2. Effect of reagent content on NPT samples.

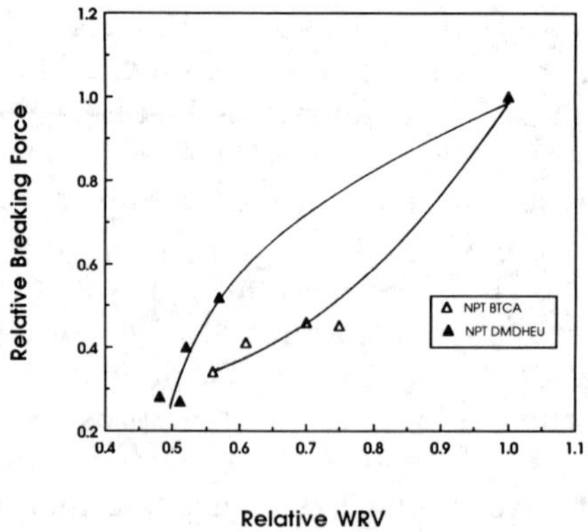

Fig. 3. Relation between strength and WRV for NPT samples.

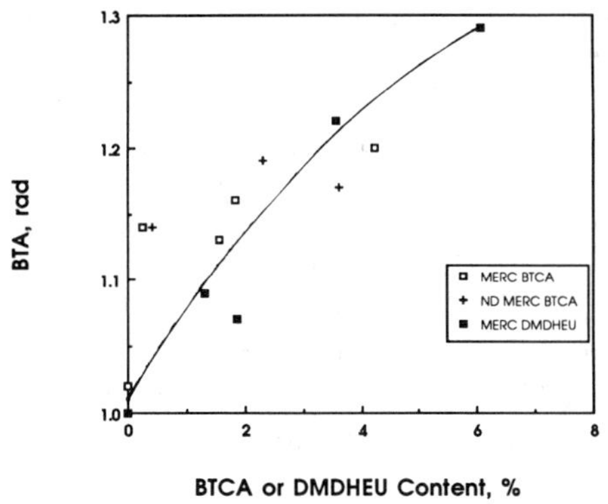

Fig. 4. Effect of reagent content on BTA of merc. samples.

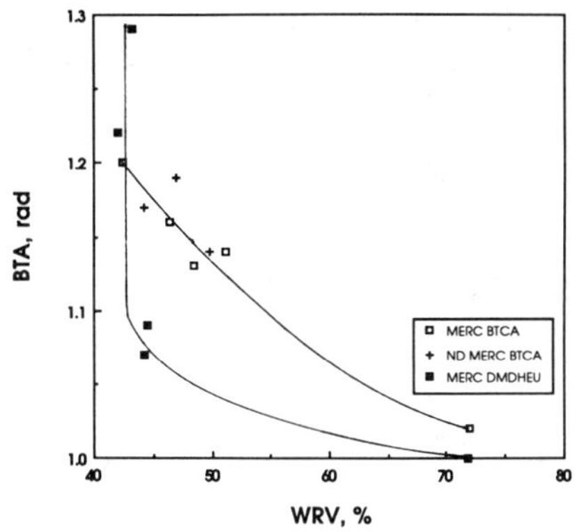

Fig. 5. Relation between BTA and WRV for merc. samples.

With a liquid ammonia preswelling for equivalent strengths and extensibilities the BTCA products had higher WRVs than their DMDHEU counterparts. Unlike NPT cotton the same relation between strength and reagent content did not appear to be present for mercerized BTCA and DMDHEU products. The DMDHEU samples had lower strengths at reagent contents between 2 and 4 percent. However both sets of samples appeared to have approximately the same strengths for equivalent WRVs. Thus it appears that the higher swelling induced by mercerization causes different fiber/crosslinking compound interactions than found when NPT cotton was treated with these reagents. Initially, the BTA values of the BTCA-treated MERC and ND MERC samples were higher than those of the DMDHEU-treated MERC samples for a given reagent content or WRV value (Figures 4 and 5). Thus it appears as in the case of the NPT samples that initially the BTCA samples were more brittle at a given WRV. It should be noted again that at high BTCA contents, the performance of the BTCA MERC sample was similar to that of the DMDHEU MERC samples of equivalent reagent content. The same was also observed for the strength of crosslinked LAT fibers. In addition, it should be noted that the properties of the BTCA treated mercerized products were independent of whether the cotton was dried or not dried after mercerization.

ACKNOWLEDGEMENT

The authors thank N. M. Morris for determining the BTCA contents by FTIR.

REFERENCES

1. C. M. Welch. *Text. Res. J.* **1988**, *58*, 480.
2. C. M. Welch and B. A. Kottes Andrews. *Textile Chem. Color.* **1989**, *21*, 13.
3. A. C. Nuessle, M. N. Fineman and E. O. J. Heiges. *Text. Res. J.* **1955**, *25*, 24.
4. S. H. Zeronian, N. R. Bertoniere, K. W. Alger, J. L. Duffin, M. S. Kim, L. K. Dubuque, M. J. Collins and Q. Xie. *Text. Res. J.* **1989**, *59*, 484.
5. S. H. Zeronian, N. R. Bertoniere, M. S. Kim and Q. Xie. *Cellulose Sources and Exploitation* (J. F. Kennedy, G. O. Phillips and P. A. Williams, Eds.), Ellis Horwood, Chichester, England 1990, pps. 203-209.
6. S. H. Zeronian, N. R. Bertoniere, K. W. Alger and Q. Xie. *J. Text. Inst.* **1990**, *81*, 310.

43

Relationships between properties of crosslinking agents for cellulose and their structures evaluated by molecular modelling methods

H. Petersen, E. Hädicke and S. Brode - BASF Aktiengesellschaft, Ludwigshafen/Rhine, Germany.

ABSTRACT

The reaction mechanism of hydroxymethylated and methoxymethylated cyclic ureas for finishing cellulose fabrics have been studied with molecular modelling methods. Quantum mechanical calculations were performed with the ab initio Hartree Fock program TURBOMOL [1] and the semiempirical program MOPAC [2].

INTRODUCTION

In the crosslinking of cellulosic fabrics and cellulosic fiber-containing blended fabrics preference is given to hydroxymethyl and methoxymethyl compounds of urea, five-membered and six-membered cyclic ureas and carbamates. Reactivity and stability towards hydrolysis are two of the most important properties of the easycare reactants. The hydroxyl groups of hydroxymethyl compounds and the methoxy groups of methoxymethyl compounds can be substituted by cellulose residues or other nucleophilic substances under the catalytic action of an acid or an acid catalyst. Specific kinetic measurements lead to the conclusion that the condensation reaction with the reactants is not subject to a specific but rather to a general acid catalysis [3], preferably in the presence of at least a small amount of water (Figure 1). The rate constant k_1 of the cleavage of the carbon-oxygen-bond is responsible for the reactivity of the hydroxymethyl or methoxymethyl compound. This is the rate-determining step. As an intermediate ureidomethyl-[carbonium-immonium]-cations are produced under the influence of an acid liberating water or methanol respectively. The ureidomethyl-[carbonium-immonium]-cations can react with the nucleophilic hydroxyl groups of the cellulose or with another nucleophilic compound, i.e. water.

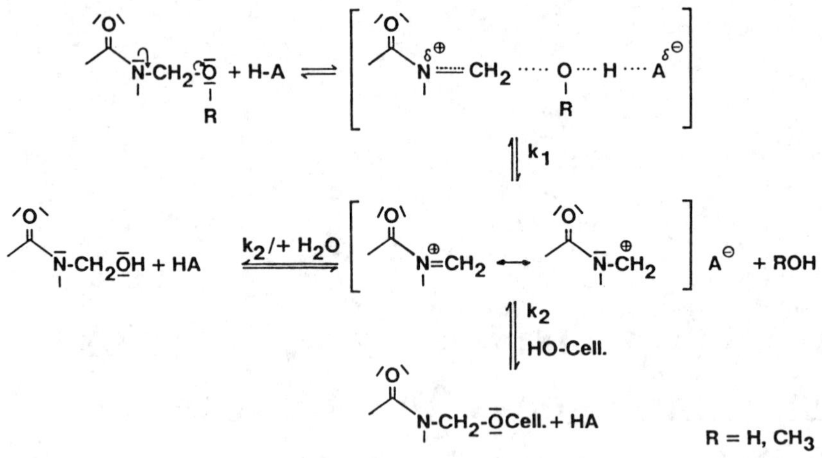

Figure 1: Reaction Mechanism

With molecular modelling techniques it was illustrated such a simulation with the acid-catalyzed cleavage of the carbon-oxygen-bond of the first methoxymethyl group of dimethoxymethyl-ethyleneurea over the transition state to the ureidomethyl-[carbonium-immonium]-cation.

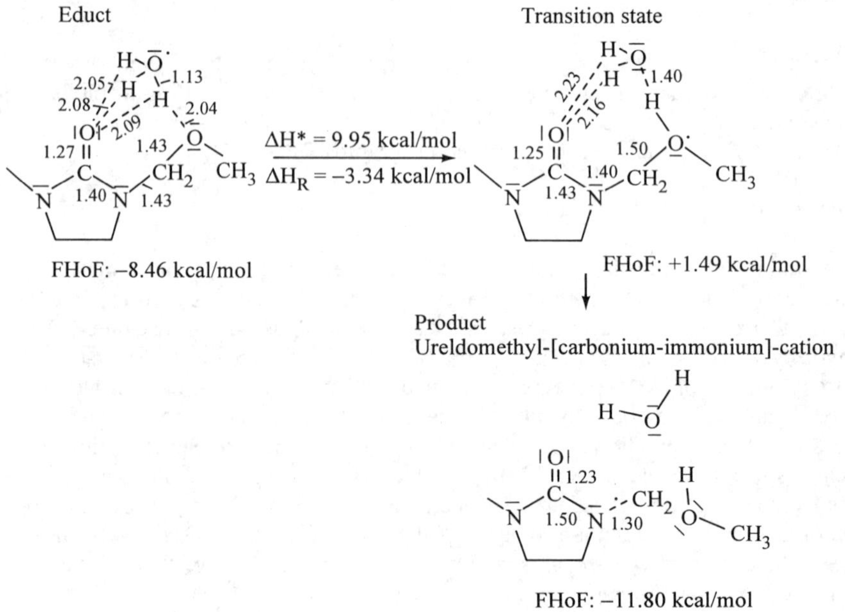

Figure 2: Reaction path of Dimethoxmethyl-ethyleneurea

Starting with an educt at which the cyclic urea forms a stable aggregate with H_3O^+ with a maximum number of hydrogen bonds to the carbonyl oxygen of the five-membered ring and the oxygen of the methoxymethyl substituent. During this reaction one has to consider the bond cleavage of the "ether-bond" and the bond formation of one H^+ of H_3O^+ with the methoxy group to methanol. These two directions are the reaction coordinates (Fig. 2). The distances of these reaction coordinates were varied from 1.3 to 0.95 Angstrom and 1.4 to 2.0 Angstrom respectively with a step size of 0.05 Angstrom to come from the educt to the ureidomethyl-[carbonium-immonium]-cation. At each point of the grid of the reaction coordinates all other bond lengths and angles were refined using the "eigenvector following"-routine by Baker [4] with no enhanced precision. As can be seen from Fig. 2 there is a remarkable change in the bond length between the carbon atom of the CH_2-group and the oxygen of the ether group from 1.43 to 1.50 Angstrom going from the educt to the transition state. The final heat of formation of the transition state was calculated with a value of + 1.49 kcal/mol. The difference between the final heats of formation between the transition state and the educt resulted with an activation energy of 9.95 kcal/mol. Very strong changes of bond lengths result from the transition state to the ureidomethyl-[carbonium-immonium]-cation. The bond length between the nitrogen atom and the carbon atom of the CH_2-group has a value of only 1.30 Angstrom. This short distance is characteristic for a nitrogen-carbon double bond and indicates the [carbonium-immonium]-cation.

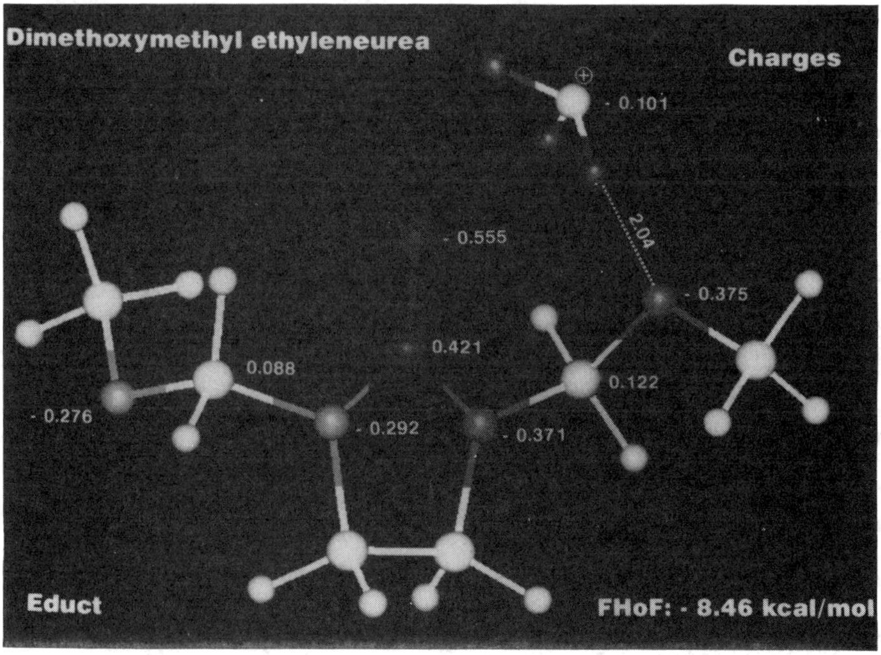

Figure 3 a: Change of Charges during the reaction of Dimethoxymethyl-ethyleneurea

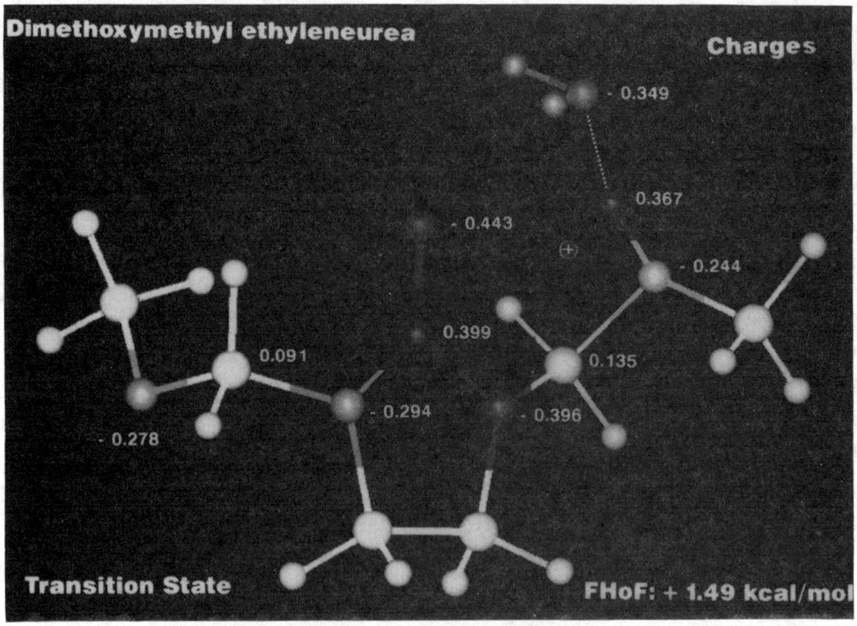

Figure 3 b: Change of Charges during the reaction of Dimethoxmethyl-ethyleneurea

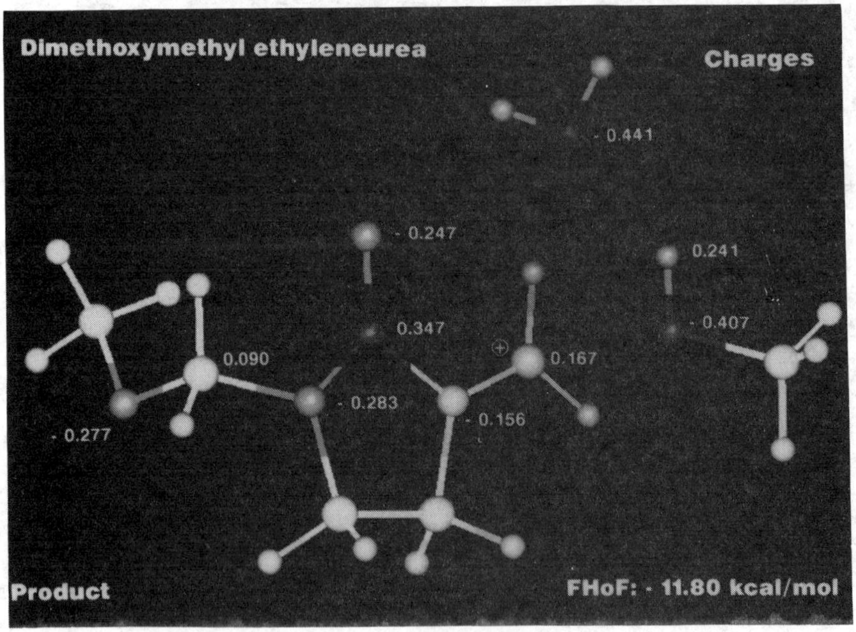

Figure 3 c: Change of Charges during the reaction of Dimethoxmethyl-ethyleneurea

At the same time the charges change at the nitrogen atom, the carbon atom of the CH_2-group and the oxygen atom of the ether group. The charge of the oxygen atom of the ether group changes from - 0.375 of the educt to - 0.244 of the transition state and for the carbon atom of the CH_2-group from 0.122 of the educt to 0.135 of the transition state (Figure 3a, 3b). The nitrogen atom becomes a more negative charge. Very strong changes of charges result from the transition state to the ureidomethyl-[carbonium-immonium]-cation. The charge of the carbon atom of the CH_2-group changes from 0.122 of the educt over 0.135 of the transition state to 0.167 of the [carbonium-immonium]-cation. At the same time the charge of the nitrogen atom changes from - 0.396 of the transition state to - 0.156 of the product (Figure 3b, 3c).

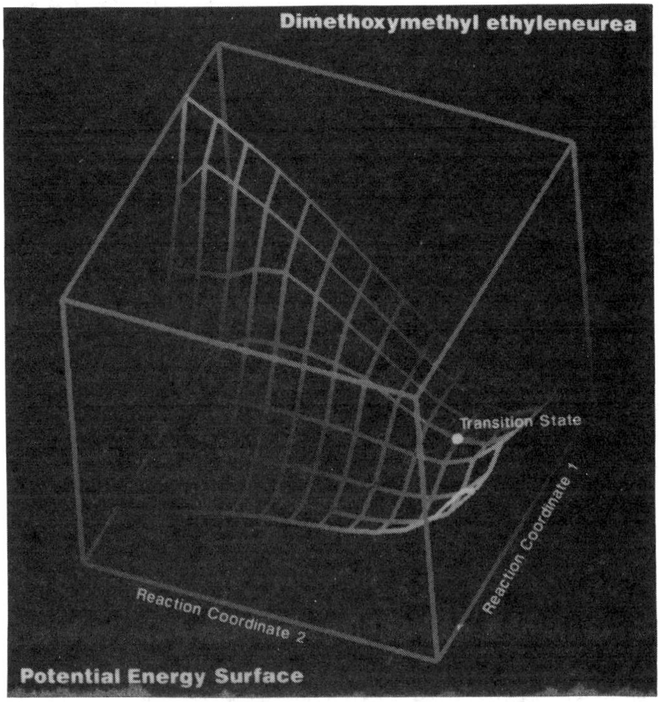

Figure 4: Potential Energy Surface

By inspection of this potential energy surface the lowest energy transition point between the educt and the [carbonium-immonium]-cation was localized following the line of the reaction path (Fig. 4). This position was refined by a saddle point calculation followed by a nonlinear least-squares-optimization by Bartel's method [5]. The transition state was confirmed to be a saddle point by force calculation. This transition point is characterized by only *one* negative eigenvalue. The other points along the reaction path have more than one negative eigenvalue. The vibration with negative eigenvalues was displayed with the modified program "VIBRATE" from Glaser and Hall [6]. The movement of atoms of this vibration direct along the reaction path. Small displacements of the structure in either direction led to the educt or [carbonium-immonium]-cation, during normal geometry optimization.

The molecule of dimethoxymethyl-4.5-dimethoxy-ethyleneurea contains two types of reactive groups. One species are the methoxmethyl groups in the positions 1 and 3 and the other species are the two methoxy groups in the positions 4 and 5. At first it was investigated with two reaction coordinates the H_3O^+-catalyzed cleavage of the carbon-oxygen-bond in the methoxymethyl group forming the [carbonium-immonium]-cation. The numbers for the activation enthalpy and the reaction enthalpy were calculated from the heat of formation values obtained from MOPAC for the educt, transition state and product geometries. The activation energy was calculated with a value of 11.14 kcal/mol (Fig. 5).

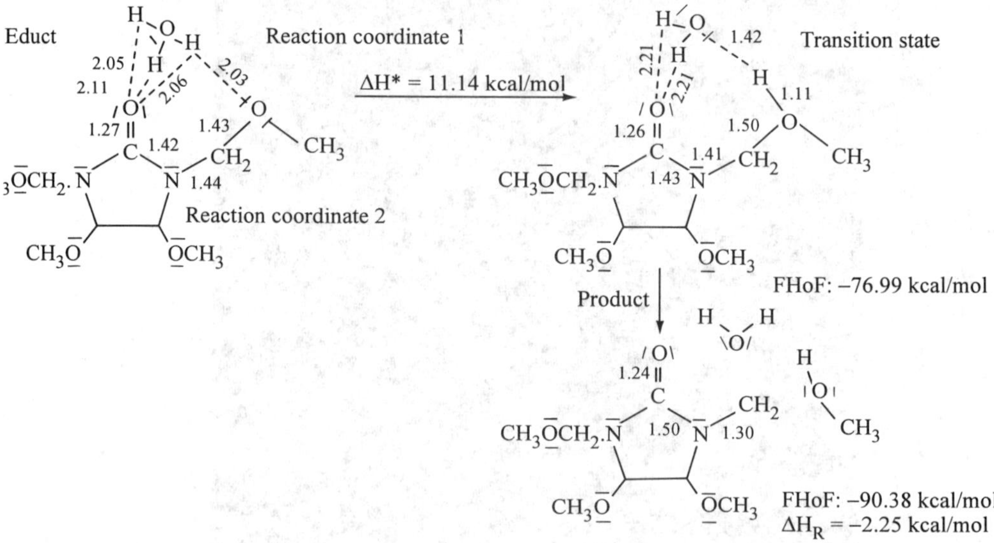

Figure 5: Reaction path of Dimethoxymethyl-4.5-dimethoxethyleneurea

There is a remarkable change of the bond lengths between the nitrogen atom and the carbon atom of the CH_2-group from 1.44 Angstrom for the educt over 1.41 for the transition state to 1.30 Angstrom for the [carbonium-immonium]-cation (Figure 5). In Figure 6a - 6c are shown the charges for the educt, transition state and the [carbonium-immonium]-cation. The charge of the carbon atom of the CH_2-group changes from 0.104 to 0.143 of the transition state to 0.175 of the product and the charge of the nitrogen atom becomes more negative from - 0.356 to - 0.419 of the transition state. The charge of the nitrogen atom changes from - 0.419 of the transition state to a more positive character with a value of - 0.176.

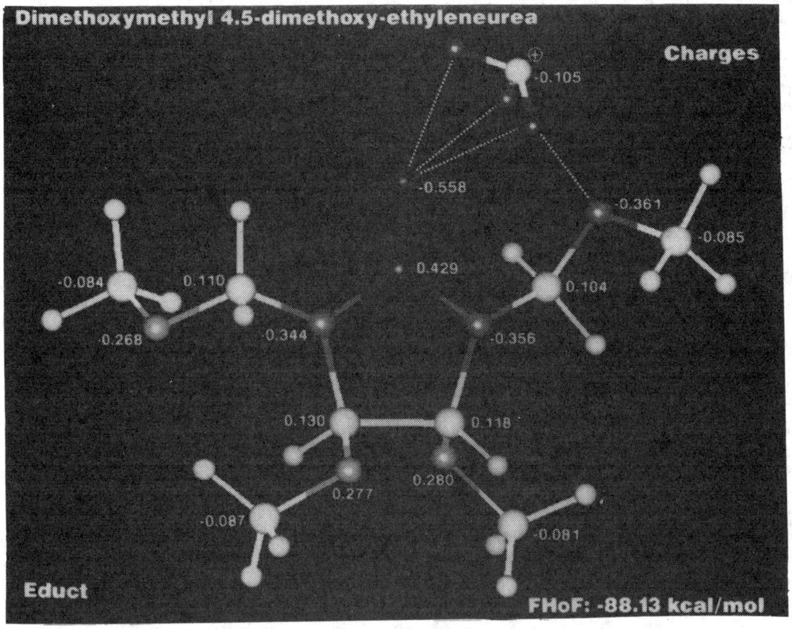

Figure 6a: Change of charges during the reaction of Dimethoxymethyl-4.5-dimethoxy-ethyleneurea

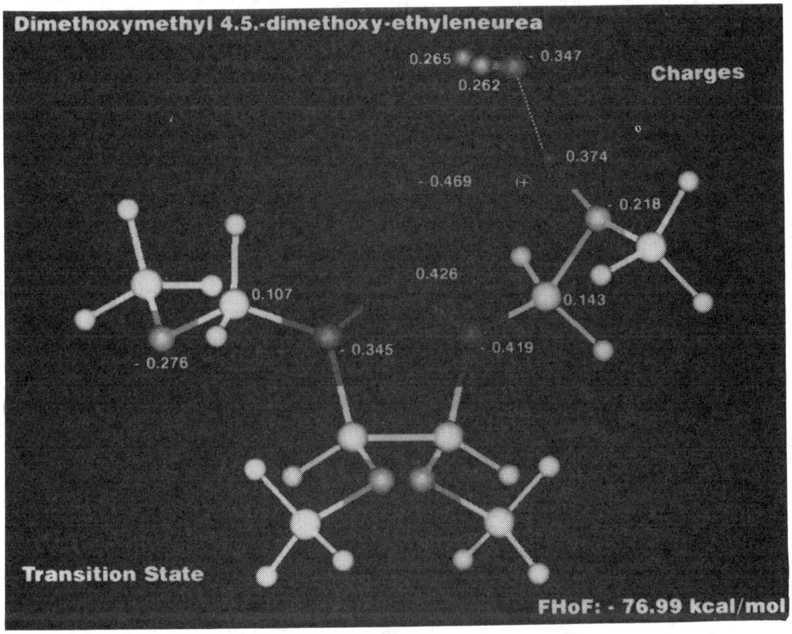

Figure 6b: Change of charges during the reaction of Dimethoxymethyl-4.5-dimethoxy-ethyleneurea

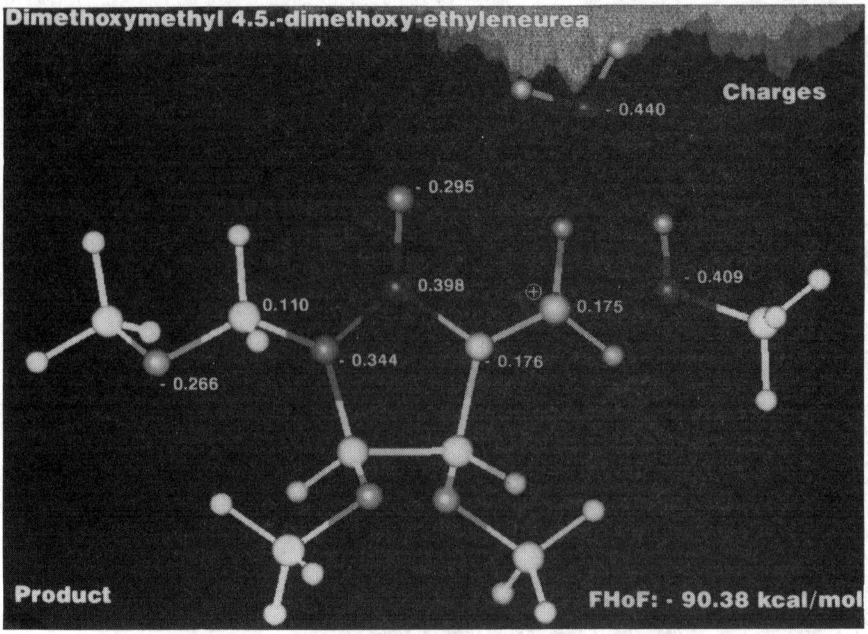

Figure 6c: Change of charges during the reaction of Dimethoxymethyl-4.5-dimethoxy-ethyleneurea

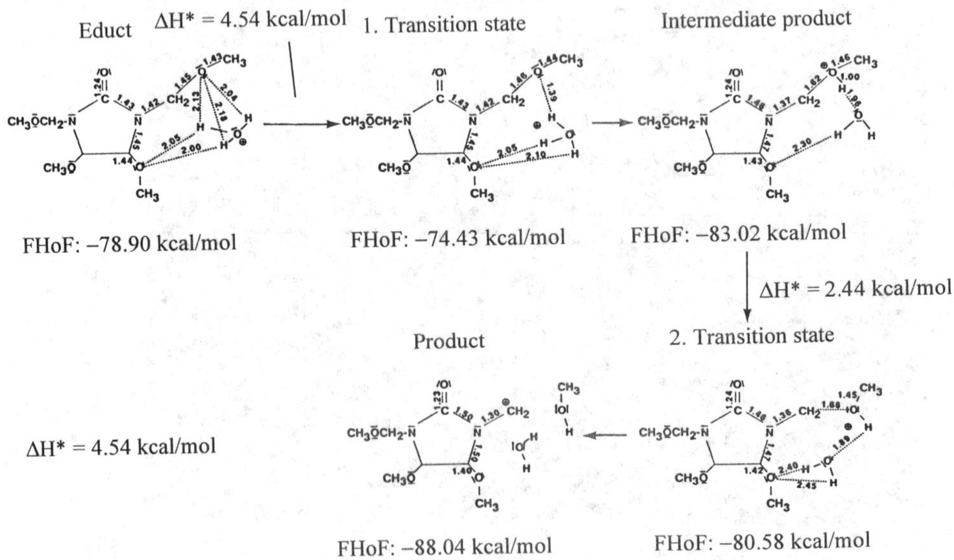

Figure 7: Reaction path of Dimethoxmethyl-4.5-dimethoxy-ethyleneurea

A further possibility for a stable arrangement of H_3O^+ with the dimethoxymethyl-4.5-dimethoxy-ethyleneurea is shown on Figure 7. The H_3O^+ forms hydrogen bonds with the methoxy group in position 4 and the adjacent methoxymethyl group. Here two reaction pathways are possible. In the first - shown in Figure 7 - methanol will be formed from the methoxymethyl group, in the second from the methoxy group (shown in Figure 10).

Starting with the educt it can be seen from Figure 7 that the bond length between the nitrogen atom and the carbon atom of the CH_2-group has a normal value of 1.42 Angstrom. There is no change in the bond length for the first transition state. The intermediate product has a bond length of only 1.37 Angstrom. This short distance is characteristic of a double bond. There is nearly no change in the bond length between the intermediate product and the second transition state. In the last step the [carbonium-immonium]-cation is formed showing a bond length of only 1.30 Angstrom.

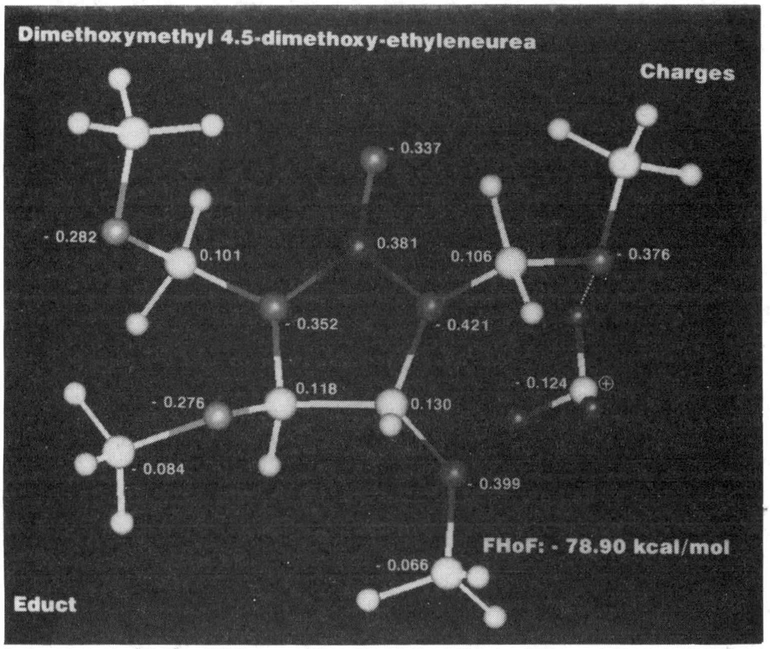

Figure 8a: Change of charges during the reaction of Dimethoxymethyl-4.5-dimethoxy-ethyleneurea

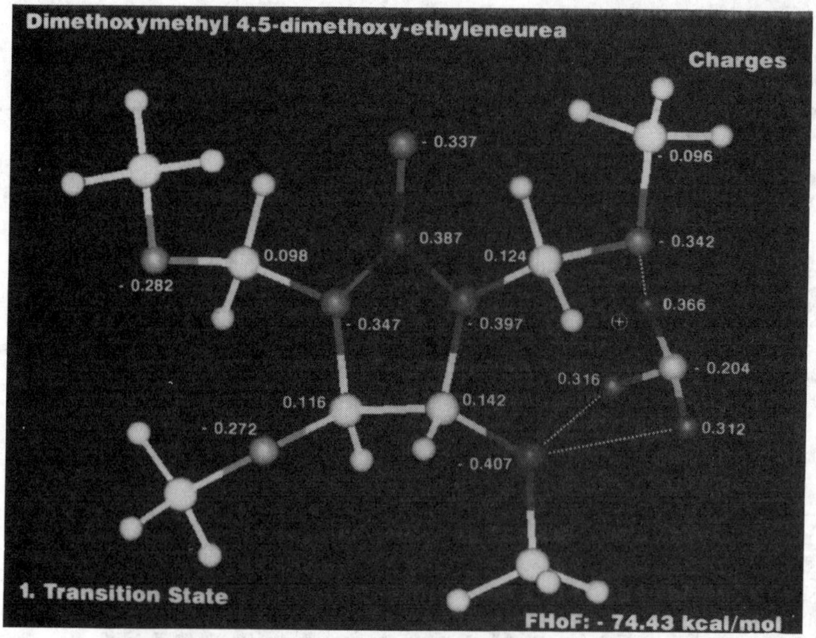

Figure 8b: Change of charges during the reaction of Dimethoxymethyl-4.5-dimethoxy-ethyleneurea

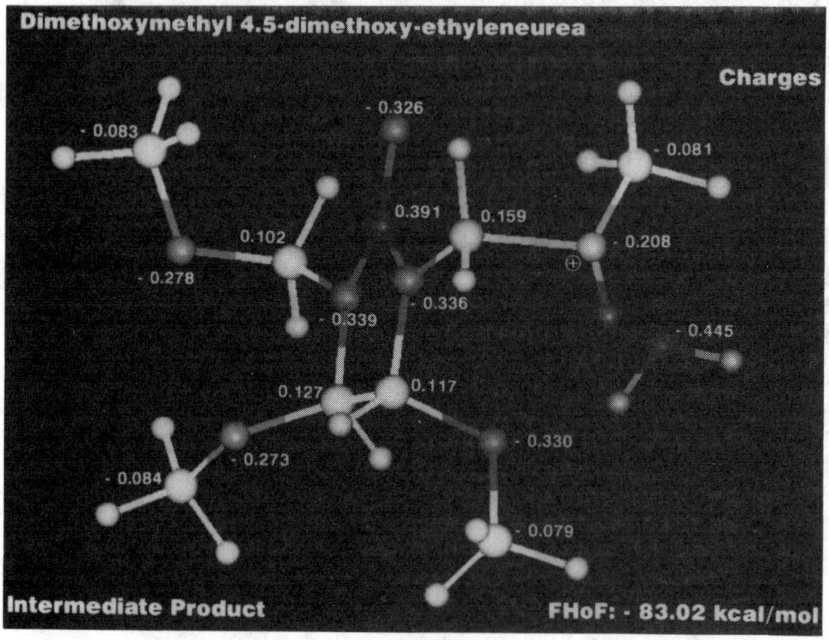

Figure 9a: Change of charges during the reaction of Dimethoxymethyl-4.5-dimethoxy-ethyleneurea

The charges at the carbon atoms of the CH_2-groups of the educt and the first transition state change from 0.106 to 0.124 (Figure 8a - 8b). There is also a change of the charges at the nitrogen atoms of the educt and the first transition state from - 0.421 to - 0.397 (Figure 8a - 8b). The intermediate product and the second transition state show the trend of the charges at the carbon atom of the CH_2-groups to a more positive charge (Figure 9a, 9b). A significant change in the charges occurs at the nitrogen atoms of the second transition state to the [carbonium-immonium]-cation from - 0.345 to - 0.174 (Figure 9b, 9c).

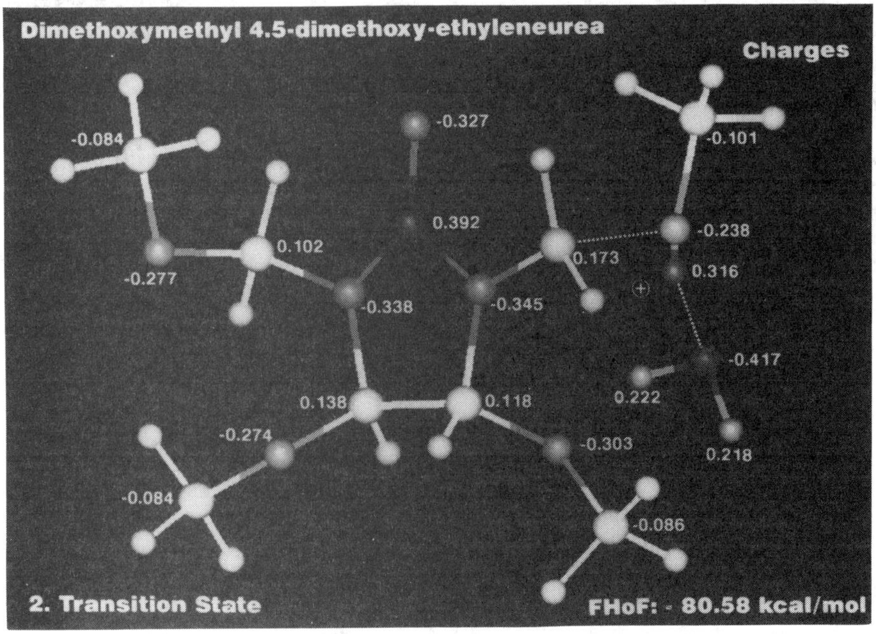

Figure 9b: Change of charges during the reaction of Dimethoxymethyl-4.5-dimethoxy-ethyleneurea

Without the formation of an intermediate there is no change in the bond length between the nitrogen atom and the carbon atom in position 4, but there is a significant change from the transition state to the [carbonium-immonium]-cation in position 4 (Figure 10). The bond length of only 1.32 Angstrom shows double bond character. Between the educt and the transition state there is nearly no change of the charges at the nitrogen atoms and the carbon atom in position 4. But from Figures 11a - 11c it can be seen that there is an obvious change of the charges at the nitrogen atoms and at the carbon atoms in position 4 with a trend to a more positive character. The result is the [carbonium-immonium]-cation in position 4. The activation enthalpy shows a value of 7.61 kcal/mol. Summarizing the results of this reaction path it can be concluded that the activation enthalpy is about 3 - 4 kcal/mol lower in comparison to the reaction path where the arrangement ov H_3O^+ and the methoxymethyl group was formed between the carbonyl and the methoxymethyl group.

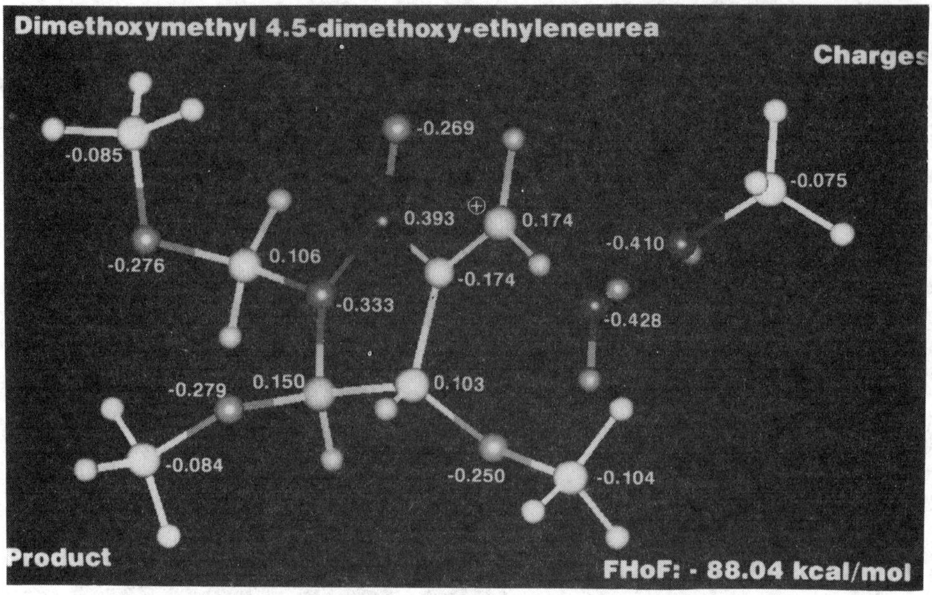

Figure 9c: Change of charges during the reaction of Dimethoxymethyl-4.5-dimethoxy-ethyleneurea

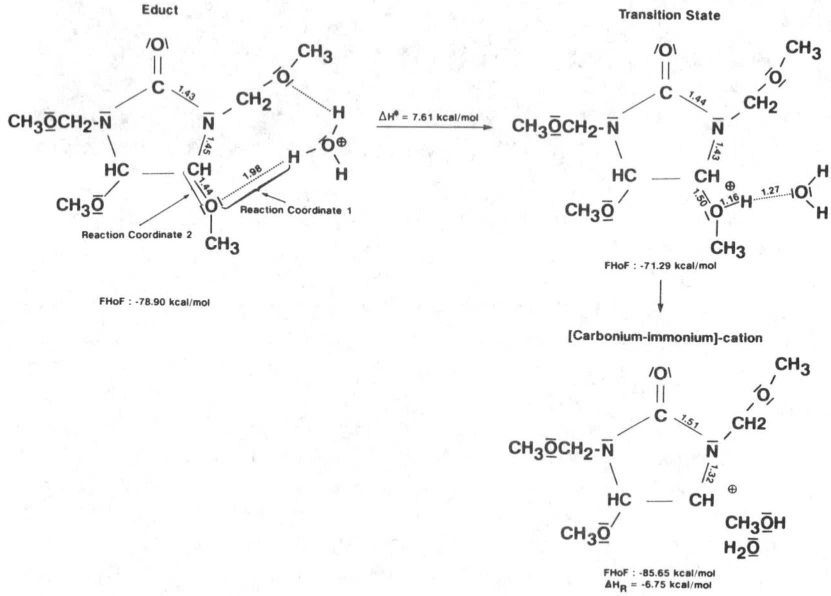

Figure 10: Reaction path for Dimethoxymethyl-4.5-dimethoxy-ethyleneurea

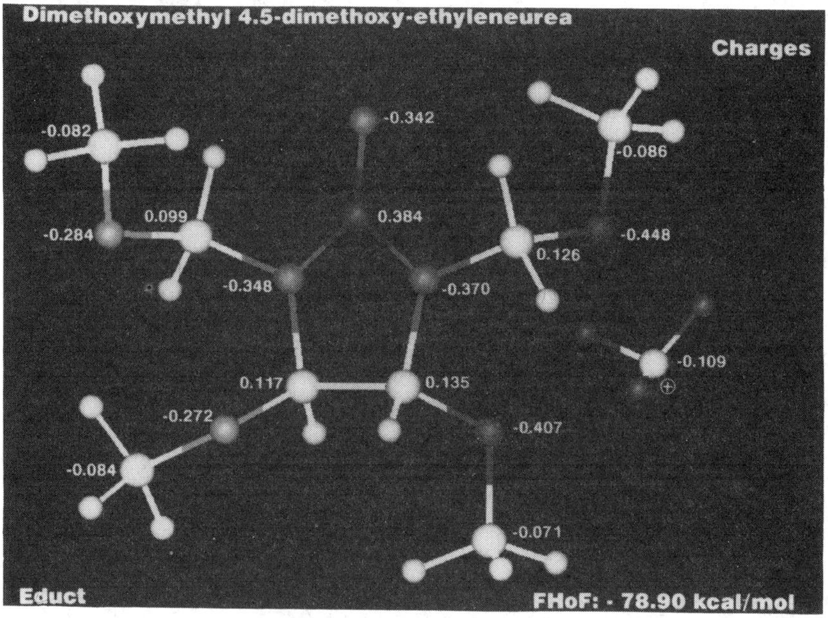

Figure 11a: Change of charges during the reaction of Dimethoxymethyl-4.5-dimethoxy-ethyleneurea

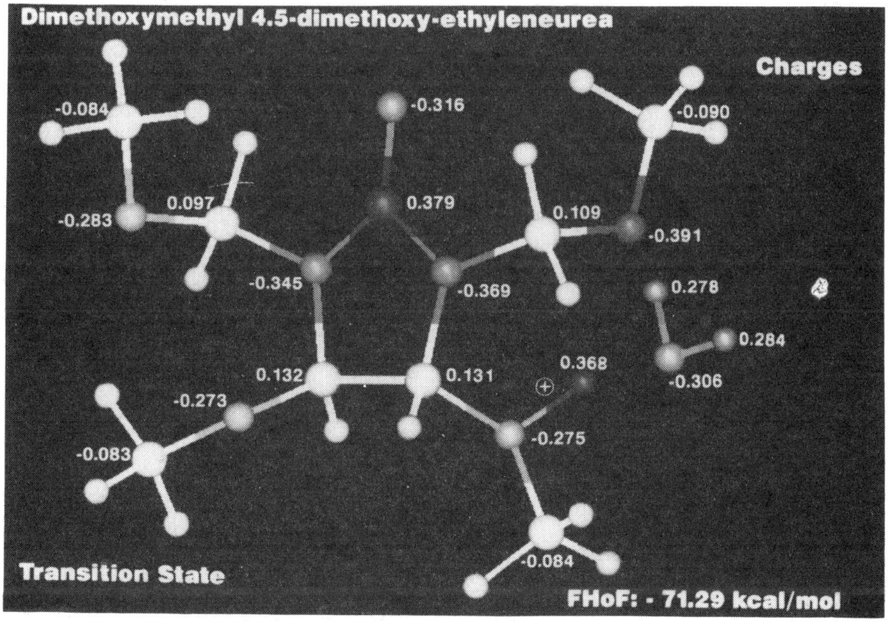

Figure 11b: Change of charges during the reaction of Dimethoxymethyl-4.5-dimethoxy-ethyleneurea

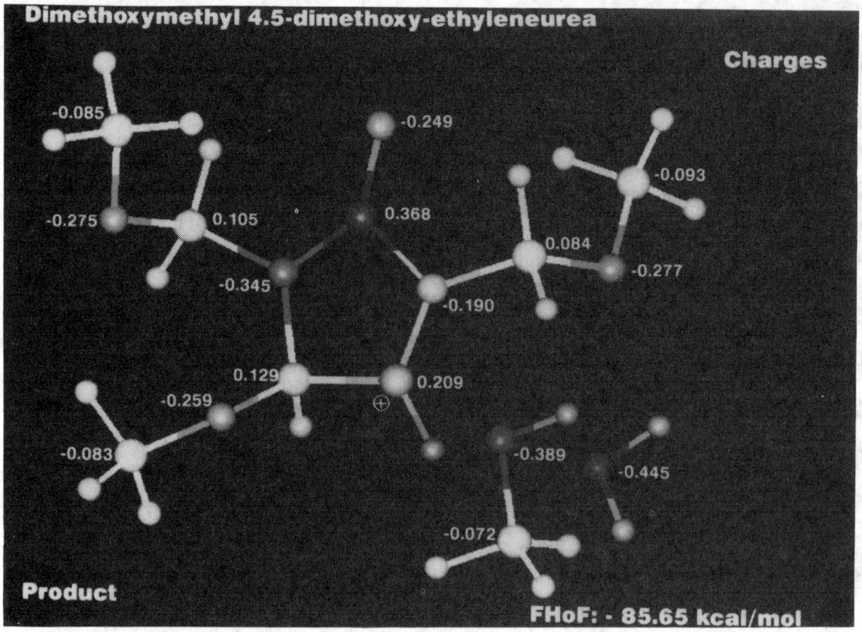

Figure 11c: Change of charges during the reaction of Dimethoxymethyl-4.5-dimethoxy-ethyleneurea

In this connection, it is a matter of interest to see by how much the reactivities of the two different species of reactive groups differ from one another (Figure 12). Summarizing all calculated results of the semiempirical and ab initio quantum chemical methods it can be concluded that the reactivities of the methoxy groups in the positions 4 and 5 are comparable with the methoxymethyl groups in the positions 1 and 3.

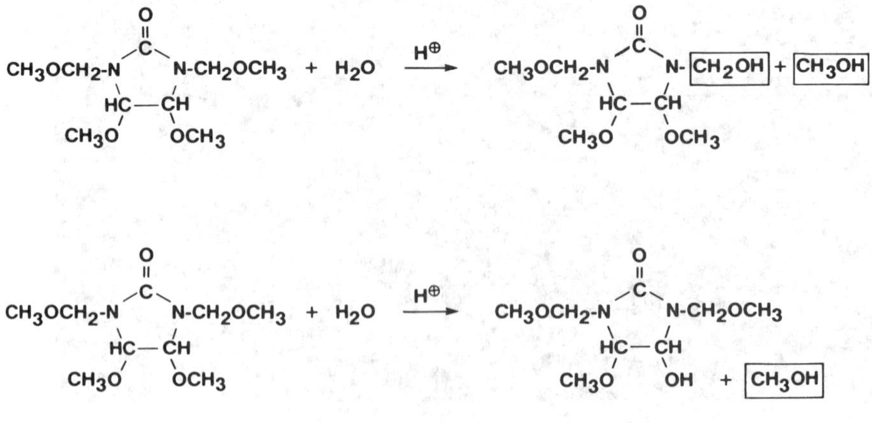

Figure 12: Analytical procedures

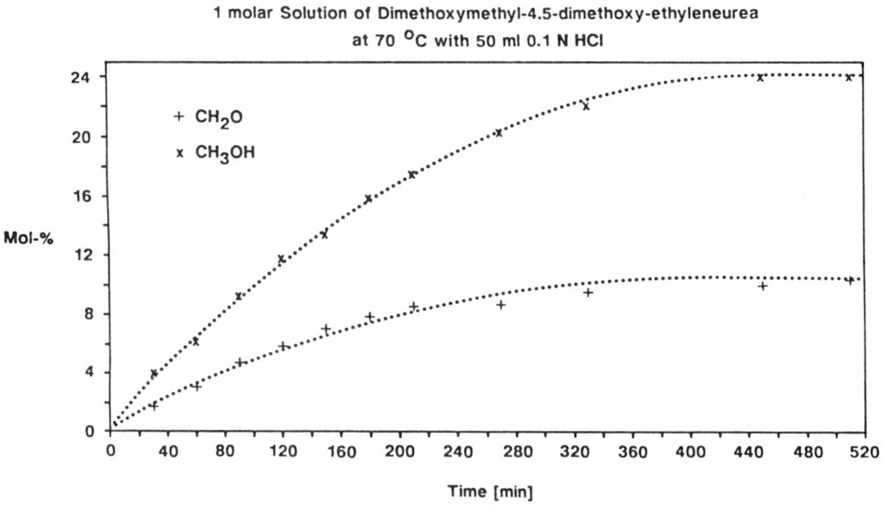

Figure 13: Measurements of reactivities

With special kinetic and thermodynamical measurements it was possible to demonstrate the equality of the reactivities of these two types of reactive groups. The reaction of dimethoxymethyl 4.5-dimethoxy-ethyleneurea with water was chosen as an example of a nucleophilic compound in the presence of an acid. The protonized N-methoxymethyl groups react with water to form hydroxymethyl groups and methanol. The protonized methoxy groups in the positions 4 and 5 react with water to the hydroxy groups and methanol. The course of the reactions could be followed by determining the content of the formed hydroxymethyl groups at various intervals by a potassium cyanide titration method. At the same intervals it is possible to detect the liberated content of methanol by a gas chromatographic method. As an example Figure 13 shows the reaction of a 1 molar solution of dimethoxymethyl-4.5-dimethoxy-ethyleneurea in water at 70° C in the presence of 50 ml 0.1 normal hydrochloric acid. It can be seen that both species of reactive groups react nearly with the same velocity. At every measured interval double the amount of liberated methanol results in comparison to the content of N-hydroxymethyl groups.

References

1) MOPAC, Version 6.0, October 1990, James P. Stewart. Frank J. Seiler Research Laboratory, United Air Force Academy, Colorado, USA.

2) Prof. Ahlrichs, University of Karlsruhe, Germany.

3) H. Petersen, Textilveredlung 5, 437 (1970); Text. Res. J. 41, 239 (1971); Colourage (India) 35 (1971); Chem. Ztg. 95, 692 (1971); H. Petersen in "Handbook of Fiber Science and Technology: Volume II, Chemical Processing of Fibers and Fabrics", pp. 48 - 327, Marcel Dekker, Inc., New York and Basel (1983).

4) J. Baker, J. Comp. Chem. 7, 385 (1986).

5) R. H. Bartels, University of Texas; Center for Numerical Analysis, Report CNA - 44, Austin, Texas (1972).

6) R. Glaser and M. K. Hall, Vibrate: A normal Mode Visualization Program, Department of Chemistry, University of Missouri, Columbia, Missouri 65211 (1991).

44

Improvements in durable press cotton fabrics from esterification crosslinking

B.A. Kottes Andrews, Brenda J. Trask-Morell and Nancy M. Morris -Southern Regional Research Center, Mid South Area, U.S. Department of Agriculture, New Orleans, Louisiana 70179, USA.

ABSTRACT

Esterification-crosslink finishing of cotton fabric for durable press is a technically successful response to promises of a possible ban on formaldehyde in textiles by U. S. regulatory agencies. Certain of the polycarboxylic acid class of agents produce smooth-drying properties equivalent to those of formaldehyde-based counterparts. The polycarboxylic acid of most interest to the textile industry is the tetracarboxylic acid, butanetetracarboxylic acid (BTCA); its expense, however, has hindered its immediate adoption. Another, more available, and more economical, polycarboxylic acid is the trifunctional citric acid; its problems include fabric yellowing and a lower level of smooth drying properties. Options for improvement of these finishes include proper catalyst choice and finishing additives. Concurrent instrumental studies have produced screening methods for crosslink efficiency and fingerprinting of catalyst systems.

INTRODUCTION

United States regulatory agencies have applied increasing pressures to lower formaldehyde in textiles beyond levels feasible with formaldehyde based agents [1]. Although the pool of formaldehyde free agents includes sulfones, acetals, aldehydes other than formaldehyde, and derivatives of glyoxal and ureas, their use is not without attendant problems [2].

At the Southern Regional Research Center we have discovered that polycarboxylic acids, catalyzed by alkali metal salts of phosphorus-containing inorganic acids, will esterify cotton to produce durable press fabrics [3,4,5]. Acids receiving most industry attention are 1,2,3,4-butanetetracarboxylic acid and citric acid which are shown below. This paper traces the developments in the field.

BTCA	CITRIC ACID
H_2C-COOH	
|	
HC-COOH	H_2C-COOH
|	|
HC-COOH	HOC-COOH
|	|
H_2C-COOH	H_2C-COOH

FIRST GENERATION ESTERIFICATION FINISHES

The first generation of smooth drying fabrics from esterification crosslinking was produced by reacting partially neutralized di-, tri-, and tetra-carboxylic acids with cotton. Because the acid salts themselves were acting as catalysts, efficient esterification was hampered and durable press levels of resiliency equivalent to that from formaldehyde-based agents were not achieved [6].

SECOND GENERATION ESTERIFICATION FINISHES

The second generation of smooth drying fabrics from esterification crosslinking was produced by reacting tri- and tetra-carboxylic acids with cotton in the presence of certain alkali metal salts of phosphoric acid. Catalysis by these weak bases supports an anhydride mechanism for crosslinking esterification of cotton [7]. Because of increased acid functionality, tetracarboxylic acids were more successful than trifunctional acids in the durable press finishing process. Use of citric acid produced an additional drawback, noticeable yellowing from high temperature curing.

THIRD GENERATION ESTERIFICATION FINISHES

Third generation catalysis for esterification crosslinking for durable press represents a departure from the 1960's technology. Catalysts from alkali metal salts of phosphinic and hypophosphinic acid formed the basis for new patents in the field [8,9,10]. Infrared spectra of the phosphorus-hydrogen bond, present in phosphite and hypophosphite anions, but not in phosphate anions, provide the means for distinguishing catalysis by alkali metal salts of those acids (phosphites and hypophosphites) from that by alkali metal salts of phosphoric acid (phosphates, pyrophosphates, metaphosphates and polyphosphates) [11].

Infrared absorbance by the phosphorus-hydrogen bond is at 2340 cm^{-1} and can be observed in sodium phosphite and sodium hypophosphite in Figure 1. There is no absorbance in this region by phosphate salts.

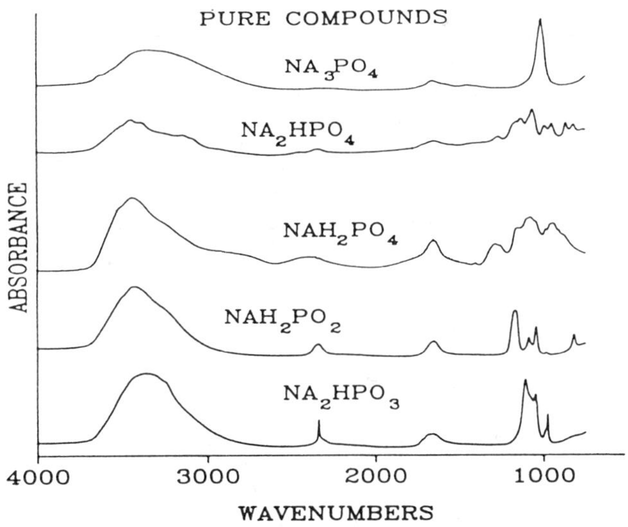

Figure 1. Infrared spectra of sodium phosphate, phosphite and hypophosphite salts [11].

Of the catalysts based on P^{+++}, hypophosphite salts proved most effective (Table I). In fact, they are superior to any other currently used catalyst for esterification crosslinking [4,12,13].

Table I. Phosphite and hypophosphite catalysis of esterification crosslinking.

Acid	Catalyst	Durable Press Rating	Whiteness (CIE)
BTCA	$Na_2HPO_3 \cdot 5H_2O$	3.5	78
	$NaH_2PO_3 \cdot 2.5H_2O$	4.0	56
	$NaH_2PO_2 \cdot H_2O$	4.5	78
Citric	$Na_2HPO_3 \cdot 5H_2O$	3.0	63
	$NaH_2PO_3 \cdot 2.5H_2O$	3.5	0
	$NaH_2PO_2 \cdot H_2O$	4.0	62
Untreated		1.0	83

Figure 2 shows the use of thermogravimetric analysis to compare catalyst efficiency among disodium hydrogen phosphate, monosodium hydrogen phosphate, and sodium hypophosphite. The catalyst which left the highest residue after heating and exhibited the slowest rate of weight loss also produced the highest level of resiliency in fabric treatments. For BTCA, this catalyst was sodium hypophosphite. Such use of thermogravimetric analysis can be used to predict catalyst efficiency [14].

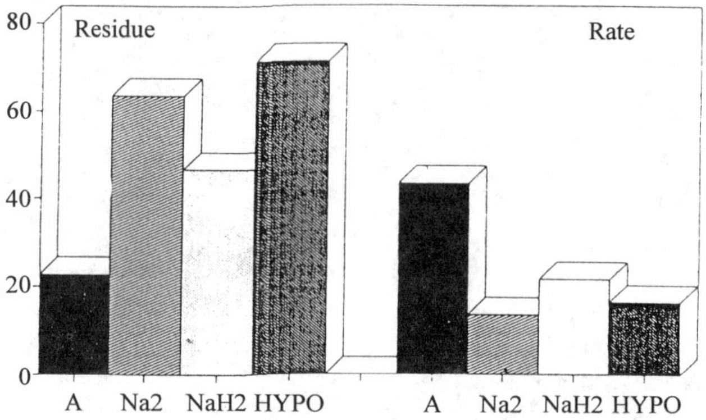

Figure 2. Comparison of % residue and rate of weight loss after controlled heating in thermogravimetric analysis of BTCA, alone (A), and in the presence of Na_2HPO_4 (Na2), NaH_2PO_4 (NaH2), and NaH_2PO_2 (HYPO) [14].

In addition, it was found that sodium hypophosphite is the only catalyst to date that produces durable press level cottons from citric acid crosslinking. The reducing capability of sodium hypophosphite improves the whiteness of the citric acid finish [5].

CAUSE OF YELLOWING IN CITRIC ACID-FINISHED FABRIC

We suspected that yellowing in citric acid-finished fabrics was due to double bond formation during the high temperatures of curing. This hypothesis was confirmed by a $KMnO_4$ spot test (Table II).

WHITENESS IMPROVEMENT IN CITRIC ACID-FINISHED FABRIC

We found that capping of the -OH group in citric acid with the tertiary amine, triethanolamine [$N(CH_2CH_2OH)_3$], protects the -OH group from abstraction and prevents formation of aconitic acid and finished fabric yellowing in treatments with citric acid [15]. When used in the salt form with 6.9% citric acid and sodium hypophosphite catalyst, triethanolamine produced finished fabrics with whiteness equivalent to that from formaldehyde-based finishing even at the lowest of three levels of catalyst.

Table II. Cause of yellowing in citric acid finished fabric.

Finish	Color change with 5% $KMnO_4$ after drying	after curing
Aconitic acid	+	+
Citric acid	-	+
BTCA	-	-

DYEABILITY OF POLYCARBOXYLIC ACID-FINISHED FABRICS

As a result of an anhydride mechanism and less than total efficiency of polycarboxylic acid reactivity with cellulose, there are unreacted polycarboxylic acid groups present in the finished fabric. These unreacted groups provide sites for attachment of basic dyes [16] so that dyeability is imparted by polycarboxylic acid in the finish. In addition, fabrics finished with a polycarboxylic acid/triethanolamine system can be after-dyed with direct, acid and reactive dyes, expanding the classes of dyes amenable to the process [17]. Thus, durable press fabrics finished by esterification crosslinking can be after-dyed even at the garment stage, allowing the short inventories of "just in time" or "quick response" marketing.

SUMMARY

In the 1960's, research by Gagliardi, et al., [6] and Rowland, et al. [7], unlocked the secret of practical esterification crosslinking of cotton and cellulosics for increased fabric resiliency. Unfortunately, the catalysts used at that time did not produce durable press level finished fabrics.

In the late 1980's, the textile industry developed a renewed interest in non-formaldehyde finishing because formaldehyde had been labeled a probable human carcinogen. Responding to this need, researchers at the Southern Regional Research Center, ARS, USDA, developed finishing systems for durable press based on polycarboxylic acids and catalysis by alkali metal salts of phosphorus-containing inorganic acids. The polycarboxylic acids of most interest to the textile industry today are butane tetracarboxylic acid (BTCA) and citric acid. The most successful catalyst is sodium hypophosphite. Problems with yellowing of citric acid-finished fabrics can be eliminated by a use of a combination of sodium hypophosphite catalysis and a triethanolamine salt additive.

Assessment of catalyst efficiency and finish performance has been accomplished by instrumental analysis of the finished fabrics. Predictor factors have been developed for acid/catalyst systems.

REFERENCES

1. United States Occupational Safety and Health Administration, Occupational Exposure to Formaldehyde, Final Rule. *Federal Register* **1987**, Vol. 52, No. 233, December 4, 1987, 46168.
2.Andrews, B.A. Kottes *Colourage Annual*, in press.
3.Welch, C,M.; and Andrews, B.A. Kottes *Textile Chem. Color* **1989**, 21(2), 13.
4.Andrews, B.A. Kottes; Welch, C.M.; and Trask-Morrell, B.J. *Amer. Dyestuff Reptr.* **1989**, 78(6), 15.
5.Andrews, B.A. Kottes *Textile Chem. Color* **1990**, 22(9), 63.
6.Gagliardi, D.D.; and Shippee, F.B. *Amer. Dyestuff Reptr.* **1963**, 52, P300.
7.Rowland, S.P.; Welch, C.M.; and Brannan, M.A.F. *U.S. Patent 3,526,048*, 1970.
8.Welch, C.M.; and Andrews, B.K. *U. S. Patent 4,820,307*, **1989**.
9.Welch, C.M.; and Andrews, B.K. *U. S. Patent 4,936,865*, **1990**.
10.Welch, C.M.; and Andrews, B.K. *U. S. Patent 4,975,209*, **1990**.
11.Morris, N.M.; Andrews, B.A. Kottes; and Donaldson, D.J. *Appl. Spectroscopy*, in press.
12.Brodmann, G.L. *Textile Chem. Color* **1990**, 22(11), 13.
13.Brotherton, D. L.; Fung, K.W.; and Addison, A.L. *Book of Pap., AATCC 1989 Intl. Conf. & Exhib.* **1989**, 170.
14.Trask-Morrell, B.J.; and Andrews, B.A. Kottes *Textile Res. J.*, in press.
15.Andrews, B.A. Kottes; Blanchard, E.J.; and Reinhardt, R.M. *Book of Abstracts, Southeast/Southwest Regional American Chemical Society Meeting* **1990**, Abstract No. 171.
16.Andrews, B.A. Kottes; Blanchard, E.J.; and Reinhardt, R.M.; *Amer. Dyestuff Reptr.* **1990**, 79(9), 48.
17.Blanchard, E.J.; Reinhardt, R.M.; and Andrews, B.A. Kottes *Textile Chem. Color.* **1991**, 23(5), 25.

45

Polymerization-crosslinking of cotton fabrics using maleic acid-containing systems

H. Choi - Louisiana Agricultural Experiment Station, Louisiana State University Agricultural Center, Baton Rouge, Louisiana 70803, USA.

ABSTRACT

Cotton fabrics were treated by aqueous baths containing carboxylic monomers, such as maleic acid, itaconic acid, and methacrylic acid, along with a free radical initiator and a catalyst for the non-formaldehyde PC treatment. The system of one-to-one mole ratio of maleic acid and itaconic acid substantially increased wrinkle resistance properties with an improved retention of mechanical properties of the treated fabrics when compared with the fabric treated by DMDHEU. The increase in Stoll flex abrasion resistance was most striking.

INTRODUCTION

In durable press finishing of cotton fabric a conventional resin, such as dimethyloldihydroxyethyleneurea (DMDHEU), released formaldehyde during the process, storage, and consumer use, and resulted in a significant loss of mechanical properties of the treated fabrics. Recent studies in the SRRC of USDA examined polycarboxylic acids such as butanetetracarboxylic acid and citric acid to solve the formaldehyde release problems [1,7-9]. Even though a high level of resiliency and stability of the finishes to machine laundering were obtained, these finishes still caused a considerable decrease in strength and abrasion resistance of the treated fabrics.

A polymerization-crosslinking (PC) treatment has been examined to alleviate the loss of strength and abrasion resistance in the durable press finished cotton fabrics and fibers [2,4,6]. Better retention in mechanical properties was obtained with the deconvolution of cotton fiber through the deposition of polymerizing resins within the fiber interior and subsequently with the longer, flexible crosslinks of cellulose through the functional groups in the resins [4,6].

In the present study, we examined the feasibility of certain monomer systems being used to develop non-formaldehyde durable press finishes of a cotton fabric while maintaining optimal mechanical strength. These monomers contain an unsaturated double bond for vinyl addition polymerization and carboxyl groups for the crosslinking reaction with cellulose through esterification.

EXPERIMENTAL

Desized, scoured, and bleached 100% cotton twill fabric supplied by Cotton, Inc. was used throughout this study. Maleic acid (MA), itaconic acid (IA), methacrylic acid (MAA), 2-hydroxyethyl methacrylate (HMA), N-vinyl-2-pyrrolidone (NVP), and poly(N-vinyl-2-pyrrolidone) with molecular weight of 40,000 (PNP) were purchased from Aldrich Chemical Company. Potassium persulfate for a free radical initiator and various catalysts for esterification reactions were all obtained from commercial sources. DMDHEU (Aerotex 900, 48% solid content) and a polyethylene emulsion softener (Cyanalube TSI) were obtained from American Cyanamide, Inc. Triton X-100 was used as a wetting agent. All chemicals used were reagent grade and used without further purification.

A conventional pad-dry-cure process [1] was used to treat cotton fabrics with a wet-pick-up of the fabric between 80-95% on weight of fabric. Whiteness index (WI) was determined before washing according to ASTM E313-67 using the Colorgard System 1000 Colorimeter from Pacific Scientific Company. The standard methods used for evaluation of performance properties of the treated fabric were as follows: durable press (DP) rating by AATCC 124-1982, wrinkle recovery angle (WRA) by ASTM D-1296-67, bending stiffness by ASTM D-1388-55, Elmendorf tear strength by ASTM D-1424-82, breaking strength by ASTM-1682-64, Stoll flex abrasion resistance by ASTM-1175-55T.

With the exception of WRA, all tests were in the warp direction.

TABLE I. Experimental results for various comonomer systems.[a]

Monomers (mole ratio)	Add -on[d] %	DP[e]	WRA[f] (°) w+f	TS[g] %	BS[h] %	AR[i] %	BL[j] cm
Mock-treated[b]	-	1.5	189	109	87	362	3.2
DMDHEU-treated[c]	4.0	4.5	278	55	45	18	3.0
MA/NVP (2:1)	7.7	3.8	229	73	63	156	2.7
MA/PNP (2:1)	7.5	4.0	240	103	58	229	2.4
MA/HMA (2:1)	8.6	3.8	233	86	61	241	2.5
MA/MAA (3:1)	8.2	3.5	229	75	65	150	2.8
MA/MAA (2:1)	7.1	3.7	241	93	90	188	2.9
MA/MAA (1:1)	5.3	3.0	207	95	67	169	3.1
MA/MAA (1:2)	3.9	3.0	203	80	69	138	3.2
MA/IA (2:1)	9.6	4.0	254	78	56	196	2.5
MA/IA (1:1)	11.6	4.4	268	76	49	214	2.7
MA/IA (1:2)	10.8	4.0	258	68	67	131	2.7
MA/IA/MAA(1:2:1)	9.5	3.7	251	87	62	188	2.8

[a]Treating bath-12% monomers, 1.5% owm $K_2S_2O_8$, $NaH_2PO_2 \cdot H_2O$ (1:1 mole ratio between monomer and catalyst), 1% softener, and 0.1% Triton X-100. Dry-100°C/10 minutes. Cure-180°C/2 minutes. Wash and tumble dry. [b]Treating bath-1% softener and 0.1% Triton X-100. Dry-100°C/10 minutes. Cure-180°C/2 minutes. Wash and tumble dry. [c]Treating bath-12% DMDHEU, 1.5% $Zn(NO_3)_2 \cdot 6H_2O$, 1% softener, and 0.1% Triton X-100. Dry-100°C/5 minutes. Cure-165°C/3 minutes. Wash and tumble dry. [d]Weight add-on(%) [e]DP rating [f]Wrinkle recovery angle [g]Tear strength retention [h]Breaking strength retention [i]Stoll flex abrasion retention [j]Bending length

VARIOUS MONOMER SYSTEMS

Since homopolymerization of MA is generally difficult to achieve, MA was copolymerized with other comonomers such as IA, MAA, HMA, and NVP as shown in Table I. The pre-formed polymer, PNP, was also used to enhance polymerization of MA by forming a polymer complex through hydrogen bonding [5].

The system of one-to-one mole ratio of MA and IA resulted in DP rating and WRA comparable with those treated by DMDHEU. At similar levels of resilience, retention in mechanical properties except breaking strength was much higher in the fabric treated by the PC technique with MA and IA. Striking effects were shown in the increase of Stoll flex abrasion resistances, which were 30% to 140% higher than the untreated fabric. Conversely, DMDHEU treated fabric resulted in only 18% of abrasion resistance of the untreated sample.

In the mock-treated sample the application of polyethylene softener alone increased abrasion resistance to more than 350% of that of the untreated control. Considering that the same amounts of softener were used throughout the treatments, the discrepancy in Stoll flex abrasion resistance was extremely considerable. These results, therefore, tend to suggest that the PC treatment of cotton fabric resulted in better retention of strength and abrasion resistances through the deconvolution of cotton fibers and the longer, flexible crosslinks.

Both MA and IA are cheap dicarboxylic monomers capable of undergoing an esterification reaction with cotton cellulose. It has been well known that both MA [7] and IA [3] lost water molecules and subsequently formed anhydrides at elevated temperatures. These anhydride intermediates underwent esterification reactions with hydroxyl groups in cellulose molecules.

On the other hand, the systems containing MA and other comonomers, such as MAA, HMA, NVP, and PNP, provided some improvement in DP rating and WRA, but the improvement was generally not comparable with that of the DMDHEU treated fabric. Substantial increases were, however, shown in the retention of tear strength and Stoll flex abrasion resistance of the treated fabric. Exceptionally high tear strength and abrasion resistance were obtained in the fabric treated by MA in the presence of pre-formed polymer (PNP). Further study would be needed to find optimal condition for this system.

MONOMER CONCENTRATION

The aqueous baths containing different concentrations of MA and IA at the same mole ratio (1:1) were used to treat the cotton fabrics as listed in Table II. Except at 6.2% monomer concentration, MA and IA treatments significantly improved DP rating and WRA comparable with those of the DMDHEU treated fabric with much higher retention in tear strength and Stoll flex abrasion resistance. Durable press properties tended to be increased with increase in monomer concentration. The results suggested that some degradation of the treated fabric occurred with 15.6% monomer concentrations due to the excess amounts of acids. Nevertheless, retention of the Stoll flex abrasion resistance was still higher than that of the DMDHEU treated sample. The variation in monomer concentration did not affect significantly on whiteness and stiffness of the treated fabric.

TABLE II. Effects of the PC Treatment on 100% Cotton Twill (MA/IA, 1:1)[a]

Monomer %owb Curing°C/min	Add-on %	DP	WRA (°) w+f	TS %	BS 5 %	AR %	BL cm	WI[b]
6.2%	5.2	3.7	242	85	62	221	2.7	46.7
8.9%	8.4	4.0	271	83	59	148	2.5	47.0
12% (160/3.5)	11.4	4.2	272	62	58	143	2.5	44.7
12% (170/2)	11.3	4.0	273	70	60	158	2.5	49.7
12% (180/2)	11.6	4.4	268	76	49	214	2.7	49.0
12% (190/2)	11.3	4.4	283	51	58	133	2.4	30.3
15.6%	13.5	4.5	290	46	54	81	2.5	45.5

[a]Treating bath-monomers, 1.5% owm $K_2S_2O_8$, 8.8% $NaH_2PO_2 \cdot H_2O$, 1% polyethylene emulsion softener, and 0.1% Triton X-100. Dry-100°C/10 minutes. Cure-180°C/2 minutes, unless otherwise specified. [b]Before washing.

CURING CONDITIONS

As presented in Table II, at curing temperatures of 160-190°C, the system containing MA and IA (1:1) resulted in good DP rating and wrinkle resistance properties. In addition, these samples resulted in better retention of the mechanical properties. As before, the increase in tear strength and Stoll flex abrasion resistance were substantial. The whiteness indices for the DMDHEU treated fabric and the untreated fabric were 48.3 and 56.3, respectively. Extended curing time (3.5 minutes) at 160°C slightly decreased the whiteness index, while the treatment at a high curing temperature (190°C) caused a considerable decrease of whiteness index of the treated cotton fabric before washing.

ACKNOWLEDGMENT

The author wishes to acknowledge financial support from Louisiana Agricultural Experiment Station, Louisiana State University Agricultural Center.

REFERENCES

1. Andrews, B.A.K.; Welch, C.M.; Trask-Morrell, B.J. *Am. Dyes. Rep.* **1989,** 36(6), 15.

2. Choi, H.; Yeh, K. *Cell. Chem. Tech.* in press.
3. Marvel, C.S.; Shepherd, T.H. *J. Org. Chem.* **1959,** 24, 599.
4. Rowland, S.P.; Stark, S.M.; Cirino, V.O.; Mason, J.S. *Textile Res. J.* **1971**, 41, 57.
5. Sato, T.; Nemoto, K.; Mori, S.; Otsu, T. *J. Macromol. Sci.-Chem.*, **1979,** A13(6), 751.
6. Shin, Y.; Hollies, N.R.S.; Yeh, K. *Textile Res. J.* **1989,** 59, 635.
7. Trask-Morrell, B.J.; Andrews, B.A.K.; Graves, E.E. *Textile Chem. Col.* **1990,** 22(10), 23.
8. Welch, C.M. *Textile Res. J.* **1988,** 58, 480.
9. Welch, C.M.; Andrews, B.A.K. *Textile Chem. Col.* **1989,** 21(2), 13.

PART 3

46

Life cycle analysis: a marketing nightmare

James Benda* and Ramani Narayan*, ** - *Chemical Engineering, Michigan State University, East Lansing, MI 48824 USA. **Michigan Biotechnology Institute, 3900 Collins Rd., Lansing MI, 48910 USA.

ABSTRACT

In order to develop new technologies, modify existing ones, or redesign products to meet new environmental criteria, one will have to do cradle-to-grave analyses of the technology. These types of analyses are holistic environmental and energy audits (accounting procedures) that focus on the entire life cycle of a product from raw material acquisition to final product disposal, rather than on one single manufacturing step or environmental emission. In this paper, we discuss the potential of life cycle analyses to develop new environmentally compatible products and processes. Specifically, we will address issues relating to disposables vs. reusables controversy based on Life Cycle Analyses (LCA). We will show that LCA cannot and should not be used to certify that a product's environmental attributes are superior to another product.

INTRODUCTION

In order to develop new technologies, modify existing ones, or redesign products to meet the new environmental criteria, Life Cycle Analysis (LCA) or Cradle to Grave Analysis of the technology has to be performed. LCA is a holistic environmental and energy audit (accounting procedure) that focuses on the entire life cycle of a product from raw material acquisition to final product disposal, rather than on a single manufacturing step or environmental emission. In today's rush by industry to make changes in product design that would enable their products to meet requirements for their ultimate disposability or recyclability, LCA assumes great importance. For example, in order to reduce the volume of waste (municipal solid waste - MSW) a scheme could be designed to collect and recycle a particular product or the product could be redesigned to biodegrade in an appropriate waste management infrastructure. However, many times, this change results in the overall process consuming more energy or producing wastes/emissions in some other step. The net result of the change towards recycling or biodegradation may do more harm than good to the environment. LCA helps quantify all these effects and permits making an informed decision on the proposed change. It would also help in identifying significant areas for improvement in energy usage and waste reduction throughout the entire process.

LCA is a tool to quantify inputs and outputs of an industrial manufacturing system. The key is in defining the system boundary and the assumptions that go in to defining the system boundary. Much work has been done by Franklin Associates in developing the research protocol and underlying assumptions in LCA [1]. Figure 1 is a generalized materials flow diagram for a complete Life Cycle Analysis.

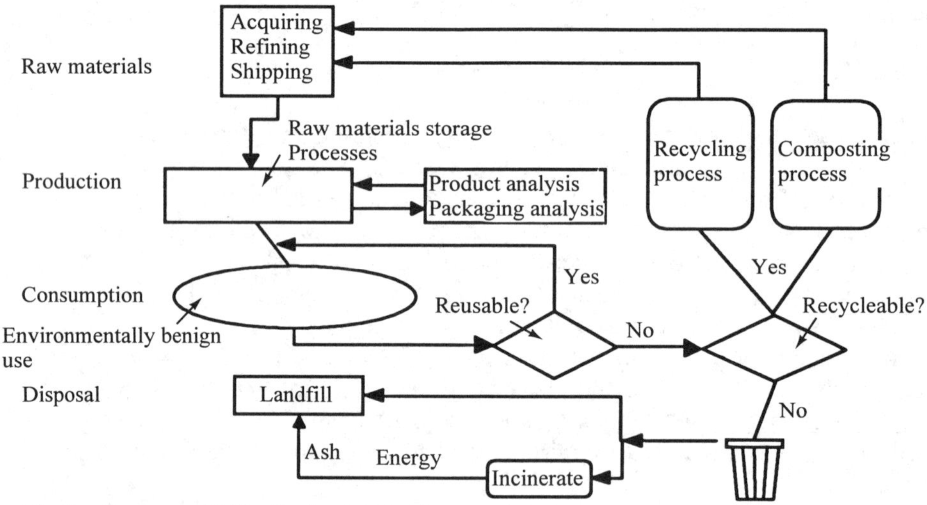

Fig 1: Stages of Life Cycle Analysis.

The analysis consists of five major steps:

1. Raw Materials Acquisition
2. Product Manufacturing / Packaging Manufacturing
3. Consumer Use
4. Recycle, Compost
5. Final Disposal.

"Since all processes produce wastes, the characteristics of the wastes and their impacts on the environment must be one of the basis for comparison of alternate products and processes." [2] Life Cycle Analysis should only be used for process improvement and product development, but not for product comparisons as has been done in the case of paper vs. polystyrene [3]. For example, two seemingly contradictory reports on reusable vs. disposable diapers were issued [2,4]. Table 1 summarizes the data from the two reports.

Table 1: Environmental Impacts of Diaper Usage (Arthur D. Little and Carl Lehrburger)
(ADL numbers compiled by Arthur D. Little and Associates and were per 85 diaper changes but have been normalized to per single diaper use; CL numbers are from Carl Lehrburger et.al. and were per 1000 diaper changes, but have been normalized also).

	Reusable Diapers		Disposable Diapers	
	ADL	CL	ADL	CL
Raw Materials				
*Renewable	0.0047 lbs	0.0046 lbs	0.254 lbs	0.216 lbs
*Non-renewable	0.038 lbs			
Water Consumption	14.05 lbs		2.3 lbs	
	1.69 gal	3.779 gal	0.278 gal	5.986 gal
Energy	(Btu)		(Btu)	
*Renewable sources	175.2		43.8	
*Non-ren. sources	752.9		230.2	
Total	928.1	2,030.6	274.0	3,455.5
Atmospheric Emissions	(*10^3)	(*10^3)	(*10^3)	(*10^3)
* Particulate Matter	3.0 lbs	0.45 lbs	0.035 lbs	1.28 lbs
* Nitrogen Oxides	1.8 lbs	1.32 lbs	0.071 lbs	1.18 lbs
* Sulfur Oxides	3.8 lbs	2.29 lbs	0.082 lbs	2.29 lbs
* Carbon Monoxide	0.4 lbs	0.81 lbs	0.094 lbs	2.76 lbs
* Hydrocarbons	1.2 lbs	0.74 lbs	0.080 lbs	1.01 lbs
Total	10.2 lbs	5.61 lbs	1.082 lbs	8.82 lbs
Waste Water Effluents	(*10^3)	(*10^3)	(*10^3)	(*10^3)
* Total Sus. Solids	0.152 lbs	1.796 lbs	0.082 lbs	1.939 lbs
* Chemical Oxy. Dmd.	0.047 lbs	4.226 lbs	----------	1.227 lbs
* Biological Oxy. Dmd.	0.141 lbs	1.887 lbs	0.035 lbs	1.5 lbs
Process Solid Waste	0.0369 lbs	0.0040 lbs	0.024 lbs	0.014 lbs
Post Consumer Waste	0.0028 lbs	0.055 lbs	0.261 lbs	0.428 lbs

Direct comparison of results show extreme contradictions, even though both research teams evaluated the same products, reusable and disposable diapers:

1) A 300% Energy difference-
   - CL: single-use diapers use 70+% more energy
   - ADL: reusables use 230% more energy
2) An 800% Water-usage difference (per diaper change basis)-

CL: single-use diapers use greater volumes of water:
reusable, ~3.8 gal, disposable, ~6.0 gal
ADL: reusable, ~1.78 gal, disposable, ~0.27 gal

3) A 14% Materials difference
CL: reusables use 72% fewer raw materials
ADL: reusables use 86% fewer raw materials

Thus, different studies of the same subject may arrive at contradictory conclusions as a result of varying assumptions and depending on the boundary conditions set for LCA. The results are not always black and white[5]. As stated earlier, the key is in defining the system boundary and the assumptions that go into this boundary. There is thus a need for an analyzing technique broad based enough to allow applicability to a wide spectrum of products within their Life Cycle boundary. The analyzing tool needs to judge the environmental merits/demerits of a product or process. This tool would be capable of drawing together reports such as those used in the above example and provide practical, sound environmental product and process information.

An Expert System has such capability of being able to make judgments based upon referencing a large knowledge base. We are developing an Artificial Intelligence (AI) system that can act as a consultant to a product developer which will allow him/her to make decisions and more easily adjust the design of a particular product depending on its eventual environmental merits.

EXPERT SYSTEMS

Expert Systems (ES) make intelligent artifacts capable of solving tough problems, process design, fabrication control, plant trouble shooting, etc. Today, ES is used only in a limited domain, that is, a specific area of application in the 'real-world' where a great deal of knowledge is available. In the engineering domains, these systems solve problems that would otherwise be solved by human 'experts' (hence, "Expert Systems") by drawing upon past experiences and 'back of the envelope calculations.' Thus, these systems must contain expertise that mimic the performance, skill and robustness of a human expert [6].

For these engineering problems, 'exact' answers are mostly unattainable, and when found are frequently 'over exact', i.e. too specific to be useful. ES try to work their way around this problem by using heuristic 'rules of thumb' which an expert would have applied in the same situation, and thus provide an accurate answer to tough problems. This is especially useful where exact answers are difficult or impossible to obtain, such as in LCA, due to various factors such as random inputs. LCA acts as a consultant to the analyzer, and for the case of Life Cycle Analysis, ES can aid in decision making of product or process design.

It is generally agreed upon that such a knowledge based system has four main components:

1) a knowledge base (to store domain specific knowledge)

2) an inference engine (to make decisions based upon the knowledge base)

3) a knowledge acquisition module (a module to facilitate the retrieval of knowledge required to adequately describe a particular domain)

4) an explanatory interface (this makes the expert system tangible to the user; the reasoning path can be explained if necessary).

If information (knowledge) needs to be input into the system, it must be provided in a 'computer-understandable' manner. The Expert System must also be able to provide the reasoning as to the development of a particular solution, such that the person solving the problem will trust the answer. This is what separates ES from conventional algorithms. Expert Systems contain the ability to explain how an answer was deduced. The development and use of such an Expert System is shown in Fig. 2.

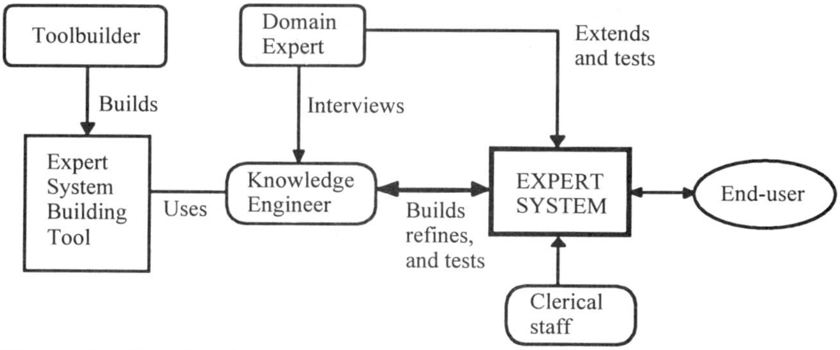

Figure 2. The development and use of an Expert System (form ref 7).

The basic elements of an expert system as described by Rich [8] are:

1) these systems derive their power from a great deal of domain specific knowledge, rather than a single powerful technique

2) in successful systems, the required knowledge is about a particular area and is well defined.

3) usually derived with the aid of one or more experts.

4) the transfer of knowledge takes place gradually through many interactions between the expert and the system.

5) the amount of knowledge required depends on the task.

Expert Systems are used for problem solving, which is a search through potential solutions, guided by heuristic rules. These point to a destination without exhausting every possible avenue, and eliminates the study of undesirable paths.

LCA determines environmental impacts of a product and its processes. Unbridled, this leads to a complicated system of solutions. However, by breaking the solution search into the five parts of a products life cycle (as mentioned in the introduction) one can then deal with each of these separately. Four of the five sections are potentially quantifiable. These are: Raw materials handling, Production, Recycling and Composting, and Disposal and can all be described as shown by Tummala and Koenig [9] using an Enterprise Model. Such models leave out the product use. In the diaper example above, the largest differences resulted in the use of the product, not the production or disposal. The use of an Expert System would help address this issue and provide a better quantitative model.

CONCLUSIONS

In the near future, an environmental audit will be required for every product and process, if not due to government regulation then due to the markets (consumers) demand. Life Cycle Analysis can provide the essential information necessary to improve each product and process environmentally and economically. However, it should not be used to compare the environmental merits or demerits of competing products. The characteristics of an Expert System can be exploited to provide a tool to perform such analyses.

REFERENCES

1) Hunt, R., Sellers, J., Franklin, W.; Franklin Associates, LTD.

2) Lehrburger, C., Mullen, J., Jones, C.V.; "Diapers: Environmental Impacts and Life Cycle Analysis." January 1991

3) Hocking, M.B., *Paper versus Polystyrene: a Complex Choice*, Science, vol. 251, 1 Feb. 1991 pages 504-505

4) Arthur D. Little, "Disposable versus Reusable Diapers: Health, Environmental and Economic Comparisons." March 1990

5) Stipp, D., Wall Street Journal "Life Cycle Analysis Measures Greenness, but Results May not be Black and White", Feb. 28 1991

6) American Laboratory, *An Information System for selection methods of Chemical Analysis*, March 1987, page 18

7) Waterman, D.A., A Guide to Expert Systems, Addison Wesley Publishing Co., Reading, Mass., 1986 page 7

8) Rich E., Knight, K.; Artificial Intelligence, 2nd ed., McGraw Hill, Inc. 1991

9) Tummala, R., Lal, Koenig, B., "Classes of Process Models and Their Characteristics Useful for the Design of Manufacturing Systems" Presented to the IEEE International Conference on Systems, Man and Cybernetics, October 1991

47

The environmental aspects of solvent-spun cellulose production and use

C.R. Woodings - Courtaulds Research, Coventry, England.

ABSTRACT

This paper provides a qualitative consideration of the life-cycle of solvent-spun cellulose fibres form cradle-to-grave. Forestry, pulping, solvent production, energy requirements, fossil reserves, effluent, and disposal issues are all considered. Data on the relative rates of biodegradation of solvent spun cellulose relative to cotton, rayon and the synthetic fibres is presented.

INTRODUCTION

The fibre industry, defined in cradle-grave terms to include such sectors as polymer making, textiles and nonwovens, and their disposal methods, provide some of the basic necessities of 20th century life. Our responsibility is to try and minimise the environmental impact of the processes by which such products are manufactured.

To do this we need to make comparisons, using the best available science, of the environmental impacts of the routes to fibrous products. We must also remember that all products made form our fibres will ultimately require disposal, and that it is our duty to ensure that the disposal methods available are environmentally sound.

Such cradle-to-grave comparisons are of course complex, difficult to carry out, and beyond the scope of this paper. Here we indicate some of the environmental factors which come into play when investing in a new fibre which could become one of the major raw materials for the textiles industry in the coming decades.

THE SOLVENT SPUN CELLULOSE PROCESS

Our new fibre making process (Figs 1&2) has one major raw material, the cellulose polymer, and one minor raw material, the amine oxide solvent. It also requires energy. If markets evolve along traditional lines, then some versions of the fibre will contain titanium dioxide dulling agents, and some will be bleached. All are likely to be finished with the surfactants needed by the subsequent conversion processes. Perhaps we should stress that the non-cellulosic components and bleaching options arise because the market demands them, not because they are an essential part of the process.

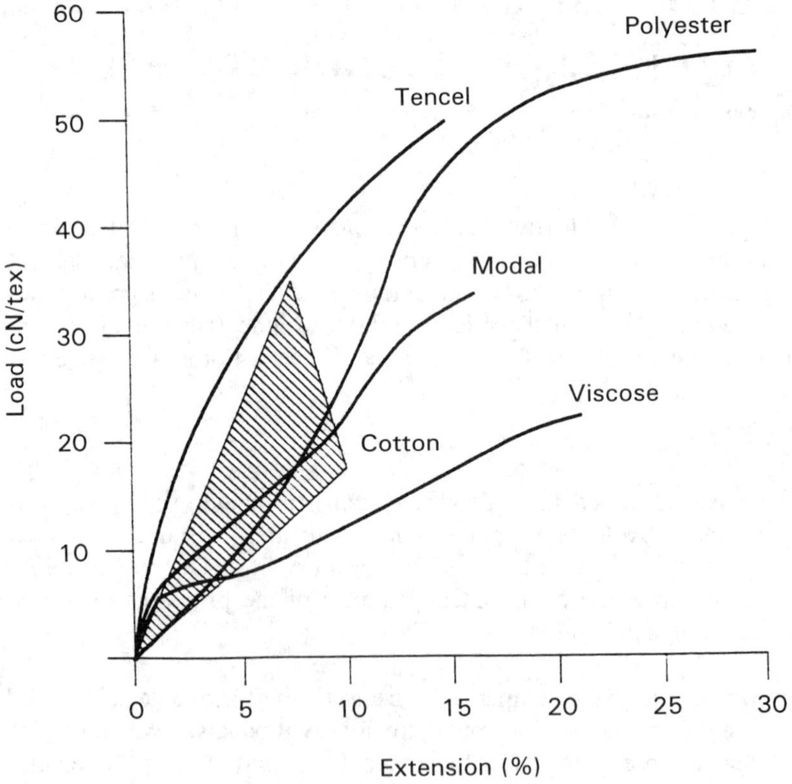

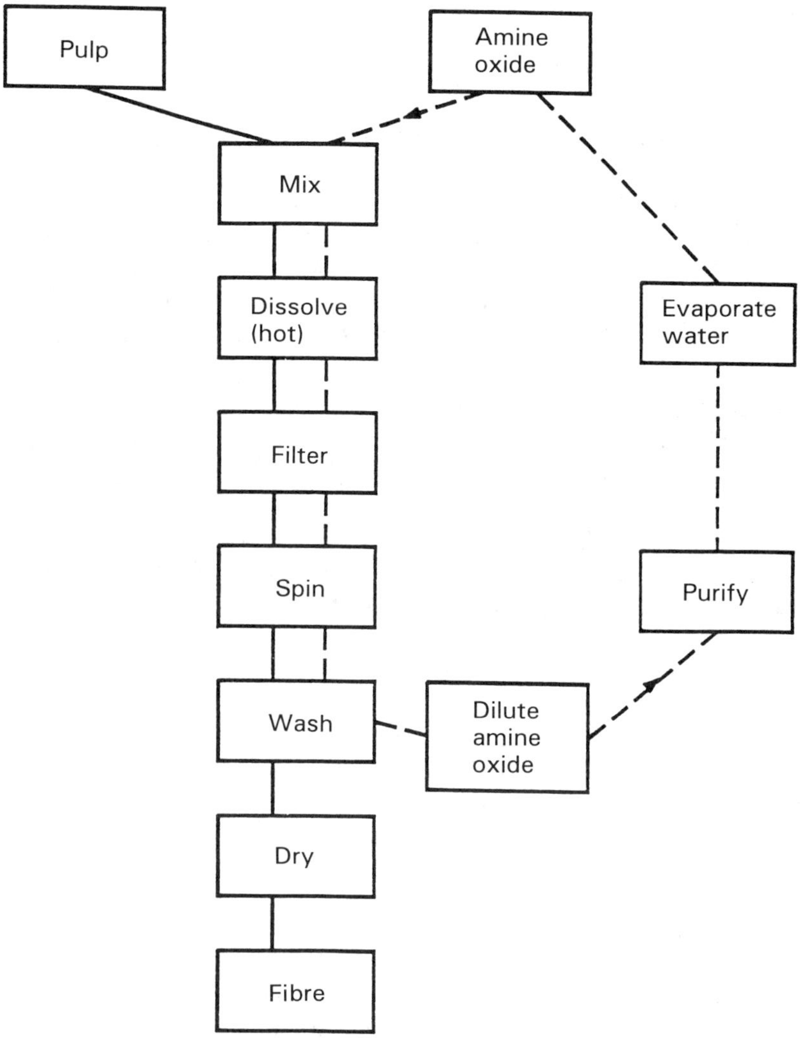

Cellulose

Cellulose is the natural polymer which makes up the living cells of all vegetation. It is the material at the centre of the carbon cycle, and the most abundant and renewable biopolymer on the planet. Rayon fibre producers have converted it from the fine short fibres which come from trees into the fine long fibres used by textiles and nonwovens for almost a century. Rayon nevertheless remains unique among the mass produced man-made fibres because it is the only one to use the natural polymer directly.

Polyesters, nylons, polyolefins, and acrylics all come indirectly from vegetation. They come from the polymerisation of monomers obtained from fossil fuels, which in turn are formed by the incomplete biodegradation of vegetation which grew millions of years ago.

The industrial grade of cellulose used to make our rayon comes from tree-farms, where specially chosen species are grown from sapling to maturity in 7-10 years. New trees grow from the stumps of the cut trees, and this happens on marginal land, generally unsuitable for food crops and without the intensive use of fertilisers or pesticides. The best farms yield in excess of 2.5 tonnes of pure cellulose per acre per year. For comparison, cotton growing at its most intensive yields about 0.35 tonnes/acre and needs good soil.

Cutting down trees is popularly regarded as an unfriendly activity, and it is therefore quite important to put the usage of trees as a raw material in the correct perspective. It has been estimated[1,2] that:

* 100 billion tonnes of vegetation grow and decay annually on land. This represents about 12% of the planets total production of vegetation, the majority being produced in the oceans.

* 12% of this land-based vegetation is in the form of wood (trees).

* Of this 12 billion tonnes of wood, a maximum of 3 billion tonnes is removed by man. Half of this is burnt, either as fuel or to clear land for agriculture. The other half is used by Industry. (Compare this with 6 billion tonnes of fossil reserves "mined" each year.)

* Of the 1.5 billion tonnes of wood used by industry, half becomes timber in saw mills, and half is used raw.

* Of the 0.75 billion tonnes used raw, half goes into construction (pit-props, telegraph poles etc) and half is converted into pulp and chipboard.

* Of this 0.375 billion tonnes, 0.29 billion tonnes of wood become wood-pulp for the paper, board, fibre, film and chemicals industry.

* A significant proportion of this 290,000,000 tonnes of pulpmill feedstock (up to 40% in some areas) comes from forest thinnings, and saw mill waste and 6% from non-pulp sources such as straw, bagasse, hemp and cotton. This feedstock yields 161,000,000 tonnes of pulp.

* About 4.5 million tonnes of this pulp output are a high quality dissolving grade for forming into fibres, films, water soluble polymers and

chemicals. Dissolving grade pulp is perhaps better described as industrial grade cellulose polymer, and should be considered alongside the polyester or nylon polymer beads which are the feedstocks of the synthetic fibre plants.

* Viscose rayon manufacture consumes 2,600,000 tonnes of this cellulose, with 2 million tonnes going into the staple fibre process.

These details are summarised in Figs 3-6.

Global Cellulose 1988

Fig 3: Plant Matter - Million Tonnes

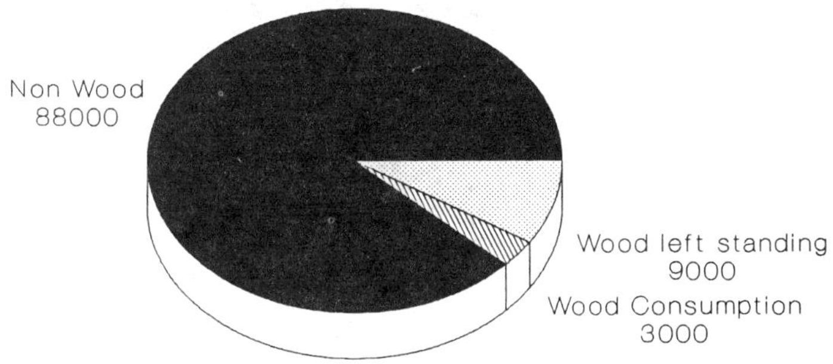

Global Cellulose 1988

Fig 4. Wood Consumption (Million Tonnes)

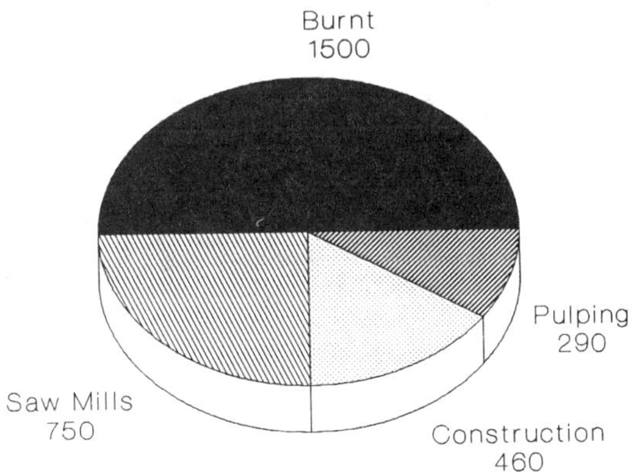

The rayon plant at Mobile currently uses about 100,000 tonnes/year of the industrial grade cellulose to make viscose rayon fibres. When the first solvent-spun plant is on stream, an additional 18,000 tonnes/year will be required.

Global Cellulose 1988

Fig 5. Pulp Use (Million Tonnes)

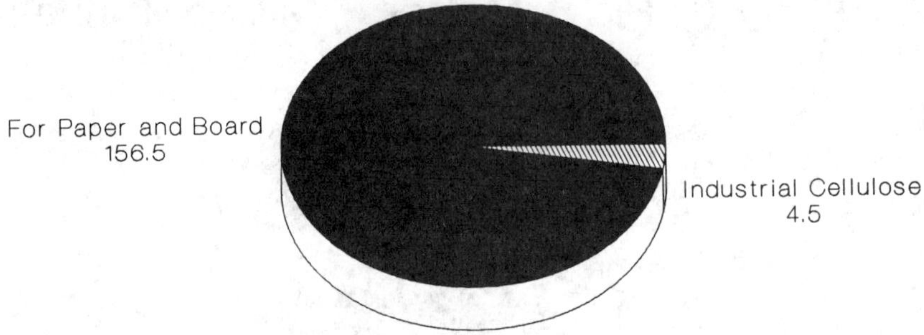

Global Cellulose (1988)

Fig 6. Industrial Cellulose (Mn Tonnes)

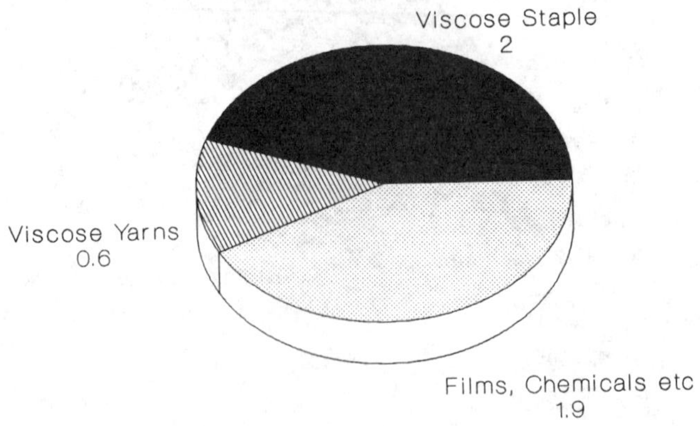

Cellulose Extraction (Pulping)

From the above figures, it can be seen that the 2.6 million tonnes of dissolving grade pulp currently manufactured to feed the rayon fibre industry represents 0.01% of the annual production of cellulose, on land, in nature, and about 0.7% of the cellulose in wood used by industry.

The solvent spun process as currently designed will use the same sources of pulp (at slightly higher levels of efficiency) as the viscose process, but concerns related to pulp mill effluents are still with us. However I think we can by now conclude that concerns over dioxins in the pulp itself, and products made from the pulp are now behind us. Our raw material and final products, both viscose and solvent-process, have been shown by independent analysis to be free of such compounds at a detection level of 0.5 parts per trillion.

With regard to the pulp mill effluent issue, our major suppliers are undertaking programmes of work to eliminate elemental chlorine at the pulp bleaching stage, and these changes should be complete before the new fibre becomes commercial in the second half of 1992. Alternative pulping sequences which eliminate all chlorine compounds from the process are being investigated for use in both viscose and solvent systems.

Amine Oxide Solvent

N-methylmorpholine N-oxide (NMMO) is the solvent used. It is manufactured by methylation and oxidation of morpholine, which comes from a reaction between diethylene glycol and ammonia.

Whilst this is the only major chemical used in the solvent spun process, its consumption is reduced to the absolute minimum by the recycling which is made possible by solvent recovery. In the plant at Grimsby, which has been operating semi-commercially for the last 3 years, we have developed techniques which now recycle virtually all of the solvent used to dissolve the pulp.

Strong NMMO solution as delivered to our process has been subjected to a series of acute mammalian toxicological evaluations with conclusion that the material would not be considered harmful by oral, dermal or ocular ingestion.

In-vitro genotoxicity studies were also carried out on dilute solutions of the NMMO solvent. The test samples were inactive in Cell transformation, Mouse lymphoma forward mutation, Primary rat hepatocyte/DNA repair, and Optimised Ames assays.

Energy

The usage of energy and the means by which it is obtained contributes a major component of the environmental impact of most complex industrial process sequences.

The methodology of assessing the energy usage of products and processes is currently the subject of much debate, and a standardised approach has yet to emerge. Not surprisingly, most of the published work on fibres was carried out during the last energy crises in the 1973-81 period, and we could find nothing in the public domain from more recent studies.

We are aware of several attempts[3-7] to assess the total energy required to make baled staple fibre from naturally occuring raw materials, wood in the case of cellulosics and oil in the case of synthetics. In general they break the fibre production sequence into monomer making, polymer making and fibre production, and while a variety of fibres are covered, only viscose rayon and polyester are mentioned in all of them. The overall picture that emerges from these early studies was that whilst the wet-spun cellulosic fibres required more energy than melt spun polyester for the fibre making step, they had no monomer energy requirement, and the "polymerisation" requirement was minimal. In the case of the very low values for rayon emerging from Lenzing and CIRFS, we think full credit was being given for the fact that the pulp mills energy needs were in fact renewable and not dependent on fossil fuels, and that pulp could be fed directly into the viscose process without incurring any transport or drying cost. In other words, the pulp mill could be driven by energy obtained from burning the parts of the tree which were not needed in the final product, and this "free" and renewable energy was not counted.

From an energy viewpoint, the solvent route to cellulosic fibres is identical to the viscose route up to the point where the cellulose enters the solvent. The energy requirement for the non-cellulosic raw materials is significantly lower in the case of the solvent route, but the solvent route will require similar energy levels in dope handling, spinning, washing and recycling. The lower water imbibition of the solvent fibre (65% versus 95%) will yield savings in fibre drying and of course in any subsequent washing and drying operations.

Overall, the solvent route will show a useful economy in this important resource when compared with viscose production on the same scale.

Fossil Reserves

Renewable resources will become increasingly important as the planet's stocks of fossilised reserves are depleted and as governments realise that biomass can provide a truly sustainable, cost-effective source of energy and materials.

The viscose route currently needs fossil reserves for energy generation but for little else. At our Mobile plant the vast majority of energy requirements come from locally available natural gas.

The solvent used in the new process is made from ethylene glycol which currently comes via ethylene from oil refineries. However, as indicated above, the recycling rate is so high that solvent usage is kept down to a few kilos per tonne of fibre, and hence fossil reserve dependence is minimal.

Gaseous Effluents

The solvent process involves direct dissolution of cellulose in a liquid which is recycled very efficiently. There are no chemical reactions and no by-products of the sort which are unavoidable in the regeneration of cellulose from the viscose route.

In the viscose process, gaseous effluent control and treatment is a fundamentally important part of the overall process and is continuously improving as the technology of the "closed-box" process evolves. The air handling and cleaning systems employed are costly and most of the emissions to atmosphere are collected and discharged through tall stacks.

The solvent process produces very little atmospheric emission. There are traces of volatile organic compounds associated with the solvent and the soft finish which will leave the plant in the normal course of ventilation. There is no need for any central air handling or emissions stack.

Liquid Effluents

The spinning and washing liquors from the viscose process are recycled to allow reuse of the sulphuric acid and zinc sulphate components wherever this is feasible from economic and environmental standpoints. Nevertheless, in common with most industrial washing and bleaching systems, large volumes of process water have to be cleaned on-site before discharge to river. As is the case with gaseous effluents, most rayon producers stay well ahead of the regulatory requirements and this means continuously working towards improved plant designs.

The solvent route uses much less water overall, and the process effluent needs significantly less treatment.

Disposability/Recycling

Cellulosic fibres are, as stated at the start of this paper, simply a tiny subset of the most abundant bio-polymer on the planet.

Like natural vegetation, they can become food for micro-organisms and higher life forms (they biodegrade) and they will burn with a rather greater yield of energy than natural vegetation.

In complete biodegradation or incineration, the final breakdown products are carbon dioxide and water, and so in the overall sense these disposal methods simply recycle the cellulose to the atmospheric components from which it was made.

It is also possible to liberate and use some of the "free" solar energy which powered the polymerisation step. In the case of incineration this is straightforward in that the free-burning cellulose can be used to generate steam etc. In the case of landfill disposal, it is now well known that slow anaerobic biodegradation occurs in all landfill sites dealing with municipal solid waste. This process generates methane from cellulose, which can, and increasingly is, being used to drive gas-turbines directly. Admittedly, this process makes only a small contribution to reducing the volume of waste in the landfill, but as fuel costs rise, this "free" and renewable energy source will become more important. If landfills are designed from the start to be anaerobic bioreactors, i.e. lined and operated with moisture addition and leachate recycling, then energy generation and the return of land to normal use can be accelerated.[9]

Tests show that the anaerobic degradation process is so fast that disposables made from cellulosic fibres are likely to disappear, yielding their energy content, in the sludge digestion process used in sewage farms.

CONCLUSION

Rayon fibres, made for a century by the direct conversion of abundant vegetable matter, have always had much to recommend them in textiles compared with synthetics made from fossil fuels. The renewability of their main raw material, their overall energy efficiency, their lack of dependence on fossil fuels, their long history of safe use in hygiene applications, and their easy disposal and natural recyclability make them strong contenders for tomorrow's textile industry also.

The new solvent route to rayon reinforces these inherent strengths by using a modern fibre production system which, being physical rather than chemical, reduces environmental impacts to a minimum. The Tencel investment, coupled with the continuous improvement of the traditional route, gives us what we believe is a winning approach to textile industry fibre supply for some time to come. The new plant is due to be on-stream at Mobile Alabama in mid 1992.

References:

1 "Eco-profiling of cellulose-based products", August 1990. (A study prepared for Courtaulds and others by Envirocell - 256 references).

2 John Emsley (Kings College London) - "Plant a tree for Chemistry", published in New Scientist, 8th October 1987.

3 Woodhead; ICI; International TNO Conference; 1976.

4 Lane and McCombes; Courtaulds; Textile Manufacturer No 1; 1979.

5 Kogler; Lenzing; EDANA AGM; Munich 1980.

6 Armstrong; Consultant; EDANA AGM; Munich 1980.

7 Marini and Six; EDANA Nonwovens Symposium Milan, 1985.

8 Unpublished Data; Interantional Committee for Rayon and Synthetic Fibres; 1982

9 Pohland and Cross, "Controlled Landfill Management - Principles and Applications"; Insight 91, Charleston, Oct 91.

48

Photodegradation of cellulosic cottons by environmental conditions

F.M. Tera and K.E. Shady - National Institute for Standards, Dokki, Giza, Egypt.

ABSTRACT

The present work deals with the photodegradation of cellulosic-cotton caused by monthly direct exposure to the environmental conditions at an urban industrial site in Egypt for one year. The deteriorations produced were assessed by different methods. Spectrophotometric measurements were used to follow the change in brightness and colour differences. Also, scanning electron microscopy was used to follow the change in surface morphology. Moreover, changes in the carboxyl end-groups, copper number and the pH values were also evaluated and discussed. Finally, the deterioration in mechanical properties expressed as losses in tensile strength and elongation % were evaluated. The results obtained were discussed in relation to the prevailing environmental conditions.

INTRODUCTION

The term "weather" refers to a complex system of many components for which there can be no actual standards. The features of weather most responsible for photodegradation[1-2] are light particularly of near UV of the range (290-315 nm), moisture in liquid or vapour form, temperature, oxygen, ozone and pollutants[3,4]. Weathering of high polymers causes breakdown of molecular structure which in turn results in loss of strength, extensibility, general durability, discolouration, fading and possibly appearance[4-6]. Direct photolysis of macromolecules such as those of cellulose[7] will occur only when light of sufficient high energy content is absorbed. This necessarily has to be capable of breaking a chemical bond in the molecule and the energy required will depend upon the nature of the bonds and the atoms involved. Considerable efforts have been made to establish whether the presence of photosensitizer is in fact necessary to promote the degradation of cellulose by light.

EXPERIMENTAL

Two cellulosic unfinished cotton fabrics of different counts and densities namely lino and duck supplied by El-Shorbagy and Misr El-Mahala Companies were used,and table (1) shows their specifications.

Table (1)

Samples	Colour	Nominal Wt.	Thread/ cm.		Linear density (Tex)	
		g/m2	Warp	Weft	Warp	Weft
Lino	Cream	90	34	34	35	35
Duck	Cream	480	20	15	100	100

Monthly, outdoor, unprotected exposure was performed for each fabric at an urban industrial site at Cairo, Egypt. The exposure was done according to the standard method[2] and started on the first of January and lasted until 31st December 1990 for set (A). Another set(B) was exposed gradually from one to 12 months.

Environmental data were recorded continuously at the nearest meteorological station to the exposure site with the cooperation of the Egyptian Meteorological Authority. The samples examined were removed simultaneously after exposure, and were conditioned in a room kept at 23°C and 65% R.H. for 24 hours before testing. Those samples which were used for colour measurement and surface morphology were purified extensively prior to testing by an ultrasonic cleaner (Branson). Reflection spectra was recorded for the exposed and the control fabric samples in the visible range (300-700nm) by a Shimadzu Double Beam UV-300 Spectrophotometer using MgO as a reference. The tristimulus values (X,Y and Z), also the brightness (L) and the colour difference values (ΔE) were evaluated according to the C.I.E. System[8].

A Joel 100S scanning electron microscope with a resolving power of 50 A° and magnification of 4000 times was used to study surface changes produced by outdoor weathering for the control samples and those exposed for one year. The samples were attached to a metal stub, then coated with gold layer of about 200 A° thick using a vacuum coating device, Edward Scattering unit S150.

Control and the monthly exposed samples of 5 cm width and 20 cm length were tested for tensile strength and elongation by Shimadzu Universal Testing Machine S500, (C.R.T.), according to the standard method of testing[2], twenty samples for each fabric were taken. The pH values of aqueous extracts of both monthly exposed and unexposed samples were measured according to the standard method of testing[4], using an Orion pH meter. Ten samples of each fabric were tested. Carbonyl and carboxy contents were determined according to the standard method[9], for the control samples and those exposed for 4,10, and 12 months.

RESULTS AND DISCUSSION

The monthly incident solar radiation energy (I.S.E) evaluated according to the World Radiation Reference in (MJ/m2) are depicted in Fig.(1) curve (a) where a cyclic seasonal variation is observed having a maximum in June and July, and a minimum in January and December. Also, the monthly levels of UV radiation[3] in (mw min/cm2) at the range (290-315nm), were estimated from the previous data and are represented in Fig.(1) curve (b).

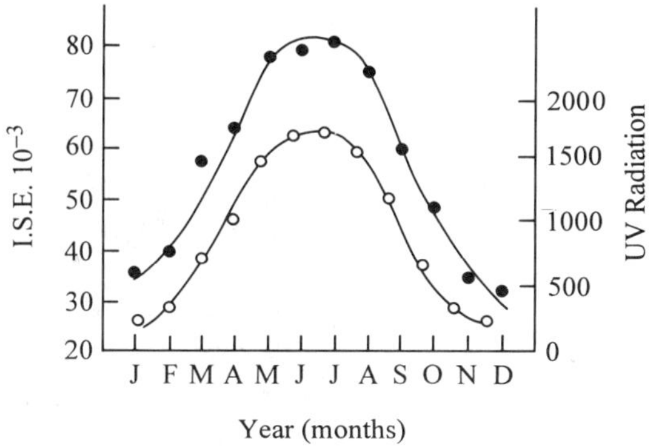

Figure 1. (a) The incidence solar energy (Mega Joule/m2),(b) The UV radiation (mw.min.cm2) versus the year months.

A similar variation is obtained with a maximum in summer time and minimum in winter. Also, a similar cyclic variation was obtained when the mean of maximum temperature or the sun hours per day were represented against the months of the year. Comparing the values of the (I.S.E.) and those of UV radiation, temperature and sun hours of (1990) with those recorded on 1983 in our previous work[3], it was found that they exceeded the latter considerably and was attributed to the noticeable changes of the earth's atmosphere and ozone concentration in the last few years. The total sun hours/month showed the fact that Egypt is characterized by long daily sunny periods, especially in summer. Also, the results of the R.H. proved to be high throughout the whole year. These severe environmental conditions proved to contribute markedly to the photodegradation of cellulosic materials[3,5,6].

Spectrophotometric studies of the samples showed a change in colour after weathering for one year, and that their creamy colour had turned white. The lino samples were deeply soiled by the exposure, even after being purified. The spectral data of table (2) revealed that considerable change in the colour difference values (Δ E) had occurred and was greater for the fine lino than that of the coarsest cotton duck. Moreover, when considering the brightness value (L), the duck showed an appreciable increase, while for lino the brightness decreased by duration of exposure.

When considering the S.E.M. photographs of the duck and lino samples Fig. (2), we may notice the change of the fibers surfaces arising from exposure to the environmental conditions. The fibers became more opaque and the surface roughness was increased and a number of depressions were formed. Moreover, the diameters of the exposed fibers were apparently enlarged. Although both fabrics were of identical chemical composition and were exposed to the same incident solar energy, yet the observed changes in surfaces were markedly different due to the different total exposed surfaces which in turn affected the absorbed radiation doses. Duck fabric proved to be more resistant and less damaged.

Figure 2. SEM photographs (a & b)for lino and duck fibres before and after one year of outdoor exposure. 3000x.

The monthly changes in mechanical properties expressed as losses of tensile strength and elongation of the fabrics arising from weathering are represented in Fig.(3). Cyclic curves resembling those of the monthly variation of the (I.S.E.) and/or the UV radiation, the sunny time and the temperature of the surroundings were observed. Also, the aforementioned losses were plotted against the monthly (I.S.E), and the estimated level of UV

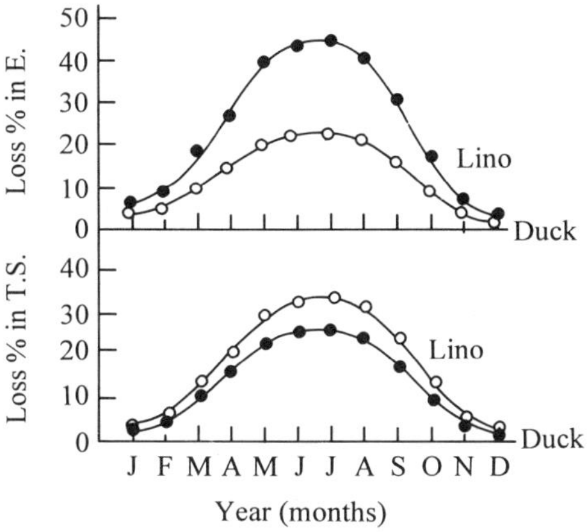

Figure 3. The loss % in tensile strength, elongation versus the months of the year for lino and duck fabrics.

radiation at wavelength (290-315 nm), a good correlation was obtained (Figs. 4,5). The results obtained clarify significantly the contribution of the environmental conditions to the photodegradation. In other words, the deterioration is affected by the duration of exposure, effective absorbed solar energy, seasons of the year, UV radiation, temperature and R.H. of the surroundings.

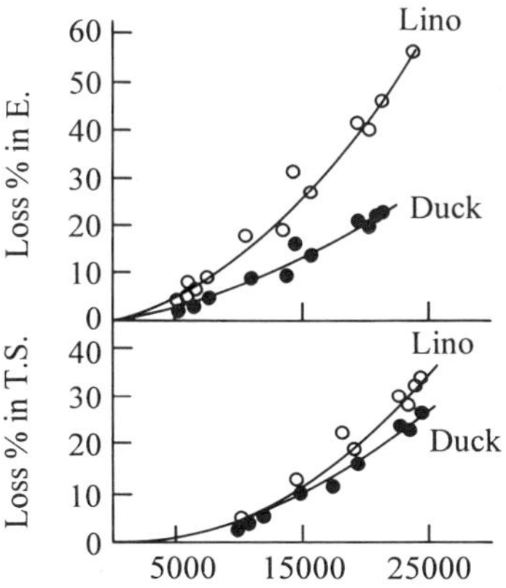

Figure 4. The loss % in tensile strength, elongation versus the monthly incident solar energy (langleys) for duck and lino fabrics.

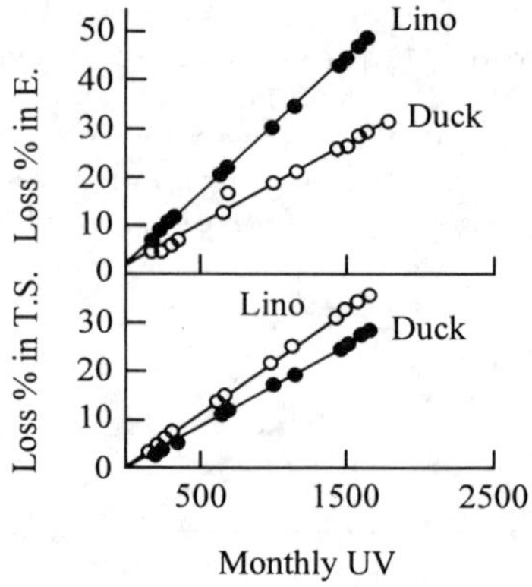

Figure 5. The loss % in tensile strength, elongation versus the monthly UV radiation (mw.min/cm2) for duck and lino fabrics.

The pH values of an aqueous extract showed also an apparent change, possibly because of the polluted atmosphere at the industrial sites in Cairo. The figure obtained indicates however that aqueous extracts of the lino exposed samples are more acidic than those of duck ones and Table (2) gives such results. Exposure caused soiling especially for the light weight lino, but both cotton samples were free from any microbiological growth.

Table (2)

Fabric	Lino				Duck			
Exposed months	0	4	8	12	0	4	8	12
Bright. (L)	78.3	75.3	73.6	72.0	78.5	80.1	83.1	84.5
Col.Diff.(Δ E)	0.0	4.4	6.8	9.0	0.0	1.9	5.6	7.4
pH values	7.5	5.7	5.5	4.9	7.7	6.2	6.0	5.9
Carboxyl cont. (m.mol)	3.3	5.2	6.9	9.4	3.2	6.7	8.6	10.4
Cu No.(g/100g)	0.2	0.4	1.0	1.5	0.2	0.5	1.1	1.6

The results of table (2) indicate that weathering of cellulosic material produces carboxyl and carbonyl groups' resulting from the photochemical degradation.

REFERENCES

1. Tera,F.M.; Shady K.E.; and Hegazi H. Kolor. Ert. 1987, 4, 80.

2. Tera, F.M.; Shady K.E.; and El-Zaher N.A. The First Egyptian-British Conf. on Biophysics, 1987, Cairo Univ., Egypt.

3. Tera F.M.; Shady K.E. Kolor Ert. 1987, 5, 153.

4. Tera F.M.; Amer M.A.; and Shady K.E. Kolor.Ert.,1987, 5, 125.

5. Little A.H.; Parsons H.L.; J. Text. Institute,1967, 58, 449.

6. Rugger G.R.; "Weathering and Environmental Effects on Polymeric Materials". Intersciences New York, 1968.

7. Norbert M.B.; and Leon Segal. "Cellulose and Cellulose Derivatives". John Wiley & Sons, Inc., Canada, 1971.

8. Wright W.D.; "The Measurements of Colour". Watts Ltd, 1964.

9. George D.; Reinhardt R.M.; and David J.; J. Text. Res. 1953, 23, 719.

49

Photodegradatin of linen by sunlight

N. Kohara, E. Nakayama and H. Toyoda - Faculty of Home Economics, Showa Women's University, 1-7-57, Taishido, Setagaya-ku, Tokyo, 154, Japan.

ABSTRACT

In order to elucidate the mechanism of degradation of cellulosic fibers, linens have been exposed to sunlight on an underglass-exposure-rack for 3 years and 5 months and changes in their properties have been investigated as a function of incident sunlight energies. The presence of carbonyl groups, especially aldehyde groups, increased remarkably in the samples after early exposure times, while the number of carboxyl groups were not observed to increase. Degree of polymerization (D.P.) decreased with an increase in light energy and became less than 1/6th of the value of the unexposed control at the incident UV light energy of ca. 900 Mj/m^2. These facts suggest that degradative reactions including oxidation took place at glycosidic bonds as well as hydroxyl groups of cellulose to cause scission of the molecular chains. Their crystallite size of longitudinal dimension decreased markedly at first and then only slowly, while the size of their lateral dimension and crystallinity index remained almost unchanged. These results indicate that the degradation did not proceed from the surface to the inner of the fibers, but proceeded longitudinally. The tensile strength decreased monotonously until ca. 4% of the value of the unexposed control.

INTRODUCTION

Cellulosic materials such as paper, vegetable fiber and wood are inevitably degraded by various factors. It is very important for their good preservation to know how and why their degradation proceeds. In this work, in order to elucidate the mechanism of the degradation of cellulose, linens were exposed to sunlight on an underglass-exposure-rack equipped with an integrating actinometer for 3 years and 5 months, and changes in their oxidation-stages, degrees of polymerization, cellulose crystallinities and mechanical properties were investigated as a function of incident sunlight energies, especially in the UV region.

EXPERIMENTAL

MATERIALS

Plain-weave linen woven with yarns of 25/1 S was used;its thickness was 0.5 mm. The sample linen had been exposed to sunlight with an IS-1 model underglass-exposure-rack equipped with a PH-11M-2AT model integrating actinometer (Suga Test Instruments Co., Ltd.) since March 1988 and collected at various exposure times.

PROCEDURES

Oxidation-stage:Oxidation stages of the samples were evaluated by comparison of amounts of aldehyde, ketone and carboxyl groups formed in the samples with those in the unexposed control. Total carbonyl group content was determined by a semicarbazide method[1] except that the reaction time of the samples and the reagents was prolonged to 192 h. The ketone group content of the samples was determined by the same method after re-oxidation with sodium chlorite[2]. The amount of aldehyde groups was calculated by subtracting the amount of ketone groups from that of total carbonyl groups. The amount of carboxyl groups was determined by a dynamic ion exchange method[3].
Degree of polymerization:Degrees of polymerization of the samples were determined as the derivatives of cellulose triphenylcarbanilates by high performance size exclusion chromatography (HPSEC). Cellulose triphenylcarbanilates were obtained by the treatment of the sample linens with phenyl isocyanate in dry pyridine according to Evans et al.[4]. HPSEC was carried out on a system made up of a Shimadzu LC-6A liquid chromatograph, a Shimadzu CTO-6A column oven, a Shimadzu SPD-6A UV spectrophotometric detector and a Shimadzu C-R3A chromatopac connected with an FDD-1A floppy disk drive. The column sets used were a series of 1.5 x 30 cm Shodex columns KF-806, -805

and -804, and a KF-800 guard column (Showa Denko K. K.). Other analytical conditions were the same as that reported by Evans et al.[4] . TSK standard polystyrene kit (Toyo Soda Manufacturing Co., Ltd.) was used as the molecular weight standards.
Cellulose crystallinity: The yarns extracted from the sample linens were used for the X-ray diffraction study. X-ray diffraction traces were obtained on a Rigaku Geiger flex RAD-IIIB X-ray diffractometer using nickel-filtered copper radiation over the angular range 5 to 60 . The crystallinity index (CrI) proposed by Segal et al.[5] was used as a measure of the degree of lateral order. Crystallite size of the cellulose in the linens was determined by measuring the width at half-maximum intensity of the 002 and 020 reflections and using the Scherrer equation[6] .
Tensile strength and elongation-at-break:All of the tensile tests were carried out at 20℃ and 65% R.H., using a UTM-4-100 model tensile testing machine (Toyo Baldwin Co., Ltd.). The gauge length was fixed at 100 mm and the stress was applied at a fixed rate of 300 mm/min. Specimens 100 mm long by 25 mm wide were used for the test, which was carried out only in the warp direction. The test was repeated for 4 times in each sample.

RESULTS AND DISCUSSION

Since March 1988, linens had been exposed to sunlight on an underglass-exposure-rack equipped with an integrating actinometer, which measured the amount of incident sunlight energies, separating them into three parts, ultraviolet (300-400 nm), visible(400-700 nm) and infrared (700-1,200 nm). The total amount of the incident sunlight energies during 1988 was 4,448 Mj/m^2, of which 6.26, 50.30 and 43.54% were the light energies in the UV, visible and IR regions, respectively. The values in the other two years were almost equal. The total amount of the incident light energies for these 3 years and 5 months was 14,352 Mj/m^2, in which 902, 7,298 and 6,152 Mj/m^2 were the incident light energies in UV, visible and IR regions, respectively. However, ca. 30% of the UV light and 10% of the others were estimated to be absorbed by the glass of the exposure-rack[7] .

OXIDATION-STAGE

The relationships between the light energies in the UV region and amounts of aldehyde, ketone and carboxyl groups formed in linens are shown in Fig.1. Carbonyl groups (>C=O + -CH=O) were increased with increasing light energies until the total incident UV light energies of ca. 400 Mj/m^2 is reached, and then decreased. On the other hand, carboxyl groups were hardly formed. At the incident UV light energies of 379 Mj/m², 129 mole equivalents of carbonyl group (95 mole equivalents of aldehyde) per 1000 glucose units were formed, which was 5 times that of the unexposed control. When all oxidations are supposed to occur only at glycosidic bonds, formation of carbonyl groups should cause scission of the molecular chains at ca.65 sites. This corresponds to the lowering of D.P. into ca. 1.6% of that of the unexposed linen. However, the actual D.P. value obtained was 42.3% of the initial value. Therefore, the results indicate that oxidation of cellulose took place at hydroxyl groups and/or ether linkages of glucose residues as well as glycosidic bonds. Decrease of carbonyl groups observed in the secondary stage may suggest that they are converted into volatile compounds as reported by Desai and Shields[8) .

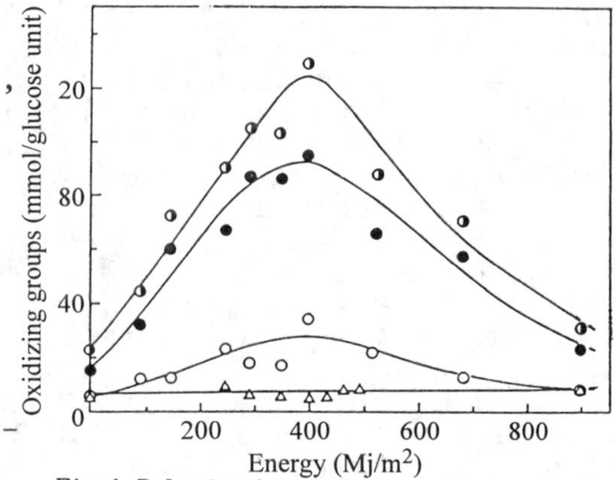

Fig.1.Relationship between oxdizing groups formed in linen and incident UV light energies.
◐:>C=O + -CH=O, ●:-CH=O, ○:>C=O, △:-COOH

DEGREE OF POLYMERIZATION (D.P.)

Fig.2 shows the D.P. of samples evaluated as their triphenylcarbanilates. The values reduced with an increase in light energies since the early time of the exposure. The value of the sample exposed for the whole period was less than 1/6 of that of the unexposed control. These results distinctly indicate that the degradative reactions including oxidation caused breakage of the molecular chains of cellulose.

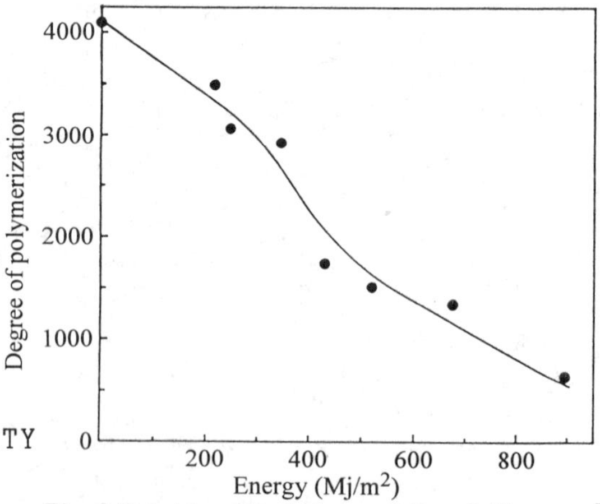

Fig.2.Relationship between D.P. of linen and incident UV light energies.

CELLULOSE CRYSTALLINITY

Fig.3 shows the changes of crystallite sizes in lateral (D_{020}) and longitudinal dimensions (D_{002}) and crystallinity indices (CrI) during the exposure to sunlight. The values of D_{020} and CrI were scarcely changed, but the value of D_{002} became lower as the light energies increased, especially in the early stages of the exposure. These results suggest that the degradative reaction probably began to occur in the crystalline region as well as in the amorphous region even in the early stages, and proceeded not from the surface to the inner of the fiber, but longitudinally.

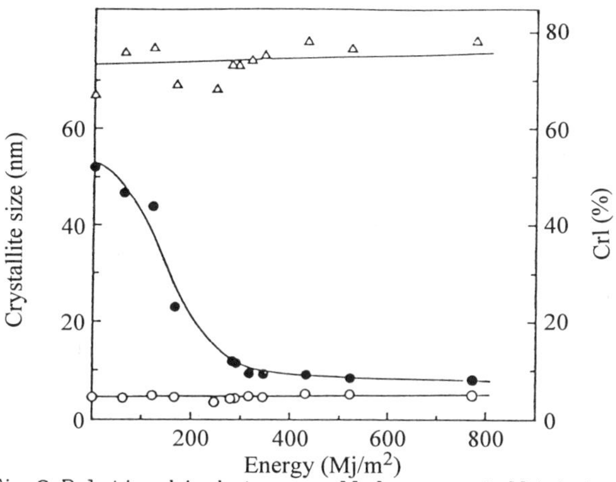

Fig.3.Relationship between cellulose crystallinities of linen and incident UV light energies.
●: D_{002}, ○: D_{020}, △: CrI

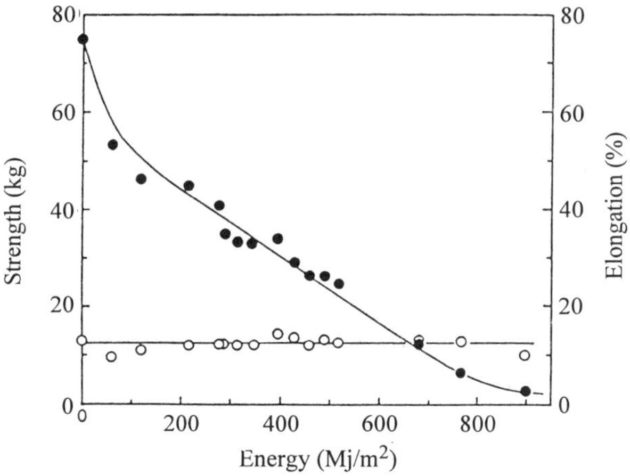

Fig.4.Relationship between mechanical properties of linen and incident UV light energies.
●: Tensile strength, ○: Elongation at break

MECHANICAL PROPERTIES

Tensile strength and elongation-at-break of the samples are shown in Fig.4. The strength began to reduce at the initial stage and became only about 1/25 of that of unexposed linen finally. In spite of such significant reduction in the strength, distinct alteration was not observed in their elongation properties.

CONCLUSION

The effect of sunlight irradiation on linens was investigated for more than 3 years by the alterations in oxidation-stages, D.P., cellulose crystallinities and mechanical properties. Formation of aldehyde groups, reduction in D.P. and lowering of tensile strength had already begun at the early stages of the exposure to sunlight. Invariance of D_{020} and decrease of D_{002} suggest that the reaction did not proceed from the surface to the inner of the fibers, but did longitudinally. Reduction in D.P. also suggests that the degradative reactions including oxidation took place at glycosidic bonds and hydroxyl groups and/or ether linkages of glucose residues in cellulose, and reaction at the former caused scission of the molecular chains.

ACKNOWLEDGMENT

The authors wish to thank T. Sano, Showa Women's University for statistical calculation of light energies.

REFERENCES

1. J. Yamamoto and T. Miyata, J. Chem. Soc. Japan, Industrial Chemistry Section, 62, 1433(1959)
2. W. K. Wilson and A. A. Padgett, TAPPI, 38, 292 (1955)
3. M. Sobue and M. Okubo, TAPPI, 39, 415(1956)
4. R. Evans, R. H. Wearne and A. F. A. Wallis, J. Appl. Polym. Sci., 37, 3291(1989)

L. Segal, J. J. Creely, A. E. Martin, Jr. and C. M. Conrad, Text. Res. J., 29, 786(1959)

J. L. Hengstenberg and H. Mark, Z. Krist, 69, 271 (1928)

Report to Agency of Industrial Science Technology, Ministry of International Trade & Industry by Japan Weathering Test Center(1984)

8. R. L. Desai and J. A. Shields, Makromol. Chemie (Basel), 122, 134(1969)

50

Effects of air quality and radiation on wood

David N.-S. Hon - Department of Forest Resources, Clemson University, Clemson, South Carolina 29634-1003, USA.

ABSTRACT

The impact of air pollutants and ultraviolet light on wood may entail chemical and physical attack. Discoloration has been observed from wood exposed to light, and also to sulfur dioxide and nitrogen dioxide with or without ultraviolet light. Electron spin resonance spectroscopy study revealed that photo-induced wood free radicals reacted readily with oxygen, sulfur dioxide and nitrogen oxide. The change in surface chemistry was monitored by a Fourier transform infrared spectrophotometer and an ultraviolet-visible spectrophotometer. Degradation of carbohydrates and lignin as well as formation of carbonyl groups at the oxidized wood surface were observed.

INTRODUCTION

The atmosphere, a gaseous envelope surrounding the earth, is divided into several zones. About 95% of the mass of the air is found in the innermost layer of the atmosphere known as the troposphere. About 95% of the volume of clean, dry air consists of nitrogen (78%) and oxygen (21%). The remaining 1% consists of small amounts of other gases such as argon and carbon dioxide. Air also holds small amounts of water. As clean air moves across the earth's surface, it picks up additional loads of chemicals produced by natural events and human activities. A series of complicated photochemical reactions of air pollutants in the atmosphere, such as oxides of sulfur and nitrogen, always leads to dry and wet acidic precipitations. The latter is popularly known as "acid rain."

During the past forty years in Europe and the past twenty years in North America, scientific evidence has accumulated suggesting that air pollution resulting from emissions of oxides of sulfur and nitrogen may have significant adverse effects on vegetation, animals and materials [1]. Wood materials in the form of siding, panelling, decking, etc. are used extensively outdoors and as a result are subject to sunlight, air pollutants and acid precipitation. Although degradation of wood used outdoors by acidic pollutants and acid rain is strongly suspected, little information is available. In this report, I would like to discuss our study on the effects of sulfur dioxide and nitrogen oxide on the wood surfaces. Synergistic action of photoirradiation was also considered.

EXPERIMENTAL

Southern yellow pine (*Pinus spp.*) was used as the raw material. High purity sulfur dioxide and nitrogen dioxide were used as the pollutants. Wood specimens with or without pollutant gases were exposed with a high pressure mercury-vapor lamp (ACE-Hanovia) through a pyrex window (λ>292 nm). A Fourier transform infrared spectroscopy (Nicolet 20 DX) and an Ultra-violet-visible spectrophotometer (Perkin Elmer Lambda 3A) were used to analyzed oxidized surface and water-soluble degradation products. An electron spin resonance (ESR) spectrophotometer (Varian E-12) was used to detect free radicals.

DISCOLORATION AND CHANGES IN SURFACE CHEMISTRY

Surface oxidation of wood during weathering involved a set of complex reactions. Cellulose, hemicelluloses and lignin participated in the reactions with ultraviolet light, air pollutants and other environmental elements. The immediate recognizable change on wood surface was the change in color.
For example, the initial pale yellow color of yellow pine was changed to dark yellow, brown, and dark brown, depending on the irradiation time, light source and the combination of light, sulfur dioxide and nitrogen dioxide. The change in color, as illustrated by color

difference, during 672 hours of exposure is depicted in Figure 1. When the wood was exposed to ultraviolet

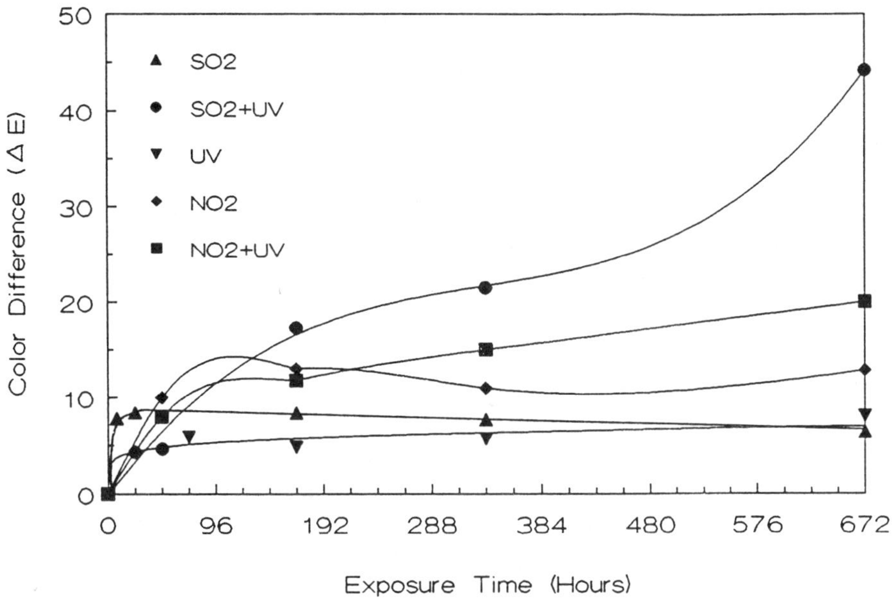

Figure 1. Color change of wood exposed to pollutant gas and ultraviolet light.

light (λ>292 nm), the change in color was readily noticed at the initial 72 hours of irradiation, i.e., 5.9 units of color difference. The color changed slowly thereafter. Sulfur dioxide itself reacted with wood surface components rapidly during relatively short period of exposure. The color appeared to be stabilized after 8 hours of exposure. The degree of color change was greater than in the specimens exposed to the ultraviolet (λ>292 nm) alone. When the wood specimens were exposed to sulfur dioxide in the presence of ultraviolet light, the change in color was the most noticeable of all. After 168 hours of exposure, the unit of change was 17. An additional 168 hours of exposure brought the change to 21 units; and after 672 hours of exposure, the unit change was 44. It appeared that nitrogen dioxide itself changed wood color more than sulfur dioxide; however, in the presence of ultraviolet light, the latter appeared to trigger more color change (see Figure 1).

The FTIR study confirmed that during photoirradiation, carbonyl groups (increased absorption at 1734 cm-1) were formed and lignin (decreased absorption at 1510 cm-1) was degraded. The degradation products generated at the surface exhibited uv absorption at 200-205 nm and 280 nm, implying that carbohydrates and lignin were both degraded. The degradation products increased as wood was exposed to either sulfur dioxide or nitrogen dioxide in the presence of ultraviolet light.

FREE RADICAL REACTIONS

Wood is a good light absorber due to the various chromophoric groups distributed at the surface. One of the consequences of light absorption is the formation of free radicals. It has been reported [2] that several radical species were generated at the wood surface when it was subject to photoirradiation:

$$\text{WOOD} \Rightarrow \text{P-Os}\bullet + \text{P-Oa}\bullet + \text{R-O}\bullet + \text{Rs}\bullet + \text{Ra}\bullet$$

where P-Os•, P-Oa•, R-O•, Rs• and Ra• represent stable phenoxy, active phenoxy, alkoxy, stable carbon and active carbon radicals, respectively. ESR studies have shown that active radicals were reactive toward oxygen, sulfur dioxide and nitrogen oxide. When wood radicals reacted with oxygen, peroxy radicals were generated. A typical ESR signal of peroxy radicals is shown in Figure 2. However, they were unstable and decomposed

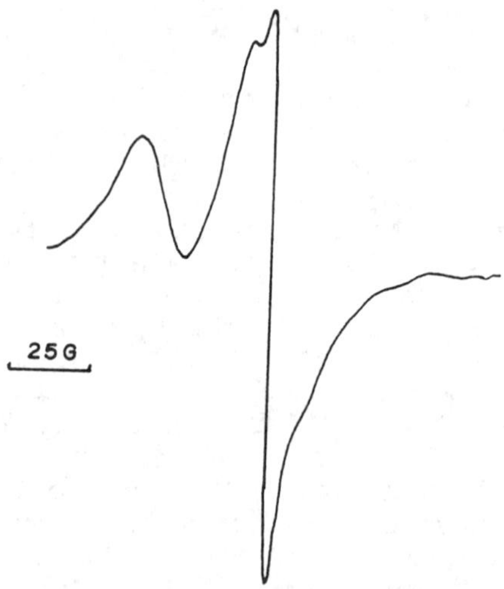

Figure 2. ESR spectrum of wood irradiated in liquid oxygen at -196°C for 60 min.

and terminated rapidly to non-radical products which normally contained carbonyl groups [3].

In common with other radical chain reactions in polymers, photooxidation of wood surfaces can be divided into three separate processes: initiation, propagation, and termination, as indicated below:

$$\text{Initiation:}\quad RH \xrightarrow{h\nu} R\bullet + H\bullet$$

$$\text{Propagation:}\quad R\bullet + O_2 \xrightarrow{k_o} RO_2\bullet$$

$$RO_2\bullet + RH \xrightarrow{k_p} ROOH + R\bullet$$

$$\text{Termination:}\quad RO_2\bullet + RO_2\bullet \xrightarrow{k_{t1}} \text{non-radical products}$$

$$R\bullet + RO_2\bullet \xrightarrow{k_{t12}} \text{non-radical products}$$

$$R\bullet + R\bullet \xrightarrow{k_{t2}} \text{non-radical products}$$

where RH represents chemical components, such as cellulose, hemicelluloses and lignin, at wood surface. Detail study of kinetics of peroxidation has been reported elsewhere [3].

When wood radicals were exposed to sulfur dioxide at -50°C, the ESR multiplet signal of wood radicals was transformed into an asymmetric singlet signal contributing to the sulfonyl, sulfinic and sulfite radicals (Figure 3). However, at room temperature, the asymmetric singlet signal was further transformed into a singlet signal of phenoxy radicals. Based on this, there is little doubt that the following reaction took place:

$$P\text{-}O\bullet + SO_2 \rightarrow \text{No reaction}$$
$$R\text{-}O\bullet + SO_2 \rightarrow R\text{-}O\text{-}SO_2\bullet$$
$$R\bullet + SO_2 \rightarrow R\text{-}SO_2\bullet$$

Likewise, ESR studies also proved that free radicals reacted with nitrogen dioxide in the following manner:

$$P\text{-}O\bullet + NO_2 \rightarrow \text{No reaction}$$
$$R\text{-}O\bullet + NO_2 \rightarrow R\text{-}O\text{-}NO_2$$
$$R\bullet + NO_2 \rightarrow R\text{-}NO_2$$

CONCLUSIONS

Wood is sensitive to the environment in which it is exposed. Sulfur dioxide and nitrogen dioxide discolored wood readily. The presence of UV light triggered more color change. Carbonyl groups were formed at the wood surface. Carbohydrates (cellulose and hemicelluloses) and lignin were degraded. Free radicals were generated at wood surface during photoirradiation. These radicals were reactive toward oxygen, sulfur dioxide and nitrogen dioxide to produce peroxy, sulfonyl, sulfinic and sulfite radicals. Nitro and nitrate ester groups were also formed.

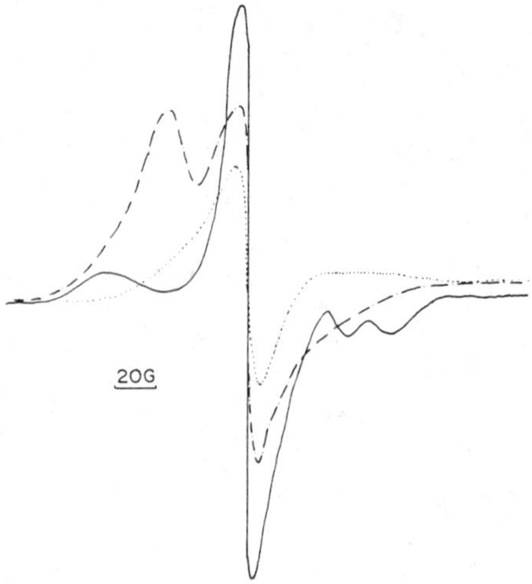

Figure 3. ESR spectra of wood free radicals. (———) wood irradiated in vacuum at -196°C for 60 min, (—·—·) photoirradiated wood in contact with sulfur dioxide at -50°C for 5 min, (·········) after warming the specimen of (b) at 25°C for 60 sec. All ESR spectra were recorded at -196°C.

REFERENCES

1. Stern, A.C.; Boubel, R.W.; Fox, D.L. *Fundamentals of Air Pollution*, Academic Press, New York, 1984.
2. Hon, D.N.-S. in Hon, D.N.-S.; Shiraishi, N. (Eds.) *Wood and Cellulosic Chemistry*, Marcel Dekker, New York, 1991. Chapter 11.
3. Hon, D.N.-S.; Feist, W.C. *Wood and Fiber Science*, in press.

51

Rapid loss of lignin at wood surfaces during natural weathering

Philip D. Evans*, Karl J. Schmalzl and Anthony J. Michell**** - *Department of Forestry, Australian National University, Canberra, A.C.T. 2061, Australia. **CSIRO Division of Forest Products, Clayton, Victoria 3168, Australia.

ABSTRACT

Thin veneers of radiata pine were exposed to natural weathering over a period of 30 days. FTIR-IR spectra of weathered veneers show that after 3 days exposure there was substantial delignification at veneer surfaces. Large losses in the Klason lignin content of veneers also occurred during weathering. Losses in the wet finite span tensile strength of veneers were more rapid than losses in dry zero span tensile strength and changes in the morphology of transverse surfaces exposed to the weather were largely confined to the lignin rich middle lamella layers. A stepwise loss in weight of exposed veneers after rainfall supports suggestions made in the literature that lignin degradation products are leached from weathered wood by rain.

INTRODUCTION

On exposure to the weather, wood is superficially degraded by light and water. An early experiment showed that after only 100 hours of natural weathering, the surface layers of hardwoods could be substantially delignified [1]. Subsequent experiments have confirmed the photo-labile nature of lignin [2], but have not examined further the early stages of surface delignification during natural weathering. This issue is addressed here by exposing thin wood veneers to the weather for periods ranging from 4 hours to 30 days. Lignin degradation is then assessed directly by Fourier Transform Infra Red Internal Reflectance Spectroscopy (FTIR-IRS) and by conventional lignin analysis. Lignin degradation is also assessed indirectly during weathering using measurements of the loss of veneer tensile strength and weight and by examining erosion of the lignin rich middle lamella region in transverse wood surfaces.

EXPERIMENTAL

Veneers approximately 85 μm thick were cut using a microtome from the radial faces of radiata pine sapwood blocks after they had been vacuum impregnated with distilled water. Veneers were batched serially into batches of 10 replicates and dried in an oven at 105 °C for 18 hours. Each batch of veneers was then weighed and the thickness of each veneer was measured with a micrometer.

Batches of veneers were then mechanically fixed to glass backing plates and exposed outdoors in 5 horizontally inclined weathering racks facing equatorially in Canberra during the summer of 1991. Batches of serially cut veneers were retained as unexposed controls. After 4 hours weathering two batches of veneers were removed, oven dried as above and weight losses calculated. Thereafter two batches of veneers were removed on a daily basis over a period of 30 days and batch weight losses after exposure were determined as above. Percentage weight losses during weathering are expressed relative to the initial oven dry weight. Tensile strength tests on weathered veneers and controls were carried out using a Pulmac paper tester. Batches of veneers exposed for 4 hours, and 1 and 2 days and thereafter for odd numbered days were tested wet with a jaw separation of 0.6 mm. Batches of veneers exposed to the weather for even numbered days were tested dry at zero span. Strength was calculated as ultimate load (kg) at failure divided by the cross sectional dimensions (mm^2) of each veneer. Percentage tensile strength losses during weathering are expressed relative to the strength of the unexposed serially cut batches of control strips.

Veneers from batches of veneers weathered over the first six days were fixed with the weathered surface against the faces of a KRS-5 prism in a RIIC TR25 frustrated multiple internal reflectance (FMIR) unit. The spectra were obtained by using a Mattson Alpha Centauri Fourier Transform Infrared Spectrophotometer equipped with a water cooled source, with a computer controlled iris and a DTGS detector. They were at a resolution of 4 cm^{-1} and were the averages of 64 scans. Background spectra were obtained with the FMIR unit in the spectrophotometer without having a sample against the crystal. Batches of veneers exposed to the weather for 4 hours, 1, 2, 3, 5, 10, 20 and 30 days and unexposed controls were also analysed for their lignin content using the 72% sulphuric acid method.

Radiata pine specimens about 5 mm square and measuring 8 mm longitudinally were soaked in distilled water for 4 days until saturated and a single edged razor blade was then used to prepare clean T.S. surfaces prior to exposure. Specimens were exposed to natural weathering for 4 and 9 days during the autumn of 1991. After weathering, controls and exposed specimens were dried and attached to aluminium stubs. They were then coated with a layer of gold/palladium alloy and examined in a Cambridge Stereoscan 360 SEM.

Information regarding the meteorological conditions encountered during the exposure trials was obtained from measurements made in Canberra by the Australian Department of Science and Technology, Bureau of Meteorology. Measurements of U.V radiation (290 to 400 nm) at the exposure site were made using a photovoltaic cell and automatically recorded by a solar powered data logger and microcomputer.

LOSS OF LIGNIN

FTIR-IR spectra of weathered wood veneers show a substantial and remarkably rapid decrease in the peak at 1505 cm^{-1} which corresponds to aromatic C=C bond stretching in lignin (Fig. 1).

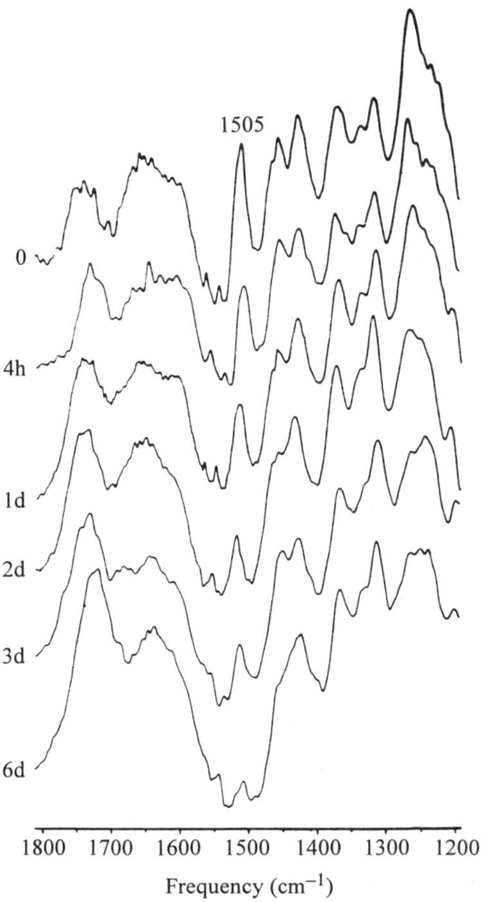

Figure 1. FTIR-IR absorbance spectra of control and weathered veneers.

At this frequency (1505 cm^{-1}) using KRS-5 at an incident angle of 45°, the average depth of light penetration would be ~ 1.2 μm assuming that the refractive index for radiata pine is similar to 1.43, the refractive index estimated for white fir [3]. Changes in peak height suggest substantial surface delignification after 3 days exposure and almost complete surface delignification after 6 days. The finding that lignin at pine surfaces is substantially degraded after only 3 days exposure is noteworthy and probably reflects the greater sophistication of the FTIR-IRS technique which makes it possible to study chemical changes taking place in only the top 1-2 μm of the wood surface. Loss of Klason lignin from veneers is also surprisingly rapid over the first 5 days of exposure decreasing from 26.2% to 20.3%, (Fig. 2).

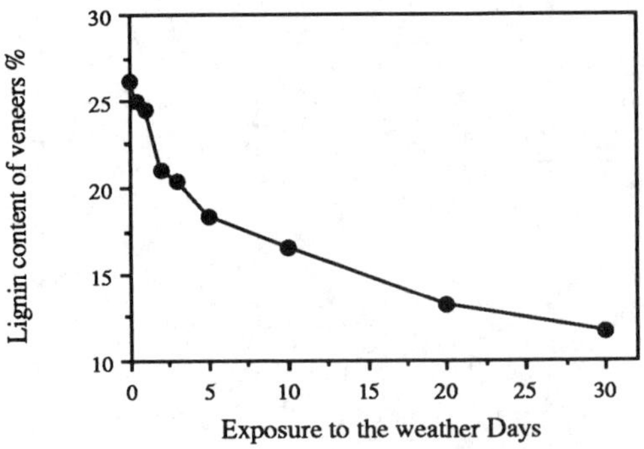

Figure 2. The loss of lignin from wood veneers during natural weathering.

LOSS OF VENEER TENSILE STRENGTH AND WEIGHT

The losses of veneer tensile strength are shown in Figure. 3. Delignification of wood is reported to reduce wet tensile strength but to have little effect on dry tensile strength [4]. Dry tensile strength particularly when testing at zero span is more dependent on the properties of cellulose [4].

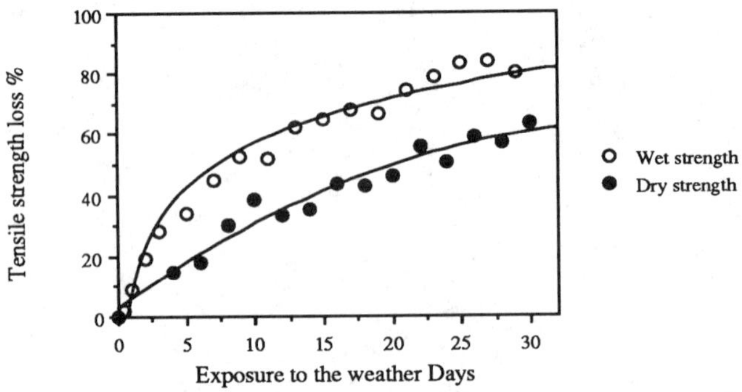

Figure 3. The loss of veneer tensile strength during natural weathering.

The pattern of veneer tensile strength loss here in which loss of wet finite span tensile strength is more pronounced initially than loss of dry zero span tensile strength indicates that lignin degradation occurs more rapidly than that of cellulose. In contrast to losses in tensile strength, weight loss of veneers during weathering occurs in a stepwise manner (Figure. 4). Substantial losses in weight appear to occur after periods of rain. For example, weight losses increase dramatically from 3.6% after 10 days exposure to 10.6 % after 11 days exposure following a period of rainfall. Previous work using artificial accelerated weathering has suggested that during weathering soluble photo-chemical lignin degradation products are leached from wood by water. The present findings using natural weathering clearly support such suggestions.

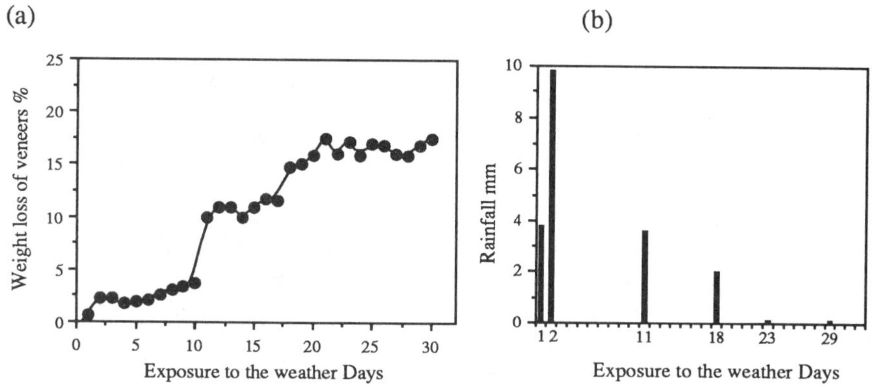

Fig 4. (a) Weight losses of veneers during weathering and (b) the pattern of rainfall during the exposure period.

MICROSCOPY OF TRANSVERSE SURFACES

Transverse wood surfaces exposed during the autumn of 1991 for 4 and 9 days received a cumulative U.V. dose equivalent to approximately 2 and 4 days of summer exposure. After 4 days natural weathering there is delamination of the cell wall and loss of some of the lignin rich middle lamella in earlywood (Figure 6a) and latewood (Figure 6b) compared to controls (Figure 5a, b). After 9 days there is further extensive erosion of the middle lamella (Figure 6c, d) which is most pronounced at the cell corners and between the radial walls where the middle lamella is thickest. The rapid degradation of the lignin rich middle lamella is consistent with the losses of lignin above (Figs. 1 and 2). Although 75% of lignin is associated with the secondary wall layers, the concentration of lignin is much higher in the middle lamella and hence it is not surprising that structural degradation is most pronounced there.

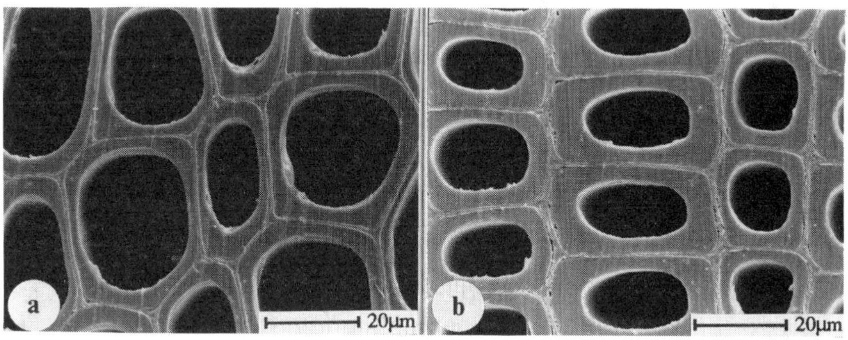

Figure 5. Scanning electron micrographs of unweathered transverse surfaces (a, earlywood; b, latewood)

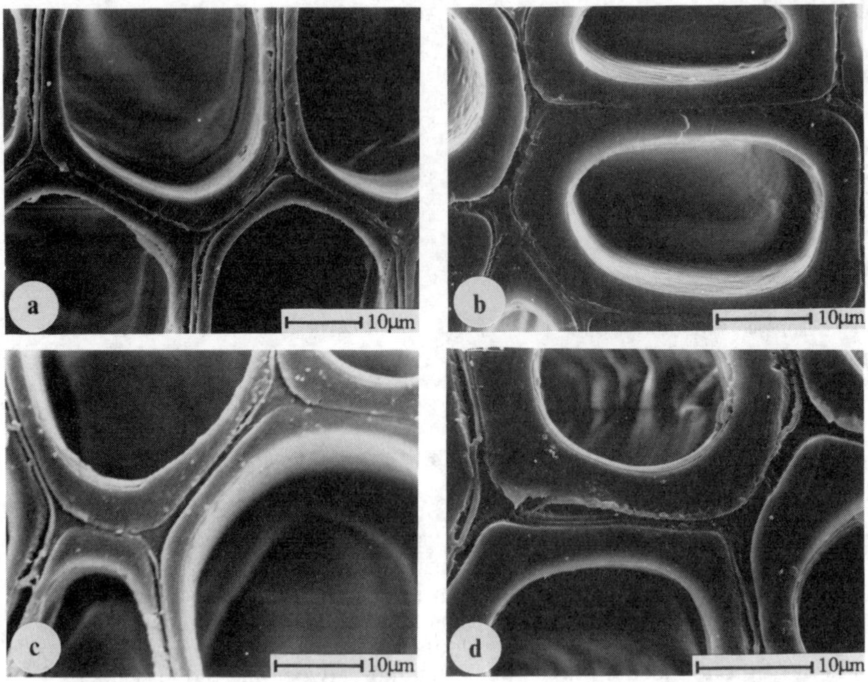

Figure 6. Scanning electron micrographs of transverse surfaces exposed to the weather for 4 days (a, earlywood; b, latewood) and 9 days (c, earlywood; d, latewood).

PRACTICAL IMPLICATIONS OF THIS RESEARCH

Acrylic latex primer adhesion to various timbers is lowered if the wood has first been exposed to the weather under summer conditions near Madison Wisconsin, USA [5]. Lower adhesion of finishes to weathered wood surfaces could be related to lignin degradation. The present study has demonstrated that after only 3 days exterior exposure under summer conditions in Canberra, most of the lignin in the top 1-2 μm of the surface of radiata pine has been lost. Future work may well be justified to evaluate the possible effect of prior short term weathering exposure on subsequent acrylic latex adhesion to radiata pine.

REFERENCES

1. Richter, G.A. *Ind. Eng. Chem.* **1935**, 27, 432.
2. Kalnins, M.A. *U.S. For. Prod. Lab. Rept.* **1966**, 57, 23.
3. Zavarin, E.; Cool, L.G.; Jones, S.J. *J. Wood Chem. Tech.* **1991**, 11, 41.
4. Mark, R.E. *Cell Wall Mechanics of Tracheids*. Yale Univ. Press, New Haven **1967**.
5. Williams, R.S.; Plantinga, P.L.; Feist, W.C. *For. Prod. J.* **1990**, 40, 45.

52

Direct liquefaction of wood extended batch reactions using recycled solvent

T. Ogi, T. Minowa, Y. Dote and S. Yokoyama - National Institute for Resources and Environment, 16-3, Onogawa, Tsukuba Ibaraki 305, Japan.

ABSTRACT

Wood was converted to heavy oil by extended batch reactions in which water or a water/2-propanol mixture was used repeatedly as the reaction solvent. Experiments were conducted in six series with different sets of reaction parameters and each series consisted of five sequential runs. When wood was treated at 275°C and 90 atm for 30 min, heavy oil yields ranged from 24 to 41%. Among the five runs, the heavy oil yields were almost the same till the third run or were the highest on the second or the third run, which indicates the effectiveness of the repeated use of reaction solvent. When both sodium carbonate and 2-propanol were added to the first run and either sodium carbonate or 2-propanol was supplemented from the second to the fifth run, heavy oil with low viscosity was obtained in high yields of 30-41%.

INTRODUCTION

In recent years, serious environmental problems, such as the greenhouse effect, have arisen. Biomass, which is the only renewable organic substance and does not affect the carbon dioxide balance, has drawn increasing attention from the standpoint of a clean alternative fuel. The development of the efficient use of biomass is greatly opted for.

In previous papers (1,2), we reported that heavy oil was obtained in yields of 40-50% by liquefying wood with sodium carbonate and water under a pressurized nitrogen atmosphere without hydrogen and/or carbon monoxide. We also found that when some organic solvents such as propanol was added to a reaction system, heavy oil with low viscosity, which was easy to separate, was obtained (3,4). The propanol did not function as a hydrogen donor and it was not consumed during liquefaction. This indicates the possibility of repeated use of propanol as a reaction solvent.

Recycling of a reaction solvent is an advantageous factor for liquefaction in a continuous run. In this report, wood was converted in water or a water/2-propanol mixture in the presence of sodium carbonate. The water or the water/2-propanol mixture was used repeatedly as the reaction solvent, and the effects of solvent recycling were examined.

EXPERIMENTAL

The wood used in all runs was Konara (Japanese oak), and it was ground to 80 mesh. The wood had a chemical composition of 77.89% holocellulose, 21.27% lignin, 2.08% alcohol and benzene extracts, and 0.38% ash, and had elemental analysis of 46.65% carbon,

	1st Run		2nd to 5th Runs	
	Addition of		Supplement of	
	Catalyst	2-Propanol	Catalyst	2-Propanol
Series I	+	−	−	−
Series II	+	−	+	−
Series III	+	+	−	−
Series IV	+	+	+	−
Series V	+	+	−	+
Series VI	+	+	+	+

Reaction conditions:
Temperature 275℃, Pressure 90 atm, Holding time 30 min.
Wood (Konara) 20g.
Catalyst (Na_2CO_3) 1.0g (5wt%)

1) Initial reaction solvent at the 1st run:
−: Water 120ml
+: Water 100ml +2-Propanol 20ml

2) Supplement of catalyst:
−: none
+: Na_2CO_3 1.0g

3) Reaction solvent from the 2nd run to the 5th run
−: Recovered aqueous layer 90ml + Water 30ml
+: Recovered aqueous layer 90ml + Water 23ml + 2-Propanol 7ml

Table 1. Reaction Parameters for Series I to VI

5.83% hydrogen, 0.21% nitrogen and 47.31% oxygen.

Experiments were conducted in six series (Series I to VI), and each series consisted of five sequential runs. Reaction parameters for Series I to VI are listed in Table 1.

In each run, the experiment was conducted using the same apparatus (stainless steel autoclave) and by a similar method described in the previous paper: wood powder was liquefied in a reaction solvent (water or a water/2-propanol mixture) in the presence of sodium carbonate at 275°C for 30 min. under 90 atm. A mixture after the reaction consisted of a tar-like product, solid residue and an aqueous layer. The tar-like product was separated from this mixture, then extracted with dichloromethane. A dark fluid or viscous material obtained after evaporation of dichloromethane was referred to as "heavy oil", and its yield was defined as below:

$$\text{Heavy oil yield (\%)} = \frac{\text{Weight of heavy oil}}{\text{Weight of wood}} \times 100$$

To clarify the effect of a reaction solvent, water (or a water/2-propanol mixture) was used in the first run, and the aqueous layer that remained in the previous run was used with a prescribed amount of additional reaction solvent (water or a water/2-propanol mixture) for the second to the fifth runs.

RESULTS AND DISCUSSION

1. Yields and properties of heavy oil

The yields of heavy oils are shown in Fig.1. The yields of heavy oils ranged from 24 to 41%, and they were almost the same till the third runs or were the highest during the second or the third runs. Then, the yields decreased, and for the fifth run they became as low as, or lower than those for the first run. The recovered aqueous layer was effective as a reaction solvent and can be used repeatedly three or four times.

The yields of heavy oils in Series III to VI were higher than those in Series I and II. In addition, heavy oils obtained in Series III to VI were more fluidized and easier to handle and separate than those obtained in Series I and II. The initial addition of 2-propanol was effective for obtaining heavy oil with low viscosity in high yields.

Comparing the yields of Series III and IV with those in Series V and VI, the latter was higher for the second and the third runs. The difference, however, became smaller and scarcely observed in the

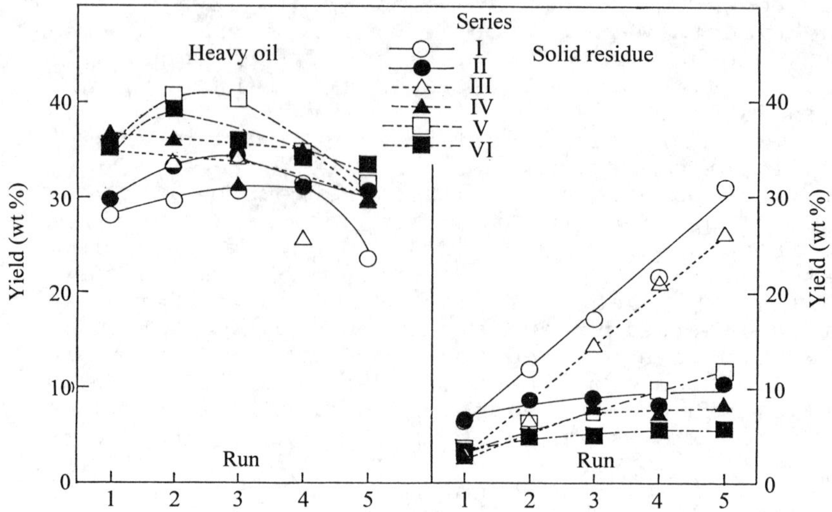

Fig.1 Yields of Heavy Oil and Solid Residue

fifth run for four series. The effect of the supplement of 2-propanol became less conspicuous with the number of runs.

To obtain heavy oil in high yield, as described in the previous papers, it is essential to add a catalyst. The supplement of sodium carbonate from the second runs was effective for obtaining heavy oil in higher yield, when only water was used in the first run in Series I and II. On the other hand, the supplement of sodium carbonate scarcely affected the heavy oil yields, when 2-propanol was added to the first run in Series III to VI.

Concerning the properties of the heavy oil, heavy oils had almost the same elemental composition (C, 67-69% ; H, 8%; O, 23-25% in average) and they had the same heating value (30 MJ/kg).

2. Yields of solid residue

The yields of solid residue are also shown in Fig.1. They changed from 3 to 31%, depending on the reaction conditions. Supplementing with sodium carbonate had a significant effect to depress the production of solid residue. The amount of solid residue increased in proportion to the number of runs and its yield exceeded 25% for the fifth run in Series I and III. Whereas, the production of solid residue was constantly suppressed in low yield in Series II, IV and VI, in which sodium carbonate was supplied during every run. The production of solid residue was also suppressed in Series V, in which 2-propanol and not sodium carbonate was used as supplement.

3. Property of an aqueous layer

The yields of heavy oil and solid residue changed depending on the reaction conditions, that is, the properties of the reaction solvent.

The pH value of an aqueous layer increased from 4.9 to 6.2 with the number of runs when sodium carbonate was added to each run. In contrast with this, the pH value decreased to 3.4 without addition of sodium carbonate. Sodium carbonate served as a catalyst for hydrolysis. It served also as a base and neutralized various kinds of acids, which were generated from wood during the reactions and accelerated undesirable reactions, such as repolymerization.

2-propanol was used as an initial organic solvent. Besides 2-propanol, however, various lower molecular compounds, which were derived from wood, were formed in the aqueous layer after the reaction. Methanol, ethanol, propanol, butanol, acetone, methyl ethyl ketone, diethyl ether, formic acid, acetic acid, and propionic acid, were identified by GC. These organic compounds, except acid, served as reaction solvents as propanol to varying extent. The total amount of organic carbon (TOC) in an aqueous layer is shown in Fig. 2. Fig. 2a indicates the results including carbon from 2-propanol, and Fig. 2b indicates the

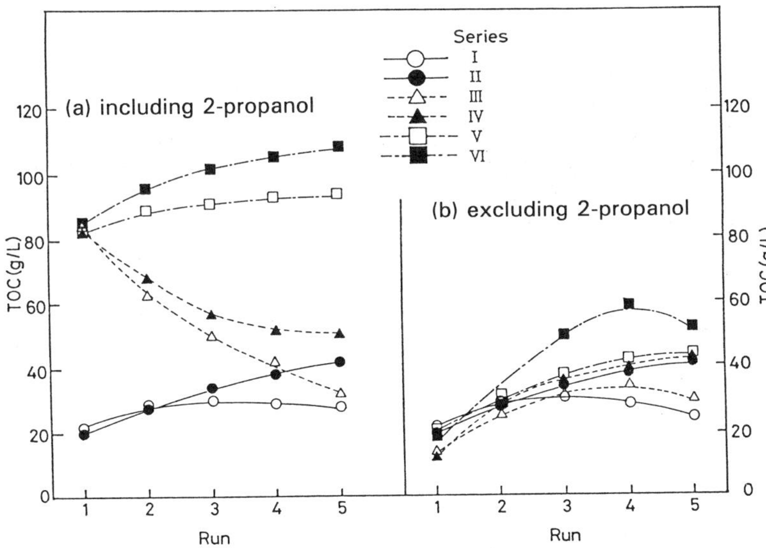

Fig.2 Total Organic Carbon in Aqueous Layer

results after subtracting the amount of 2-propanol from the total amount of organic carbon in (a).

Fig. 2a shows that the TOC in Series II increased, while that in Series III decreased with the number of runs. For the fifth run, the reverse took place, that is, the TOC in Series II was larger than that in Series III. As was mentioned, heavy oils obtained in Series III to VI were more fluid than those in Series I and II, due to the addition of 2-propanol to the first run. However, heavy oil obtained in Series II became gradually more fluid with the number of runs. On the other hand, heavy oil obtained in Series III gradually became sticky. The reason of this phenomenon was explained by this TOC change. The organic compounds produced from wood, except organic acids, had an effect similar to 2-propanol, and the total amount of organic compounds had an effect on the fluidity of the heavy oil.

CONCLUSION

When wood was directly liquefied in water or a water/2-propanol mixture in the presence of sodium carbonate, the aqueous layer was effective when used repeatedly as a reaction solvent.

In the case of using only water to the first run, the resulting heavy oil became more fluid with the number of runs when sodium carbonate was supplemented in each run.

When using a water/2-propanol mixture in the first run, heavy oils with low viscosity were obtained. Their yields were 30 to 41% with the highest in the first or the second runs. Supplementing either 2-propanol or sodium carbonate was necessary to maintain low viscosity of the heavy oils and to suppress any solid residue formation.

REFERENCES

1. Ogi, T., Yokoyama, S., and Koguchi, K., J. Japan Petroleum Institute, 28, 239 (1985)
2. Ogi, T., Yokoyama, S., and Koguchi, K., Chem. Lett., 1199 (1985)
3. Ogi, T., Yokoyama, S., Koguchi, K., and Minowa, T., J. Japan Petroleum Institute, 32, 21 (1989)
4. Ogi, T., Yokoyama, S., Minowa, T., and Dote, Y., J. Japan Petroleum Institute, 33, 383 (1990)

53

Thermal reaction pathways of syringol

C.P. Masuku - Department of Chemical Engineering, Laboratory of Industrial Chemistry, Helsinki University of Technology, SF-02150 Espoo, Finland.

ABSTRACT

The objective of this investigation is to elucidate the thermal reaction pathways of both the alkyl and aromatic C-O bonds which is a key reaction in thermochemical conversion of lignocellulosic materials. Thermolysis of syringol (2,6-dimethoxyphenol), a common aromatic unit in lignins have been studied under inert atmosphere at 250° to 325 °C for reaction times ranging from 2 to 6 h. The following were observed:

* cleavage of the alkyl C-O bond
* cleavage of the aromatic C-O bond
* ring methylation and O-methylation reactions
* steric and electronic effects in the molecule enhance the cleavage of these bonds
* the role of the ortho-oxygen is very significant

Reaction pathways and reaction mechanisms are discussed.

INTRODUCTION

Though so diverse in nature, wood contains three major chemical components: cellulose (50%), hemicellulose (25%) and lignin (25%) [1]. In the wood pulping process, cellulose is isolated leaving behind a by-product "waste", containing mainly lignin, along with various pulping chemicals in solution (spent liquor).

There have been numerous studies and proposals on the utilization of these pulping by-products. One area of continuing interest has been thermochemical liquefaction to lower molecular weight materials. However, this has achieved only limited success because of the numerous unselective reactions which give a large number of products in low individual yields. More knowledge is needed on the controlling fundamental thermal reaction paths and mechanisms in order to tailor the process towards desirable products in good yields.

Lignin is made of phenylpropane units of 4-coumaryl, guaiacyl and syringyl types that are linked together by different types of ether and carbon-carbon bonds. Over 3/4 of the constitutive units of lignin contain aryl alkyl ether groups [2,3]. In particular, hardwood lignin contains mainly syringyl units whereas softwood lignin contains largely guaiacyl type structures.

The present paper describes the thermolysis of syringol (2,6-dimethoxyphenol), a typical lignin model compound.

EXPERIMENTAL

2,6-Dimethoxyphenol was obtained from Aldrich Chemie GmbH and used as received. All the other chemicals were commercial pure or analytical grade and were used without further purification.

Thermolysis was carried out in 1 ml sealed boronsilicate ampoules (Duran 50) under argon inert atmosphere heated in a fluidized sand bath. The procedure is described in detail elsewhere [4]. The sand bath was maintained at the desired temperature (±2 °C) throughout the reaction period. The reaction temperatures used were 250, 275, 300 and 325 °C. Reactions were performed for 2 to 6 h. The experiments were run in a single set without any repetition to determine the reproducibility of the results. The inner surface of the ampoules do not exhibit wall effects and hence were not deactivated before use. The gaseous products were not analysed because there was no provision for gas sampling.

The condensed liquid material was dissolved in 2 ml acetone and analysed using a HP 5880 A gas chromatograph carrying a flame ionisation detector and a fused silica capillary column SE-30 type (25 m long and 0.32 mm i.d). 4-Heptanone was added as internal standard. The temperature programme was set initially at 70°, then 10°/min to 230°, and 10 min at 230°. The injection port was kept at 250° and the detector at 300°. The carrier gas was hydrogen at 3 ml/min and a split ratio of 3:100 was used.

In the GC chromatograms, products were divided into monoaromatics and dimers. The monoaromatics were identified using authentic compounds as well as literature data [5] and further confirmed by mass spectrometry. Dimeric products were not characterized and were calculated as a sum.

Mass spectrometric analysis were performed on a JEOL DX 303/DA 5000 instrument combined with a HP 5790 A gas chromatograph equipped with a similar capillary column and operating under the same conditions as described above for GC.

RESULTS AND DISCUSSION

The results for three temperatures are presented in Tables 1-3. The selectivities of formation were normalised to 100% of the conversion.

The major products formed are 3-methoxycatechol, dimethoxymethylphenol, methylguaiacol and guaiacol. The minor products include 1,2,3-trimethoxybenzene, 1,2,3-trihydroxybenzene, 2,3-dimethoxyphenol, 3-methoxyphenol, 3-methoxymethylphenol, 3-methoxymethylcatechol, veratrole, methylveratrole, catechol, methylcatechol, phenol, methylphenol and dimethylphenol.

Table 1. Thermolysis of 2,6-dimethoxyphenol at 275°

Reaction time (h)	2	3	4
Products (selectivity* %)			
3-Methoxycatechol	42.7	45.6	47.2
Guaiacol	5.4	6.6	7.6
Methylguaiacol	32.9	24.8	21.7
Dimethoxymethylphenol	2.7	3.4	4.0
1,2,3-Trimethoxybenzene	2.8	2.3	1.8
1,2,3-Trihydroxybenzene		0.7	2.1
3-Methoxymethylcatechol	1.2	1.1	2.2
3-Methoxyphenol	1.6	1.1	0.6
3-Methoxymethylphenol		0.4	0.5
Catechol	1.9	2.5	1.6
Methylcatechol	1.6	2.3	2.9
Unidentifieds	7.1	9.2	7.8
Conversion (%)	**11.0**	**20.8**	**27.5**

(*) normalised to 100%

The number of products generally indicates the number of reaction paths. And, in so far as the products mentioned above are concerned, the reaction pathways described in the following are accountable to the formation of the products. Although the major reaction pathway can be clearly distinguished, the other reaction pathways occur inevitably thereafter concurrently with or consecutively to the major reaction pathway.

Cleavage of the O-CH_2 bond

The dominant thermal reaction pathway in the thermolysis of 2,6-dimethoxyphenol is the cleavage of the alkyl C-O bond. Of the bonds in 2,6-dimethoxyphenol, the O-CH_2 bond has the lowest enthalpy of dissociation and hence it is not suprising that its scission is thermodynamically favoured more than any other

Table 2. Thermolysis of 2,6-dimethoxyphenol at 325°

Reaction time (h)	2	3	4
Products (selectivity* %)			
3-Methoxycatechol	46.8	38.1	10.6
Guaiacol	10.5	15.9	35.3
Methylguaiacol	19.9	14.9	12.2
Dimethoxymethylphenol	4.4	4.2	3.0
1,2,3-Trimethoxybenzene	1.0	0.6	0.5
2,3-Dimethoxyphenol	2.8	5.8	7.0
3-Methoxymethylcatechol	3.0	2.8	2.8
Catechol	2.4	3.1	3.6
Methylcatechol	1.6	1.2	1.5
Veratrole	1.7	1.8	2.6
Methylveratrole	0.7	1.7	2.4
Phenol		0.4	3.4
Cresol		0.9	3.9
Unidentifieds	5.1	8.5	11.2
Conversion (%)	**35.4**	**51.7**	**82.8**

Table 3. Thermolysis of 2,6-dimethoxyphenol at 300°

Reaction time (h)	2	3	4	5	6
Products (selectivity* %)					
3-Methoxycatechol	47.5	49.3	47.6	46.2	41.7
Guaiacol	8.5	9.3	9.4	11.5	15.3
Methylguaiacol	26.9	21.6	17.4	15.3	14.4
Dimethoxymethylphenol	3.2	3.7	4.5	4.4	4.2
1,2,3-Trimethoxybenzene	1.5	1.5	1.7	0.9	0.7
1,2,3-Trihydroxybenzene			0.3	0.5	0.9
2,3-Dimethoxyphenol	2.9	2.1	2.7	3.8	3.4
3-Methoxymethylcatechol	2.9	3.2	3.7	3.4	2.7
3-Methoxyphenol		0.5	1.6	1.3	0.9
3-Methoxymethylphenol			0.5	0.5	0.6
Catechol	1.5	2.5	3.0	3.8	5.6
Methylcatechol		0.5	0.7	1.1	1.6
Cresol		0.4	0.4	0.6	0.8
Dimethylphenol		0.7	1.5	1.4	1.7
Veratrole			1.5	3.0	3.3
Unidentifieds	5.1	4.7	3.5	2.3	2.2
Conversion (%)	**18.9**	**25.9**	**32.7**	**41.6**	**52.2**

bond. This is consistent with previous works on the thermolysis or pyrolysis of aryl alkyl ethers [6-13]. In a companion paper [4] of the thermolysis of 4-propylguaiacol between 325° and 400°, the formation of 4-propylcatechol was accompanied by an apparent first order activation energy of 228 kJ/mol. Parkhurst et al. [6] have estimated the bond energies of the various C-O bonds in guaiacol as follows: O-CH_2 247 kJ/mol, CH_2 O-C_5 H_4 356 kJ/mol and HO-C_5 H_4 415 kJ/mol. And using the Benson method, the bond dissociation energy of the C_5 H_4 OCH_1-H bond has been calculated to be 459 kJ/mol [14].

Cleavage of the O-$C_5$$H_4$ bond

Cleavage of the O-C_5 H_4 bond can occur as a demethoxylation reaction or a dehydroxylation reaction. As the enthalpy of bond dissociation of the aromatic bonds (both C_5 H_4 -OCH_2 and C_5 H_4 - OH) are rather high, it is reasonable to assume that these reactions proceed to a lesser extent in the early stages of the thermolysis reaction and become important only at higher reaction temperatures.

Ring methylation

Ring methylation is quite diverse in the thermolysis of 2,6-dimethoxyphenol. This stretches from methylation of 2,6-dimethoxyphenol itself to that of the products of the accompanying reactions. The substituent groups in the benzene ring activate the unoccupied positions and render the ring susceptible to substitution.

O-Methylation

O-Methylation constitutes a minor reaction pathway in the thermolysis of 2,6-dimethoxyphenol. This occurs as a result of both intermolecular (1,2,3-dimethoxybenzene) and intramolecular (2,3-dimethoxyphenol) methyl group shift.

Mechanisms

There are obviously several mechanisms underlying the set of reaction paths discussed above. Thermal decomposition of aryl alkyl ethers, according to current views, proceeds primarily via a free radical reaction mechanism. The free radical mechanism is auto-initiated by scission of the weak alkyl C-O bond into phenoxy and methyl radicals. This is followed by propagation steps (bimolecular and disproportionation) of radical stabilization to form neutral products and new radicals. Finally, termination steps occur by radical combination to form dimers and gaseous products. Evidence for the formation of a gaseous product is shown by the fact that opening of the ampoules was accompanied by a bang. Giabian and Corley [15] have also noted that a methyl radical has a tendency to abstract hydrogen to yield methane or another methyl group to yield ethane, rather than couple with other radicals present in the system. The mechanism of thermal decomposition of 2,6-dimethoxyphenol to form the major products can be depicted as shown in equations 1-5.

An interesting observation here is the phenomenon of methyl radical versus hydrogen radical abstraction. Methylguaiacol formed according to equation (5) is formed in significant amounts already at lower reaction temperatures. On the other hand, guaiacol formed according to equation (4) is found in smaller amounts, until at higher temperatures. This is in agreement with the results of Afifi et al. [10,11] who have shown that in the thermolysis of anisole, the abstraction of a methyl radical is preferred to that of a hydrogen radical at low reaction temperatures.

For reasons not apparent at the moment, the ipso-rearrangement previously noted in most thermal decompositions of aryl alkyl ethers is suppressed.

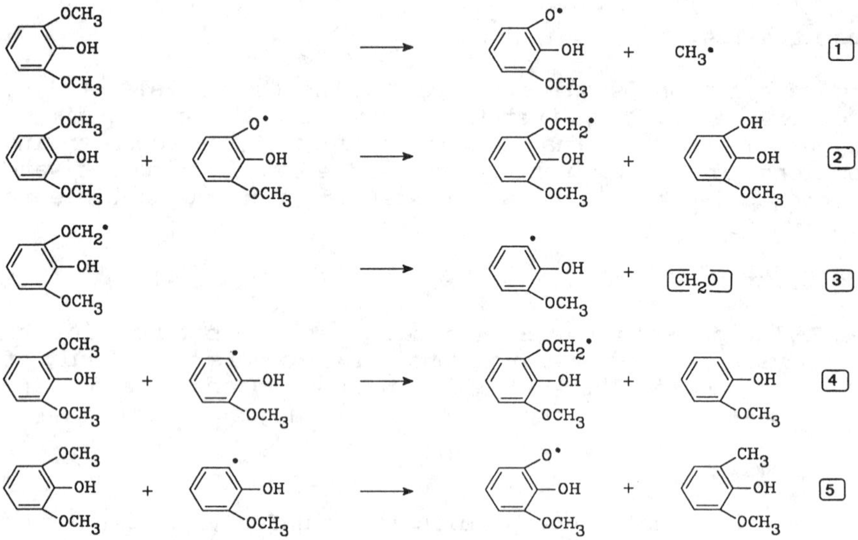

Another probable mechanism that may prevail under the present conditions is the pericylic group transfer mechanism proposed by Klein and Virk [12,13] for guaiacol pyrolysis. In this mechanism, the ortho-oxygen group (methoxyl) to the hydroxyl group plays a vital role. At a particular orientation these two groups produce a 6-electron ($\sigma\pi$) configuration intermediate. This intermediate eliminates a methyl group to form an enol type compound that eventually forms a stable catechol compound. Note that it is not feasible for both methoxyl groups to be involved simultaneously in this type of reaction. These observations are consistent with other findings on ortho-oxygen substituted phenols such as guaiacol and substituted guaiacols [4,6-9,12,
13]. Klein and Virk [12,13] have conducted pyrolysis of 2,6-dimethoxyphenol and shown that the major reaction pathways are similar to those of guaiacol.

It has also been suggested that hydroxy-substituted aryl alkyl ethers can undergo an intramolecular rearrangement resulting in the formation of a C-C linkage. According to Ceylan and Bredenberg [7] 2-cresol can be formed from guaiacol via an enolic form of the 2-hydroxyl group. This mechanism, in addition to the radical one, would account for the high yield of methylguaiacol found in the present case.

The strength of the alkyl C-O bond in the methoxyl group is influenced by steric and electronic effects in the molecule. The unshared electron pairs on one of the oxygen atom can overlap with the p-orbital of the adjacent carbon atom. This displacement of π-electrons in turn produces a perturbation of the π-electrons in the benzene ring. The end result is that the alkyl C-O bond in the methoxyl group is weakened.

In conclusion, it is quite clear from the presented discussion that the thermal reaction paths identified for 2,6-dimethoxyphenol are mainly due to free radical mechanisms. The very limited number of major products found suggests that it may be possible to direct thermal liquefaction along perhaps only one or two initial reaction paths.

REFERENCES

1. Fengel, D.; Wegener, G., Wood;chemistry, ultrastructure, reactions. Walter de Gruyter, Berlin 1984, 613p.
2. Freudenburg, K. in Freudenburg, K.; Neish, A.C. (Eds.) Constitution and biosynthesis of lignin, Springer-Verlag, New York, 1968, p.47.
3. Sakakibaru, A. Wood Sci. Technology **1980**, 14, 89.
4. Masuku, C.P. Acta Polytechnica Scandinavica, Chemical Technology and Metallurgy Series **1991**, 196, 27p.
5. Faix, O.; Meier, D.; Fortmann, I. Holz als Roh-und Werkstoff **1990**, 48, 281.
6. Parkhurst, H.J. Jr.; Huibers, D.T.A.; Jones, M.W. Prepr. Pap.-Am.Chem.Soc., Div.Pet.Chem. **1980**, 25, 657.
7. Ceylan, R.; Bredenberg, B-son. Fuel **1982**, 61, 377.
8. Vuori, A.I. Acta Poyltechnica Scandinavica, Chemical Technology and Metallurgy Series **1986**, 176, 31p.
9. Bredenberg, J.B-son.; Huuska, M.; Vuori, A.I. in Volborth, A. (Ed.) Coal Science and Technology 10. Coal Science and Chemistry, Elsevier, Amsterdam, 1987, 1.
10. Afifi, A.I.; Hindermann, J.P.; Chornet, E.; Overand, R.P. 10th Canadian Catalysis Symposium, Kingston, **1986**, 16p.
11. Afifi, A.I.; Hindermann, J.P.; Chornet, E.; Overand, R.P. Fuel **1989**, 68, 498.
12. Klein, M.T.; Virk, P.S. Prepr.Pap. - Am.Chem.Soc. Div. Fuel Chem. **1980**, 25, 180.
13. Klein, M.T.; Virk, P.S. Report MIT-EL 81-005, Cambridge, Mass., Massachusetts Institute of Technology, 1981, 99p.
14. Benson, S.W. Thermodynamic Kinetics, Wiley and Sons, Chichester, UK, 1976, p. 19.
15. Giabian, M.J.; Corley, R.C. Chem.Rev. **1973**, 73, 441.

54

Environmentally acceptable conversion technology for the biochemical utilization of lignocellulosics

A. Blazej and M. Kosik - Slovak Technical University, 812 37 Bratislava, CSFR.

1 Introduction

Many different types of technologies have been developed on a pilot scale for the conversion of cellulosics into suitable biosubstrates [1]. They involve:

i/ Pretreatment and enzymatic hydrolysis.

ii/ Dilute acid hydrolysis/or use of the low quantity of acid (H_2SO_4).

iii/ Strong acid hydrolysis (HF; HCl; H_2SO_4).

iv/ Thermolytical saccharification; Fig.1.

For the evaluation of all these procedures (Fig.1) we have evaluted the principles of environmentally acceptable technologies which respect the general, so-called "Technological principles".

	saccharification		product-biosubstrat
	enzymic-various		D-glucose
	pretreatment		
cellulose	chemical	diluted acid	D-glucose
		strong acid	oligomers - D-glucose
	thermolytic		glycosans

Fig.1 The main types of conversion technologies.

Principles of environmentally acceptable technologies for processing of lignocellulosics.

I. The maximal utilization of all components of phytomass (material balance).

II. The technological optimalization and level of technology.

III. The use of the highest driving force of conversion.

IV. The achievement of required quality of products.

V. The minimal energy consumption.

VI. The minimal environmental impact on human environment.

VII. Economical feasibility to secure sufficient profitable value-added products.

No single cellulose conversion technology has been developed which could be acceptable from the viewpoint of all the principles suggested. On top of this there is an excess of competitive starch containing agricultural products not only in Europe but also in many other parts of the world. This is the main reason why research activity should be still focused on this field.

Let us discuss more particularly the efficiency of individual conversion procedures (shown in Fig.1) according to the above-mentioned environmentally acceptable technological principles.

2 Pretreatment and Enzymatic Hydrolysis

The enzymatic action is limited by an accessibility of cellulosics when the hydrolysis is running under a very low driving power. This cannot be sufficiently improved either by various pretreatment methods. In addition, the inhibition and regression are negatively affected by the rate and extent of hydrolysis. Two different ways are attractive for research in this field:

- improvement of enzymes activity,
- diminution of inhibitive actions.

In our laboratory, a number of phenolic compounds have been tested for the inhibition/activation effects in experiments with cellulases and xylanases [3,4]. The most efficient activator is guaiacol which increases the enzymatic hydrolyzability of Whatman cellulose by 38 %. D.J.Senior et al. [5] have shown the similar activation effect of soluble lignin derivatives on xylanases at the low concentration treatment.

In the numerous experimental studies of prehydrolysis treatments the possibility to improve the rate and extent of enzymatic hydrolysis has been shown [6-8]. However, there are also results pointing to the evidence of the formation of carbonyl and carboxyl groups which interact with enzymes [9]. This is another reason why we cannot be very optimistic and expect the development of the new acceptable technology of phytomass pretreatment and enzymatic hydrolysis.

3 Dilute Acid Hydrolysis

The basic principles of the processes using a dilute acid are still being applied. The major industrial plants (more than 40) have recently been operated in the Soviet Union territory, producing mostly ethanol, furfural and fodder yeasts. All the dilute-acid hydrolysis technologies are not economically feasible and they run with government support.

The profitability of dilute acid hydrolysis depends on:

- relatively low yield of reducing sugars (50-65 % from theory),
- high energy consumption (25-42 % of manufacturing costs),
- pollution problems (with the non-regenerated acid and soluble lignin fraction).

The achievement of particular values of these

parameters depends also on the selected type of reactor: batch, percolator, cocurrent, countercurrent and plug-flow [10-12].

The highest yield of monosaccharides can be obtained only at the lower concentration of products using all dilute acid technologies, Fig.2.

A very important criterion of dilute acid hydrolysis is the selectivity (φ_M) [12]:

$$\varphi_M = \frac{k_i C_P - k_2 C_M}{k_1 C_P} = 1 - \frac{k_2}{k_1} \cdot \frac{C_M}{C_P}$$

k_1 - hydrolysis rate constant, k_2 - degradation rate constant, C_M - conc. monosaccharides, C_P - conc. polysaccharides.

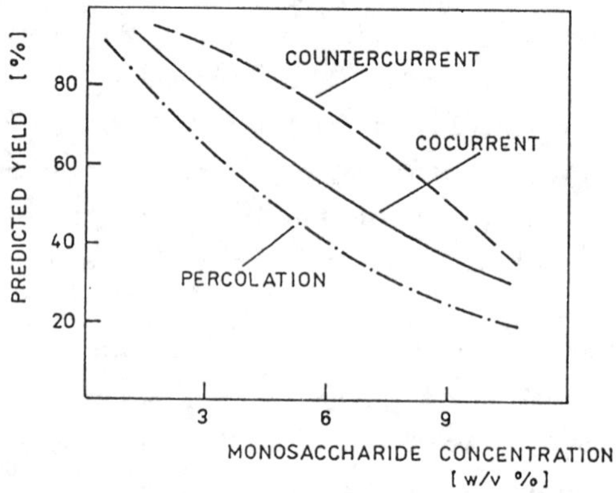

Fig.2 Plots of the sugar yield and concentration in hydrolysates (1% H_2SO_4) for different types of reactor [11,12].

The use of high-hydromodule thermochemical treatment (a plug-flow and chemi-explosion reactor) can improve the profitability of hydrolytic production [10,11], but the parameters achieved, sugar yield and energy consumption will still be in the above-described ranges. Wright and D'Agincourt [13] have calculated the per cent of the production of operating costs in the case of the application of a plug-flow reactor instead of the percolation technology. Nevertheless, the problem of the rapid and effective separation of a lignin residue and its utilization has not been solved. The hydrolyzed lignin contains a part of the acid used (15-20 %) and the rest of the acid (75-20 %) should be removed from hydrolyzates. The amount of the acid adsorbed per 1 g of phytomass depends on the acid concentration and is almost constant: 0.017 g per 1 g of phytomass, if the concentration of H_2SO_4 is 0.5 M or higher [14].

4 Strong Acid Hydrolysis

Although the concentrated acid (65-80 % H_2SO_4, 42 % HCl, HCl/g, HF/g) gives the 75 % or higher conversion of potential, in the phytomass present glucan to oligomers, which are recovered as monosaccharides by dilute posthydrolysis, the large amount of acid used must be recovered efficiently in order to have an economical process and to meet the environmental standards [15-22]. As it is summarized in Table 1, HF has several potential advantages when compared to other strong acid technologies [21,22]. Despite the high HF recovery, the process can be hardly acceptable from the point of view of safety and environmental needs. According to our experience it is rather difficult to achieve the 95 % recovery of HF and solve the problems of the posthydrolysis of oligomers formed.

Table 1 Parameters of Strong Acid Hydrolysis

Acid	Yield of monosaccharides	Recovery of Acid % from total used amount	References
H_2SO_4 (65-80 %)	75-96	80	15,16,17
HCl (g)	85-90	90-95	18-20
HF (g)	90-95	95-98	21,22

5 Thermolytic Saccharification

The pure crystalline cellulose can be rapidly depolymerized into glycosans at 300 to 480 °C and at the reduced pressure, giving the maximal yield of 70 %. The 1,6-anhydro-β-D-glucopyranose is the main product of the depolymerization reaction [23,24]. The glycosan conversion products can be produced by the thermolysis of hemicelluloses and starch in a different yield. The main advantage of thermolytic conversion lies in the possibility of applying a technology which is free of acid. The hydrolysis of glucosans into monosaccharides can be completed by strong cationic resins or by enzymes. There exists also the possibility to use the small amount of phosphoric acid which should be converted into ammonium phosphates at the end of posthydrolysis [6], Fig.3. Our experiments with aureobasidium pollutants CCI-2771-14 [25] have shown good growing parameters on using glycosans as a single carbon source. Therefore a direct biochemical conversion of glycosans into the fodder yeast should have a priority, all the possibilities shown in Fig.3 being acceptable from the point of view of environmental criteria.

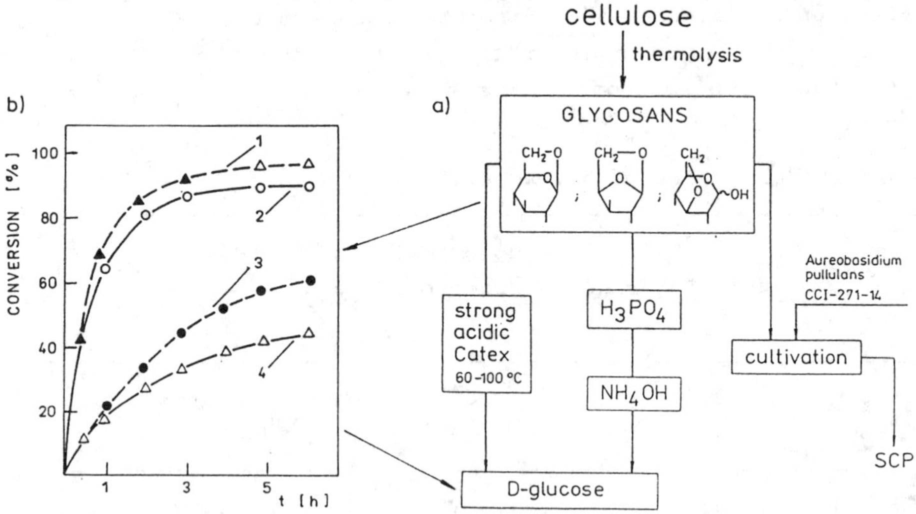

Fig.3 Thermolytic saccharification; a - thermolysis of cellulose into glycosans; b - conversion of glycosans into D-glucose; 1 - 10% H_2SO_4, 2 - 10% H_3PO_4, 3 - 4% H_3PO_4, 4 - Dowex 50 (Catex).

6 Conclusions

Various types of technologies for the conversion of cellulosics into biosubstrates have been compared according to the criterua of technological principles. The conclusions of these comparisions are presented in Table 2.

Table 2 Evaluated parameters of compared technologies

Parameter	Pretreatment + Enzym.hydrol.	Dilute acid	Strong acid	Thermolytic saccharification
Mass-convers.	+	++	+++	++
Level of technology	++	+ (+)	+	++
Driving forse	+	++ (+)	++	+++
Quality of product	++	+	+++	+
Energy consumpt.	+	+	++	+++
Environmental impact	+++	+	+	++
Economic. feasibility	+	+ (+)	+	++
	11	9 -(12)	13	15
	1	2	3	4

(+) - high hydromodul

In the case of cellulose conversion to glucose we have evaluated the respective technologies in accordance with seven principles.

I. Maximum mass conversion

up to 50 % +

up to 80 % ++

> < 80 % +++

II. Level of technology

conventional +
advanced ++
Hi-tech +++

III. Driving force of cenoversion

T - minutes for 50 % conversion +++
hours for 50 % conversion ++
days for 50 % conversion +

IV. Required quality of product

nonfermentable sugars > 5 % + < 5 % -
impurities > 3 % + < 3 % -
inhibition doses for cellulase (ID_{50})
10 hours - ID_{50} > +
10 hours - ID_{50} < -

V. Energy consumption - conventional 100 %

energy saving 25 % +
50 % ++
< 50 % +++

VI. a/ Environmental impact

cancirogenity compounds ---
toxicity compounds (irreversible concentration) --
nonbiodegradability compounds < 10 % -

b/ Environmental level of technology

- "end of pipe" technology +
- low waste ++
- close cycle system +++

VII. Economical feasibility

profit up to 10 % +
10-20 % ++
< 20 % +++

The order of conversion technology according to the technological principle criteria is: 4 > 3 > 2 > 1.

7 References

1. Blažej A., Košík M.: Degradation Reaction of Cellulose and Lignicellulose; Cellucon 87, Chapt. 8, 1989, 97.

2. Bretsznajder S. et al.: General Principles of Chemical Technologies, WN-T, Warszawa, 1973.

3. Pekarovičová A. et al.: Cellulose Chem. Technol. 23, 1989, 225.

4. Pekarovičová A. et al.: Chim. drevesiny (Riga), No 1, 1989, 115.

5. Senior D.J., Mayers P.R., Saddler J.N.: Biotechn. Bioengineer. 37, 1991, 274.

6. Blažej A. at al.: Bioconversion of Lignocell. in Theory and Practise, Cellulcon 88, 1988,

7. Fan L.T., Charpuray M.M., Lee Y.H.: Cellulose Hydrolysis, Springer Verlag, 1987.

8. Brownell H.H., Saddler J.N.: Biotechn. and Bioengineer. 29, 1987, 228.

9. Pekarovičová A. et al.: Drevársky výskum 106, 1985, 9; 108, 1986, 51.

10. Greenwald C.G., Nystrom J.M., Lee L.S.: Biotech. and Bioengineer Symp. No 13, 1983, 27.

11. Wright J.D. et al.: SERI Review 7 (12), 1985, 1-7.

12. Markoš J. et al.: Chem. průmysl, 39/64, 1989, 568.

13. Wright J.D., D'Agincourt C.G.: Biotechn. and Bioengineering No 14, 1984, 105.

14. Ullal V.G., Matharasan R., Grossmann E.D.: Biotechn. and Bioengineer. 14, 1984, 69.

15. Oshima M.: Wood Chem. Process Eng. Aspects, n.y. 1965.

16. Karlivan V.P.: New Aspect of the Production of Chemical from Biomass, CHEMRAWN, Toronto, July 10-13, 1978.

17. Tsao G.T. et al.: Process Biochemistry, Sept./October, 1982, 34.

18. Kusama J., Ishi T.: Wood Saccharification by HCl (g), Kogyo Kagaku Zasshi 69 (3), 1966, 469.

19. Chalov N.V.: Hydrolysis of Vegetable Raw Materials, USSR-patent 119 491, 1959.

20. Beardsmore A.I.: ICI Agricult. Div. Contract BO5084-UK, 1984.

21. Selke S.M. et al.: Ind. Eng.Product Res. Develop. 21, 1982, 11.

22. Selke S.M. et al.: Wood Agriculture Residues, 1983, 329.
23. Shafizadeh F., Cochran T.G., Sakai Y.: AICHE Symp. Ser. 75, 1979, 184.
24. Košík M. et al.: Czechoslov. pat. 215514, 1989.
25. Kocková-Kratochvílová A., Hrmová M., Košík M., Šturdík E., Biely P.: Czechoslov. pat. 245339, 1989.

55

Biological and thermochemical pretreatment of water hyacinth for improved biogas production

Vikram Patel and Datta Madamwar - Department of Biosciences, Sardar Patel University, Vallabh Vidyanagar 388 120, Gujarat, India.

ABSTRACT

Water hyacinth was subjected to various pretreatments and used as substrate in anaerobic digestion for biomethanation. Pretreatment in general increased the solubility of biomass and improved gas production.

INTRODUCTION

Rapid industrialization and population growth have increased the global energy requirement over the recent decade. Due to growing demand for energy there is an increased interest in the production of fuels from renewable sources. Water hyacinth (*Eichhornia crassipes*) has a high productivity coupled with an excellent pollutant removal and methane generation potential that makes it amenable to utilization in integrated pollution control and energy conversion system in rural areas. In practice, however, a problem arises due to the nature of the organic content in this material.

Like other biomass residues, water hyacinth is lignocellulosic in chemical composition. The degree of their conversion being dependent on the degree of their association with lignin, a natural binder which itself is virtually totally recalcitrant[1]. To improve the biodegradability of lignocellulosic materials, pretreatments to decrease the assocation with the carbohydrate portion may be affected, thereby increasing the substrate accessibility.

Water hyacinth, a lignocellulosic material, can be pretreated physically, chemically or biologically to increase its biotechnological substrate value[2]. Therefore it was felt necessary to study the effect of biological and thermochemical pretreatments on water hyacinth with the objective of improving both the performance stability and the rate of decomposition, with improved gas production.

MATERIALS AND METHODS

Resources - All the chemicals used were of analytical grade. Water hyacinth was obtained locally, it was dried at 60°C and powdered to 50 mesh. Fresh cattle dung was always obtained locally.

Microorganisms - *Heterobasidion annosum*; *Pleurotus ostreatus* and *Trametes hirsuta* were obtained from Institute of Microbial Technology, Chandigarh, India and were maintained on a medium containing yeast extract (5.0 g/l), Glucose (10.0 g/l), Agar (15.0 g/l), pH 5.8 at 30°C. They were subcultured once in two months.

Pretreatment - In a heat treatment 50 mesh water hyacinth powder in a 10% (W/V) slurry were treated at temperatures of 50, 100 and 121°C (with steam) for 1 hour and cooled.

In a chemical treatment 50 mesh water hyacinth powder in a 10% (W/V) slurry were treated with 2% NaOH (W/V), 2% $Ca(OH)_2$ (W/V) and 2% H_2O_2 (V/V) for 24 hours.

After the treatments, the samples were cooled down to room temperature and used as such for loading in to the anaerobic digesters. Whereas after chemical treatment samples were washed with water to neutral pH and used for anaerobic digestion. Care was taken to adjust the total solid (TS) as 7% (W/V) before loading in to digester.

In a biological treatment three delignifying fungi - *Heterobasidion annosum*, *Pleurotus ostreatus* and *Trametes hirsuta* were used. Inoculum was prepared by harvesting cell mass from one week old slant in a sterile distilled water containing small amount of Tween 80. Water hyacinth powder (100 g) was mixed with 200 ml of sterile distilled water in a 1000 ml Erlenmayer flask. The flaks were inoculated with the above inoculum and incubated for 4 weeks at 30°C. After the treatments, the samples were used as such for loading in to anaerobic digesters. Care was again taken to adjust the TS concentration to 7% (W/V) before loading into the digester.

Anaerobic digestion - Several bench scale anaerobic digesters were used. Each vessel consisted of a 10 litre glass reaction bottle, having working volume of 6 litre and containing 7% (W/V) of TS. The digesters were intermittently stirred with a magnetic stirrer and maintained at 37 $\pm$ 1°C. Gas was collected and measured by displacement of saturated acidified salt solution, making due correction for atmospheric pressure and temperature. The digesters were fed on a semi continuous basis : once per day using a freshly prepared mixture of water hyacinth biomass or treated powdered (50 mesh) water hyacinth and cattle dung in the ratio of 7:3 (W/W), with a retention time (RT) of 8 days (where loading rate was 8.75 g TS/litre of digester/ day). This was found to be most suitable from the initial study[3]. Prior to feeding, an equal quantity of sludge was withdrawn from the bottom of the digester.

A fresh digester was always started by preparing a mixture of the treated or untreated water hyacinth and cattle dung in the ratio of 7:3 (W/W) to give a final TS concentration of 7% (W/V) and using a 10% inoculum from the running biogas digester of the same type. A steady state condition was decided upon by a fairly constant mean gas production rate. Experiments were carried out in triplicate for each treatment and average data are presented.

Analysis - Gas composition was analysed by a CIC gas liquid chromatogram with stainless steel Chromosorb 2 column and thermal conductivity detector. Feed and effluent slurry were routinely analysed for pH, BOD, COD, volatile acids etc. as per standard procedures[4].

RESULTS AND DISCUSSION

Table 1 summarizes the effect of various pretreatments on chemical composition of water hyacinth slurry prior to feeding the digesters. Furthermore the percentage of solubilization as measured by the ratio of SCOD to COD is also presented in Table 1. The results indicate that heat treatment as low as 50°C produce an increase of 26.98% on COD solubility. The treatment at 121°C increased COD solubility by 39.77% indicating that solubilization of organic matter increases with heat

TABLE 1 :Effect of pretreatments on COD solubilization of water hyacinth slurry (10% (W/V)

Treatment		COD		Volatile acids
	Total (g/l)	Soluble (g/l)	% Solubi- lization	(g/l)
Control	51.20	8.19	16.00	1.05
Physical : 50°C	49.91	13.47	26.98	1.02
100°C	51.91	19.60	37.75	1.10
121°C	46.71	18.58	39.77	1.15
Chemical: H_2O_2	53.12	17.74	33.39	1.38
NaOH	52.00	29.72	57.15	2.98
$Ca(OH)_2$	48.79	26.34	53.98	2.11

treatment[3]. Treatment with H_2O_2, NaOH and $Ca(OH)_2$ showed a different effect. In general the alkali addition showed a more pronounced effect. The alkali effect was probably due to the solubilizatin of lignin This treatment is also responsible for delignification, lignin would otherwise be a major obstacle. The compositional changes in the treated solids can be advantageous or disadvantageous from the digestibility point of view. The solubilization of lignin constitute advantage in the sense that it liberates the remaining solid phase carbohydrate and increases its porosity while on the other hand, the soluble lignin derivative may exert a toxic effect.

The biodegradability of pretreated water hyacinth biomass was tested in a semicontinuous anaerobic fermentation in terms of its gas production and its methane content. Table 2 shows the steady state effluent and gas data from the digesters fed with heat chemical and biologically treated water hyacinth slurry. Pretreatment of water hyacinth by H_2O_2 did not enrich methane content in the digester gas. This is also supported by higher volatile acid content in the steady state pretreated water hyacinth digesters.

In general process stability is evidenced by lower amounts volatile acids[5-8].However, in this system, in spite of an increase in volatile acids, the process seems to be highly favourable with a high yield. During the steady state period a mixed population develops the ability, to convert particular organic molecules[9] to methane through population selection or enzyme induction.

TABLE 2 : Stedy state data of anaerobic digestion of water hyacinth (treated) - cattle dung.

Treatment	Gas*	Mathane (%)	BOD (g/l)	COD (g/l)	VA (g/l)	COD removal (%)
Control :	1.01	64	17.2	27.3	1.35	62.01
Physical :						
50°C	1.08	64	16.9	26.5	1.28	62.09
100°C	1.12	64	16.6	25.8	1.23	64.34
121°C	1.18	65	16.0	24.7	1.16	65.59
Chemical :						
H_2O_2	1.17	64	18.2	25.1	1.20	65.04
NaOH	1.51	66	14.8	22.9	1.10	68.10
$Ca(OH)_2$	1.43	66	15.0	23.4	1.19	67.40
Biological:						
H. annosum	1.59	67	12.3	24.9	0.73	65.32
P. ostreatus	1.73	68	11.4	24.0	0.71	66.57
T. hirsuta	1.59	67	13.6	25.2	0.87	64.90

*1/1 of digester / day

The treatment of alkali gave enhanced gas production. Maximum enhancement (over 50%) was achieved with the treatment of NaOH. In addition this treatment was also responsible for higher methane content in the digester gas. Alkali treatment showed greater increase in digestibility and was found to be advantageous to anaerobic fermentation for higher yield. This is attributed to the fact that alkali treatment decreases the association of lignin with carbohydrate and hence increases the substrate accessibility and anaerobic biodegradability.

Methane produciton rates from the biologically treated slurry are the highest since both the gas production rates and methane percentge are higher than those from heat and chemically treated slurries. This might be due to delignification of lignocellulose present in the waste. Biological processes proceed under mild conditons, give higher product yields and fewer side reactinos and demand less energy and less reactor resistance to pressure and corrosion. But at the cost of higher retention time.

Volatile acid contents were low in treated digesters as compared to the untreated digester. It appears that alkaline (and biological) treatment did enhance the methane forming step of the digestion process which often involves the degradation of fatty acids, in addition to solubilization of water hyacinth biomass.

The values of COD are presented in Table 2. It shows that gas production increases with COD reduction[10]. This can be explained as follows: substrate stabilization, which is important in the digestion of wastes, is related to COD reduction. The only way, in which COD reduction can occur in an anaerobic digestion process, is through the removal of organic material from wastes, such as by evolution of methane and carbon dioxide. Hence the total gas production is directly proportional to the COD reduction. The values of effluent BOD in a digester with pretreated water hyacinth biomass were low, indicating greater biodegradation with high process performance. This also indicates that microbial degradatin of organic matter was at a higher rate in the treated water hyacinth digester than in the control. Thus in general pretreatment improves the digester performance.

ACKNOWLEDGEMENT

Thanks are due to the DNES, New Delhi, India for financial support.

REFERENCES

1. Pavlostathis, S. F.; Gossett, J. M. Biotechnol. Bioeng. 1985, 27, 443.
2. Muller, H.W.,; Trosch, W. Appl. Microbiol. Biotechnol. 1986, 24, 180.
3. Madamwar, D. B.; Patel, A., Patel, V.; J. Ferment. Bioeng. 1990, 70, 340.
4. APHA. 1976. Standard Methods for the Examination of Water and Waste Water. 14th Ed. American Public Health Association, Washington DC.
5. Stuckey, D. C.,; McCarty, P. L. Biotechnol. Bioeng. Symp. 1978, 8, 213.
6. Fraser, M. D. 1977. Proc. Symp. Fuels from Biogas and Wastes. Inst. of Gas Technology, Chicago.
7. Varel, V. H.; Hashimoto, A. C.; Chen, Y. R. Appl. Environ. Microbiol. 1980, 32, 420.
8. Madamwar, D. B.; Mithal, B. M. Biotechnol. Bioeng. 1986, 28, 624.
9. McCarty, P. L.; Young, L. Y.; Cossett, J. M., Stuckey, D. C.; Healy, J. B. 1976. In Proc. Seminar on Microbial Energy Conversion, Germany, Oct. 1979.
10. Tosum, A.; Selcuk, N.; Soyupak, S. Chem. Eng. Technol. 1989. 12, 6.

56

Cellulases and ligninases of basidiomycetes

Z.R. Ahmedova - Institute of Biochemistry and Physiology of Microorganisms, Russian Academy of Sciences, Pushchino, Moscow Region, 142292, Russia.

ABSTRACT

Screening of wood-degrading fungi has been performed to select active strains decomposing lignocellulose substrates. The active cultures of basidiomycetes *Panus tigrinus, Pleurotus ostreatus, Fomes fomentarius* and *Inonotus hispidus* have been selected. Enzyme formations have been noted to depend on the structure of lignocellulose component of the medium and cultivation time. It has been shown that all the above cultures have cellulolytic activities, and the cultures of *P.tigrinus* and *P.ostreatus* also have laccase, as well as the greatest ligninolytic activities. Fungi *F.fomentarius* and *I.hispidus* reveal no laccase activities. It has been established that a potential producer of cellulases is the fungus *I.hispidus,* and that of hemicellulases and ligninases is a basidiomycete *P.ostreatus.*

INTRODUCTION

Cellulose is a plant structure component resistant to various effects. This resistance is provided by another structure component - lignin. The best method for wood cellulose and lignin degradation is their enzymatic decomposition by microorganisms.

Literature data show that relatively complete wood decomposition occurs with white rot fungi (mostly basidiomycetes and some ascomycetes), brown rot fungi (basidiomycetes), and soft rot fungi (ascomycetes and imperfect fungi). However, in Kirk and Shimada's opinion, natural degradation of lignocellulose wastes occurs mostly with white rot fungi [1]. Brown rot fungi utilize mostly polysaccharides, thus loosening lignin structure and creating favourable conditions for the functioning of while rot fungi which are producers of oxidation enzymes depolymerizing lignin, such as laccase and peroxidase.

Despite a lot of investigations carried out at present, possible ways of lignin oxidation by certain microorganisms and participation of certain enzymes in this process are still unclear, the whole complex of enzymes in cultures has not been studied fully, and universal cultures promoting complete degradation of lignocellulose substrates are little described.

In this connection, the present investigation was aimed at selection of new microbial strains having complete sets of corresponding enzymes, and study of the formation of cellulolytic, xylanolytic, and ligninolytic activities of selected cultures.

METHODS

The work used more than 264 cultures from culture collections of the Institute of Microbiology and the Laboratory of mycology, Uzbek Republic Academy of Sciences, and the cultures isolated from decaying plant field wastes from soils of Uzbek Republic regions where cotton is sown. As a result of screening by Bavendamm [2], the following cultures were selected: *Panus tigrinus, Pleurotus ostreatus, Fomes fomentarius, Inonotus hispidus.*

Cultivation of fungi was carried out in submerged culture on a specially chosen nutrient medium, containing various plant wastes: depleted bagasse of cotton seed ground oilcake, cotton stems, rice hull, hemp-mallow chaff, and beer wort in different concentrations. Fungi were grown in 500 ml Ehrlenmeyer flasks with 200 ml of the nutrient medium for 3-30 days.

For enzymatic analysis, the culture liquid filtrate was used, samples being taken every 3 days. Protein content in the samples was determined by the Lowry method [3].

Endoglucanase activity was determined by viscosimetry (by decrease of high-polymer CMC viscosity), and by formation of reducing substances as the result of degradation of low-polymer CMC [4].

Cellobiase activity was determined by 1 mkM of decomposed cellobiose for 1 min at 40°C [4]. Xylanase activity was determined by the amount of reducing sugars formed in the result of enzymatic xylan hydrolysis [4].

Laccase activity was determined by syringaldazine oxidation [5]; peroxidase activity by dianizidine oxidation [6]; ligninase activity by veratryl alcohol oxidation [7].

RESULTS AND DISCUSSION

The study of formation of cellulolytic, xylanolytic and ligninolytic activities (see Table) in selected fungi *P.tigrinus, P.ostreatus,*

Table. Maximum activity of cellulolytic, xylanolytic (A) and ligninolytic (B) enzymes of some basidiomycetes during their cultivation on media with plant wastes.

A

Substrate		Enzyme activity in culture liquid, units/ml					
Cultures		Xylanase	Cultivation time, days	Endoglucanase	Cultivation time, days	Cellobiase	Cultivation time, days
1. *P.tigrinus*	I*	8.27	3	3.27	3	4.9	21
	II	9.38	3	4.38	15	3.23	18
	III	4.02	12	2.98	12	4.63	18
	IV	4.07	15	2.47	9	3.77	3
2. *P.ostreatus*	I	5.62	3	4.34	9	4.16	3
	II	9.97	3	5.4	6	6.27	9
	III	9.59	3	4.24	9	6.45	12
	IV	9.35	6	3.94	15	5.66	3
3. *F.fomentarius*	I	7.52	6	3.4	6	3.7	6
	II	10.07	3	4.08	18	2.3	9
	III	5.98	12	4.20	9	3.72	12
	IV	9.17	3	3.3	12	5.67	15
4. *I.hispidus*	I	8.71	9	6.72	6	8.02	9
	II	6.83	12	8.09	9	9.18	6
	III	9.92	12	6.7	9	5.0	9
	IV	8.67	12	9.08	6	5.02	9

*I - medium with depleted bagasse of cotton seed ground oilcake, II - medium with cotton stems, III - medium with rice hull, IV - medium with hemp-mallow chaff

B

Substrate		Enzyme activity in culture liquid, units/ml					
Cultures		Laccase	Cultivation time, days	Peroxidase	Cultivation time, days	Ligninase	Cultivation time, days
1. *P.tigrinus*	I	53.0	15	17.97	15	36.0	18
	II	30.8	18	9.14	15	2.11	24
	III	32.3	9	10.53	15	1.44	18
	IV	28.0	12	6.95	15	2.41	21
2. *P.ostreatus*	I	108.2	3	67.1	24	24.2	27
	II	90.5	9	42.5	18	12.8	6
	III	65.9	6	15.9	12	16.0	24
	IV	47.6	18	37.6	21	13.4	12
3. *F.fomentarius*	I	-	-	34.34	12	21.7	12
	II	-	-	14.00	9	1.57	18
	III	-	-	7.83	9	6.0	12
	IV	-	-	9.1	6	1.3	21
4. *I.hispidus*	I	-	-	3.7	6	16.3	12
	II	-	-	3.2	12	3.0	15
	III	-	-	4.1	9	1.4	24
	IV	-	-	-	-	0.7	27

F.fomentarius and *I.hispidus* on media with various plant wastes has shown that during the growth on lignocellulose wastes the above fungi secrete into the culture medium cellulolytic and ligninolytic enzymes, depending on the structure of ligninocellulose component of the medium and cultivation time.

Thus for example, we have revealed that the highest endoglucanase and cellobiase activities are expressed by the fungus *I.hispidus.* Maximum formation of cellobiase is observed when the fungus is cultivated on the medium with cotton stems (9.18 units/ml) (see Table), and that of endoglucanase - on the medium with hemp-mallow chaff (9.08 units/ml) on the 6th day of the producer growth (see Table). The fungus *Fomes fomentarius* shows high xylanase activity on the medium with cotton stems on the 3^{d} day of growth (10.7 units/ml). Usually the formation and accumulation of xylanase activity in nearly all the media anticipates those of other enzymes (see Table), which seems to be the result of the presence in substrate of xylans - easily hydrolyzed hemicellulose regions accessible for enzymes. Anticipating accumulation of cellobiase activity, when fungi *P.tigrinus* and *P.ostreatus* are grown on the medium with hemp-mallow chaff, is an exception. Probably, this substrate contains low-molecular oligosaccharides inducing the formation of cellobiase in the medium. The similar accumulation of cellobiase and endoglucanase activities during growing the fungi on the medium with bagasse can be explained by the presence in substrate of oligosaccharides, together with polysaccharides.

The experiments have shown that the favourable medium for the formation of ligninolytic enzymes in all tested cultures is usually a medium with depleted bagasse of cotton seed ground oilcake. Results given in Table show that fungi *P.tigrinus* and *P.ostreatus* have high laccase activities on this medium, as well as on the medium with cotton stems. Formation of ligninolytic enzymes occurs at both exponential and stationary phases of culture growth. Compared with laccase and peroxidase, ligninase is synthesized significantly later - on the 18-27th day of growth (Table, Appendix B). In the literature there is information on cellulose being a co-metabolizing substrate for the biosynthesis of enzyme complexes [8]. It is not to be ruled out that the expression of high laccase and peroxidase activities during the exponential phase of the fungi growth is associated with the effect of cellulose present in plant substrates.

Investigations of lignin-degradative ability of the fungi have shown that the culture of *F.fomentarius* and *I.hispidus* are not able to oxidize syringaldazine which was a substrate for laccase activity expression in the experiments. Moreover, *I.hispidus* shows very low peroxidase activity on all media, and in the medium with hemp-mallow chaff this enzyme has not been revealed (Table).

It should be noted that in all nutrient media used the protein content increased during the culture growth, and the quantity of biomass started to decrease gradually on the 15th day of cultivation. Maximum accumulation of biomass on all media falls on 6-9th day of producers growth.

Results of our study indicated that total enzymic activities of cellulases, hemicellulases and ligninases of the assayed fungi for the production of the assayed enzymes are assigned like the following:

xylanase: *P.ostreatus* » *I.hispidus* » *F.fomentarius* » *P.tigrinus*
cellobiase: *I.hispidus* » *P.ostreatus* » *P.tigrinus* » *F.fomentarius*
endoglucanase (by degradation of high-polymeric CMC): *I.hispidus* » *P.ostreatus* » *F.fomentarius* » *P.tigrinus*

endoglucanase (by degradation of low-polymeric CMC): *I.hispidus* » *P.tigrinus* » *P.ostreatus* » *F.fomentarius*
peroxidase: *P.ostreatus* » *F.fomentarius* » *P.tigrinus* » *I.hispidus*
laccase: *P.ostreatus* » *P.tigrinus*
ligninase: *P.ostreatus* » *P.tigrinus* » *F.fomentarius* » *I.hispidus*,
which shows that a potential producer of cellulases is the fungus *I.hispidus*, and that of xylanases and ligninases is *P.ostreatus*.

Thus, in our investigation it has been established that the local strain of basidial fungus *P.ostreatus* (isolated from soils of cotton-sowing parts of Tashkent Region, Uzbek Republic) is a universal culture having a wide set of enzymes: cellulases (endoglucanases, cellobiases), hemicellulases (xylanases), and ligninases (laccases, peroxidases, ligninases), which take an active part in degradation of lignocellulose plant wastes.

The fungus *I.hispidus* can also be successfully used as a producer of cellulases for fungal biomass and protein production.

REFERENCES

1. Kirk,T.; Shimada,M. *Biosynthesis and biodegradation of wood components*. Acad.Press, **1985**, 579.
2. Bavendamm,W. *Zbl. fur Bakt.*, II abt. **1928**, *75*, 15.
3. Lowry,O.H.; Rosebrough,N.J.; Farr,A.L.; Randall,R.J. *J.Biol.Chem.* **1951**, *193*, 1, 265.
4. Feniksova,R.V.; Tiunova,N.A.; Rodionova,N.A.; Martynovich,L.I.; Kudryashova,T.I. In: *Methods of modern biochemistry*. Moscow: Nauka, **1975**, 23 (in Russian)
5. Harkin,J.U.; Obst,J.R. *Experienta*, **1975**, *29*, 391.
6. Gudkova,L.V.; Degtyar,R.G. In: *Enzymes in medicine, food industry and agriculture*. Kiev: Naukova Dumka, **1968**, 172 (in Russian).
7. Tien,M.; Kirk,T.K. *Science* **1983**, *221*, 4611, 661.
8. Golovleva,L.A.; Ganbarov,Kh.G.; Skryabin,G.K. *Mikrobiologiya* **1982**, *51*, 4, 543 (in Russian).

57

Characterization of cellulases activity and electrophoretic analysis from *Ganoderma resinaceum* Boud. Apud Patouillard extract

M.C. Sàágua, Mª M. Carolino and N. Teixeira Rodeia - Departamento de Biologia Vegetal, Faculdade de Ciências, 1700 Lisboa, Portugal.

ABSTRACT

The species studied was *Ganoderma resinaceum* with cotton cellulose as carbon source, in a mineral medium. Incubation was at 28ºC for 36 days. Extracellular extracts were concentrated by ultra-filtration and the fraction analysed for both endoglucanase and ß-glucosidase activities (5270 U x 10^{-3} and 1242 U x 10^{-3} respectively). The fractions eluted with enzymatic activity were chromatographed through DEAE Sephadex A-50 by ionic charge and two enzymes were separated .The maximum endoglucanase stability was in the crude extract between 50 and 55ºC and between 50 and 65ºC in the isolated fractions. Also the K_m were evaluated (to glucose; 1.18% or 0.4% in the crude extract or in the isolated fractions respectively) and V_{max} (to glucose; 0.19 μmol min^{-1} or 0.124 μmol min^{-1}).The different fractions with each enzymatic activity were joined, lyophilized, resuspended and submitted to electrophoresis 7.5% PAGE/ 0.1% SDS at pH 8.3 using a tris-glycine 25mM buffer. Two bands were respectively revealed, one corresponding to a ß-glucosidase (M.W. between 67,000 and 43,000) and the other one to an endoglucanase (M.W. between 20,000 and 14,400).

INTRODUCTION

We have always called attention to the importance of the study of cellulases produced by fungi, which would be used in industrial fermentation and the food industry. We have, therefore, studied the cellulasic enzyme complex from *Ganoderma resinaceum* Boud. Apud. Pat., in order to purify different enzymatic components with a view to studying the mechanism of the action of cellulases on cellulose. We have now used cotton cellulose as the only carbon source to induce *G.resinaceum* to excrete cellulases to the extracellular medium.We obtained one ß-glucosidase with action on ortho-nitrophenyl ß-D-glucoside (ONPG) and one endoglucanase with action in carboxymethylcellulose. For this paper we researched the differentiation of cellulases using the knowledge we have about cellulosic substrate specificity.We then studied the mechanism of enzymatic action of the cellulases, namely of endoglucanases. This involved the utilization of cellooligosacharide substrates, with a polymerization degree of up to 6, our objective being to determine the glucosidic bond which is specifically broken by different endoglucanase.

MATERIAL AND METHODS

MATERIAL: One strain of *G. resinaceum*, cultured during 36 days at 28° C; the reagents necessary in the different methods used.

METHODS:

1) Norkrans and Hammarström (1963) medium modified (Melo, 1977) - culture media and endoglucanase activity.
2) Nelson (1944), Somogyi (1952) - CMCase activity. Effect of temperature on the stability of enzymes - Dixon & Webb, 1964.
3) Wood (1985) - ß-glucosidase activity.
4) Lineweaver & Burk (1934) - K_m and V_{max} evaluation of the different enzymes on different substractes.
5) Dixon (1953) - K_i evaluation of the different enzymes on different substractes.
6) Davies (1964). Electrophoretical analysis.
7) Parr (1983) - Kinetic properties of the ß-glucosidase and endoglucanase.

RESULTS AND DISCUSSION

In the Tables I and II are inserted the total activity and yield from one ß-glucosidase and one endoglucanase from *G. resinaceum* both before and after purification by ultra-filtration.

TABLE I - Purification of one ß-glucosidase from *G.resinaceum*.

Step	Volume (ml)	Total activity (unitsx10^3)	Yield (%)
Culture filtrate	150	450	100
Ultrafiltrate	12	5270	94

TABLE II - Purification of one endoglucanase from *G.resinaceum*.

Step	Volume (ml)	Total activity (unitsx10^3)	Yield (%)
Culture filtrate	150	110	100
Ultrafiltrate	12	1243	72

The kinetic properties of one endoglucanase and one ß-glucosidase of the cellulase system from *G. resinaceum* as exhibited in Figures 1-8 appear to be typical of cellulases in general, Stutzenberger (1988). The assay conditions were pH 5.5 and 50°C respectively.

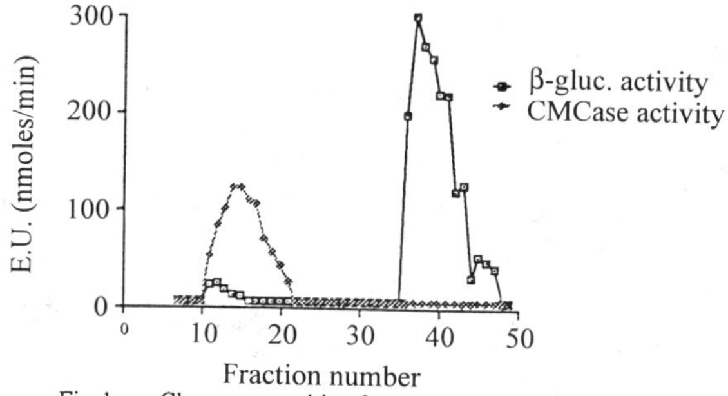

Fig.1 - Chromatographic fractionation of the ultrafiltered extract through one DEAE - Sephadex A - 50 column.

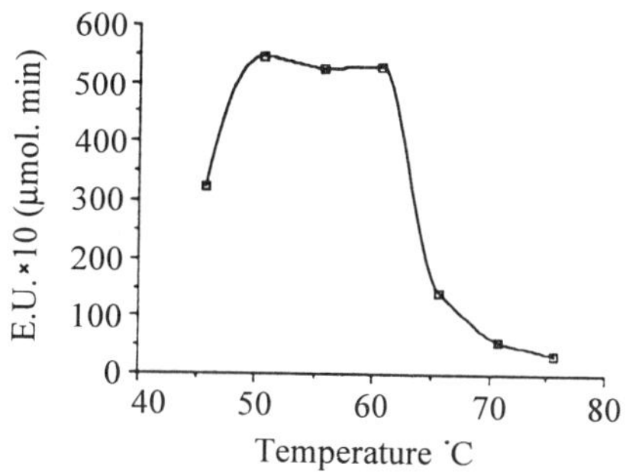

Fig.2 - Optimal temperature endoglucanase activity in the crude extract.

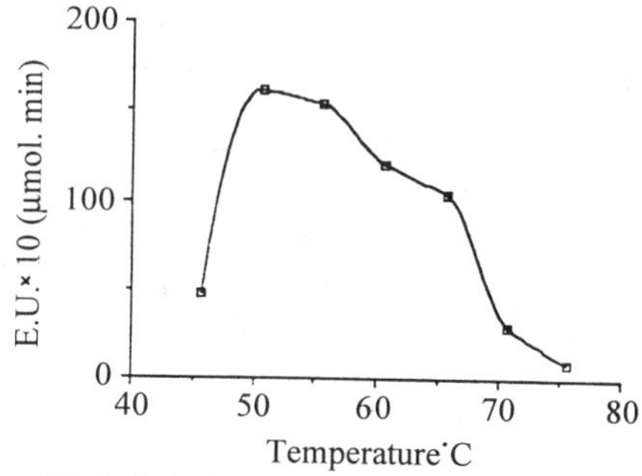

Fig.3 - Optimal temperature endoglucanase activity in the fraction isolated by chromatography.

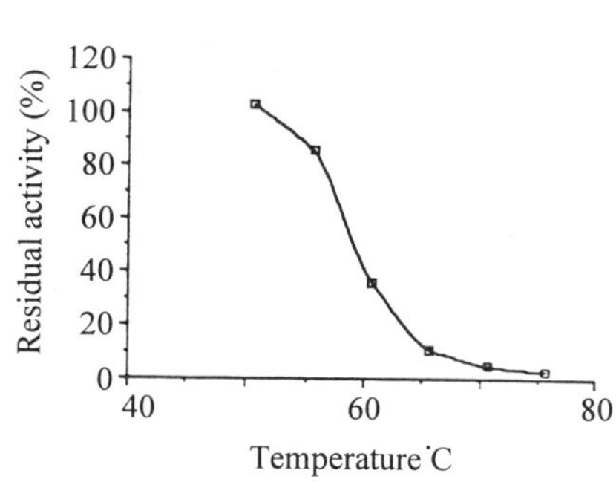

Fig.4 - Effect of temperature on the enzymatic stability of the endoglucanase in the crude extract.

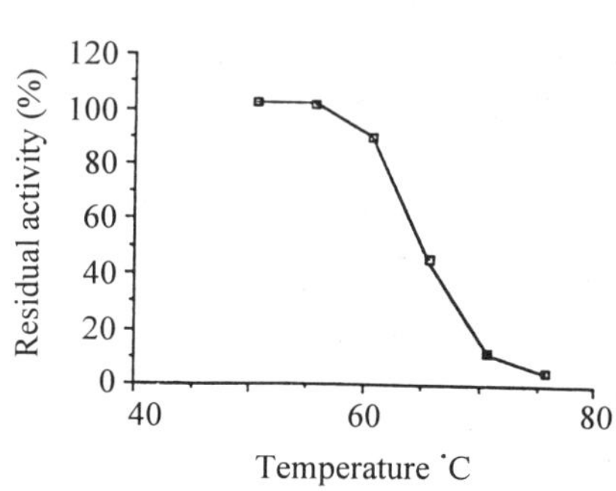

Fig.5 - Effect of temperature on the enzymatic stability of the endoglucanase in the isolated fraction.

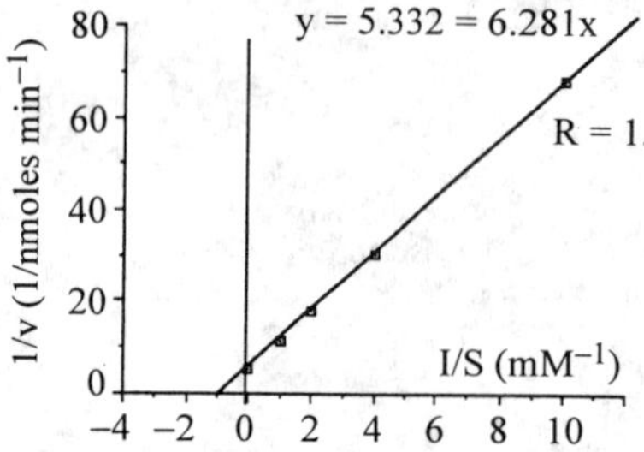

Fig.6 - Graphic representation of Lineweaver-Burk to determine the kinetic constants (Km and Vmax) of the endoglucanase present in the crude extract. Km = 1.18 %; Vmax = 0.19 µmol / min

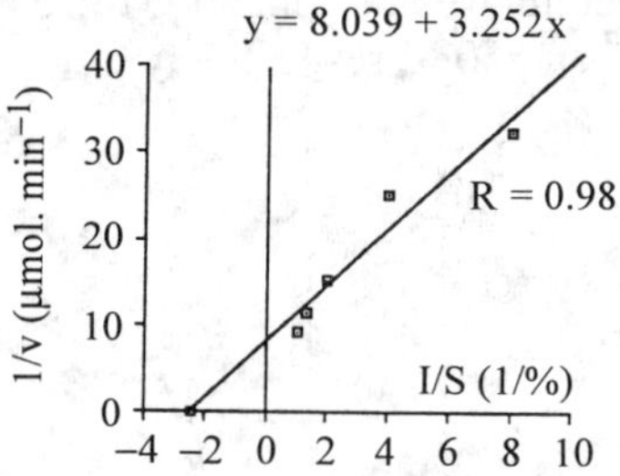

Fig.7 - Graphic representation of Lineweaver-Burk to determine the kinetic constants (Km and Vmax) of the endoglucanase present in the isolated fraction Km = 0.4 %; Vmax = 0.124 µmol / min

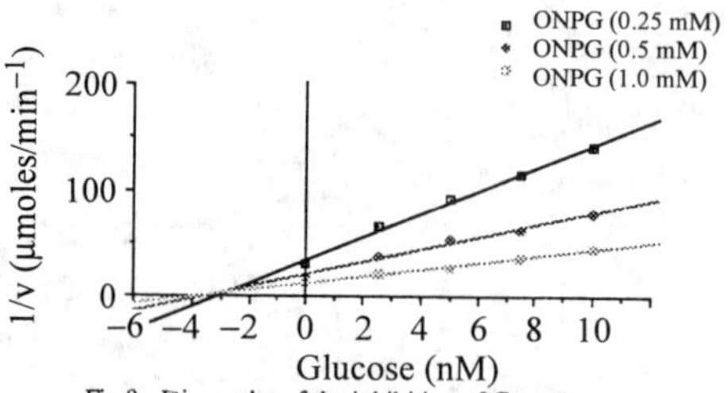

Fig.8 - Dixon plot of the inhibition of Ganoderma resinaceum ß-glucosidase by D-glucose. Conditions: 0.2M sodium acetate buffer, pH 4; temperature 50 oC. Ki = 2.8 mM.

The effect of the presence of D-glucose on the initial rate of reaction of ß-glucosidase with ONPG as substrate shown as a Dixon plot, indicating that D-glucose acts as a competitive inhibitor with an inhibition constant (k_i) of 2.8mM under the specified reaction conditions. The results show that preparation of cellulases from *G.resinaceum* are particularly effective in catalysing the hydrolysis of CMC or the ONPG substrates to produce glucose.

As the K_m is an inverse measure of the affinity of the enzyme to the substrate, we can affirm that the *G.resinaceum* endoglucanase in an isolated form presents a higher affinity to the CMC than the enzyme present in the crude extract . We do not present any photography of electrophoresis because the bands were barely perceptible. Although our results concord with Wood,1988 - one ß-glucosidase from *Trichoderma koningii* has a molecular weight of 39,800 however Shepherd *et al.* 1988 determined a M.W.of 85,000 for one ß-glucosidase of *Thermoascus aurantiacus* and Dias, 1981 evaluated one ß-glucosidase of *Fusarium graminearum* with a M.W. of 305,00. In relation to the endoglucanases Lindner (1988) mentions a M.W. of 54,000 or 45,000 (with a Biogel P-150 or Biogel P-100); this value was different from that obtained by us, but Ulker & Sprey (1990) verified a position of the 25,000 endocellulase in *Trichoderma reesei* which is a value nearer that determined by us.

CONCLUSIONS

1 - Extracellular extracts have endoglucanase and ß-glucosidase activities -Table I and Table II.

2 - Ultrafiltration promotes a good concentration of those enzymes - Table I and Table II.

3 - Chromatography of concentrated samples analysed through DEAE-Sephadex-A-50 separated one endoglucanase and one ß-glucosidase - Fig.1.

4 - The optimum temperature for endoglucanase activities in the crude extract and in the isolated fraction is about 50°C - Figs 2 and 3.

5 - The endoglucanase is stable at a temperature within the range of 50-60°C either in the crude extract or in the isolated fraction; but above this temperature it quickly denatured - Fig.4 and 5 .

6 - The endoglucanase purified on DEAE Sephadex A-50 presents a K_m value of 0.4 % and a V_{max} value of 0.124 µmol/min for carboxymethylcellulose. However those values in relation to the crude extract are 1.18 % and 0.19 µmol/min respectively - Figs 6 and 7.

7 - On the basis of a Dixon plot the hydrolysis of ONPG appeared to be competitively inhibited by D-glucose - K_i = 2.8mM - Fig.8.

8 - The different fractions which show cellulase activity when joined, lyophilized, resuspended and submitted to a 7.5 PAGE / 0.1% SDS electrophoresis at pH 8.3 using a Tris-glycine 25mM buffer, revealed 2 bands. One corresponds to a ß-glucosidase with a Rf of 0.318, M.W. between 67,000 and 43,000. The second corresponds to an endoglucanase with a Rf of 0.618, M.W. between 20,000 and 14,400.In both cases the bands were compared with one kit standard of low molecular weight between 94,000 and 14,400 daltons (phosphorilase b, albumin, ovalbumin, carbonic anhydrase, trypsine inhibitor, and α-lactalbumin).

9 - *G. resinaceum* could be a good source of cellulase systems since high levels of ß-D-glucosidase and endoglucanase can be produced extramycelially in a single reaction.

ACKNOWLEDGEMENT

This research was financed with a grant from JNICT and had a financial help from INIC enabling me to participate in Cellulose 91 Conference, New Orleans, Louisiana , 2-6 December 1991.

REFERENCES

Bhat, K . M.; May, A . J.; Clayessens, M. C.; Wood, T . M. - **1990** - Study of the mode of action and site specificity of the endo (1->4)-ß-D-glucanases of the fungus Penicillium pinophilum with normal 13H-labelled reduced and chromogenic cellooligosacharides.*Biochemistry Journal* , 266 : 371 - 378.

Davies, B . J.- **1964** - Disc electrophoresis.II. Method and application to human proteins. *Annals of the New York Academy of Sciences*, 121 : 404-427.

Dias, M.C.L.- **1981** - Isolation and some properties of a ß-glucosidase from Fusarium graminearum. *Ciência Biológica* 6 : 279 - 286.

Dixon, M.- **1953** - *Biochemistry Journal* , 55 : 170.

Dixon, M.; Webb, E. C. - **1964** - Enzymes.2nd edition. Longmans. London.

Lineweaver, H.; Burk,D.-**1934** - The determination of enzyme dissociation constants.*Journal of American Chemical Society* , 56 : 658-666.

Lindner, W. A. - **1988** - Carboxymethylcellulase from Sclerotium rolfsii. *Methods in Enzymology* , 160 : 376 -382.

Lugtenburg, et al. - **1975** - *FEBS Letters* , 58 : 254-258.

Melo, E.M.P.- **1977** - Contribuição para o estudo de celulases de Rosellinia necatrix. Relatório de Estágio Científico - Faculdade de Ciências de Lisboa.

Nelson, N. - **1944** - A photometric adaptation of the Somogyi method for the determination of glucose. *Journal of Biological Chemistry* , 153 : 375-380.

Norkrans, B.; Hammarstrom, A.-**1963**-Studies on growth of Rhizina undulata Fr.and its production of cellulose and pectin decomposing enzymes. *Physiologia plantarum* , 16 (1) : 1-10.

Parr, R.S.-**1983** -Some kinetic properties of the ß-glucosidase (cellobiase) in a commercial cellulase product from Penicillium funiculosum and its relevance in the hydrolysis of cellulose. *Enzyme Microbiology Technology*, 5: 457-462.

Shepherd, G. M.; Cole, L. A. & Tong, C. C. - **1988** - Cellulases from Thermoascus aurantiacus. In: *Methods in Enzymology*, 160 (A) : 300-307. Wood, A.W.; Kellog ,T.S. (eds). Academic Press , Inc.London

Somogyi, M. - **1952** - Notes on sugar determination.*Journal Biological Chemistry* 195 : 19-23.

Stutzenberge, F. - **1988** - Production and activity of Thermonospora curvata cellulases on protein extract lucerne fibers. *Applied Microbiology Biotechnology* , 28 : 387 - 393.

Wood,T.M. - **1985**- Aspects of the biochemistry of cellulose degradation. In: Cellulose and its derivatives, chemistry, biochemistry and applications, 173-178. Kennedy, F., Phillips,G.O., Wedlock, D.J.; Williams, P.A. (eds). Edward Arnold, London.

Wood, M. T. - **1988** - Cellulase of Trichoderma koningii. In: *Methods in Enzymology*, 160(A): 210-234 . Wood, A.W.; Kellog,T.S.(eds). Academic Press, Inc.London.

Yeoh, H. H.; Tan, T .K.; Tian, K. E. - **1984** - Cellulolytic enzymes of fungi isolated from wood materials. *Mycopatologie* , 87 : 51-55.

58

Complexity of glycoside hydrolases from *Trichoderma viride* A 10

Pei-Ji Gao, Yin-Bo Qu and Zu-Nong Wang - Institute of Microbiology, Shandong University, Jinan, 250100, P.R. China.

ABSTRACT

At least 37 components of glycoside hydrolases obtained from Trichoderma viride A 10 have been detected and purified by ion-exchange on DEAE-Sephadex, SE-Sephadex, and CM-Bio-gel A. Most of them showed a single band on polyacrylamide gel electrophoresis. Their characteristics including molecular weight, isoelectric point, 200~600 nm absorption spectrum, glucosyl content and substrate specificity were determined.

According to their substrate specificity activities, they were divided into four groups: glucosidase, endo-glucanase, glucan cellobiohydrolase and other polysaccharide hydrolases.

All six glucosidases could hydrolyze (1-4) and (1-2) linked β-D-glucosidic bonds and some had activity for (1-3) and (1-6) β-D-glucosidic bonds, (1-3), (1-4), and (1-6) α-D-glucosidic linkages and some could hydrolyze (1-4) linked β-D-xyloside efficiently. Similarly all 17 endoglucanases were identified with high activity against glucoside linkages in amorphous cellulose (CMC, swollen cellulose), and nine of which had activity for (1-4)-β-D-xylan, seven for (1-4)-α-D-glucoside bonds in starch, and two for (1-3, 1-4)-α-D-glucosidic linkages in lichenin.

INTRODUCTION

Cellulose, hemicellulose and lignin are the main constituents of the plant cell wall and a large spectrum of glycoside hydrolases are needed for conversion of these fractions. Important groups of such hydrolases are cellulase, xylanase and ligninase. Many saprophytic fungi such as Trichoderma spp.[1], Penicillium pinophilu [2], Fusarium sola [3] and Phanerochaete chrysosporium [4] etc. are able to produce a wide variety of polysaccharide hydrolases. Fractional studies on culture filtrates of the fungi mentioned above have demonstrated that there are three major types involved in hydrolysis of native cellulose. They are: endo-β-1,4-D-glucanase (1,4-β-D-glucan 4-glucanohydrolase, EC 3.2.1.4), cellobiohyrolase (1,4-β-D-glucan cellobiohydrolase, EC 3.2.1.91) and β-D-glucosidase (β-D-glucoside glucohydrolase, EC 3.2.1.21). All of those enzymes are generally found to exist in multiple forms[5], some of which have cross-substrate specificities [6]. This is a most interesting phenomenon in enzyme synthesis and substrate hydrolysis, particularly with regard to defining the optimal preparations of the enzyme necessary to gain efficient hydrolysis of raw materials. This paper reports on the fractionation of a crude enzyme preparation from Trichoderma viride A 10 and characterization of all the components which could be detected and purified to homogeneity.

MATERIALS AND METHODS

Enzyme preparation and activity assay

The crude enzyme powder used was a commercial product produced by precipitation of the culture filtrate of Trichoderma viride A 10 by ethanol and then dried in vacuo.

One unit of enzyme is defined as the amount of protein which catalyses the release of one micromole of substrate per minute (expressed as glucose or other aldose) under assay conditions.

Endo-D-glucanase activity assay: A reaction mixture containing 2 ml 1% polysaccharide suspension in 0.02 M sodium acetate buffer, (pH 4.8) and 0.5 ml enzyme solution was incubated at 40° C, and the reducing sugar was measured by addition of Nelson-Somogyi reagent [7]. The incubation time varied depending on the type of substrates used: cotton linter pulp, 24h; cellulose powder, Avicel, Phosphate-swollen cellulose and chitin, 4h; xylan, pectin, lichenin and laminarin, 2 h ; starch, 1 h ; CMC and HEC, 0.5 h.

Glucosidase activity assay: 2- or 4-nitrophenyl glucoside and similar glycosides of oligosaccharides were selected as substrates for determining glucosidase activity. The reaction mixture contained 1 ml glucoside substrate (0.5%) in 0.02 M sodium acetate buffer, pH 4.8 and 0.5 ml enzyme solution. After incubation for 0.5h at 40° C, reaction was stopped by addition of 0.1 ml 10% $NaNO_3$ solution. The liberated nitrophenol was measured at 400 nm. When the oligosaccharide glycoside were used, the reducing sugars released were determined by the method of Nelson-Somogyi.
Cellobiohydrolase activity (CBH) was estimated by the method described by Deshpande et al [8] using 4-nitrophenyl β-D-celluloside as substrate.

Column chromatography

Chromatography was carried out on DEAE-Sephadex A 25 (Pharmacia), SE-Sephadex C-50 (Pharmacia), DEAE-Biogel A (Bio-Rad) or CM-Biogel A (Bio-Rad) in a column (2.5 x 25 cm) at 10--15° C and fractions were collected with an automatic fraction collector (type BS-100, China) and analyzed for enzyme activities. An ionic strength gradient mixture was employed as necessary to provide a continuous linear gradient with respect to acetate buffer, citrate buffer or sodium chloride. Prior to rechromatography, each pooled fraction was concentrated in an ultrafiltration cell using a PM 10 membrane, dialysed in an appropriate buffer and then applied to a column equilibrated with the same buffer. Protein was detected by U.V. at 280 nm.

Polyacrylamide gel electrophoresis

Homogeneity of each purified component was analyzed by polyacrylamide gel disc-electrophoresis [9]. Analytical isoelectric focusing was performed in 7.5% gels containing 1% ampholyte (LKB, Sweden) in the pH range of 3.5 to 10.0. The total carbohydrate content of the purified component was determined by the phenol-H_2SO_4 method [10].

RESULTS

Separation and purification of all detectable components

A summary of the procedure used for purification of the enzyme complex is given in Fig.1.
About 15 g crude cellulase powder was dissolved in 50 ml 0.02 M sodium acetate buffer (pH 5.8), and then

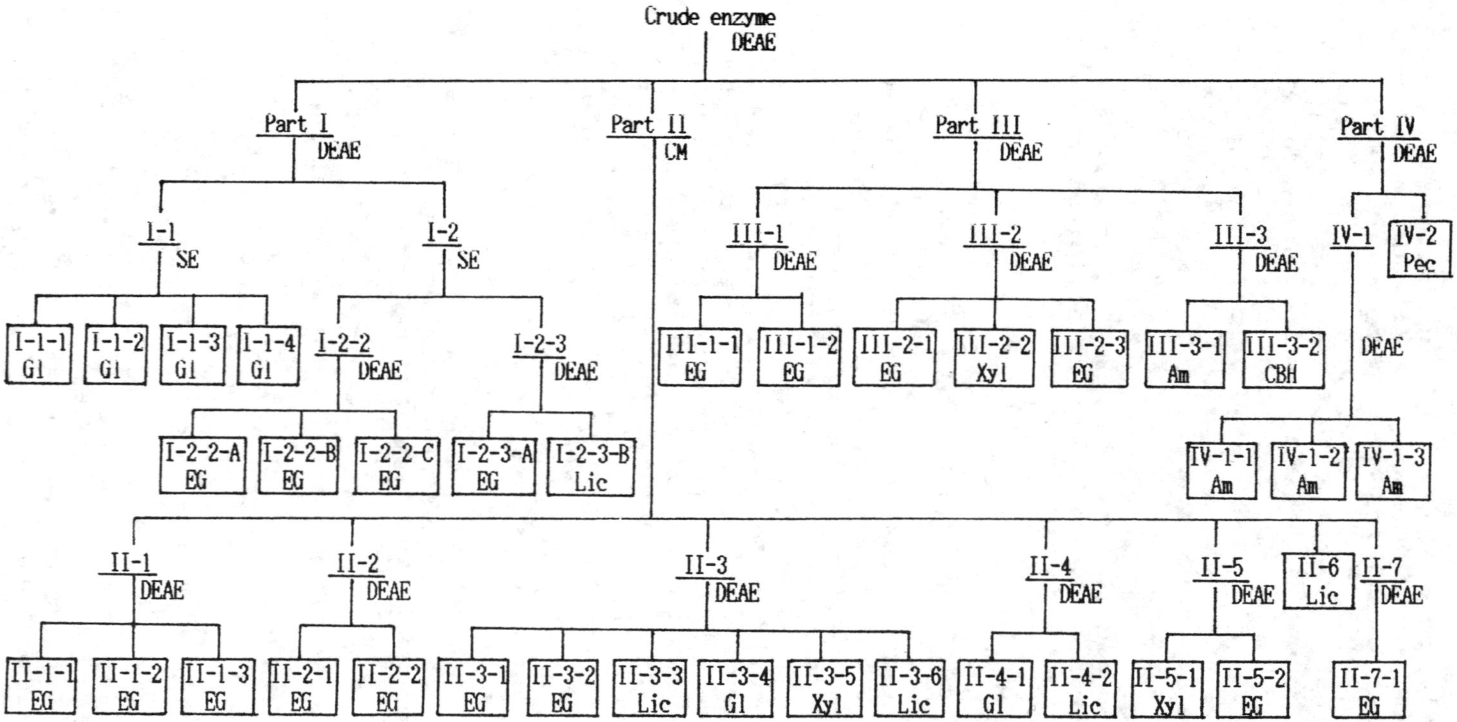

Fig.1. Schematic representation of the procedure used for the purification of the enzyme complex.
Am, amylase; CBH, exo-glucan cellobiohydrolase; EG, endo-glucanase; Gl, glucosidase; Lic, lichinase; Pec, pectinase; Xyl, xylanase; DEAE, DEAE-Sephadex A-25 or DEAE-Biogel A; CM, CM-Biogel A; SE, SE-Sephadex C50.

centrifuged at 10000 g for 20 min. The supernatant was applied to a column of DEAE-Sephadex A 25 equilibrated with the same buffer. The column was washed with the initial buffer until the first peak was eluted and then a linear gradient (0 - 0.8M NaCl) was used. The crude enzyme preparation was fractionated into four protein fractions (part I, II, III and IV). Each fraction of these four parts was applied to a DEAE-Sephadex A or CM-Biogel A column for further separation and then the component obtained was further purified by rechromatography as shown in Fig.1. In the procedure used for separation and purification of the enzyme complex from T.viride A 10 at least 40 distinct protein components were obtained. Of those, 37 compoments, each appeared as a single protein peak on disc-electrophoresis in polyacrylamide gel.

Classification of all purified components

According to their substrate specificity the 37 enzyme components were classified into four main groups: glucosidase, endoglucanase, glucan cellobiohydrolase (CBH) and other glycoside hydrolases. A total of 17 endo-glucanase components were isolated which were all active on CMC, cellulose powder and phosphate swollen cellulose, but only some of which were active on HEC, xylan and starch and two (I-2-2-A and I-2-3-A) showed some activities towards lichenin. The substrate specificities of the other purified components belonging to amylase, endo-xylanase, lichenase, pectinase and CBH, were also tested with starch, xylan, lichenin, pectin, 4-nitrophenyl cellobioside and CMC as substrates. Of the three xylanases, the one component II-3-5 showed activity towards xylan only, the others II-5-1 and III-2-2 were highly active on xylan , but also active on starch and lichenin. However, of the five lichenases tested, only the component I-2-3-B showed some activity towards xylan as well as towards lichenin. No cross-substrate activity were found with the amylases (III-3-1, IV-1-1, IV-1-2 and IV-1-3), pectinase (IV-2) and CBH (III-3-2).

DISCUSSION

The multiplicity of enzymes, especially for endo-glucanase from Trichoderma spp has been well stutied[2]. In the present study, we have identified 17 endoglucanase components and most of them have been purified to homogeneity as checked by PGA electrophoresis.They differ in molecular weight, isoelectric point, substrate specificities and sugar content. In another similar study [11], we have obtained six CBH and eight endo-glucanase components from the culture filtrate of Penicillium decumbens. Most of them showed homogeneity in the gel electrophoresis with or without SDS, and six of the components each showing a wide band on the electrophoresis were further fractionated respectively into two to six subcomponents by CM-Bigel A ion-exchange chromatography. Similarly, Fageratam and Pettersson[12] have reported that about 30 components are readily detectable from a crude filtrate of Trichoderma reesei QM 9414 by using immunoelectrophoresis. Sprey and Lamber [13] concluded that to obtain complex-free enzymes further purification with 2-dimensional techniques seems to be needed. It may be speculated that electrophoretically homogeneous components(in PAG) might become heterogeneous if the separation is performed with a higher resolution technique. However, there is considerable discussion on the actual number of enzymes in their original form needed for solubilizing cellulose. It is evident that only a very simple enzyme system is required to solubilize crystalline cellulose [3]. Therefore it seems possible that the multiple components of endo-glucanase are derived from proteohydrolytic modification and different glycosylation during cultivation, purification and storage. We feel that during the several successive purification steps used the amount of endoglucanase components is increased. One of the most intricate phenomena in the cellulase-xylanase system is the substrate cross-specificity activity. i.e. some of the homogeneous endo-glucanase components exhibit xylanase activity, and cellulase activity were found in some homogeneous xylanase components [14,15]. In the present work, the 17 endo-glucanase components tested all showed activities towards (1-4) glycoside linkages (for CMC, cellulose powder, phosphate swollen cellulose, avicel and lichenan). Nine of them (I-2-2-A, 1-2-2-B , I-2-3-A, II-1-1, II-1-2, II-1-3, II-5-2, III-1-1 and III-1-2) were found to have xylanase activity. Seven components (II-1-2, II-1-3, II-3-1, II-5-2, III-1-1, III-1-2 and III-2-3) have the ability to hydrolyze the (1-4) α-D-glycosidic linkages substrate (starch). Thus, this cellulolytic fungus can produce many kinds of glycoside hydrolases, some of which have the cross-substrate activities. This might be of help for the improvement of a total biomass conversion process.

REFERENCES

1. Sprey,B.; Lambeet, C. FEMS Microbiol. 1983,18,21.
2. Wood,T.M.; McCrae, S.I. Carbohyd. Res. 1982, 140, 111.

3. Wood,T.M. Biochem. J. 1969, 115,457.
4. Kirk, T.K. Croan, S. and Tiem, M. Emzyme Microb. Technol. 1986, 8, 27.
5. Sprey, B.; Lambeet, C. FEMS Microbiol Lett. 1984, 23, 227.
6. Sadana, J.C., Lachke, A.H. and Patie, R.V.. Carbohyd. Res. 1984, 133,297.
7. Nelson, N. J. Biol. Chem. 1952, 153, 376.
8. Deshpande, M.V., Eriksson, K.E. and Pettersson,L. G. Anal. Biochem. 1984, 138, 481.
9. Mikes, O. in Laboratory Handbook of Chromatographic and Applied Methods. pp 649. John Wiley & Sons, New York, 1979.
10. Herbert, O., Phipps, P. J. and Strange, R. E. in Methods of Microbiology 5B, pp 265. Acadmic Press, London and New york, 1971.
11. Qu Y., Gao P. and Wang Z. Acta. Microbiol. Sinica. 1988, 28, 121.
12. Fageretam, L.G.; Pettersson, L. FEBS Lett. 1979, 98, 363.
13. Sprey, B.; Lambert,C. FEMS Microbiol Lett. 1987, 23, 227.
14. Toda, S.; Suzuki, H. J. Ferment. Technol. 1971, 49, 499.
15. Deshpande, V.; Lachke, A.; Mishra, C.; Keskar, S. and Rao, M. Biotechnol. Bioeng. 1986, 28, 1832.

59

Adsorption and interactions of cellobiohydrolase I from *Trichoderma reesei* on microcrystalline cellulose

L. Kuutti*, J. Pere*, B. Henrissat, H. Chanzy**, T. Reinikainen*, L. Laaksonen*, M. Srisodsuk* and T.T. Teeri*** - *VTT, Biotechnical Laboratory, P.O. Box 202, SF-02151 Espoo, Finland. **Centre de Recherches sur les Macromolécules Végetáles (CNRS), BP 53X, F-38041, Grenoble, France.

ABSTRACT

The adsorption and interactions of cellobiohydrolase I (CBH I) from *Trichoderma reesei* on microcrystalline cellulose have been studied using transmission electron microscopy, molecular modelling and genetic engineering. Electron microscopic observations confirm that the enzyme operates by reducing the lateral dimensions of the individual microcrystals. Using a specific gold-labelled antibody it was shown that CBH I binds randomly along microcrystals without obvious preferred binding sites. A surface model of crystalline cellulose was constructed based on a crystal structure proposed for *Valonia ventricosa* native cellulose [4]. Interactions between the optimized crystalline cellulose and the cellulose binding domain (CBD) of CBH I were explored using molecular dynamics calculations carried out in vacuum. The binding of the CBD to a relatively large portion of a linear polysaccharide was shown to involve numerous individually weak interactions. Single amino acid mutations and larger deletions introduced to the CBD and the interdomain linker peptide of CBHI were analysed to get further insight into the molecular mechanism of this important enzyme.

INTRODUCTION

The filamentous fungus *Trichoderma reesei* produces an efficient set of enzymes capable of hydrolysing crystalline cellulose. One of these enzymes, cellobiohydrolase I (CBH I), binds strongly to highly crystalline cellulose. CBH I is composed of two structurally and functionally distinct domains which are linked by a long O-glycosylated linker peptide. The core domain contains the active site and the tail domain interacts with crystalline cellulose. However, the mechanism of the almost irreversible adsorption of CBH I on the surface of cellulose is not yet fully understood and its catalytic mechanism is still speculative. We have used transmission electron microscopy, molecular modelling and molecular genetic studies to visualise and explore the adsorption and interactions of CBH I with crystalline cellulose surfaces.

EXPERIMENTAL

Microcrystalline *Valonia macrophysa* cellulose was prepared for electron microscopy as described previously [1]. Native CBH I was purified essentially as described in [2]. The cellulose sample was incubated with CBH I (0.16 mg/mg cellulose) for 24 h whereafter drops of the suspension were placed on copper grids, negative stained with 1 % phosphotungstic acid and examined with a Jeol 100CX transmission electron microscope operating at 80 kV. The monoclonal antibody specific to CBH I was purified by a protein-G column (MabTrap G, Pharmacia LKB Biotechnology Inc., Sweden) and conjugated to 10 nm gold particles as in [3]. *Valonia* microcrystals were incubated with CBH I for 1 h, washed twice with buffer, labeled with the antibody-gold probe for 2 h at room temperature and negatively stained as described above.

The molecular model of crystalline cellulose used for the molecular dynamic calculations was based on the coordinates of cellulose I by Gardner and Blackwell [4]. The surface model constructed is composed of five parallel sheets and every cellulose chain was made of 12 momoner units. The model was first optimized by minimizing the intra- and intermolecular energies to remove some bad van der Waals contacts between successive sheets. The minimization algorithm used the conjugate gradient method and all atoms were allowed to move freely.

The whole bottom sheet of the surface model and the ends of every chain were kept fixed in the dynamic simulations to get a stable crystalline surface. The timestep for the simulation was 1 femtosecond. The trajectory history of the coordinates of every atom in the system was automatically recorded every 50 steps during the dynamics calculation for a total simulation of 15 picoseconds. The MD simulations were performed in the NVT emsemble. The Discover program by BIOSYM [5] was used in all simulations.

The point mutations and deletions were introduced to the *cbh1*-gene by PCR [6]. The mutated genes were transformed into a CBHI-deficient strain of *T. reesei* as described [7,8] and the mutated proteins purified as in [2]. The adsorption and activity of the mutated proteins were assayed using microcrystalline cellulose from *Acetobacter xylinum* according to [8].

RESULTS

Adsorption of CBHI onto crystalline cellulose. Cellulose microcrystals from *Valonia macrophysa* appear as smooth and well defined rod-like elements in the microscope (fig.1a). After incubation with CBH I the lateral dimension of the

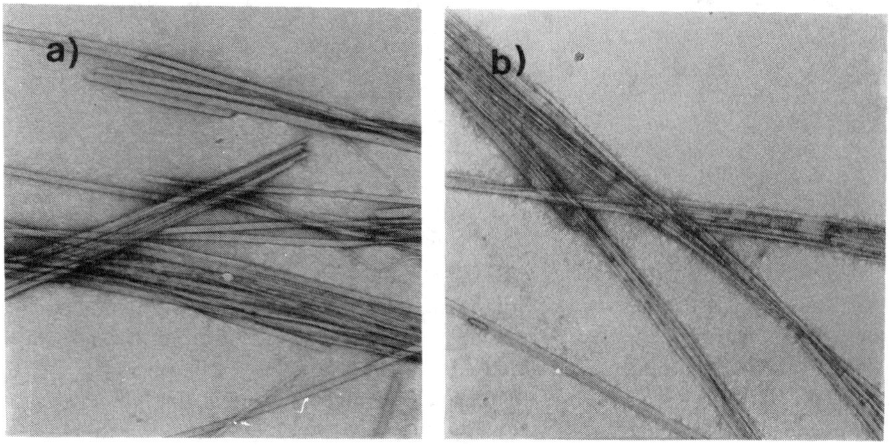

Figure 1. *Valonia* microcrystals a) prior to CBHI treatment and b) after hydrolysis by CBH I for 24 h (37 000x).

microcrystals diminishes while there is no change in their longitudinal dimension (fig.1b). As seen in fig.2 immuno-gold labelling of the bound enzyme reveals no preferred binding sites of CBH I, the enzyme binds randomly along the microcrystals.

Surface model calculations. In the absence of a better structure of cellulose the model of Gardner and Blackwell served our purpose to model the surface of crystalline cellulose. The NMR structure of the cellulose binding domain (CBD) of CBH I has been determined by Kraulis *et al.* [10]. The peptide folds into a unique wedgelike shape with two relatively flat faces, one of which is mainly formed by three conserved tyrosine residues. The tyrosines are predicted to be essential for binding the enzyme onto crystalline cellulose [11]. The CBD of CBH I was oriented over the cellulose surface as suggested by Henrissat *et al.* [12]. The flat face of tyrosine residues was placed over the exposed glycosidic oxygens of the surface chain.

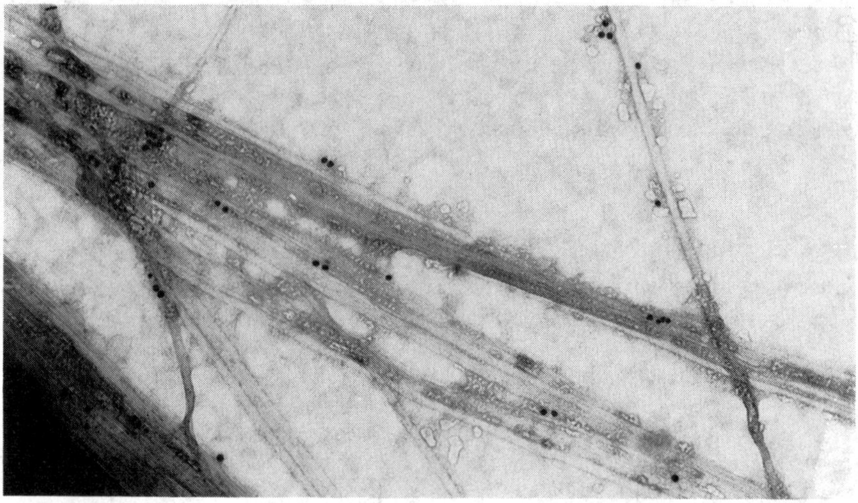

Figure 2. Adsorption of CBH I onto *Valonia* microcrystals as revealed by specific labelling with an antibody-gold conjugate (52 000x).

During the molecular dynamic simulations the motions of the cellulose atoms on the surface were restricted to small local fluctuations in the direction against the fiber axis, as can be expected. In the beginning of the MD simulation the CBD rose to a more stand up position and stayed there for the whole period of the simulation (fig. 3). The structure of the CBD remained almost unchanged during the simulation. The flat face of the tyrosine residues moved rather freely over the exposed glycosidic oxygens of the surface chain. One Leu (489) and one Gln (495) residue moved freely on a large area over the surface. The residues His465, Tyr466, Leu489, Asn490, Tyr492 and Tyr493 seemed to lift up the cellulose chains slightly during the simulation. In spite of the large-scale motion of some residues, the CBD of CBH I remained practically at the same position over the exposed glycosidic oxygens of the surface chain.

Adsorption and activity of mutated CBHI. The binding and activity of CBHI on crystalline cellulose were studied by introducing specific amino acid changes on both of the flat surfaces of the CBD. Mutation of a Tyr residue (Tyr622) on the hydrophilic face and another mutation (Pro607) on the more hydrophobic face both reduced the binding and activity of CBHI to the level of the isolated catalytic domain lacking the CBD and the interdomain linker peptide [8]. The results confirm the role of tyrosine residues in the binding events and strenghten the hypothesis that both surfaces of the wedge-shaped CBD are required for efficient interaction of CBHI with crystalline cellulose.

Internal deletions of the interdomain linker peptide reducing the distance between the catalytic and the cellulose binding domain of CBHI did not significantly affect the binding behaviour of CBHI on Avicel [13]. Interestingly, however, the activities of the deletion mutants were significantly lower than that of the wild type CBHI on crystalline cellulose. Based on these results it seems likely that tight binding is not a sufficient prerequisite for efficient hydrolysis of crystalline cellulose. It is possible that the interdomain linker peptide mediates specific interactions of the catalytic and the binding domain or, as suggested by Warren and co-workers [14], optimizes the orientations of the two domains relative to the cellulose surface.

Figure 3. CBH I cellulose binding domain as modelled on crystalline cellulose.

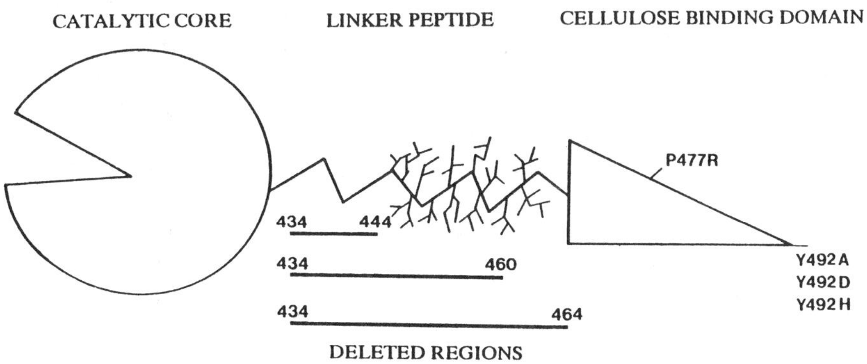

Fig. 4. Schematic representation of the domain structure of CBHI. The point mutations and deletions introduced are indicated.

CONCLUSION

CBHI is the key enzyme in the degradation of highly ordered crystalline cellulose by *T. reesei.* In this study we show that the enzyme binds randomly along the cellulose surface and operates by thinning the cellulose microfibrils. Our results on molecular dynamic simulations partially confirmed by analysing site specific mutations on the cellulose binding domain reveal a number of interactions determining the specificity and tightness of binding of CBHI on crystalline cellulose. Moreover, significant reduction of activity but not of adsorption of CBHI upon gradual shortening of the interdomain linker peptide suggests that the two domains may undergo specific interactions during crystalline cellulose hydrolysis.

REFERENCES

[1]Chanzy, H.;Henrissat, B.; Vuong, R. & Schülein, M. FEBS Lett. 1983, 153, 113-117.

[2]van Tilbeurgh, H.; Bhikhabai, R.; Pettersson, L. & Claeyssens, M. FEBS Lett. 1984, 169, 215-218.

[3]Roth, J. The Colloid Gold Marker System for Light and Electron

[4]Gardner, K. & Blackwell, J. Biopolymers 1974, 13, 1975-2001.

[5] Discover: Molecular simulation program,BIOSYM Technologies Inc., San Diego, USA

[6] Higuchi, R. Recombinant PCR. *In* Innis, M.; Gelford, D.; Sninsby, J. & White, T.(eds.) PCR Protocols, Academic Press, 1989, 177-183.

[7] Uusitalo, J.; Nevalainen, H.; Harkki, A.; Knowles, J. & Penttilä, M. J.Biotechnol.,1991, 17, 35-50.

[8] Penttilä, M.; Nevalainen, H.; Rättö, M.; Salminen, E., & Knowles, J. Gene,1987,61, 155-164.

[9] Reinikainen, T.; Ruohonen, L.; Nevanen, T.; Laaksonen, L.; Kraulis, P.; Jones, A.; Knowles, J. & Teeri T.T. Proteins, 1992, in press.

[10] Kraulis, P.; Clore, G.; Nilges, M.; Jones, T.; Petterson, G.; Knowles, J. & Gronenborn, A. Biochemistry, 1989, 28, 7241-7257.

[11] Claeyssens, M. & Tomme, P. Structure-function relationships of cellulolytic proteins from *T.reesei. In* Kubicek, C. *et al.* (eds.) *Trichoderma reesei*:Biochemistry, Genetics, Physiology and Application. Proc. Tricell 1989, Royal Society of Chemistry, 1989,1-11.

[12]Henrissat, B.; Vigny, B.; Buleon, A. and Perez, S. FEBS Lett. 1988, 231, 177-182.

[13] Srisodsuk, M.; Reinikainen, T; Ruohonen, L & Teeri, T.T. Manuscript in preparation.

[14] Shen, H.; Schmuck, M.; Pilz, I.; Gikes, N.; Kilburn, D. Miller, R. Jr. & Warren, A. J. Biochem., 1991,266,11335-11340.

60

Mechanism of enzymic hydrolysis of cellulose and its activation and inhibition by phenolics

A. Blažej and M. Košik - Slovak Technical University, Faculty of Chemical Technology, Radlinskeho 9, 812 37 Bratislava, ČSFR.

ABSTRACT

In the paper three-dimensional structures of cellulose and at least three conformations of β-D-glycosidic bonds conditioned by distinct intramolecular hydrogen bonds are discussed. The enzymic hydrolysis is elucidated on the basis of stereospecificity of these conformers.

The second part of the paper is devoted to the iteraction of enzymes and phenolics. Both the formation and the mechanisms of polyelectrolyte complexes are described on the basis of the polyelectrolyte theory.

INTRODUCTION

The mechanism of polysaccharide hydrolysis depends on electronic structure (charges) of the reaction centre (glycosidic bonds). The second important question is the three-dimensional conformation of a glycosidic bond.

ELECTRONIC STRUCTURE OF GLYCOSIDIC BONDS

Quantum chemical calculations were applied for the electronic structure of model systems (1), using the semiempirical INDO/2 method (Intermediate Neglect of Differential Overlap). This method is capable to describe the poperties connected with the electronic structure of not very large organic molecules. Fig. 1 shows the structure and numbering of the relevant atoms in the model compounds of a glycosidic bond. Table 1 represents the calculated values of charges for considered carbon and oxygen atoms of the model compounds. The highest positive charges are located on C_1 and the maximum electron densities (negative charges) on O_1 and O_2 atoms (1,2). Both oxygen atoms are nearly equal for all eight model systems despite the different character of compunds. The bons C_1 - O_2 and O_1 - C_1 are strongly polarized and acceptable for heterolytic (ionic) cleavage.

THE INFLUENCE OF COMPOSITION AND CONFORMATION

The conformation adopted by a polysaccharide chain is influenced by the primary structure (chemical composition of polysaccharide) and its microenvironment. Each chemical bond which is rotatable in a molecule has a priori three conformational possibilities. For a glycosidic bond with two rotatable bonds there can be 3^2 creates 3^4 conformers. The real molecule itself has to seek out its one favoured conformation under given conditions. A realistic cellulose chain model has attracted attention during the last 20 years. The conformational three-dimensional structure cellulose macromolecule is preserved in oligomeric model compounds as well. Therefore the conformations of cellobiose, methyl β-cellobioside and higher

Table 1. Calculated values of charges for relevant carbon and oxygen atoms of 2-ethoxyethanol, 2-substituted tetrahydropyrans, glucose and cellobiose (1,2).

		ATOM CHARGES			
COMPOUND	r	O_1	C_1	C_2	O_2
2-ethoxyethanol	1	-0.285	0.229	0.223	-0.311
	80	-0.335	0.283	0.290	-0.368
H-O-THP	1	-0.316	0.382	-0.004	-0.321
Et-O-THP	1	-0.311	0.383	-0.009	-0.295
iPe-O-THP	1	-0.316	0.386	0.001	-0.305
cyHe-O-THP	1	-0.318	0.389	0.001	-0.309
Glucose	1	-0.317	0.356	0.144	-0.304
	80	-0.375	0.448	0.162	-0.367
Cellobiose	1	-0.308	0.353	0.148	-0.292

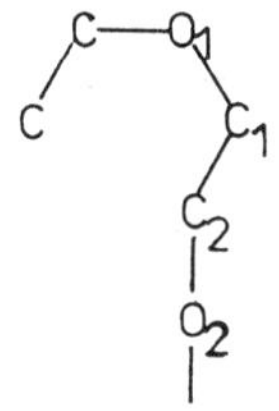

2-ETHOXYETHANOL

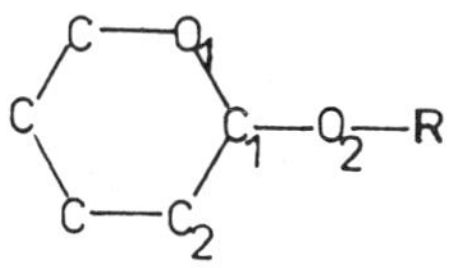

2-HYDROXYTETRAHYDROPYRAN (H-O-THP)
2-ETHOXYTETRAHYDROPYRAN (ET-O-THP)
2-ISOPENTOXYTETRAHYDROPYRAN (iPe-O-THP)
2-CYKLOHEXOXYTETRAHYDROPYRAN (cyHE-O-YHP)

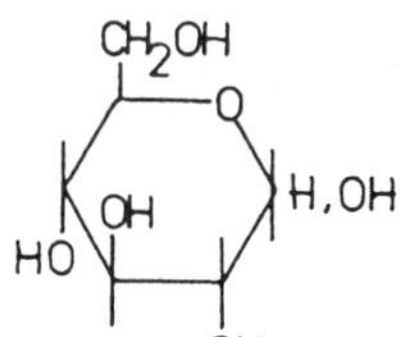

D-GLUCOSE

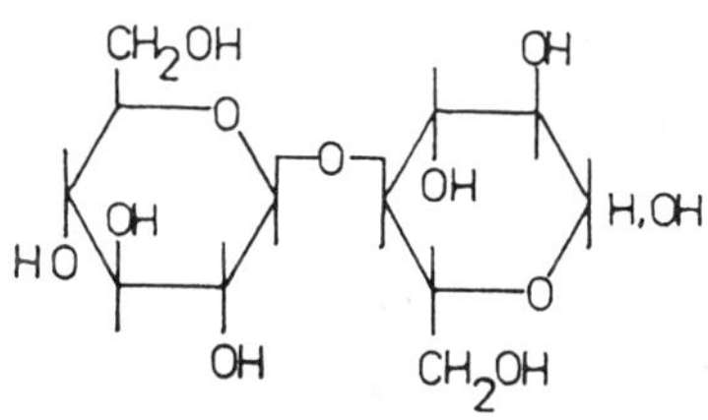

CELLOBIOSE

METHYL β-D-CELLOBIOSIDE

Fig.1. Model compounds of a glycosidic bond.

cellooligosaccharide have been studied. The conformational equilibria of five β-cellobiose conformers were investigated in various solvents by Tvaroška (3). He used the PCILO quantum-chemical method. The results (Table 2) obtained indicate that the preponderant conformer C_1 in the aqueous solution is similar to the one adopted by β-cellobiose in the crystalline form. Theoretically, the allowed steric conformers of cellobiose can be projected using X-ray and electron diffraction, CP-MAS NMR spectroscopy and computer modelling with the PCILO and MM2 algorithm (4). From the structure of the cellobiose model the presentation of two main different conformers can be concluded. Polarity of a solvent contributes to the conformation stability of cellulose by solvatation energy. The chain flexibility of cellulose increases with decreasing of the solvent polarity.

ENZYMIC HYDROLYSIS

The enzymic hydrolysis of cellulose has been reviewed and discussed in many publications. The factors influencing the rate and extent of the enzymic hydrolysis of cellulose are well known (enzyme concentration, the susceptibility of the substrate to enzyme, inhibition of the cellulose, etc.), but these effects have not been so far adequately elucidated.

The fungus *Trichoderma reesei* is one of the most interesting cellulolytic systems for the future application of cellulose hydrolysis. The unusual complexity of this system which contains 18 different enzymes (5) with a different substrate specificity (6) is the subject of great interest. For instance, research of the hydrolysis of β-1,4 oligosaccharides by the non-specific endo 1,4-β-D-glucanase isolated from *T. reesei* QM 9414 has shown how many competitive

Table 2. Calculated molar fractions x_i of stable conformers of cellobiose and calculatedmean values of dihedral angle (ϕ), dipole moment (μ), linkage rotation (λ), and coupling constant (J) in various solvents (3).

SOLVENT		CONFORMER C1	C2	C3	C4	C5	(μ)	(λ)	(ϕ)	(J)
Cellobiose in the solid state		60.0	34.1	2.9	2.0	1.0	4.54	72.8	33.1	3.65
Dioxane		65.4	30.6	2.2	1.2	0.6	4.57	66.0	34.6	3.73
Pyridine		68.7	28.1	2.1	0.8	0.3	4.60	61.3	36.1	3.61
Acetone		72.2	25.0	1.9	0.6	0.3	4.61	55.9	37.9	3.58
Methanol		75.3	22.3	1.8	0.4	0.2	4.63	51.2	39.4	3.51
Dimethylsulfoxyde		75.5	22.1	1.8	0.5	0.2	4.63	50.7	39.5	3.50
WATER	10 C	83.3	15.4	1.1	0.1	0.1	4.67	38.8	42.8	3.32
	25 C	82.2	16.2	1.3	0.2	0.1	4.66	40.3	42.6	3.33
	50 C	80.3	17.5	1.8	0.3	0.1	4.65	42.8	42.3	3.35
	70 C	78.8	18.6	2.1	0.3	0.2	4.64	44.8	42.0	3.36

Table 3. Possible mechanisms of cellotriose degradation (6); ● D-glucosyl residue ^{3}H-labelled at the reducing end.

Possible mechanism	Schematic representation
Unimolecular hydrolysis	
Shifted binding	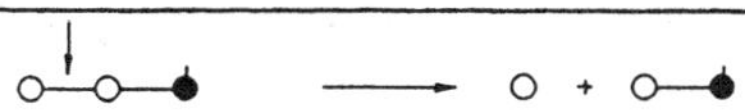
Glucosyl transfer	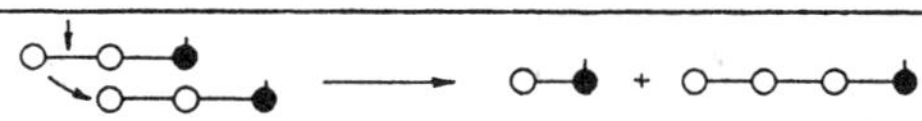
Cellobiosyl transfer	
Condensation	

biochemical processes take place in the bioconversion of polysaccharides (7). The possible mechanisms of cellotriose degradation is shown in Table 3.

In the case of cellotriose the hydrolysis proceeds on both glycosidic linkages. In D-glucosyl-D-xylose oligosaccharides the xylosidic bonds are preferentially hydrolyzed. The mixed substrates of both types of oligosaccharides are attacked on same active site of enzymes.

It has been observerd that D-glucosyl and cellobiosyl transfers occur either as primary products leading to subsrate resynthesis or as another cellotriose molecule. Transglycosylation reactions catalyzed by RG I in the mixure lead to new heterooligomers composed of G-glucosyl and D-xylosyl residues.

ACTIVATION AND INHIBITION OF ENZYMIC HYDROLYSIS

The catalytic activity glycanase enzymes can be inhibited by hydrolytic products (glucose, cellobiose), 2-furaldehyde and 5-hydroxymethyl-2-furaldehyde. These enzymes can be activated or inhibited by natural phenolics or degradation products of lignins or by a lignosulphonic acid (8).

We have studied the activation and inhibition effects of various phenolic compounds during the enzymic hydrolysis of cellulose (Whatman no 1) by celluloses from _T. reesei_ QM 9414 (9). The positive effect is enhanced in the phenolic compounds shown in Table 4. In the case of phenolics the activation capacity increases in the order shown Fig. 2. The interaction between EG from _T. reesei_ QM 9414 and phenolics was evaluated through the circular dichroism (CD) (10).

The inhibition effect of some lignin degradation products was studied on the biomass production by

Table 4. Analysis of sugars after 24 hours of enzymic hydrolysis of Whatman cellulose (HPLC analysis).

Additive	Glucose (G_1) ($mg \cdot mL^{-1}$)	Cellobiose (G_2) ($mg \cdot mL^{-1}$)	$G_1 + G_2$ ($mg \cdot mL^{-1}$)
Control	3.45	5.44	8.89
Phenol	2.79	3.93	6.72
Pyrocatechol	3.64	5.89	9.53
Pyrogallol	4.04	8.30	12.34
2,6-dimethoxy-phenol	4.64	8.79	13.44
Guaiacol	5.77	9.19	14.96
Syringaldehy-de	4.33	3.33	7.66
Vanillin	3.83	4.20	8.03
p-Hydroxy-benzoic acid	3.10	5.44	8.54
Veratric acid	5.04	5.96	11.00
Syringic acid	3.37	6.12	9.49
Vanillic acid	4.54	7.36	12.90

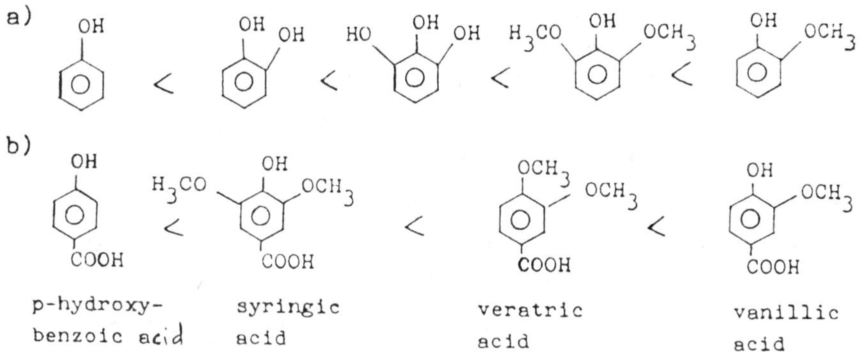

Fig.2. Various phenolic compounds and their effect on cellulose enzymic hydrolyzability:
a) phenol<pyrocatechol<pyrogallol < 2,6-dimethoxy-phenol < guaiacol
b) phenolcarboxyl acids.

Candida sp. The growth inhibition for 20 lignin monomer model compounds during the Candida sp. cultivation is shown in Table 5 (11).

The interaction of enzymes and phenolics can be elucidated by the formation of protein-ligand complexes. The mechanisms of these interactions are shown in Fig.3 . Enzyme as a polyelectrolyte protein has the polyampholyte behaviour, being in pH below the isoelectric point (IP) the protein behaves as polycation, over IP as polyanion. In the case of carboxylic groups in phenolics or sulphate groups in lignosulphonic acid the electrovalent bonds along with $R-NH_3^+$ groups of the side chains of basic amino acids of the enzyme are formed (lysine, histidine and arginine). The phenolic hydroxyl groups are bonded to the peptide bonds of the backbone of polypeptide chains by hydrogen bonds. The present aldehyde or quinone groups are bonded to the side chain of the amino acid of amino groups (being over IP the colvalent bonds are formed). The bonding of phenolics to protein influences the tertiary and quarternary enzyme structure. Formation of the polyelectrolyte enzyme-phenolic comlex affects, due to allosteric effects, the flexibility and conformation transitions of the enzyme during its "working cycles" in the course of hydrolysis. Impacts on the conformation changes influence the binding of the enzyme to a substrate and are channeled into the active site of the enzyme as well. In the polyelectrolyte enzyme-phenolic complexes the character of interaction depends on pH of the solution, IP of the enzymes and pK of the phenolics and on their concentration. These kinds of interaction are reversible processes.

In the case of covalent bonding between the enzyme and phenolics (aldehydes, quinones) irreversible inhibition processes take place. In both cases

Table 5. 50% and 100% Inhibition Dose (ID_{50} and ID_{100}) of phenolics at 8th and 24th hours of cultivation C.utilis.

	Time 8 h		*Time 24 h*	
Compound	ID_{50} [mM]	ID_{100} [mM]	ID_{50} [mM]	ID_{100} [mM]
Vanillyl alcohol	19·1	> 20·0	> 20·0	> 20·0
Guaiacol	2·2	43·2	14·6	20·9
Conipherylalcohol	> 20·0	> 20·0	> 20·0	> 20·0
Phenol	4·9	42·0	13·0	20·4
2,6-Dimethoxyphenol	3·1	24·6	12·6	20·9
Acetovanillon	1·2	1·2	7·5	11·4
Pyrocatechol	1·2	3·0	7·1	10·9
Pyrogallol	< 1·2	< 1·2	< 1·2	< 1·2
Syringic acid	> 20·0	> 20·0	> 20·0	> 20·0
Gallic acid	6·6	> 20·0	> 20·0	> 20·0
p-Hydroxybenzoic acid	6·9	19·9	12·6	21·8
Vanillic acid	2·8	16·2	9·4	24·8
Salicylic acid	< 1·2	< 1·2	< 1·2	< 1·2
Cinnamic acid	< 1·2	< 1·2	< 1·2	< 1·2
Benzoic acid	< 1·2	< 1·2	< 1·2	< 1·2
3,4-Dimethoxybenzaldehyde	< 1·2	< 1·2	6·3	12·0
Syringaldehyde	1·4	7·7	7·9	13·9
p-Hydroxybenzaldehyde	2·4	8·0	7·3	10·0
Vanillin	1·3	8·1	5·1	10·5
o-Vanillin	< 1·2	< 1·2	< 1·2	< 1·2

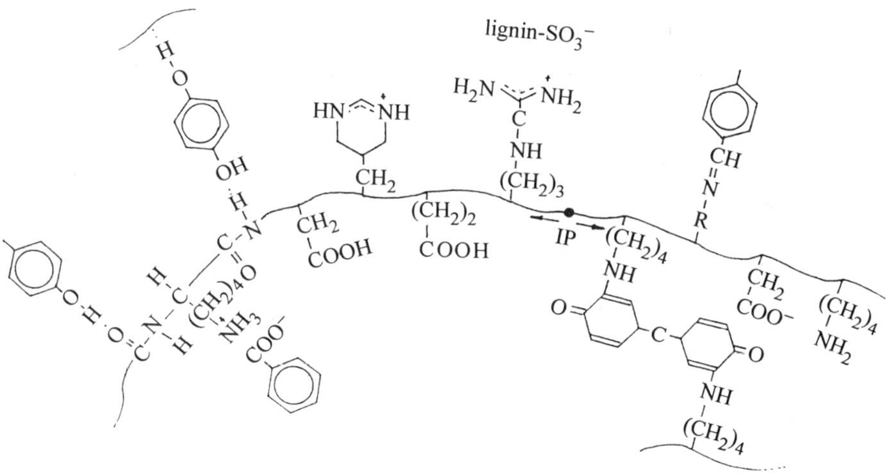

Fig.3. The polyelectrolyte enzyme-phenolic complexes.

cross-linkages on the quarternary structure can be formed, which contributes to the conformation stability of the enzyme and influences its flexibility. All these interactions between the enzyme and phenolics have positive or negative impacts on the bonding of the enzyme to a substrate and on the enzymic catalytic activity.

CONCLUSION

On the basis of the electronic model system structures and of the conformational analysis of cellulose oligomers the three non-equivalent glycosidic bonds present in cellulose can be distinguished. Specificity of the cellulolytic system is not given by a cellulose substrate but by the conformation and stereospecificity of glycosidic bonds (their secondary structure). The activation and inhibition of cellulolytic enzymes by phenolics are influenced by the creation of polyelectrolyte complexes between enzymes and phenolic compounds. The interaction can be presented by non-covalent bonds, (reversible complexes) and covalent bonds (irreversible complexes).

REFERENCES

1. Brezová V.: Study of the Selected Problems of Photocatalyse of Organic Compounds. PhD. Thesis. Slovak Tachnical University, Bratislava, 1990.
2. Brezová V., Breza M. and Čeppan M.: Photocatalytic Degradation of Cyclic Acetals in Aqueous Titanium Dioxide Suspension. Chem. Papers, Bratislava (in print).
3. Tvaroška I.: Biopolymers, 23, 1951 (1984).
4. Tvaroška I.: Theoretical Conformational Analysis of Saccharides. D.Sc. Thesis. Slovak Academy of Sciences, Bratislava, 1986.
5. Biely P. and Markovič O.: Biotechnology and Applied Biochemistry, 10, 99 (1988).
6. Blažej A.: In: Inagaki H. and Phillips G.O.: Cellulosics Utilization. Elsevier Applied Science, London and New York, 1988, pp. 173-174.
7. Biely P., Vršanská M. and Claeyssens, M.: Eur. J. Biochem., 200, 157 (1991).
8. Blažej A. and Košík M.: Phytomass as a Raw Material for Chemistry. Ellis Horwood Limited, Chichester, 1992 (in print).
9. Blažej, Košík M., Pekarovičová A. Špilda I. and Biely P: In: Kennedy J.F., Phillips G.O. and Williams P.A.: Lignocellulosics Utilization Chichester, 1989, p.301.
10. Pekarovičová A., Lužáková B., Bystrický S., Pekarovič J. and Košík M.: Cellulose Chem. Technol., 23, 225 (1989).
11. Mikulášová M., Vodný Š. and Pekarovičová A.: Biomass, 23, 149 (1990).

61

Microbial degradation of impure cellulosics by *Aspergillus fumigatus* IMI 255091

D.A. John Wase - Laboratory for Environmental Biotechnology, Biochemical Engineering Division, School of Chemical Engineering, The University of Birmingham, Edgbaston, Birmingham B15 2TT, UK.

ABSTRACT

Progress to date in the production of a range of enzymes by *Aspergillus fumigatus* IMI 255091 on hay and straw and capable of degrading such impure substrates is reviewed. Early work was aimed at a low-technology process in the air-lift fermenter, due to shear problems in conventional mechanically agitated vessels. Subsequently, the production of a wide range of enzymes was investigated. It was established that they are generally of high thermal stability and inhibited by ammonium ions. Immobilisation/solid state culture are currently viewed as potential methods to realise increased yields.

INTRODUCTION

The potential of *Aspergillus fumigatus* strains for degrading cellulosic or lignocellulosic materials has, over the years, been investigated by a number of research groups. This organism is well known for its cellulose-degrading abilities, and strains appear to be distributed very widely over the earth's continents. It is particularly prevalent in Britain, especially in the wetter areas along the western coasts, where a considerable amount of time has been expended in examining its human pathogenicity [1], and its ability to multiply and sporulate in great profusion in damp grass, straw and hay. A most aggressive coloniser, perhaps due to its prolificity in spore production, it has even been shown to occur in great abundance in the air surrounding Birmingham's School of Chemical Engineering, several miles within Birmingham's somewhat urban city boundary [2]. Our original interest in cellulase degradation arose from the large amounts of straw being burned in Britain, and we therefore looked for an organism which was robust, could possibly be grown on the farm in very simple equipment, and which degraded straw. The organism of choice which we isolated was *Aspergillus fumigatus* IMI 255091 [3].

INITIAL TESTS

This isolate was therefore exposed to a range of substrates, and it soon began to emerge that it was of scientific interest. For instance, it was evident that it grew very well in shaken flasks on a variety of cellulosic and lignocellulosic substrates: indeed, it was noteworthy that it rapidly consumed a cotton-wool plug which was accidently included one day [4]. However, it failed to thrive in a conventional sparged and mechanically agitated vessel equipped with a Rushton turbine. An investigation of this problem led to the conclusion that the organism was shear-sensitive, but would grow well and produce good yields of cellulolytic enzymes in a low-shear device like an air-lift fermenter [5], in which yields were maximal at maximum gas hold-up.

From a practical point of view, this result was significant. At the time we were trying to establish if it were possible to devise a relatively low technology process in which such an organism could produce cheaply a range of potent cellulases for immediate use (such as on stubble, to assist its degradation), and an air-lift device readily translated itself to a couple of plastic drain-pipes. Further work along these lines revealed that it was unnecessary to grind the substrate too finely; for example, with the average farm straw-chopper in mind, it was established that the penalty of using straw in 1 cm lengths was a drop of some 10% in yield, which was tolerable [6].

However, shortly after that time, interest began to shift to the scientific, as the range of cellulolytic activities which this organism would produce on impure substrates such as straw began to be appreciated. Some of these are listed in Table I.

TEMPERATURE STABILITIES

A further discovery was that many of these activities were optimal at elevated temperatures, and, even under these conditions, possessed a high degree of stability. In this connection, a much earlier study in which the stability of the

TABLE I. Activities synthesised by *A. fumigatus* [7].

Endoglucanase	Cotton Activity	β-Mannosidase	Xylanase
β-Glucosidase	α-Glucosidase	α-Galactosidase	FPA
β-Glucuronidase	α-Mannosidase	β-Galactosidase	β-Xylosidase
α-L-Arabinosidase			

crude broth was investigated showed that crude broths could be aerated and agitated with a standard Rushton disc turbine at 500 rev. per min for as long as 24h at room temperature with no loss in β-glucosidase and endoglucanase activities [5]. However, it was much later when it occurred to us that stability at elevated temperatures might be equally as good. Table II shows assay temperatures (an indication of the temperature for optimum activity), for a range of cellulase-producing microorganisms. Comparisons show clearly that *Aspergillus fumigatus* enzymes are to the fore in this respect.

TABLE II. Optimum assay temperatures (°C) for a range of enzymes derived from a variety of organisms.

Microorganism	CMC	G	X	Xyl	FPA	CA
Thermomonospora sp. [8]	70	55	N	N	N	65
T. viride ITCC 1433 [9]	55	60	N	N	N	50
Cellulomonas sp. [10]	N	N	40	40	N	N
Sporotrichon thermophile [11]	65	72	N	N	N	55
Aspergillus terreus GN1 [12]	60	N	N	N	55	N
Thielavia terrestris ATCC 26917 [13]	70	70	60	60	60	60
T. reesei TD B6 [14]	60	65	60	55	55	45
Penicillium sp. PD 20 [14]	55	60	60	55	55	45
Sporotrichon cellulophilum ATCC 20494 [14]	60	60	60	70	60	50
Schizophyllum commune [15]	N	50	N	50-52	45	N
A. fumigatus fresenius [16]	60	N	N	N	N	55-65
A. fumigatus fresenius [17]	70	70	N	N	65	N
A fumigatus IMI 255091 [7]	60-70	70	70	60	60	50

CMC - Endoglucanase; G - β-Glucosidase; X - β-Xylosidase; Xyl - Xylanase; FPA - Filter Paper Activity; CA - Cotton Activity; N - Not available.

An additional piece of information concerns overall temperature stability with time. Typical data at 70°C [7] are shown in Figs 1, 2 and 3, again demonstrating the stabilities of the activities. As these indicate, it is often not worth attempting to calculate half-lives even at temperatures as high as 70°C for many of these activities. It would thus seem that the use of an impure substrate with this particular isolate promotes production of a wide range of activities of generally very high thermostability.

INHIBITORY EFFECTS

Around the time that this work was being done, Holland and Wase [18] showed that some rather odd inhibitory effects observed previously were due to the presence of ammonium ions in the medium.

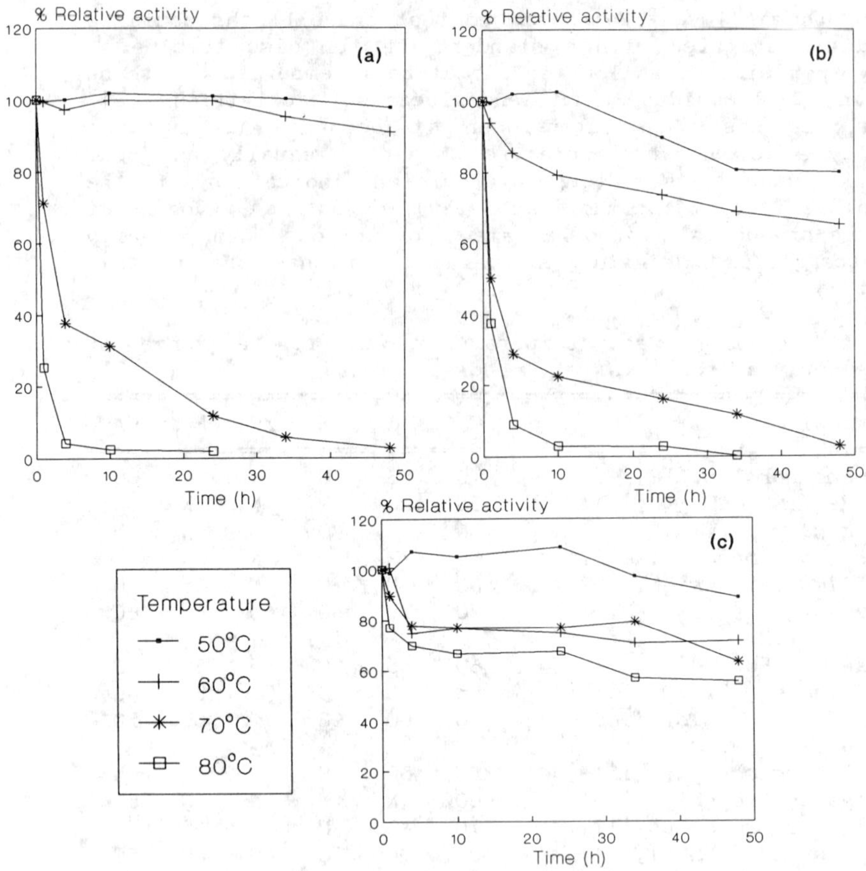

Figures 1, 2 and 3. Thermostabilities of (a) endoglucanase; (b) cotton activity and (c) α-galactosidase from *A. fumigatus.*

For instance, the production of the whole range of cellulolytic activities was affected; preliminary results suggested that some of the enzymes themselves were also inhibited. Further investigation by Hamilton [19] of a number of these enzymes has established the precise nature of this inhibition, whilst comparisons with an alternative strain used by the Coventry group have shown that the inhibitory effects of ammonium ions were not evident in their strain (Table III). Such a control mechanism is of interest: possibly it is tied to the organism's ability to proliferate on damp vegetation. In this connection, Hamilton [20] has investigated a further substrate, hay. This is much higher in nitrogen than straw, and, in general terms, high medium nitrogen levels are often associated with poor cellulase yields. However, in this particular case, enzyme yields on hay were actually higher than those on straw, whilst, in another approach, this unusual strain has proved relatively insensitive to a range of broth peptone levels. All of this evidence suggests that this organism has unusual control over its nitrogen metabolism which it might well be possible to exploit in the future.

SOLID STATE FERMENTATION

Initial work by Kim [7] in this area has shown clearly on a small scale that the use of semisolid methods of cultivation leads to enhanced yields of enzyme for almost all the enzyme activities listed in Table I. At the time, he was unaware of the inhibitory effects of ammonium ions, and it is likely that further enhancement of activities is possible. This has led us to investigate immobilisation of this strain in polyurethane pads, where enhancement of activities of up to double that of standard cultures was obtained. This result ties in with similar results recently obtained by Kent-Kirk [21], in tests with *Phanaerochaete chrysosporium*, and Evans [22], who examined a range of organisms for ligninolytic abilities.

These findings, in turn, have led us to examine the potential of semisolid and solid state cultures for cellulase production: currently results are being obtained for a modified Z-blade reactor [23]. It is evident that there still exists considerable potential for increasing the titres of fungal cellulases by means of such techniques.

TABLE III. Influence of 500 mM ammonium ions on the characteristics of enzymes from two different strains of *Aspergillus fumigatus*.

Enzyme	Substrate*	Activity from *A. fumigatus* IMI 255091 [3]	Activity from *A. fumigatus* Fresenius [16]
Endoglucanase	CMC	Non-competitive	No inhibition
β-Glucosidase	PNPG	Non-competitive	No inhibition
β-Glucosidase	MU-glc	Non-competitive	
Exocellulase	MU-cell	Competitive	

*CMC, Sodium salt of carboxymethylcellulose; PNPG, 4-*p*-nitrophenyl-β-D-glucopyranoside; MU-glc, 4-methyl-umbelliferyl-β-D-glucopyranoside; MU-cell, 4-methyl-umbelliferyl-β-D-cellobiopyranoside.

CONCLUSIONS

It is clear that microorganisms which produce cellulases in vivo rely on a variety of induction and control mechanisms, and it is therefore not altogether surprising that there should be significant differences between the strategies adopted by those organisms primarily interested in high lignin situations, and those which colonise low lignin vegetation. What is perhaps more surprising is to find that our particular strain, one of the latter, should differ so widely in this respect from a close relative. Nevertheless, it has a clear ability to elaborate a broad spectrum of astonishingly assorted activities of excellent temperature tolerence and stability on such substrates, and containing useful amounts of β-glucosidase and extracellular activities. For many practical, agricultural and industrial purposes, the potential of a low technology process, producing a wide range of enzymes in a mixture such as this, is considerable for a variety of applications. The curious method of control of this enzyme family by ammonium ions inhibition is unusual and worthy of further investigation: indeed, it is probable that this could shed light generally on the mechanisms of induction and expression of such activities.

Finally, it is of interest that constraining the organism by immobilisation or semi-solid/solid state fermentation appears to promote substantial increases in yields. Future work will clearly be aimed at the exploitation of this discovery.

REFERENCES

[1] Hacking A (1982) Internal Communication between ADAS and Welsh Office
[2] Veale D (1984) Internal Report, Safety Unit, University of Birmingham
[3] Wase DAJ & Vaid AK (1983) Process Biochem **18**(5) [Dec] 35
[4] Raymahasay S (1984) PhD Thesis, University of Birmingham
[5] Wase DAJ, McManamey WJ, Raymahasay S & Vaid AK (1985) Biotech Bioeng **28** 1166
[6] Bonakdarpour B, Kim SW, Wase DAJ & McManamey WJ (1987) In Wood and Cellulosics: Industrial Utilisation, Biotechnology Structure and Properties Ed Kennedy JFK, Philips GO & Williams PA; Ellis-Horwood Chichester p243
[7] Kim SW PhD Thesis, University of Birmingham
[8] Hagerdal B, Ferchak J, Pye EK & Forro JR (1979) Biotech Bioeng **22** 1601
[9] Herr D (1980) Biotech Bioeng **22** 1601
[10] Rickard PAD & Laughlin TA (1980) Biotech Lett **2**(8) 363
[11] Margaritis A & Creese E (1981) Biotech Lett **3**(9) 471
[12] Garg SK & Neelakantan S (1982) Biotech Bioeng **24** 737
[13] Margaritis A & Merchant R (1983) Biotech Bioeng Symp **13** 299
[14] Durand H, Soucaille P & Tiraby G (1984) Enz Microb Technol **6** 175
[15] Steiner W, Lafferty RM, Gomes J & Esterbauer H (1987) Biotech Bioeng **30** 169
[16] Stewart JC & Parry JB (1981) J Gen Microbiol **125** 33
[17] Grajek W (1986) Biotech Lett **8**(8) 587
[18] Holland TM & Wase DAJ (1991) In Lignocellulosics Science Technology Development and Use Ed Kennedy JFK Philips GO & Williams PA; Ellis-Horwood Chichester p143
[19] Hamilton LA & Wase DAJ (1991) Process Biochem **26** 287
[20] Hamilton LA & Wase DAJ (1992) This volume
[21] Kent-Kirk T (1991) Personal communication
[22] Evans CS (1991) Personal communication
[23] Erb CK (1992) This volume

62

Biological degradation of lignocellulose using *Pleurotus ostreatus* strains obtained by genetic manipulation

R. Ramirez, H. Leal, G. Eger*, F. Padilla and M. Alcántara - Departamento de Alimentos y Biotecnologia, Facultad de Quimica, División de Ingenieria, UNAM, Cd. Universitaria 04510, México D.F.

ABSTRACT

Pleurotus ostreatus is an edible mushroom which consumes lignin in nature although it also causes a simultaneous degradation of polysaccharides. Therefore it was treated with MNNG for production of acellulolytic (C-) mutants. With this treatment, 240,000 colonies were recovered and from them, one total and three partial (C-) mutants were obtained. The total and one partial (C-) mutants were improved by breeding with wild type strains. Segregation studies suggest that 5 genes control expression of the (C-) character in the total (C-) mutant, whilst five or six genes are involved in the partial (C-) mutant. Pairings of (C-) monokaryons from each progeny produced always (C-) dikaryons, while pairings between monokaryons (C-) x (C+) or (C+) x (C+) yielded only 7% of (C-) dikaryons. When the (C-) monokaryons from both progenies were cross paired only few (C-) dikaryons were produced (4%). The new (C-) dikaryons were more vigorous, and one of them showed a reduced cellulose consumption, 23% compared to 44% of the wild type strains.

INTRODUCTION

It is well known that the most common lignocellulosic by-products are composed of cellulose, 35 to 60%, hemicellulose, 10 to 30% and lignin, 4 to 25%. Lignin is the main obstacle for making polysaccharides available and only very few organisms are able to degrade it. *P. ostreatus* is an edible mushroom which grows in nature on dead trees and is artificially cultivated on different types of substrates. Since this fungus is able to degrade lignin though also metabolizes polysaccharides, this study aimed to develop strains of *Pleurotus ostreatus* with a reduced capacity of cellulose consumption to produce a selective degradation of lignin.

MATERIALS AND METHODS

Fungal strains

The *P. ostreatus* dikaryons 8x3 is a subculture bred from progenies of dikaryons FX(868x381) and Somycel 3004 (Eger *et al*, 1976). This strain shows a vigorous growth, a huge capability for primordia formation and certain selectivity for lignin consumption (Aguilera *et al*, 1982). Its two component neohaplonts (nh1 and nh4), were obtained by Leal and Eger (1980). Dr. Eger provided also the 16 "wild type" monokaryons which were paired with acellulolytic mutants.

Growth media

Malt extract agar, whole (MA) and deproteinized (DPMA) were used for mycelium growth, matings and storage of fungal cultures (Eger, *et al*, 1976).

Media for recovery of mutants

The two layer medium used by Eriksson (1981) was modified for this study. The mineral medium of the first layer was set to pH 4.5 and eventually substituted by DPMA (0.5%). For the second layer, 8 ml of cellulose medium (also set to pH 4.5) were sterilized in test tubes and supplemented with 50 μl of sodium deoxicolate (0.4 g/l) and 10 μl of thiamine (0.1 mg/l).

Media for characterization of mutants

Three sterile filter paper pieces (0.5 x 0.5 cm) were distributed on the right side of a petri dish with meat peptone agar (0.1%). Agar rods (0.5 cm ϕ) were cut from the edge of a growing colony, one rod was placed on each filter paper and three more rods were distributed on the opposite side of each petri dish. Acellulolytic strains develop the same mycelium growth (diameter) on both sides of the petri dish.

Production of acellulolytic mutants (C-)

A mycelial suspension with ca. 10^9 fragments/ml was mixed with an equal volume of 0.02% N-metyl-N'-nitro-N-nitroso guanidine (MNNG) and allowed to react for 60 minutes at room temperature. Mutagen action was stopped by dilution with 4 volumes of sterile distilled water. Thereafter, 1 ml of this solution was spread over the first layer of the recovered medium and then the second layer was poured. Plates were incubated at 28°C for 14 days and colonies which did not form a clearing zone in the cellulose medium were recovered as possible mutants.

Genetic improvement by breeding

Production of dikaryons by pairing of monokaryotic strains, and recovery of their progenies were performed according to Eger (1978).

Lignocellulosic substrates for degradation tests

Wheat straw and a mixture of wheat straw-maize straw (1:1) were used as lignocellulosic substrates. These were soaked in water overnight and then pasteurized at 85°C for two hours. For each strain, 8 kg of susbtrate were prepared and inoculated with 3% of grain spawn after cooling down to 25°C. Inoculated substrates were distributed in eight bags of 1 kg each one and incubated at 25°C for 50 days.

Chemical analysis

Degraded and undegraded substrates were dried prior to analytical determinations. In order to measure the loss of lignin and polysaccharides, ash content had to be quantified (AOAC, 1980). For determination of polysaccharides a total hydrolysis with sulfuric acid followed by quantification of xylans (Tracey, 1950) and glucans (Klein, 1953) was performed. Lignin analysis was carried out according to Effland (1977).

RESULTS AND DISCUSSION

Previous to production of mutants, various wild type dikaryons were firstly tested. The dikaryotic strain 8x3, which showed a good growth and a more selective degradation of lignin was selected. For production of acellulolytic mutants the monokaryotic mycelium from the two component neohaplonts of dikaryons 8x3 was used.

In an initial mutation experiment, survival rates were determined every 10 minutes. The recommended lethallity rate of 99.99% was achieved after 60 minutes of mutagen action. For production of acellulolytic mutants, 17 mutation experiments were performed and about 240,000 survivors were recovered. From them, 434 strains were isolated as possible mutants. In the first 10 experiments the medium described by Eriksson was used. For the last 7 experiments, the first layer of this medium was supplied with DPMA at 0.5% (Table 1).

TABLE 1. PRODUCTION OF ACELLULOLYTIC MUTANTS (C-) OF *Pleurotus ostreatus*.

RECUPERATION MEDIA	MUTATION EXPERIMENTS (NUMBER)	STRAINS	
		SURVIVORS	MUTANTS
CMC-Minerals	10	194,000	370
CMC-DPMA (0.5%)	7	41,125	64
Total	17	235,125	434

For characterization of acellulolytic mutants, the 434 possible mutants were tested twice more on the same two layer medium yielding only 20 colonies unable to clear the cellulose medium (Table 2). In order to confirm the acellulolytic character of the remaining 20 mutants, they were tested again on the same two layer medium but using now two different concentrations of DPMA. After this test one total and three partial acellulolytic mutants were recovered (Table 3).

TABLE 2.- SUCCESSIVE TESTS ON TWO LAYER MEDIUM

TESTS	REMAINING (C-) STRAINS
Recuperation media	434
First	52
Second	20

TABLE 3. EFFECT OF COMPOSITION OF THE FIRST LAYER AND INCUBATION TIME ON DETECTION OF (C-) STRAINS.

FIRST LAYER	TIME (days)	(C-) STRAINS
Minerals	5	20
	10	19
	20	7
DPMA (0.3%)	5	13
	10	7
	20	6
DPMA (0.5%)	5	8
	10	6
	20	1 / 3

* Partial acellulolytic character (C-/+)

These 20 mutants were also evaluated on filter paper medium, resulting in six acellulolytic mutants after this test (Table 4).

TABLE 4 DETERMINATION OF (C-) CHARACTER ON MEDIA WITH FILTER PAPER AS CARBON SOURCE.

MEDIUM	(C-) STRAINS
Water agar	20
Pepton (0.05%)	7
Pepton (0.10%)	6

Since the total and the partial acellulolytic mutants showed a very scanty growth in comparison to wild type strains, genetic improvement was required. For this purpose, each mutant was paired with vigorous compatible wild type monokaryons.

From their progenies, 479 monokaryons were isolated from the total acellulolytic mutant and 597 from the partial mutant. In both progenies, the quantities of acellulolytic monokaryons were evaluated to identify the number of genes controlling the acellulolytic character in each mutant. These results indicated that there are five genes controlling the acellulolytic character in the total mutant, while in the partial mutant, five or six genes are controlling this character.

The acellulolytic monokaryons from each progeny were classified according to their mating types. Compatible monokaryons were then mated, yielding acellulolytic dikaryons in all cases (Figure 1).

FIGURE 1 PRODUCTION OF (C-) DIKARYONS

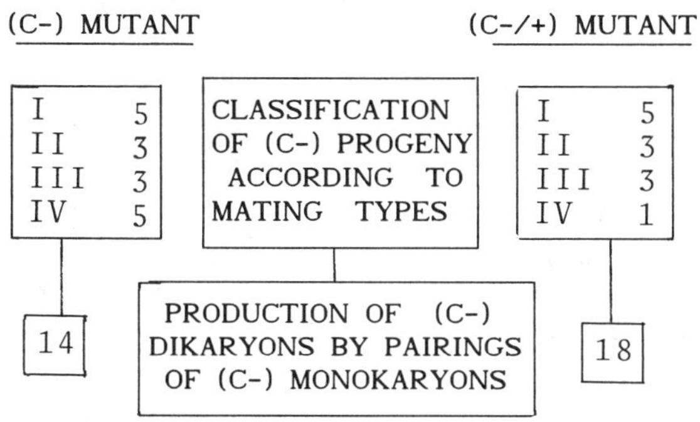

The monokaryotic progeny from the total acellulolytic mutant was used to perform pairings of acellulolytic monokaryons with cellulolytic monokaryons. Similarly, cellulolytic monokaryons were mated with each other. In both cases, 7% of all the dikaryons showed the acellulolytic character (Table 5).

TABLE 5 PRODUCTION OF (C-) DIKARYONS BY COMPLEMENTATION THROUGH CROSS MATING OF THE PROGENY FROM THE (C-) MUTANT

PAIRINGS	DIKARYONS		
	TOTAL	(C-)	(C+)
(C-) x (C+)	65	61	4
(C+) x (C+)	86	80	6

In a second stage of genetic improvement, the acellulolytic monokaryons from both progenies were cross-paired, resulting in this case, only 4% of acellulolytic dikaryons (Figure 2).

FIGURE 2 GENETIC IMPROVEMENT BY BREEDING THE (C-) PROGENY FROM BOTH MUTANTS (SECOND STAGE)

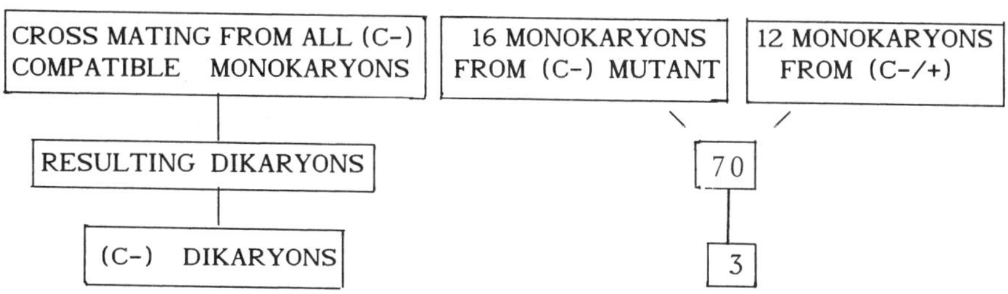

Finally, acellulolytic dikaryons produced from both stages of genetic improvement were grown on lignocellulosic materials. After 50 days incubation, the loss of lignin and polysaccharides was measured. Both types of acellulolytic dikaryons consumed glucans to a lower extent than the wild type strain. However, some strains showed also a reduced consumption of lignin.

The dikaryons from the second stage of genetic improvement, nevertheless, presented some advantages like a reduced consumption of cellulose, a more vigorous growth and a high consumption of lignin, similar to that of wild type strains (Table 6).

TABLE 6 LOSS OF LIGNIN AND POLYSACCHARIDES AFTER 50 DAYS OF INCUBATION WITH ACELLULOLYTIC DIKARYONS FROM BOTH STAGES OF GENETIC IMPROVEMENT

STRAINS	LOSS (%)		
	LIGNIN	GLUCANS	XILANS
C- Dikaryons (1st. Genetic Improvement)[1]			
Control	45 ± 1.1	45 ± 2.5	56 ± 1.4
Ia	31 ± 2.9	16 ± 0.7	52 ± 0.9
Ib	28 ± 0.4	22 ± 1.8	9 ± 1.6
Ic	34 ± 3.4	29 ± 0.6	59 ± 0.2
Id	47 ± 0.9	30 ± 0.7	53 ± 0.9
Ie	36 ± 1.0	33 ± 2.0	61 ± 0.6
If	49 ± 0.8	38 ± 0.8	56 ± 0.9
C- Dikaryons (2nd. Genetic Improvement)[2]			
Control	60 ± 0.5	44 ± 0.7	49 ± 1.7
IIa	52 ± 3.9	23 ± 0.5	35 ± 1.8
IIb	53 ± 1.4	31 ± 0.9	28 ± 0.9
IIc	63 ± 1.3	43 ± 0.2	55 ± 1.5

Substrates: 1 = Wheat straw
2 = Wheat straw-Maize straw

REFERENCES

- Aguilera, A., Hernández, M. and Ramírez B. 1982. BSc. Thesis, UNAM Facultad de Química.
- A.O.A.C. Official Methods of Analysis, 1980, 11th Edition.
- Effland, J.M. 1977, TAPPI 60 (16): 143-144.
- Eger, G., Eden, G. E. and Wissing, E. 1976. Theoret. Appl. Genet. 47: 155-63.
- Eger, G. 1978, Biology and Cultivation of Edible Fungi. Chang and Hayes (Ed.) Academic Press, New York.
- Eriksson, K. E. 1981, Pure and Applied Chem., 53: 33-34.
- Klein, B. and Weissman, M. 1953, Anal. Chem. 25(5): 771-74.
- Leal, H. and Eger, G. 1980. Abstract of the VIth. International Fermentation Symposium. London, Canada.
- Tracey, M. U. 1950, Biochem. 47: 436-43.

63

Lignin degradation: basic research progress, and applications in soil remediation and biopulping

T. Kent Kirk - Forest Product Laboratory, U.S. Department of Agriculture, Forest Service, One Gifford Pinchot Drive,Madison WI 53705, USA.

ABSTRACT

The lignin-degrading enzyme system of basidiomycetous white-rot fungi has been studied intensively in recent years. Research with *Phanerochaete chrysosporium* indicates that the extracellular part of the system is comprised of the enzymes lignin peroxidase, manganese peroxidase, glyoxal oxidase and certain metabolites. Lignin is fragmented by this system, and the plethora of degradation products are taken up by the hyphae and further metabolized by the intracellular system, which has received very little research attention. The structural complexity and heterogeniety of lignin demands that the degrading enzyme system (extra- and intra-cellular) be non-specific. Research has shown in fact that it is so non-specific that it also degrades a variety of manmade compounds, including polycyclic aromatics, some polychlorinated biphenyls and dioxins, DDT, and many chlorinated phenols. With pentachlorophenol (PCP) as a model substrate, we have studied the possibility of using *Phanerochaete* spp. to remediate contaminated soils. Successful laboratory results led to a promising field study in the summer of 1989. *Phanerochaete chrysosporium* and other white-rot fungi are also being studied as pretreatment agents for mechanical pulping of wood, with the objectives being to reduce energy requirements, increase paper strength properties, and decrease pollution. Results show that these objectives can be met.

INTRODUCTION

Cellulose and the other structural components of wood are decomposed to

CO_2 and water as wood is decayed in nature. Such decay is largely accomplished by the higher basidiomycetous fungi that inhabit wood and the litter of the forest floor. The key to their ability to decay wood is their capacity to degrade lignin. Many strongly cellulolytic microbes, including the well studied fungus *Trichoderma reesei* and the bacterium *Cellulomonas fimi,* have essentially no effect on the cellulose in wood or other lignified tissues because they cannot penetrate the lignin barrier. During the past 10 years rapid progress has been made in understanding how lignin is degraded by the basidiomycetes. At the same time, efforts have been made in a number of laboratories to use that understanding in practical applications, including bioremediation and biopulping. This paper provides an overview of the current understanding of the biochemistry of lignin biodegradation, and briefly summarizes research on the use of ligninolytic fungi in bioremediation and biopulping.

THE LIGNIN-DEGRADING ENZYME SYSTEM

Lignin-degrading enzymes were discovered, and subsequently characterized, in the ligninolytic fungus *Phanerochaete chrysosporium* (1-5). Research on lignin biodegradation began in earnest in the 1970s, when ^{14}C-lignins were prepared and used to delineate which groups of microbes are able to mineralize lignin (decompose it to $^{14}CO_2$). The higher basidiomycetous fungi were found to be the most proficient, and *P. chrysosporium* was chosen for detailed study for a number of reasons. Subsequent research with it described the culture parameters important for lignin mineralization, and showed that one or more steps are "secondary metabolic" events, triggered by limitation for certain nutrients. At the same time, the chemical changes that occur in the lignin polymer during decay by ligninolytic fungi were described (6). In the early 1980s, specific degradative reactions accomplished by ligninolytic cultures of *P. chrysosporium* were described using synthesized "dimeric" lignin model compounds; it was also found that certain polymeric dyes are decolorized. Cell-free enzyme preparations that catalyzed dimer cleavage and dye decolorization were reported in 1983, and subsequently the peroxidases responsible were described.

Evidence indicates that these enzymes, now termed *lignin peroxidase* and *manganese peroxidase*, are responsible for the initial fragmentation of the lignin polymer (Figure 1). The extracellular H_2O_2 required to activate the peroxidases is supplied by glyoxal oxidase, and perhaps by intracellular sugar- and alcohol-oxidases. Veratryl alcohol (3,4-dimethoxybenzyl alcohol), found in the extracellular milieu of ligninolytic cultures, is a secondary metabolic product synthesized *de novo*. As depicted in Figure 1, lignin peroxidase (LiP) is first oxidized by peroxide. In turn, the oxidized LiP oxidizes aromatic nuclei in lignin (phenolic and non-phenolic) by one electron, generating cation radicals. These react spontaneously with nucleophiles (primarily water), and with molecular oxygen. The result is an "enzymatic combustion" in which C-C and C-O linkages are cleaved, depolymerizing the polymer and opening aromatic rings. A plethora of aromatic and aliphatic products are thereby formed, and these enter the hyphae where they are further oxidized by as yet undescribed enzymes.

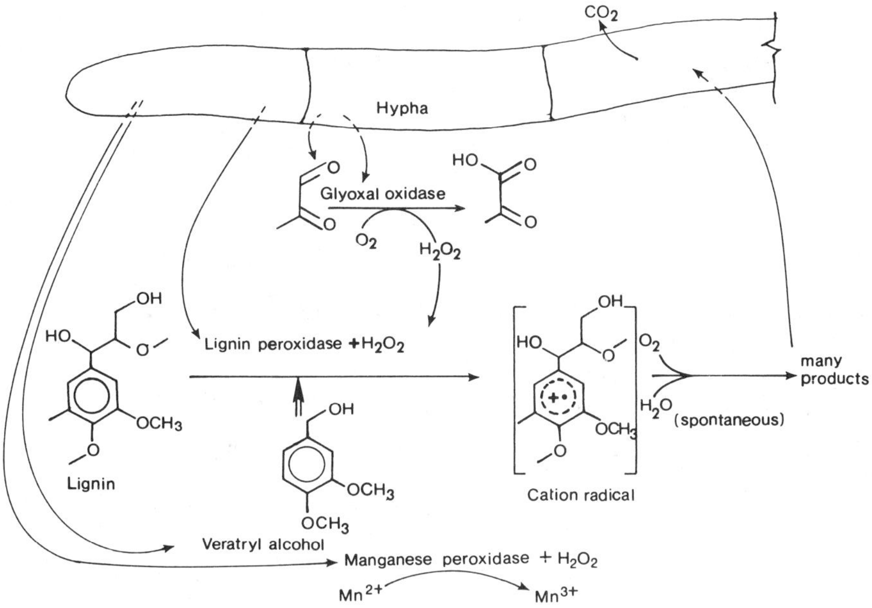

Figure 1. Scheme depicting the lignin-degrading system of *Phanerochaete chrysosporium.*

Veratryl alcohol, itself a substrate for LiP, apparently acts as a "mediator," or electron relay, between LiP and its substrates. Manganese peroxidase (MnP) catalyzes the oxidation of Mn^{2+} to Mn^{3+}, which in turn apparently oxidizes lignin. The exact function of MnP, and the chemistry of its oxidation of lignin is not yet clear, and for that reason it is shown at the bottom of Figure 1. In *in vitro* reactions, in which Mn^{3+} is stabilized by lactate or other organic acid, the oxidation potential of Mn^{3+} is too low to react at significant rates with non-phenolic units in lignin (7). Reaction only with phenolic units would not be expected to seriously degrade lignin. Even so, recent compelling evidence points to a significant role for MnP in lignin degradation (see below). It seems probable that uncomplexed Mn^{3+}, which has a higher oxidation potential than Mn^{3+}stabilized by organic acids, might be responsible for significant degradative reactions *in vivo*, despite its instability. Further research, now underway in several laboratories, should soon resolve the issue. Some evidence suggests that laccase, a phenol-oxidizing blue copper oxidase, is involved in lignin degradation in some fungi, although it is not produced by *P. chrysosporium* and certain others.

General characteristics of lignin peroxidase of *Phanerochaete chrysosporium* are given in Table 1. The enzyme has been discovered in most but not all ligninolytic basidiomycetes, which raises the question of whether it is required for lignin degradation by all fungi. It is possible, however,

Table 1. General characteristics of lignin peroxidase of *Phanerochaete chrysosporium.*

- Secondary metabolic (transcriptionally regulated)
- Extracellular
- Protoporphyrin IX heme
- Multiple acidic isoenzymes (structural genes)
- Glycosylated (N- and O-)
- M_r = 39-42 kDa
- oxidizes aromatic nuclei by one electron

that the techniques used to look for LiP activity have been equivocal. As a case in point, apparent structural genes for LiP have been detected by Southern blot analysis in the fungus *Ceriporiopsis subvermispora*, even though LiP activity has not. In any event, evidence that LiP plays a role in lignin degradation includes the following: its expression is temporally associated closely with lignin depolymerization and mineralization by *P. chrysosporium*; it is regulated, as is the overall lignin-mineralizing system, by nutrient nitrogen or nutrient carbon; its action results in reactions that are known to be involved in lignin degradation; addition of exogenous LiP to active cultures enhances lignin degradation; a defined mutant with no LiP had much decreased lignin-mineralizing ability (8); and LiP has been shown partially to depolymerize isolated lignins (9; K. Hammel and M. Tien, personal communication). That extracellular peroxide is required for lignin degradation has been established, which clearly implicates peroxide-requiring enzymes.

General characteristics of manganese peroxidase (MnP) are given in Table 2. The enzyme has been found in all examined ligninolytic basidiomycetes. It is often the dominant extracellular enzyme, as it was found to be in extracts of wood being decayed by *Phanerochaete chrysosporium* (10). Additional

Table 2. General characteristics of manganese peroxidase of *Phanerochaete chrysosporium*

- Secondary metabolic (transcriptionally regulated)
- Extracellular
- Protoporphyrin IX heme
- Multiple acidic isoenzymes (structural genes)
- Glycosylated
- M_r = 46 kDa
- Oxidizes Mn^{2+} to Mn^{3+}
- Oxidizes reducing agents, with reduction of O_2

evidence that it is important for lignin degradation is inherent in its recently demonstrated ability partially to depolymerize lignin (in the presence of H_2O_2

and Mn^{2+}) (11; Hammel and Tien, personal communication). Even so, the enzyme is not always detectable in lignin-degrading cultures (12), which leaves some uncertainty as to its necessity. It should be noted, however, that LiP, in the presence of H_2O_2, O_2, and the metabolites veratryl alcohol and oxalate, oxidizes Mn^{2+} to Mn^{3+} (13).

General characteristics of glyoxal oxidase (GLOX) are given in Table 3 (14). The enzyme substrates glyoxal and methyl glyoxal are secreted by ligninolytic cultures, and their concentrations are sufficient to account for a physiological role in peroxide generation for lignin degradation (14).

Table 3. General characteristics of glyoxal oxidase of *Phanerochaete chrysosporium*

- Secondary metabolic
- Extracellular
- Contains colorless copper
- Two acidic, allelic isoenzymes
- Glycosylated
- $M_r = 68$ kDa
- Oxidizes (methyl)glyoxal by two electrons, reducing O_2 to H_2O_2.

One of the most interesting features of GLOX is that it is inactive in the absence of a peroxide-using enzyme. Thus, the concentration of extracellular peroxide generated by GLOX is apparently never in excess, which is in accord with the fact that both LiP and MnP are sensitive to excess H_2O_2.

Until very recently, one of the enigmas in research on the biochemistry of lignin biodegradation was the inability to construct an *in vitro* enzyme/metabolite system that would depolymerize lignin; indeed, lignin peroxidase was shown to *polymerize* lignin. However, Hammel and Moen showed recently (9) that crude LiP, in the presence of veratryl alcohol and H_2O_2, partially fragmented synthetic lignin. Hammel et al. (personal communication) have now demonstrated *in vitro* depolymerization of unmodified synthetic lignins by purified LiP or MnP, in the presence of the appropriate cofactors. Failure of earlier workers -- and even some recent workers (15) -- to demonstrate depolymerization apparently was due to use of incorrect reaction conditions.

Identification of lignin-degrading enzymes made it possible to investigate the molecular genetics of of the lignin-biodegrading system. Such studies have now been done with several fungi, but *Phanerochaete chrysosporium* has been the most thoroughly studied. With that organism, complex structural gene families have been found. To date six *lip*, two *mnp*, and one *glox* have been cloned and sequenced. Most of the LiP genes are clustered; at least five *lip* and one *mnp* are found on a single dimorphic chromosome, and three *lip* are within a 30 kb domain (16). All of the genes contain introns and multiple glycosylation sites. Further detail is found in recent reviews and other papers (17, 18).

BIOREMEDIATION

A consideration of the structure of lignin made it apparent from the start of the research that the ligninolytic system would be found to be extracellular, non-hydrolytic, and non-specific. The above brief discussion shows that these are in fact characteristics of the system. Just how non-specific the system is, however, became apparent when it was found that ligninolytic cultures were able to mineralize lignin that had been subjected to kraft pulping and chlorine bleaching treatments (19). This observation led to studies of the degradation of industrial "chlorolignins," and to development of bench-scale fungal bioreactors for degrading such polymers in bleaching effluents (20, 21). (Chlorolignins are quite resistant to conventional biological waste treatments.) In connection with the studies of the degradation of chlorolignins by the fungi, it was discovered that low molecular weight chlorinated aromatics also present in the effluents are also degraded. Surprisingly, it was then shown that numerous manmade chemicals bearing little structural relation to lignin are also degraded by the fungi (Table 4). This observation has prompted several research groups to explore the potential of the ligninolytic fungi in bioremediation. At the Forest Products Laboratory, studies on the remediation of contaminated soils by the fungi have been underway for almost six years, and have been done in collaboration with the Environmental Protection Agency. The focus has been on the wood-preservative pentachlorophenol (PCP). Pentachlorophenol contaminates hundreds of wood treatment sites in the U.S. Laboratory work initially defined the factors affecting growth of lignin-degrading fungi from wood chip inoculum into soils, and PCP degradation in soils (22). Certain strains of *Phanerochaete sordida* and *P. chrysosporium* were found to be the best of several studied fungi (23). Results with these and other fungi showed that PCP was rapidly depleted, primarily by conversion to non-volatile products (22). The nature of the products, whether they were soil-bound or extractable was greatly influenced by soil type. The primary extractable transformation product was pentachloroanisole. However, in most cases most of the transformation products were not extractable. Losses via volatile products were negligible.

A field study was conducted in 1989 with the above two *Phanerochaete* species. The site, near Oshkosh, Wisconsin, was an old storage tank area used for storage of a commercial PCP formulation. Plots of 1 m^2 were established, and involved various control and inoculated sets. Inoculum consisted of wood chips thoroughly grown through with the fungi. Results (24) were essentially the same as those obtained in the laboratory.

Inoculation of soil containing 250-400 $\mu g\ g^{-1}$ PCP resulted in an overall decrease of 88-91% of the PCP in the soil in 6.5 weeks. A small percentage (8-13%) of the decrease was a result of methylation; other products were not detected in soil extracts. Thus, if losses of PCP via volatilization were negligible (as in the laboratory), then the PCP was depleted via mineralization and, primarily, by conversion to nonextractable soil-bound products. A second field study to investigate the ability of ligninolytic fungi to remediate soil contaminated with both PCP and and creosote is underway at a closed pole treatment facility in Mississippi.

Table 4. Some manmade compounds degraded by *Phanerochaete chrysosporium* and other ligninolytic fungi.

PAH's	Chlorophenols
Benzo(a)pyrene	Pentachlorophenol
Phenanthrene	2, 4-Dichlorophenol
Anthracene	
Fluoranthene	Chlorinated Biphenyls (PCB's)
Benzo(b)fluoranthene	
Benzo(k)fluoranthene	3, 4, 3', 4-Tetrachlorobiphenyl
Indeno(ghi)pyrene	
Benzoperylen	Chlorinated Dibenzo (p) dioxins
Fluorene	
	2, 3, 7, 8-TCDD
Insecticides	Azo and Heterocyclic Dyes
DDT	Orange II
Lindane	Tropaeolin
2, 4, 5-T	Congo Red
Chlordane	Azure B
Dieldren	
	Triphenylmethane Dyes
Chlorinated Anilines	
	crystal violet
4-Chloroaniline	pararosaniline
3, 4-Dichloroaniline	cresol red
	bromphenol blue
Explosives	ethyl violet
	malachite green
2, 4, 6-Trinitrophenol (TNT)	brilliant green

At this time, it would seem that ligninolytic fungi could be used in a practical treatment of soils contaminated with certain chemicals. It is not yet known, however, whether the fate of PCP or other chemicals is environmentally acceptable. (It would seem that there is a need for standard tests for efficacy of remediation technologies in general; identifying and quantifying all of the transformation products, although desirable, is not always practical, especially if processes involving "enzymatic combustion" with formation of numerous products are involved.)

BIOPULPING

One of the most attractive ways to deal with environmental contamination is to avoid it in the first place. In many industries, biological processes have the potential to replace chemical processes.

Biopulping is the treatment of wood chips with a lignin-degrading fungus prior to pulping. It addresses the pollution associated with chemical and chemimechanical pulping processes, and in addition two other problems facing the pulp and paper industry: i) the high energy consumption required in making mechanical or chemimechanical pulps, ii) the relatively low strength properties of paper made from mechanical pulps. In April of 1987 a cooperative research program on biopulping was established, involving the Forest Products Laboratory, the University of Wisconsin Biotechnology Center, and 21 pulp and paper and related companies. The overall objective of the 5-year Consortium research effort is to evaluate the scientific and technical feasibility of using a fungal pretreatment with mechanical pulping. The Biopulping Consortium is divided into six closely coordinated teams. A fungal team screens species and strains of fungi, optimizes the fungal pretreatment, and provides fungus-treated chips for the pulp and paper team to evaluate. The latter team determines energy consumption required for pulping, and measures pulp and paper properties. An engineering and scale-up team is looking at the fungal pretreatment as an engineered solid substrate fermentation, and is working with the fungal research team to determine critical parameters. An enzyme team seeks to determine which of the extracellular enzymes secreted during fungal pretreatment are beneficial for pulping and which are not, with emphasis on the lignin- and cellulose-degrading enzymes. Research of the molecular genetics team is basic, aimed at developing information that ultimately will allow improved strains to be developed. Supporting the other teams is an information team that uses sophisticated computer search strategies to screen the scientific and technical literature.

The Consortium teams have made good research progress, described in over 40 research papers and one patent. Major accomplishments include the following. Strains of several fungi that give superior results have been identified, with several strains of the little-known *Ceriporiopsis subvermispora* being the best to date. Approximately 200 biopulping runs (2 kg scale) with aspen and pine woods have demonstrated significant energy savings and in most cases improvements in paper strength properties (Table 5). Major isoenzymes secreted during successful fungal pretreatment have been identified; one MnP isoenzyme is dominant in aspen during pretreatment with *P. chrysosporium* (10). Considerable information has been obtained about the structure and organization of the genes for key lignin- and cellulose-degrading enzymes, and a versatile transformation system has been developed. The engineering parameters that would likely be most critical in a commercial fungal pretreatment operation have been identified, with aeration being a major one. Preliminary economic evaluations indicate that energy savings of only 25% during mechanical pulping would enable a substantial capital investment in fungal pretreatment; in the laboratory, energy savings usually exceed 25%. The information team has provided abstracts of hundreds of relevant papers, and conducted many tailored searches for the researchers. The Consortium has involved approximately 35 researchers and an investment of approximately $3.5 million. A new consortium is planned to start at the termination of the present one in April of 1992, with emphasis on scale-up and economic evaluation.

Table 5. Yield, paper properties and energy consumption during biomechanical pulping.

Chip treatment	Yield %	Burst index ($kPam^2/g$)	Tear index (mNm^2/g)	Tensile index (Nm/g)	Pulping Energy (Whr/kg)
Aspen (hardwood)					
RMP (control)1/	--	.92	2.03	29.5	2746
*Phanerochaete*2/	86	2.05	4.53	52.8	1477 (46%)3/
Loblolly pine (softwood)					
RMP (control)1/	--	.58	2.74	22.3	2404
*Ceriporiopsis*2/	94	.82	4.75	21.8	1167 (51%)3/

1/ RMP, "Refiner mechanical pulp," produced in a disk refiner mechanical pulping apparatus. Control pulp had no pretreatment.

2/ RMP was made from chips pretreated with *Phanerochaete chrysosporium* or *Cerioporiopsis subvermispora.*

3/ Percent energy saved by fungal pretreatment in comparison to controls.

REFERENCES

1. Buswell, J.A.; Odier, E. *CRC Critical Reviews in Biotechnology* **1987**, *6,* 1.

2. Eriksson, K.-E.; Blanchette, R.A.; Ander, P. *Microbial and Enzymatic Degradation of Wood and Wood Components*, Springer-Verlag KG, Berlin, **1990**, 407 p.

3. Gold, M.H.; Wariishi, H.; Valli, K. in Whitaker, J.R.; Sonnet, P.E. (Eds.) *Biocatalysis in Agricultural Biotechnology, ACS Symposium Series No. 389*, American Chemical Society, Washington, DC, **1989**, 128.

4. Kirk, T.K.; Farrell, R.L. *Ann. Rev. Microbiol.* **1987**, *41*, 465.

5. Tien, M. *CRC Critical Reviews in Microbiology* **1987**, *15,* 141.

6. Chen, C.-L.; Chang, H.-m. in Higuchi, T. (Ed.) *Biosynthesis and Biodegradation of Wood Components*, Academic Press, Inc, Orlando, FL, **1985**, 535.

7. Popp, J.L.; Kirk, T.K. *Arch. Biochem. Biophys.* **1991**, *288*, 145.

8. Boominathan, K; Balachandra Dass, S.; Randall, T.A.; Kelley, R.L.; Reddy, C. A. *J. Bacteriol.* **1990**, *172*, 260.

9. Hammel, K.E.; Moen, M.A. *Enz. Microbial Technol.* **1991**, *13*, 15.

10. Datta, A.; Bettermann, A.; Kirk, T.K. *Appl. Environ. Microbiol.* **1991**, *57*, 1453.

11. Wariishi, H.; Valli, K.; Gold, M.H. *Biochem. Biophys. Res. Comm.* **1991**, *176*, 269.

12. Perez, J.; Jeffries, T.W. *Appl. Environ. Microbiol.* **1990**, *56*, 1806.

13. Popp, J.L.; Kalyanaraman, B.; Kirk, T.K. *Biochemistry* **1990**, *29*, 10475.

14. Kersten, P.K. *Proc. Nat. Acad. Sci.* **1990**, *87*, 2936.

15. Sarkanen, S.; Razal, R.A.; Piccariello, T.; Yamamoto, E.; Lewis, N.G. **1991**, *266*, 3636.

16. Gaskell, J.; Dieperink, E.; Cullen, D. *Nucl. Acid Res.* **1991**, *19*, 599.

17. Pease, E.; Tien, M. in Dordick, J.S. (Ed.) *Biocatalysis for Industry*, Plenum Press, New York, NY, **1991**, 115.

18. Kersten, P.; Cullen, D. in Kinghorn, J.R.; Turner, G. (Eds.) *Applied Molecular Genetics of Filamentous Fungi*, Blackie and Sons, Glasgow, 1992, in press.

19. Lundquist, K.; Kirk, T.K.; Connors, W.J. *Arch. Microbiol.* **1977**, *112*, 291.

20. Yin, C.-F.; Joyce, T.W.; Chang, H.-m. *J. Biotechnol.* **1989**, *10*, 67.

21. Messner, K.; Jaklin-Farcher, S.; Ertler, G.; Blaha, A. *Forum Mikrobiol.* **1988**, *11*, 492.

22. Lamar, R.T.; Glaser, J.A.; Kirk, T.K. *Soil Biol. Biochemi.* **1990b**, *22*, 433.

23. Lamar, R.T.; Larsen, M.J.; Kirk, T.K. *Appl. Environ. Microbiol.* **1990a**, *56*, 3519.

24. Lamar, R.T.; Dietrich, D.M. *Appl. Environ. Microbiol.* **1990a**, *56*, 3093.

64

Mass transfer performance and enzyme production in air-lift and bubble column fermenters

C.K. Erb, W.J. McManamey and D.A.J. Wase - Laboratory for Environmental Biotechnology, Biochemical Engineering Division, School of Chemical Engineering, University of Birmingham, Edgbaston, Birmingham, B15 2TT, UK.

ABSTRACT

Higher coefficients of oxygen mass transfer and fractional gas holdup were observed in a bubble column fermenter, than in a geometrically similar air-lift fermenter. However, in fermentation of the cellulolytic fungus, *Aspergillus fumigatus* IMI 255091, higher maximal titres of β-glucosidase, endoglucanase and filter paper activity were obtained approximately 2 days earlier in the air-lift fermenter than in the bubble column.

INTRODUCTION

A bubble column is the simplest non-mechanically mixed fermenter: a gas-sparged vessel with a large height-to-diameter ratio (Figure 1a). Circulation occurs by 'gulf-streaming', during which a volume of liquid carrying a high concentration of gas bubbles rises through the centre of the vessel, whilst fluid carrying fewer gas bubbles passes down the outside of the vessel.

Air-lift fermenters are modified bubble columns designed for the growth of shear sensitive organisms, enabling better growth and product formation than stirred tank reactors [1]. They comprise main hydrodynamic zones - a riser, into which gas is sparged; the downcomer, parallel to the riser; and a gas disengager, an area separating the two (Figure 1b).

Circulation in air-lift fermenters is caused by a difference in gas holdup between the riser and the downcomer, leading to a difference in bulk fluid density, which induces circulation around the fermenter. This form of circulation improves bulk mixing within the fermenter, over bubble columns [2]. Although there has been much recent attention paid to air-lift fermenters, the advantage of this device over the mechanically simpler bubble column is unclear.

The advantages of an air-lift fermenter over a bubble column are said to be that oxygen mass transfer is more efficient, that shear is lower, that bulk mixing is improved and that the relative velocity of the bubbles to the fluid is lower and thus there is less tendency for bubble coalescence to occur. The fermenter in which an organism is grown has a great effect upon enzyme titres and thus selection of a vessel with such favourable characteristics is vital to efficient fermentation.

The organism which was used in this study, *Aspergillus fumigatus* is a very common contaminant of lignocellulosics such as sugar cane, bagasse and straw. However, it is an infrequent human pathogen, chiefly to individuals exposed to large numbers of spores, which possibly explains why the organism has not been as widely studied as its cellulolytic capabilities might have indicated; nevertheless, it is relatively safe in fermentation. The strain used in this work, *A. fumigatus* IMI 255091, was developed in this laboratory [3] and has high cellulolytic and hemicellulolytic activity, as well as some lignolytic activity [4].

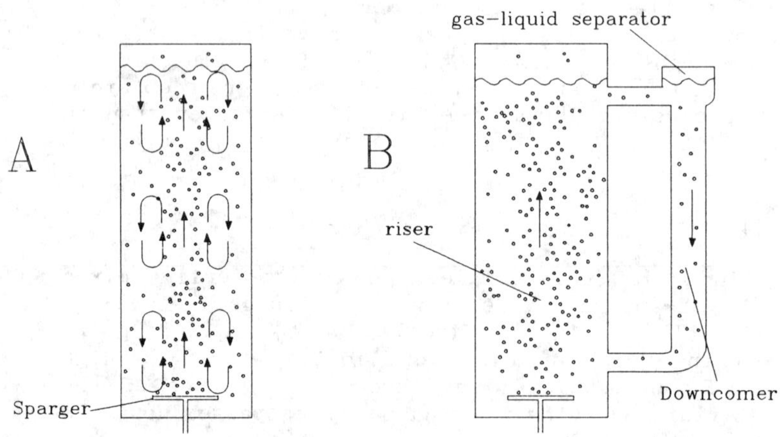

Figure 1: A bubble column (A) and an air–lift fermenter (B).

The aim of this work was to investigate the differences between an air-lift fermenter and a bubble column of similar geometry. Few published data exist for systems other than water or simple non-Newtonian systems and, clearly, the application of these data to fermentation systems comprising a non-Newtonian gas-liquid-solid system is limited [5].

METHODS

Values of fractional gas holdup and the volumetric mass transfer coefficient (K_La) were determined for the fermentation medium containing 1% (w/v) ground hay at a variety of gas flow rates by the volume expansion method and the dynamic method [6]; including the modifications of Koizumi and Aiba [7]) respectively. A galvanic oxygen probe was used to measure K_La and at least 3 replicates (normally 5) were made of each measurement, all of which were carried out equilibrated at 30°C.

The fungus was grown in the two vessels in a working volume of 11 dm^3, using 1% (w/v) ground hay as the carbon source in a medium based upon that used by previous workers [8]. The fermenters were operated at 30°C and daily samples were taken, which were filtered and the culture dry weight determined. The samples were assayed for extracellular protein [9]; β-glucosidase, measuring nitrophenol release from nitrophenyl β-D-glucopyranoside; endoglucanase, using carboxymethyl cellulose, based upon Ghose [10]; and exoglucanase (filter paper activity), based upon Ghose [10]. The latter two involved the measurement of glucose liberation by the dinitrosalicylic acid method [11].

RESULTS AND DISCUSSION

The gas holdup and K_La values from both vessels show that there was a rapid increase in both gas holdup (Figure 2a) and K_La (Figure 2b) with increasing gas flow rate, to maximum values at approximately 10 dm^3/min. At higher gas flow rates in the bubble column, gas holdup values decreased with increasing gas flow rate, up to approximately 12 dm^3/min, after which there was a slight increase again with increasing gas flow rate. The values from the air-lift fermenter were lower, with no obvious peak in gas holdup apparent. Values of K_La remained relatively static at gas flow rates in excess of 9 dm^3/min.

These data indicate that oxygen mass transfer is more efficient in the bubble column than in the air-lift fermenter and that both fermenters exhibit maximal oxygen mass transfer at approximately the same gas flow rate, 9 dm^3/min, the value used in the subsequent fermentations.

The results from both of the fermentations show an early increase in endoglucanase (Figure 3a) and filter paper activity

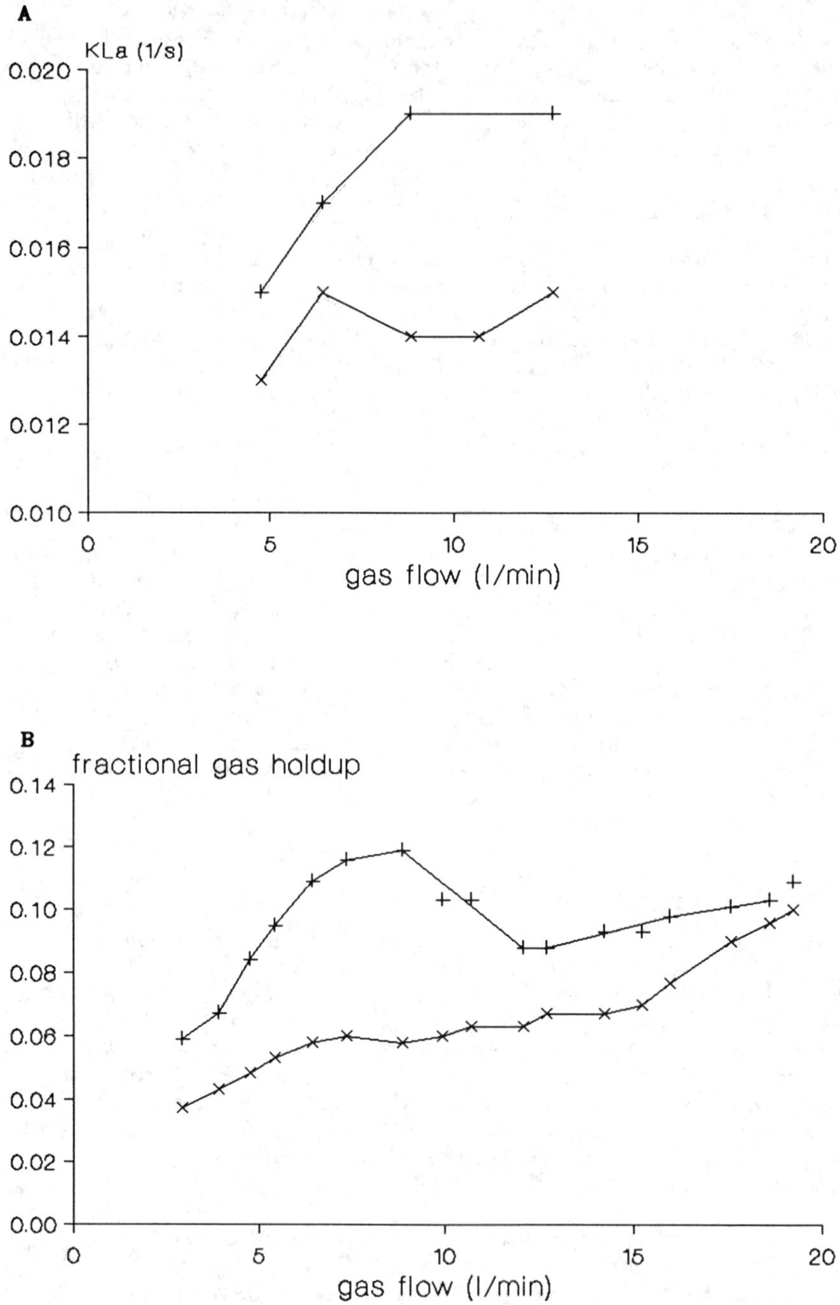

Figure 2: *Graphs showing KLa (A) and gas holdup (B) values for fermentation medium obtained in an air-lift fermenter (X) and a bubble column (+).*

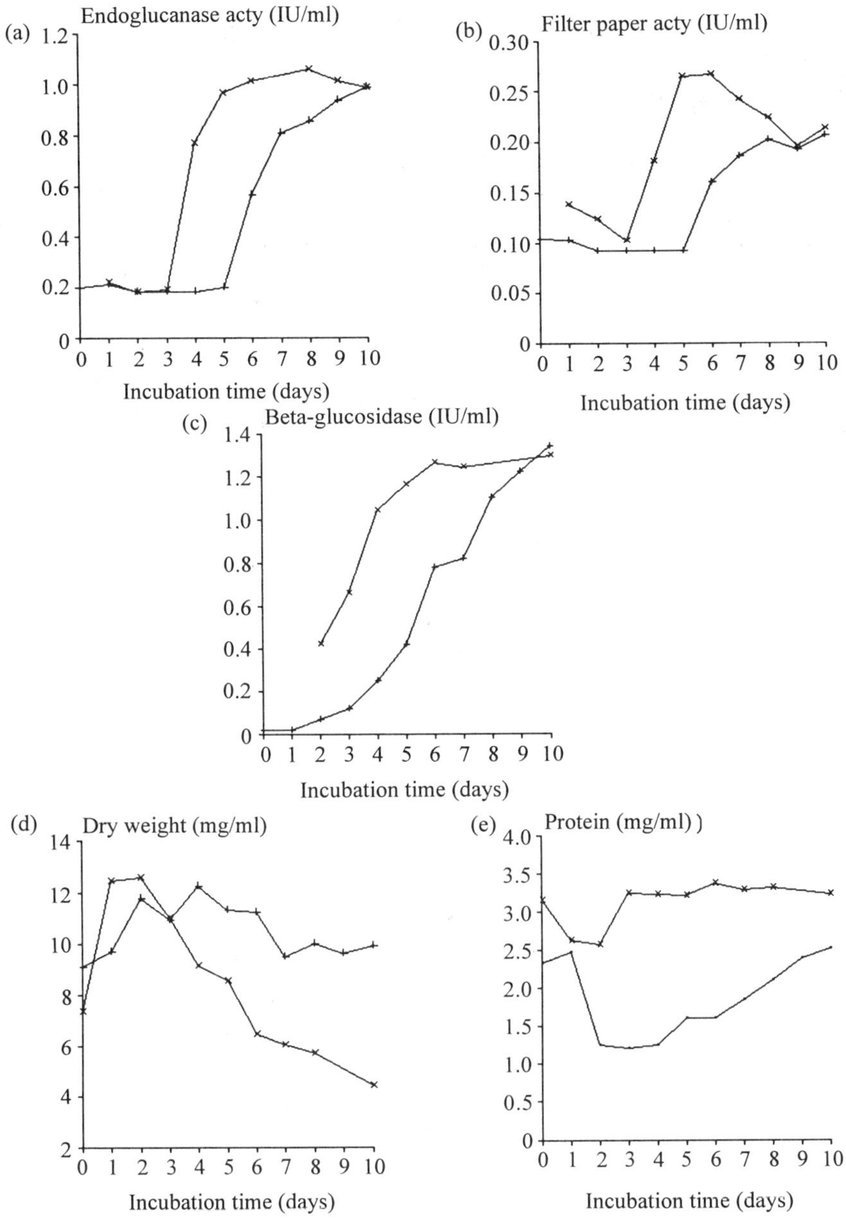

Figure 3: *Graphs showing filter paper (A), endoglucanase (B), and β-glucosidase activities* (C), *extracellular protein (D) and culture dry weight (E) during fermentation of* Aspergillus fumigatus IMI 255091 *in an air-lift fermenter (X) and a bubble column (+).*

(Figure 3a), accompanied by a slower increase of β-glucosidase activity (Figure 3c). This indicates degradation of the long-chain cellulose molecules by endo- and exo-glucanases and subsequent digestion to glucose by the β-glucosidase.

Culture dry weights from both fermenters increased early in fermentation (Figure 3e), accompanied by a decrease in extracellular protein (Figure 3d), suggesting consumption of the peptone which was present in the medium. This was followed by a gradual decrease in culture dry weight during fermentation and an increase in extracellular protein, which implies enzyme production resulting in degradation of the substrate.

The air-lift fermenter exhibited a much greater decrease in culture dry weight and higher extracellular protein, indicating higher enzyme concentrations and more complete degradation of the substrate. The cellulase titres began to rise rapidly after a lag of 3 days in the air-lift fermenter, compared to a lag of 5 days in the bubble column. In addition to this, the maximal titres of cellulase were higher in the air-lift fermenter than in the bubble column.

Thus the air-lift fermenter is a more efficient bioreactor in terms of cellulase yield from *A. fumigatus* than a bubble column of similar geometry, in spite of apparently less efficient oxygen mass transfer in the air-lift fermenter than in the latter.

REFERENCES

1. Wase, D. A. J., McManamey, W. J., Raymahasay, S. and Vaid, A. K. Biotech. Bioeng. 27 1166-1172 (1985).
2. Popovic, M. K. and Robinson, C. W. JAIChE 35 393-405 (1989).
3. Vaid, A. K. PhD. Thesis, University of Birmingham (1982).
4. Holland, T. M. PhD. Thesis, University of Birmingham (1989).
5. Allen, D. G. and Robinson, C. W. Biotech. Bioeng. 34 731-740 (1989).
6. Bandyophadyay, B., Humphrey, A. E. and Taguchi, H. Biotech. Bioeng. 9 533-544 (1967).
7. Koizumi, J. and Aiba, S. Biotech. Bioeng. 26 1131-1133 (1984).
8. Kim, S. W. PhD. Thesis, University of Birmingham (1989).
9. Lowry, O. H., Rosebrough, A. L. F. and Randall, R. J. J. Biol. Chem. 193 265-275 (1951).
10. Ghose, T. K. (ed.) Pure and Appl. Chem. 59 (2) 257-268 (1987).
11. Miller, G. L. Anal. Chem. 31 426-428 (1959).

The authors would like to acknowledge the funding of this work by the Science and Engineering Research Council.

65

Immobilisation of the cellulolytic fungus *Aspergillus fumigatus* IMI 255091

Lesley A. Hamilton and D.A. John Wase - Laboratory for Environmental Biotechnology, Biochemical Engineering Division, School of Chemical Engineering, University of Birmingham, Edgbaston, Birmingham, B15 2TT, UK.

ABSTRACT

Shake-flask experiments have shown that actively growing mycelium of the cellulolytic fungus *Aspergillus fumigatus* IMI 255091 is passively immobilised by entrapment in small reticulated foam pads (captor pads). It is clear from scanning electron micrographs that mycelial growth is abundant throughout the whole of the support matrix. The cellulolytic and hemicellulolytic enzyme titres produced under these conditions were greater than that produced by free mycelial growth of *A. fumigatus*.

INTRODUCTION

Our studies of a specially isolated strain of *Aspergillus fumigatus* (IMI 255091) have revealed that it produces all three major cellulolytic enzymes (endoglucanase, exocellulase and β-glucosidase) in abundance, together with substantial amounts of a wide spectrum of other cellulose- and hemicellulose-degrading enzymes on a variety of substrates.

As these enzymes are intended to degrade insoluble substrates such as wheat straw, hay and paper mill wastes they are secreted by *A. fumigatus* into the growth medium. However, before such an enzyme system can be profitably utilized, it is often necessary to separate the organism from the fermentation broth containing the extracellularly excreted enzymes [1, 2]. Using *A. fumigatus* as a model organism, the process of autoentrapment in small reticulated polyurethane foam pads (captor pads) has been examined in order to assess the extracellular secretion of cellulases together with the separation of fungal mycelium from the substrate.

MATERIALS AND METHODS

CAPTOR PADS

Captor pads exploit the ability of microorganisms to adhere to surfaces and colonise cavities of inert materials.

These small polyurethane reticulated foam pads (1 cm^3) have the following advantages [3]:

1. They are simple to use.
2. The soft foam has a porosity of 97%.
3. They are stable in liquid nutrient media.
4. They do not not suffer from mechanical degradation.
5. They do not suffer from chemical degradation.
6. It is the least cytotoxic method.

ORGANISM

Aspergillus fumigatus IMI 255091 was maintained on malt extract agar slopes and incubated at 37 ^{o}C for 7 days.

IMMOBILISATION METHOD AND FERMENTATION

Passive entrapment of *A. fumigatus* was carried out in 250 mL shake-flasks containing 100 mL of culture medium and incubated in an orbital shaker (200 rpm, 30 ^{o}C) [4]. To each shake-flask 4 captor pads of 1 cm^3 and known weight were added prior to sterilisation (121°C, 15 psig for 15 minutes).

Immobilisation and subsequent induction of cellulases were carried out by the addition of a 1 mL spore (4 x 10^8 mL^{-1}) suspension of *A. fumigatus* using the following two methods:

Method 1- immobilisation in which shake-flasks containing captor pads and the cellulosic source was inoculated directly from the spore suspension.

Method 2- shake-flasks containing malt extract and captor pads but no cellulosic substrate was inoculated from the spore suspension and incubated for 48 h. After this period the medium was drained from the flask and fresh sterilised medium[4] containing the carbon source for the induction of the cellulases was added to the shake-flasks.

CELLULOSIC SOURCE

The cellulosic source used in all experiments was hay, ground in a laboratory mill (Christy and Norms Ltd., Chelmsford, size 8 inches) fitted with a sieve (13 holes of 1mm inside diameter per inch).

ANALYSES

The contents of the shake-flasks were filtered through Whatman glass fibre discs and the dry weight of immobilised cells

ANALYSES

The contents of the shake-flasks were filtered through Whatman glass fibre discs and the dry weight of immobilised cells determined by drying in an oven at 120 °C for 24 h. The filtered broths were analysed for cellulolytic and hemicellulolytic enzyme activity whilst immobilisation was followed by visual/photographic inspection of the shake-flasks together with scanning electron microscopy [5].

IMMOBILISATION AND CELL GROWTH

The passive entrapment of *A. fumigatus* was successful in all shake-flasks containing captor pads regardless of whether immobilisation was carried out by method 1 or method 2. When the medium was inoculated with a spore suspension of *A. fumigatus* the mycelium had an affinity for the captor pads which resulted in a completely transparent medium. That is, immobilisation occurred with no free cell growth at all.

Autoentrapment of the fungus was allowed to proceed for the 48 h, normally employed for the production of a free cell inoculum culture, and was followed by determining the dry weight of the entrapped mycelium. Figure 1 displays changes in immobilised biomass and free cell biomass over 48 h. Up to 40 h, the amount of free biomass formed at any time was roughly double that of the entrapped biomass. However, towards 48 h the free cell biomass fell whilst entrapped biomass continued to increase.

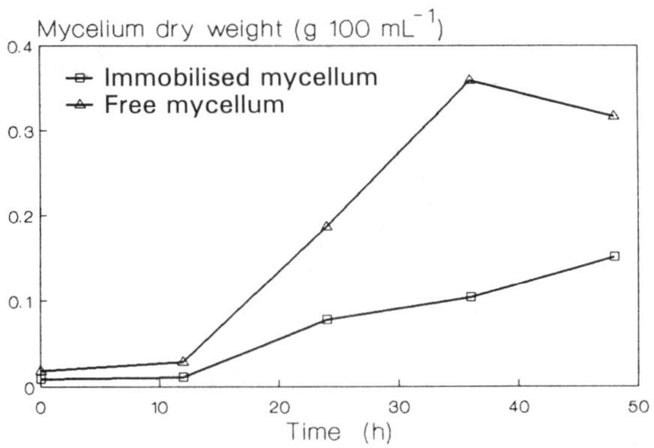

Figure. 1: Growth of *A. fumigatus* as free cell biomass and immobilised cell cultures

Scanning electron micrographs (Plates 1 and 2) also show that colonisation of the pad occurred throughout the support matrix. Thus, the large pore size of the captor pad can be considered as providing a stable environment in which sustained growth of the organism was promoted.

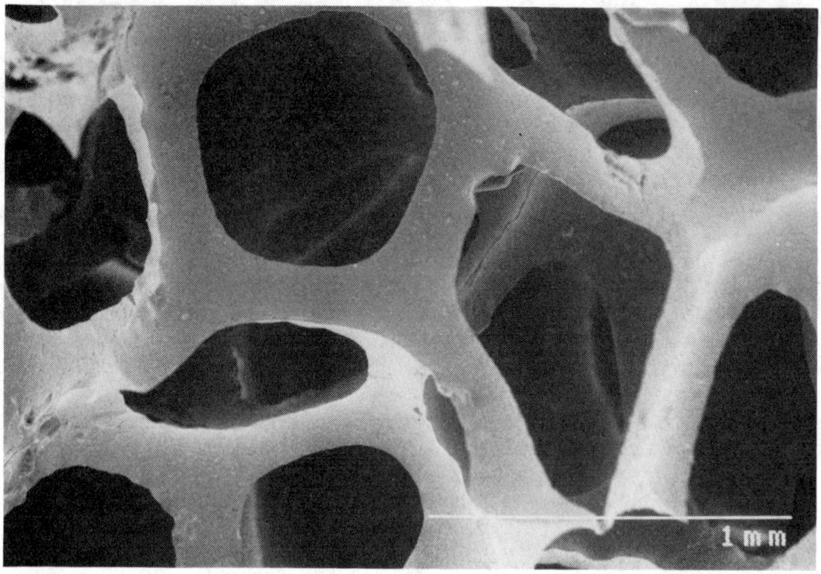

Plate 1: Outside surface of a captor pad prior to inoculation by *A. fumigatus*

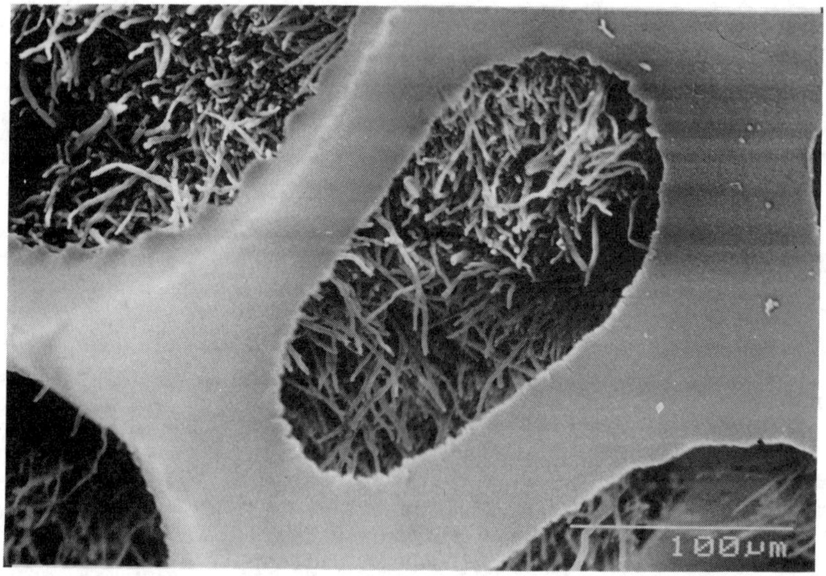

Plate 2: Section through the centre of the captor pad 48 h after inoculation with *A. fumigatus*

Direct inoculation of shake-flasks (method 1) containing the captor pads as well as the cellulase-inducing carbon source, hay, was also successful in terms of immobilisation and growth of the organism. However, the presence of both a nitrogen and carbon source in the medium resulted in the over-growth of the mycelium within the captor pad in a shorter period of time than with method 2. On the other hand, the viability of the organism is clearly evident in this immobilised form such that captor pads containing mycelium could be re-used in repeated batch culture or as a continuous culture.

IMMOBILISATION AND CELLULOLYTIC ENZYME PRODUCTION

The culture broths obtained on day seven of the fermentation, showed an abundance of cellulolytic/hemicellulolytic enzyme activity. Figures 2 and 3 show that all the enzymes assayed (endoglucanase, exocellulase, β-glucosidase and xylanase) were present in significant quantities. Moreover, when the amount of hay was increased the enzyme titre also increased, and that this increase was true for both of the methods of immobilisation tested.

A comparison of the enzyme titres with those produced by the growth of free mycelium is shown in Figure 4. For all of the cellulolytic enzymes (endoglucanase, exocellulase and β-glucosidase), the enzymic activities were approximately the same or even slightly higher than those obtained with free mycelium. On the other hand the titre of xylanase (a hemicellulolytic enzyme) was approximately double that of the free mycelium culture. Fukuda *et al.* (1985) also found improved cellulase titres with *Trichoderma reesei* immobilised in biomass support particles [6]. It was suggested that the lower specific growth rate of the immobilised mycelium may result in an increased synthesis and secretion of cellulases.

CONCLUSION

In conclusion, captor pads provide an environment which shields the mycelium from external factors such as any shearing effects. They allow the free diffusion of cellulolytic enzymes outwards and nutrients inwards, evident by the good growth that takes place throughout the matrix of the support. They also promote excellent yields of enzyme.

Recently a bioreactor has been designed for use with immobilised *A. fumigatus* incorporating features of a downflow fluidised/expanded bed reactor. Thus this reactor configuration for the production of cellulases and of simple sugars from more complex cellulosic/hemicellulosic substrates has enormous potential. The optimisation of this reactor and its application for the degradation of waste vegetation forms the basis of future work.

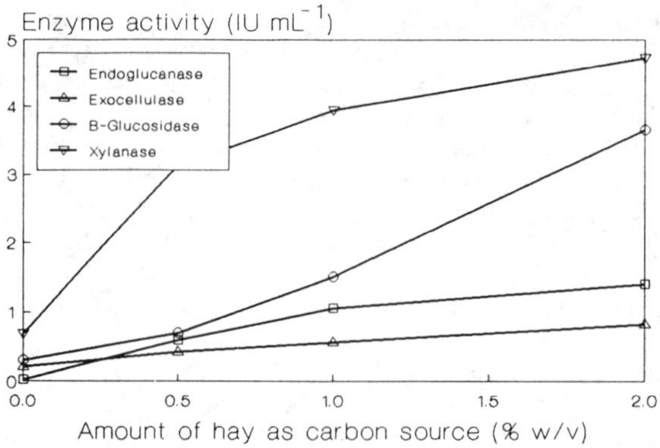

Figure 2: Enzyme activities for immobilised biomass of *A. fumigatus* by method 1 for various amounts of hay.

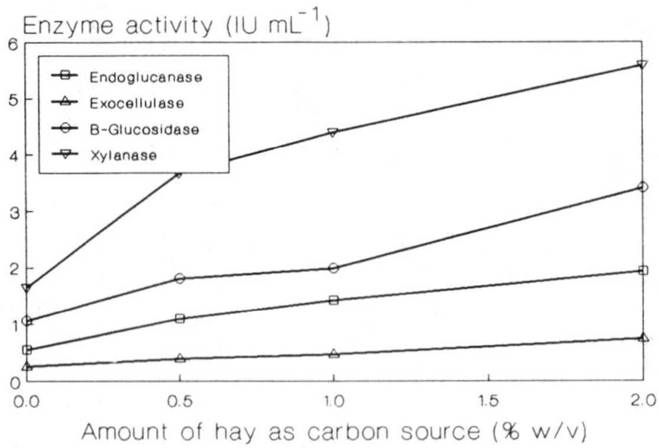

Figure 3: Enzyme activities for immobilised biomass of *A. fumigatus* by method 2 for various amounts of hay.

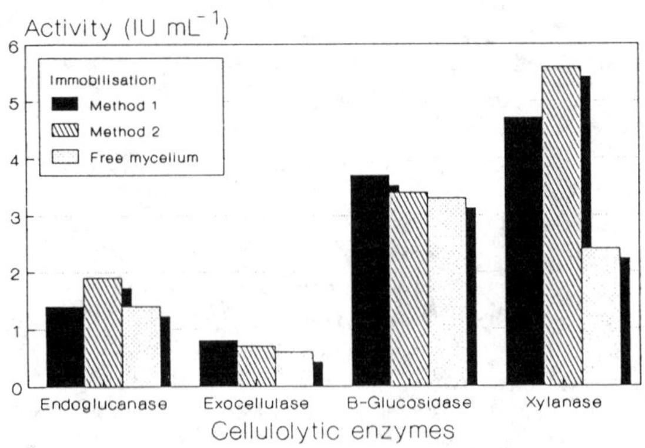

Figure 4: Enzyme activities for *A. fumigatus* in free cell and immobilised cell cultures for methods 1 and 2.

ACKNOWLEDGEMENTS

We are most grateful to Simon-Hartley Ltd., Stoke-on-Trent, for their gift of captor pads, and to Paul Adams of Simon-Hartley Ltd. for his technical help.We are also grateful to S.E.R.C. (The Science and Engineering Research Council), U.K. for the provision of the grant to carry out this work

REFERENCES

[1] Wase, D.A.J. and Raymahasay, S. (1985). In "Cellulose and its derivatives". Eds. J.F. Kennedy, G.O. Phillips, D.J. Wedlock and P.A. Williams. Ellis Horwood, Chichester, U.K. Ch.49

[2] Hamilton, L.A. and Wase, D.A.J. (1990). In "Cellulose sources and exploitation - industrial utilization, biotechnology, and physico-chemical properties." Eds. J.F. Kennedy, G.O. Phillips and P.A. Williams. Ellis Horwood, Chichester, U.K. Ch. 52.

[3] Cooper, P.F., Walker, I., Crabtree, H.E. and Aldred, R.P., (1985). In "Process engineering aspects of immobilised cell systems". Eds. C. Webb, G.M. Black and B. Atkinson. Inst. Chemical Engineers, Oxford, U.K, Ch. 11.

[4] Holland, T.M. (1989). Ph.D. Thesis, University of Birmingham, Birmingham, U.K.

[5] Kim, S.W. (1989). Ph.D. Thesis, University of Birmingham, Birmingham, U.K.

[6] Fukuda, H., Webb, C. and Atkinson, A.K., (1985). In "Process engineering aspects of immobilised cell systems". Eds. C. Webb, G.M. Black and B. Atkinson. Inst. Chemical Engineers, Oxford, U.K. Pp 309-315.

66

The dynamic mechanical behaviour of fungally degraded wood

C.J. McCarthy, C. Birkinshaw, J.T. Pembroke and M. Hale* - University of Limerick, Limerick, Ireland. *University College of North Wales, Bangor, Wales.

INTRODUCTION

An initial investigation[1] has shown that dynamic mechanical analysis (DMA) is a potentially useful technique for quantifying the mechanical changes which occur when wood is subject to biologically-induced degradation. It is common to follow fungal decay by strength loss and weight change measurement, however accurate and reproducible mechanical testing requires relatively large samples and the inhomogeneity of fungal decay processes then make it difficult to precisely relate the strength loss and the weight change. As DMA uses a smaller sample, (60 x 12 x 3mm in this work), and as the response arises from the sample as a whole it may help to resolve this difficulty.

With viscoelastic materials the application of a simple harmonic stress results in a phase difference between stress and strain. In the DMA technique this stress is applied by forcing the material to vibrate either at its natural frequency or at some preselected frequency. The tangent of the phase angle, tan ∂, measures the internal friction or damping and goes through a maximum at a thermodynamic transition. The resistance to deformation can be split into the shear storage modulus associated with the elastic response, and the shear loss modulus associated with the viscous response. Because of its potential ability to separate the individual component contributions in heterogeneous systems the technique has found extensive application in analysis of composite materials In addition to structure property investigation with whole wood[2,3], DMA work has included investigation into wood-resin composites to allow assignment of component contributions to overall mechanical behaviour[4], and investigations directed at predicting wood fibre response during pulping and in paper[5,6].

In the first investigation[1] *Pinus sylvestris* was subject to degradation by the brown rot fungus *Coniophora puteana* using soil block culture, however the white rot, *Phanerochaete chrysosporium* was found not to flourish under the relatively dry soil block conditions and so agar culture was adopted for this investigation. Here *Pinus sylvestris* has been subject to attack by the brown rot *Coniophora puteana* and the white rot *Phanerochaete chrysosporium* under identical conditions for periods up to 15 weeks. The objective was to compare directly the actions of the two fungi on dynamic mechanical properties, flexural properties and weight loss, and thus further evaluate the applicability of the DMA technique.

EXPERIMENTAL

Samples of *Pinus sylvestris* sapwood were machined to dimensions of 60 x 12 x 3 mm with the long direction parallel to the grain and the width in the direction of radial growth, and sterilized by steam autoclaving. The fungi were cultured on a 2% malt agar medium after BS6009. Test specimens were separated from the agar by a feeder strip and were only introduced when the feeder strip was covered with fungal hyphae. Incubation was at 22°C and incubation periods between one and 15 weeks were used. Control specimens were incubated under sterile conditions.

A Du Pont 983 resonant frequency instrument was used to measure the dynamic mechanical response between -150°C and 200°C at a heating rate of 5°C min^{-1}. The oscillation amplitude was 0.1mm and samples were flexed perpendicular to the grain direction and perpendicular to the direction of radial growth. Before testing the degraded samples were dried at 105°C for 24 hours. All results are the average of at least three individual tests.

Three point bending was carried out after BS2782 part 3, method 335A, using an Instron 4302 universal testing machine. The traverse speed was 0.5mm min^{-1} and the span length 50mm. The same specimen cross section was used as in the DMA work and again all results are the average of three or more tests..

RESULTS AND DISCUSSION

Figure 1 compares the dynamic response of degraded materials with an undegraded control. With both fungi the shear storage modulus of the wood decreased across the temperature range examined, whilst tan ∂ remained relatively unchanged. The result displayed in the figure is typical. The main events in both degraded and undegraded materials are the low temperature peak around -50°C corresponding to the onset of movement of bound water between hydrogen bonding sites, and at high temperature the increase in thermoplasticity as the glass transition temperatures of the structural polymers is reached[7,8].

Figures 2 and 3 show the change in shear storage modulus with time of decay for the two fungi while Table 1 details the weight loss compared with the 25°C shear storage moduli. Figures 4 and 5 display the changes in flexural strength and modulus measured in the three point bending experiments. With both fungi the fall in modulus with time of decay is approximately linear, suggesting a rapid and uniform penetration of the relatively small samples. Modulus declines more quickly with the brown rot *Coniophora puteana*, despite the somewhat lower rate of weight loss. This implies differences in the chemistry or location of attack and is consistent with the observations of other workers[9].

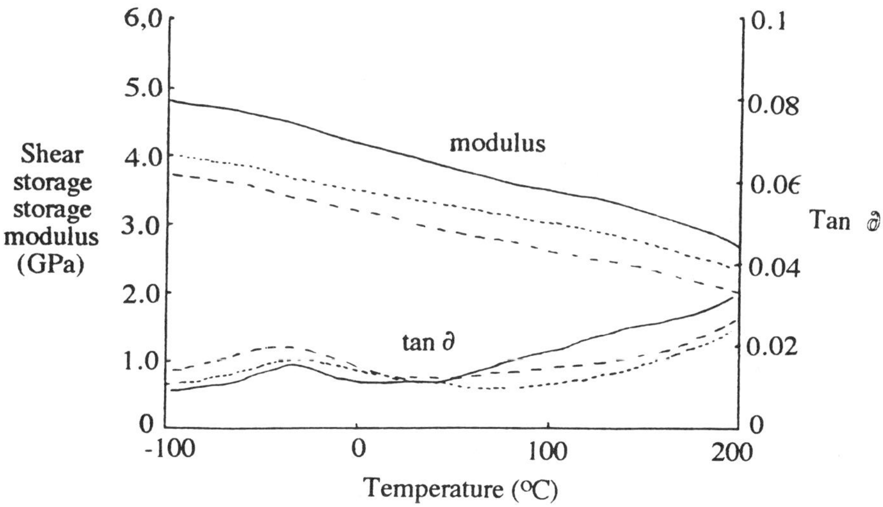

Figure 1 Dynamic mechanical response of *Pinus sylvestris.*, —— control sample, – – – – following eight weeks decay by *Coniophora puteana*, and -·-··· following eight weeks decay by *Phanerochaete chrysosporium*.

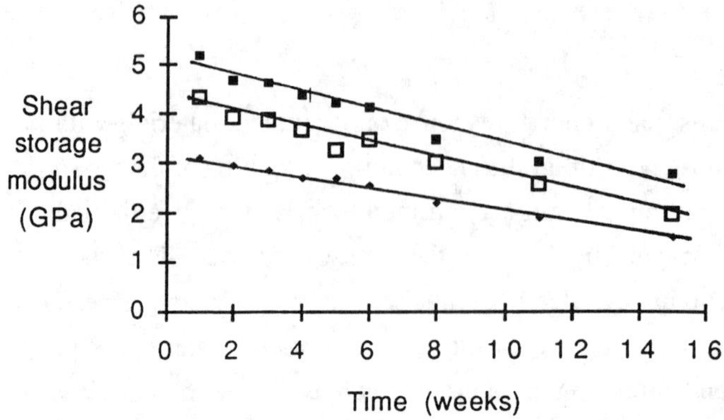

Figure 2. Effect of *Coniophora puteana* on on shear storage modulus at 75ºC ▪, 25ºC □ and 200ºC ◆.

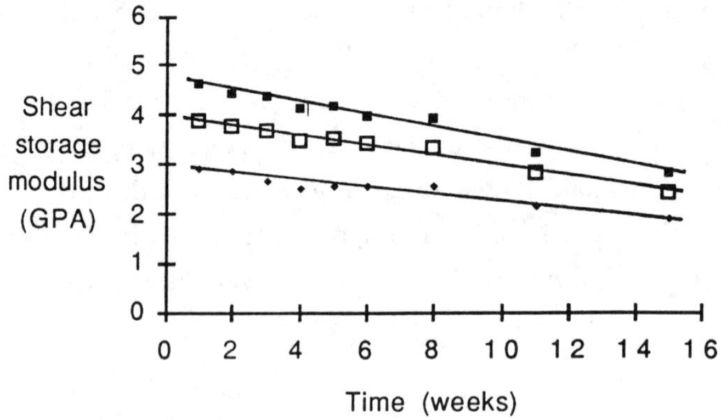

Figure 3. Effect of *Phanerochaete chrysosporium* on shear storage modulus at -75ºC ▪ ,25ºC □ and 200ºC ◆ .

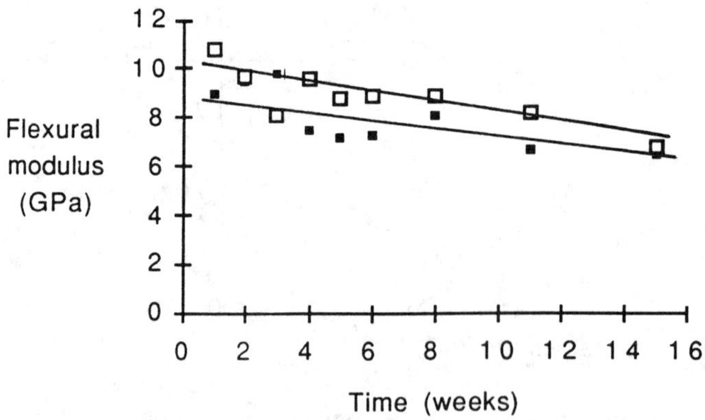

Figure 4. Effect of *Coniophora puteana* ▪ and *Phanerochaete chrysosporium* □ on flexural modulus.

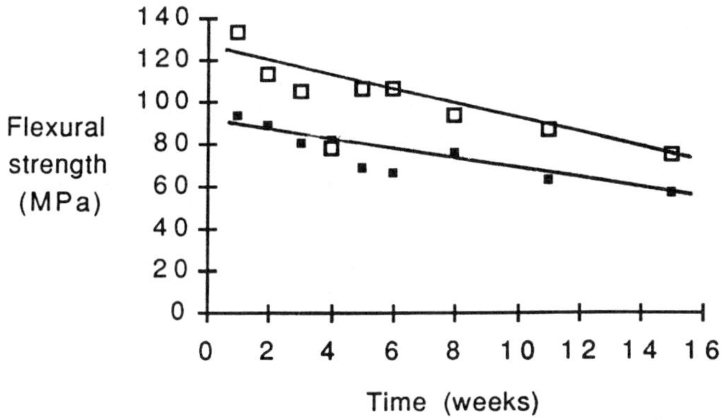

Figure 5. Effect of*Coniophora puteana* ■ and *Phanerochaete chrysosporium* □ on flexural strength

Table 1 Retention of shear storage modulus at 25° as a percentage of uninfected controls and percent weight loss.

Time (week)	Weight loss (%)	% Shear storage modulus retention *C. Puteana*	*P. Chrysosporium*
1	2.25	97.7	
	2.90		86.84
2	4.83	99.6	
	3.46		85.5
3	5.95	87.6	
	3.92		82.41
4	6.14	82.5	
	4.16		78.2
5	8.14	73.9	
	6.75		79.7
6	7.74	77.7	
	9.08		76.8
8	8.36	68.1	
	14.07		75.4
11	14.18	57.9	
	20.03		64.0
15	17.72	44.0	
	26.16		54.9

The three point bending results also display a loss of strength with both cultures and again show *Coniophora puteana* to be the more aggressive agent. In these experiments the scatter of the results is greater, however, and trends are less well defined. In DMA the result comes from the response of the sample as a whole and is thus less influenced by minor structural irregularities.

The strength and stiffness changes are consistent with the known[10] mode of attack of the two types of fungi. With brown rot the S2 and S3 layers are extensively hollowed out with loss of cellulose and hemicellulose. Any loss of material from the S2 layer, with its relatively shallow helix angle, can be expected to have substantial effects on mechanical stiffness. With white rot fungi the hyphae lie along the cell lumen promoting a more progressive degradation through the S3 and S2 layers, and will thus have less immediate effect on stiffness and strength for a given weight change.

CONCLUSIONS

The relative action of the two fungi assessed by the DMA technique is consistent with published information on their mode of attack After fifteen weeks the brown rot *Coniophora puteana* has effected a 56% reduction in shear storage modulus of *Pinus sylvestris* with a 16% weight loss, whilst *Phanerochaete chrysosporium* reduced the shear storage modulus by 45% with a 27% weight loss.

The work has supported the earlier observation[1] that dynamic mechanical properties are a sensitive and reliable indicator of the effects of fungal decay in wood. Changes in dynamic modulus are similar to those obtained from three point bending, but are subject to less experimental scatter. The procedure outlined here may be of more general applicability in the study of degradation and preservation of wood.

REFERENCES

1. C.L. McCarthy, C. Birkinshaw, J.T. Pembroke & M. Hale, Biotechnol. Techniques, **1991**, *5*, 493.
2. C. Birkinshaw, M. Buggy & G.G. Henn, J. Mat. Sci. Letters, **1986**, *5*, 898.
3. S.S. Kelley, T.G. Rials & W.G. Glasser, J. Mat. Sci., **1987**, *22*, 617.
4. C. Birkinshaw, M. Buggy & A. Carew, J.Mat. Sci., **1989**, *24*, 359.
5. D. Atack, Philosoph. Mag. **1981**, *A43*, 619.
6. L. Salmen, J. Mat. Sci., **1980**, *19*, 3090.
7. G.M. Irvine, Wood Chem., **1984**, *67*, 118.
8. L. Salmen & E.L. Back, Tappi, **1977**, *60*, 137.
9. R.W. Kennedy, Forest Prod. J., **1958**, *8*, 308.
10. T.L. Highly, Document IR6/WP/1319 US Forest Products Laboratory, **1987**.

67

Products of oxidation of cellulose by hydrogen peroxide during fungal degradation

R.G. Veness and C.S. Evans - Division of Biotechnology, The Polytechnic of Central London, London W1M 8JS, UK.

ABSTRACT

In the degradation of cellulose by wood-rotting fungi, oxidative reactions occur in addition to hydrolysis. Identification of products formed when enzyme systems from white-rot basidiomycetes reacted with Avicel have shown similar compounds to those produced by treating glucose with peroxide free-radicals generated by Fenton's reagent. It is concluded that a peroxide free-radical mechanism is the most likely cause of the oxidation. Products identified from the oxidation of Avicel by hydrogen peroxide confirm a theoretical model degradative pathway.

INTRODUCTION

Extracellular hydrogen peroxide has been implicated in the breakdown of cellulose [1], and a suggested mechanism by wood-rotting basidiomycetes was the reaction of extracellular hydrogen peroxide with Fe^{2+} to produce free radicals in a Fenton's-type reaction [2]. We have reported that peroxide free radicals were the probable cause of cellulose oxidation by <u>Trichoderma reesei</u> [3]. These experiments showed that

the oxidised products resulting from the reaction of 'Celluclast' (commercial enzyme preparation from T.reesei) with Avicel were similar to those produced by the action of Fenton's reagent on glucose [3]. Similar products were also identified from the reactions of extracellular enzymes from some white-rot basidiomycetes with Avicel, the oxidation appearing to be independent of oxygen concentration in the reaction mixtures [4].
This paper presents a theoretical pathway for the reaction of hydrogen peroxide and associated free-radicals on cellulose which could account for the range of oxidation products observed previously.

EXPERIMENTAL

The conditions for oxidation of Avicel by the extracellular enzymes from the white-rot fungi Coriolus versicolor, Heterobasidion annosum, Armillaria mellea and Poria contigua, have been described previously [3,4]. The products of oxidation were converted to their trimethylsilyl esters (TMS) [5] and analysed by gas chromatography and mass spectrometry on a Jeol Chromopak gas chromatograph fitted with a 0.25mm x 25m wall coated open tubular column with CPSil 5 liquid phase, coupled to a Jeol JMS-DX300 mass spectrometer operating at 70eV. A Jeol JMA-2000 data system was used for mass assignment and data presentation.

Glucose was oxidised with Fenton's reagent producing peroxide free radicals as reported previously [3]. Products were analysed by GC-MS as described above.

RESULTS AND DISCUSSION

Extracellular enzymes from all the white-rot fungi tested reacted with Avicel producing oxidised products which were detected as TMS derivatives by GC. Table 1 shows the compounds that were identified in all the reaction mixtures. Many of these products were identified by MS as identical to those produced by the action of peroxide free radicals on glucose. Fenton's reagent produces free radicals by the following scheme:

$$Fe^{2+} + HOOH \longrightarrow Fe^{3+} + HO^{\bullet} + OH^{-}$$

The hydroxyl radicals produced abstract hydrogen atoms and so act as oxidants. In dilute solution at low temperatures, the following products have been reported

TABLE 1. Identification of products derived from (A) the treatment of Avicel with fungal enzyme extracts, and (B) the reaction of glucose with Fenton's reagent.

Products	A	B
Glyceric acid	+	+
Glyceraldehyde	+	+
Glycollic acid	+	-
Erythritol	+	-
Erythrose	+	-
α-Xylose	+	+
ß-Xylose	+	+
Fucose	+	+
3-O-Methylglucose	+	-
Glucurono lactone	+	+
Mannose	+	-
Methyl α-D mannoside	+	-
Glucaro lactone	+	+
Glucono-1,5-lactone	+	+
Glucono-1,4-lactone	-	+
Arabinose	+	+
α-Glucose	+	+
ß-Glucose	+	+
Glucaric acid	+	+

to be derived from D-glucose: D-arabino-hexo-2-ulose which is further oxidised to glycollic acid, glyoxylic acid and D-erythronic acid [5]. Other products that may be produced include D-arabino-hexulosonic acid, D-erythro-hexo-2,3-diulosonic acid, glyceric acid, and oxalic acid.

A scheme for the oxidation of glucose by hydroxyl free radicals is shown in Figure 1. This scheme is initiated by the reaction of the hydroxyl free radical on the C2 carbon of glucose, causing dehydrogenation which leaves the sugar as a radical species. Further reaction of this sugar with hydrogen peroxide regenerates the hydroxyl radical and forms two hydroxy groups on C2, which being an unstable molecule decomposes to the 2-keto derivative of glucose with the elimination of water. Subsequent attack by hydroxyl radicals could then occur at either C1 producing an ulosonic acid, or at C3 yielding a 2,3-diketo-sugar which would decarboxylate to give carbon dioxide and a 2-keto sugar with one less carbon in the chain. Cleavage could also occur between C2 and C3 resulting

FIGURE 1.
Theoretical pathway for the oxidation of glucose by peroxide free radicals.

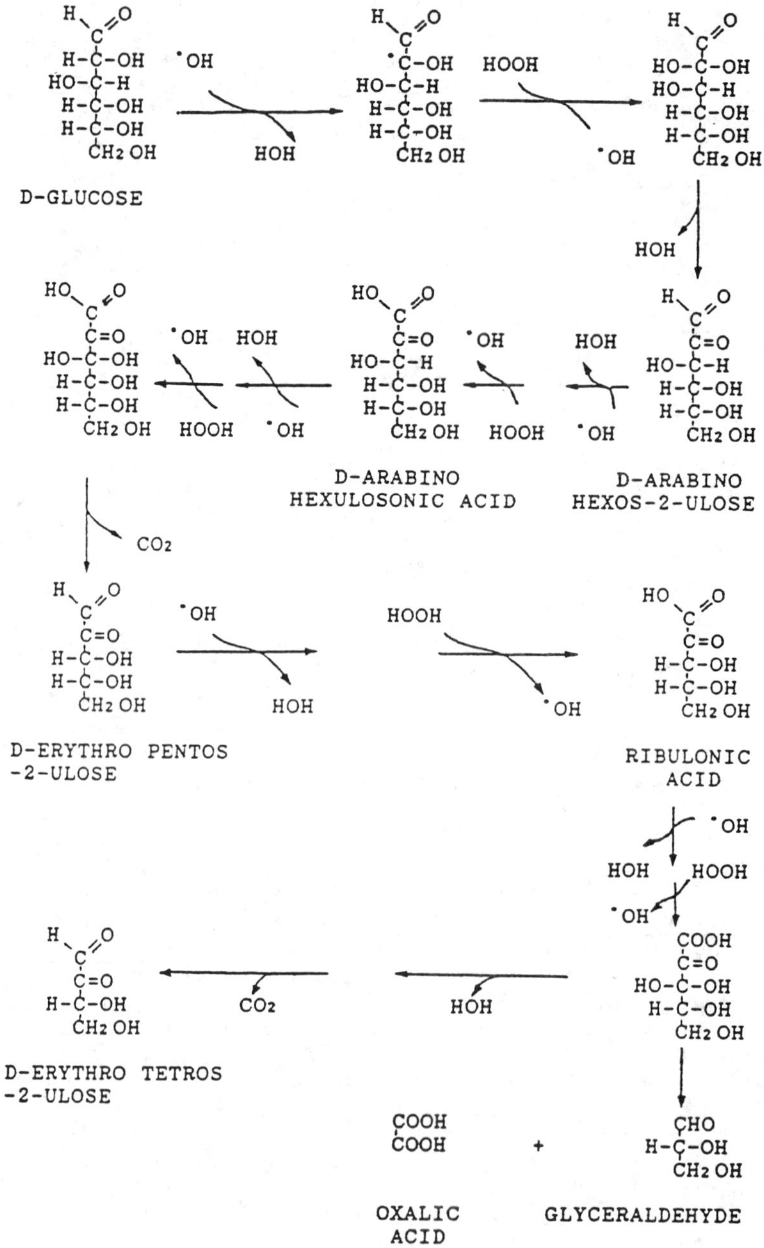

in the formation of oxalic acid and an aldose with two carbons less than the original glucose. If the stereospecificity of the sugar was low, hydroxyl radicals could attack C1 first so producing an -onic acid.

Hydrogen peroxide can also react with Fe^{3+} regenerating Fe^{2+} by the reaction [5]

$$Fe^{3+} + HOOH \longrightarrow Fe^{2+} + H^{+} + {}^{\bullet}OOH$$

The hydroperoxyl radical formed is a less effective oxidant than the hydroxyl radical, but it is known to be an active species in the Ruff degradation of an aldonic acid as shown in Figure 2. This degradation will decarboxylate gluconic acid to arabinose in a reaction which generates a hydrogen free radical which can abstract hydroxide groups giving rise to deoxyarabinose.

FIGURE 2. The Ruff degradation of an aldonic acid.

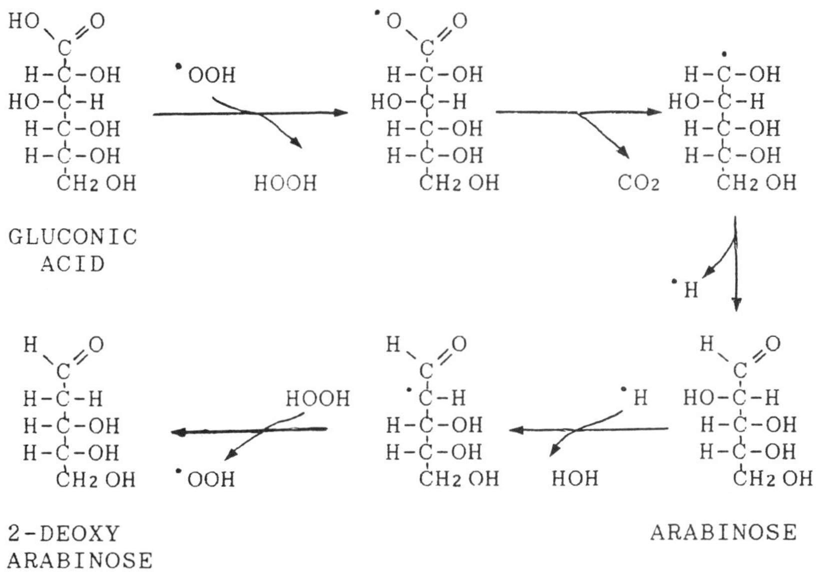

Theoretically therefore a wide range of neutral sugars, sugar acids, ketose sugars and keto-acid sugars could be produced from a single aldose. The products identified from glucose treated with Fenton's reagent included glyceraldehyde, xylose, fucose, glucuronic acid lactone, gluconic acid lactone and aldaric acid which confirmed the reactions shown in Figure 1. In addition the finding of neutral sugars and fucose suggested that the Ruff degradation also occurred with the subsequent production of hydrogen free radicals.

It would seem probable therefore that similar products identified from the treatment of Avicel with enzyme systems from white-rot fungi would be produced by these proposed schemes.

REFERENCES

1. Halliwell, G. Biochem. J. 1965, 95, 270.
2. Koenigs, J.W. Arch. Microbiol. 1974, 99, 129.
3. Veness, R.G., Evans, C.S. J.Gen. Microbiol. 1989, 135, 2799.
4. Veness, R.G., Evans, C.S. in Kennedy, J.F., Phillips, G.O., Williams, P.A. (Eds.) Cellulose: Sources and Exploitation. Ellis Horwood, Chichester, 1990, 409.
5. Green, J.W. in Pigman, W., Horton, D.(Eds.) The Carbohydrates: Chemistry/Biochemistry Acad. Press, New York, 1980, Vol. 1B.

68

Changes in pore structure and cell wall pore volume in wood decayed by brown- and white-rot fungi

Douglas S. Flournoy and Jennifer A. Paul - United States Department of Agriculture, Forest Service, Forest Products Laboratory[1], One Gifford Pinchot Drive, Madison, WI 53705-2398 USA.

ABSTRACT

Sweetgum (*Liquidambar styraciflua* L.) wood blocks were decayed by *Postia (=Poria) placenta* or *Phanerochaete chrysosporium* in soil-block cultures. Decay was terminated at various weight losses, and the cell wall pore volumes available to probes of various molecular weight and diameter were determined. The volume in sound wood attributable to cell wall pores is 0.35 ml g^{-1}. In brown-rotted samples, the volume of pores in the cell wall increased steadily to 0.7 ml g^{-1} at 35% weight loss. New cell wall pore volume was accessible to low molecular weight probes but not to molecules of $M_r \geq 6{,}000$. Within experimental error, no pores of > 20Å were observed in sound wood or >38Å in brown-rotted wood. Most of the new cell wall pore volume created by removal of components during decay was in the pore size range of 12Å to 38Å. Our results are consistent with the hypothesis that the initial depolymerization of cellulose, characteristic of brown rot, is caused by a diffusible agent. The molecular diameter of the agent is apparently in the range 12Å to 38Å and it causes erosion and thus enlargement of the pores to which it has access. In the white-rotted wood, cell wall pore volume increased to 0.6 ml g^{-1} at 40% weight loss and maximum pore diameter increased to > 50Å. Most of the cell wall pore volume increase resulted from the creation of pores of 20-50Å diameter. Analysis of loss of major wood components as a function of weight loss revealed that lignin, cellulose, and hemicellulose were removed at approximately equal rates. Under our experimental conditions,

---

[1]The Forest Products Laboratory is maintained in cooperation with the University of Wisconsin.

ligninolytic enzymes have access to only a small portion of the new cell wall pore volume, even after extensive decay.

INTRODUCTION

Brown-rot fungi remove the carbohydrate fraction of wood without substantially modifying the lignin. In contrast to this, white-rot fungi remove both the lignin and the carbohydrate fractions of wood [1]. The white-rot fungi may remove both lignin and carbohydrate simultaneously [2] or may preferentially remove the lignin component [3].

Presumably, degradative enzymes gain access to their respective substrates in the cell wall at the polymer surfaces exposed to the lumen and through pores of suitable diameter. This is generally considered to be the diameter of the enzyme if the shape of the molecule in question is spherical and the length of the minor axis if it is ellipsoid. Most pores in sound wood are to small to allow passage of enzyme sized molecules (median pore size ~10Å), and the cell wall pore volume available through pores of large diameter (~100Å) is only a very small percentage of the of total cell wall pore volume [4, 5].

It has generally been supposed that microbial action on sound wood results in an opening up of the wood structure and creation of pores. In the simultaneous rot of sweetgum by *Polyporus* (*Coriolus*) *versicolor* Cowling [1] noted "a distinct increase in porosity of the secondary walls and compound middle lamella...." In contrast to this, however, Highley and co-workers [6, 7] concluded that simultaneous rot by the white rot fungus *Ganoderma applanatum* does not open the microstructure of wood.

The purpose of this work is to examine changes in cell wall micro-pore structure as it is deteriorated by wood detroying fungi. The results of such a study may provide answers to questions regarding the accessibility of degradative proteins to their respective substrates. In the present study we have examined changes in sweetgum wood as it undergoes white-rot decay by *Phanerochaete chrysosporium* and brown-rot decay by *Postia (=Poria) placenta* .

MATERIALS AND METHODS

The fungi used in this work were *Postia (=Poria) placenta* (Fr.) Cke. (strain Madison 698) and *Phanerochaete chrysosporium* Burds. (strain ME-461). Sweetgum (*Liquidambar styraciflua* L.) blocks were decayed and the pore volume available to probes of various molecular weight and diameter were determined as previously described [8,9].

Chlorite holocellulose was prepared, intrinsic viscosities determined, and degrees of polymerization of the holocellulose were calculated as previously described [8]. Samples of white-rotted wood were analyzed for klason lignin, glucose, xylose, mannose, arabinose, and galactose and loss of components during decay were calculated as previously described [8].

RESULTS

Pore volumes in sound wood that were accessible to solutes were determined to vary from ~1.0 ml g^{-1} for the largest probes to 1.35 ml g^{-1} for water (data not shown). From these results, it is apparent that sound sweetgum wood contains ~1.0 ml g^{-1} of volume readily accessible to all of the probes. We attribute this volume to lumens, pits, and other large openings. Therefore, the cell wall pore volume was determined to be 0.35 ml g^{-1}. A plot of probe diameter as a function of inaccessible volume (Figs. 1, 2) indicates that molecules larger than ~20Å in diameter are completely excluded from the cell wall of water-swollen sound sweetgum. In the absence of fungi, the cell wall pore volume, the accessible volumes obtained for the probes, and the maximum pore diameter all remained constant over 4 weeks.

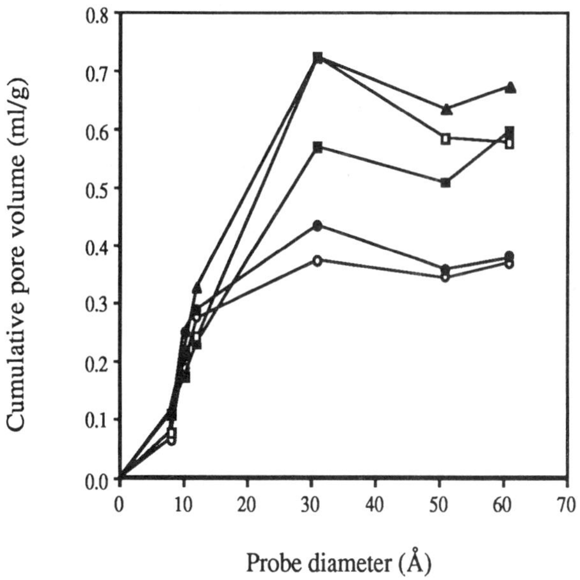

Figure 1. Inaccessible volumes in sound sweetgum wood and in sweetgum wood decayed by *Postia (=Poria) placenta* to the indicated weight losses: 0% (●), 6.9% (○), 20.3% (■), 27.6% (□), and 35% (▲).

We next determined the changes in cell wall pore volume during decay by the brown-rot fungus *Postia (=Poria) placenta*. The average DP of the holocellulose decreased gradually from 1,200 in sound wood to approximately 200 in wood decayed to 35% weight loss (data not shown). The inaccessible pore volume data based on size exclusion studies was plotted as a function of probe diameter (Fig. 1). As the probe diameter increases molecules are increasingly excluded from the cell wall. Plateaus in inaccessible volume are reached with the three largest probes for all samples. The plateaus represent fiber saturation points (total cell wall pore volumes) for the samples. It can be seen that the cell wall pore volume increases from 0.35 ml g^{-1} in sound wood to 0.7 ml g^{-1} in wood decayed to 35% weight loss. The three largest probes did not gain entry to the cell walls despite the decay. The point at which the curves plateau indicates the maximum pore diameter in that sample. Our results indicate that the maximum pore diameter in our decayed samples never exceeded 38Å.

Plots of inaccessible volume as a function of weight loss for the three low molecular weight probes show no change on decay (data not shown). Inaccessible pore volume did not increase for molecules ≤12Å in diameter during decay. Thus, newly created pore volume accessible to water is also accessible to these probes. The values for inaccessible volume for the three largest probes for each sample were averaged and plotted as a function of weight loss. In the case of the three largest probes, those ≥38Å in diameter, inaccessible volume for each of these increased as the wood decayed (data not shown), indicating that these molecules are excluded from newly created cell wall pore volume. It follows that any new pore volume created by brown-rot decay results from either enlargement of existing pores that are ≥ 12Å in diameter or from the creation of new pores between 12Å and 38Å in diameter.

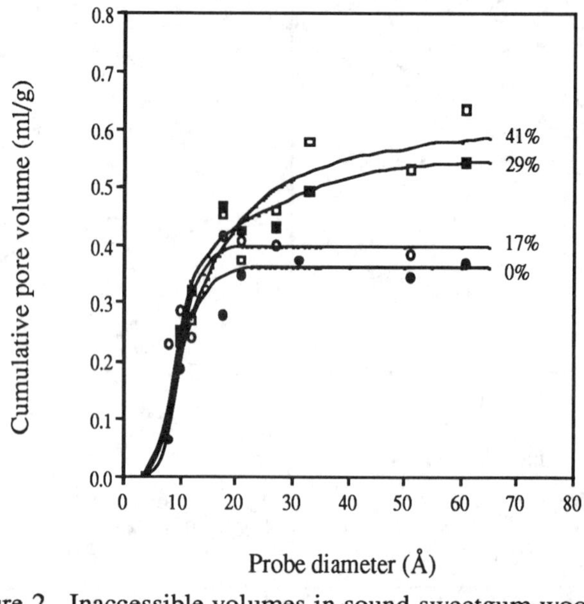

Figure 2. Inaccessible volumes in sound sweetgum wood and in sweetgum wood decayed by *Phanerochaete chrysosporium* to the indicated weight losses: 0% (•), 17% (○), 29% (■), and 41% (□).

We also examined the effect of white-rot on the cell wall pore volume and pore size distribution. Analysis of loss of major wood components as a function of weight loss revealed that lignin, cellulose, and hemicellulose were removed at approximately equal rates under our experimental conditions (data not shown). The inaccessible pore volume data for wood decayed by *Phanerochaete chrysosporium* is shown in Figure 2. It can be seen that the cell wall volume increased from 0.35 ml g^{-1} in sound wood to 0.6 ml g^{-1} in wood decayed to 40% weight loss. At low weight losses (17%), sharp plateaus are observed. At higher weight losses, these plateaus are reached at larger probe diameters, indicating a broadening in the range of pore size distribution. The maximum pore diameter is found from the point at which these curves plateau. In sound wood, this value was found to be ~20Å. The maximum pore diameter increases with increasing weight loss in white-rot decayed wood (Fig. 2).

To see more clearly the change in pore size distribution as a result of white-rot decay, the inaccessible volumes for sound and white-rot decayed wood were normalized and plotted as a function of probe diameter (Fig. 3). From this figure three points can be made: i) the increase in cell wall pore volume results from the creation of pores of 15 to 60Å diameter, ii) the median pore diameter, the pore diameter at which one-half of the pore volume is contained within larger pores and one-half is contained within smaller pores, increases from ~10Å in sound wood to ~15Å in wood decayed to 40% weight loss, and iii) the maximum pore diameter increases from ~20Å in sound wood to > 60Å in wood decayed to 40% weight loss.

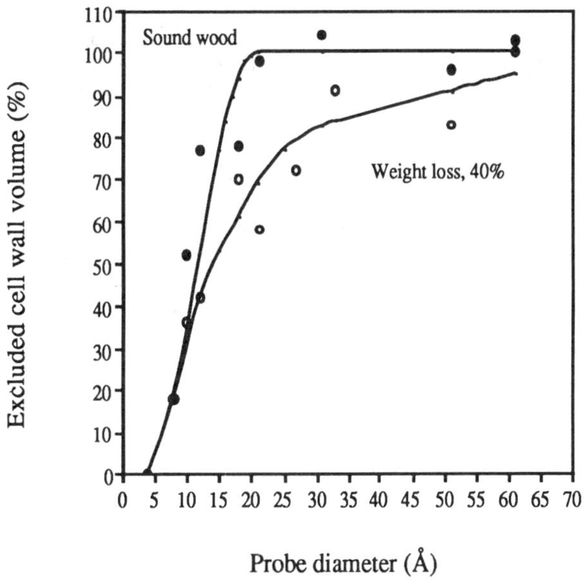

Figure 3. Normalized curves for excluded volumes in sound (•) and *Phanerochaete chrysosporium* decayed (○) sweetgum wood.

DISCUSSION

The purpose of this study was to examine changes in cell wall pore volume and pore size distribution in wood during decay by brown- and white-rot fungi. Changes in the pore structure of wood during decay processes have apparently never been examined. The results of such a study may provide insite into questions regarding the accessibility of degradative proteins to their respective substrates.

The sweetgum wood that we used had been dried, and as reported earlier [8], its maximum pore diameter was expected to be small. Our finding of a cell wall pore volume of ~0.35 ml g^{-1} are in line with previous studies which report cell wall pore volumes of the order of ~0.4 ml g^{-1} [10, 11, 12, 13, 14].

A portion of this study focused on changes taking place in the cell wall during a period of decay in which a brown-rot agent penetrates the cell walls of wood, cleaving virtually all of the cellulose in the amorphous regions of the microfibrils. The data obtained in this study indicate that molecules ≥38Å in diameter are excluded from the walls of sound sweetgum wood, as well as from the walls of the wood after it had been decayed to various losses in total weight. In addition, molecules of ≤12Å diameter gained access to all newly created cell wall pore volume. In our decayed samples, the cellulose was depolymerized to its "limit" DP of 200. Therefore, it is presumably pores of diameter 12Å to 38Å through which the cellulose-depolymerizing agent must pass to gain access to cellulose cleavage sites.

We also examined changes in cell wall structure as a result of white-rot decay. Previous studies on the effect of component removal on the pore structure of wood have revealed that the selective chemical removal of lignin leads to a substantial increase in the fiber saturation point, broadens the distribution range of pores, and increases the median pore diameter substantially [10]. We have found that the simultaneous removal of the three main components through decay by the white-rot fungus *P. chrysosporium* to 40% weight loss increases the cumulative pore volume of the cell wall only modestly (Fig. 2), and the median pore diameter increases only slightly (Fig. 3). We also detected a broadening in the range of pore size distribution. Most of the increase in pore volume, at 40% weight loss, resulted from the creation of pores of 20 to 50Å diameter. Most conventional cellulases will be excluded from such pores due to their large size [1, 4, 5, 15]. This is true of the ligninases as well whose molecular dimensions can be estimated at 47Å if their shape is considered to be spherical and 40 by 43 by 60Å if their shape is considered to be ellipsoid. Manganese peroxidase would also be expected to be excluded from most of the cell wall.

Several recent studies have examined the penetrability of wood decay enzymes into the cell wall by the use of transmission electron microscopy and immuno-gold labeling [16, 17, 18]. Those studies have concluded that lignin peroxidase (M_r ~42,000) is incapable of penetrating the walls of sound wood. The areas of the walls that have been heavily decayed by white-rot fungi, which degrade lignin and hemicelluloses somewhat selectively, were penetrated by the enzyme [18]. This suggests that selective biological lignin removal, like alkaline delignification in pulping, opens up the cell walls. Work is currently under way in this laboratory to quantify these effects in microbial delignification.

ACKNOWLEDGEMENTS

This work was supported by USDA Competitive Grants No. 86-FSTY-9-0212 and 90-37291-5753. We thank Les Ferge for expert technical assistance in preparing the samples of decayed wood. We are also indebted to A.M. Scallan and D. Caulfield for their helpful advice.

REFERENCES

1. Cowling, E.B. Technical Bulletin No. 1258, U.S. Department of Agriculture, Washington, D.C., 1961, 79 pp.
2. Kirk, T.K.; Highley, T.L. *Phytopathology* **1973**, *63(11)*, 1338-1342.
3. Kirk, T.K.; Moore, W.E. *Wood and Fiber* **1972**, *4(2)*, 72-79.
4. Cowling, E.B.; Brown, W. in Gould, R.F. (Ed.) *Advances in Chemistry Series 95, Cellulases and Their Applications*, American Chemical Society, Washington, D.C., 1969, p. 157-187.
5. Cowling, E.B.; Kirk, T.K. *Biotechnol. & Bioeng. Symp. No. 6* **1976**, 95-123.
6. Murmanis, L.; Palmer, J. G.; Highley, T.L. *Wood Sci. Technol.* **1985**, *19*, 313-321.
7. Murmanis, L.; Highley, T.L. *Wood Sci. Technol.* **1987**, *21*, 101-109.
8. Flournoy, D.S.; Kirk, T.K.; Highley, T.L. *Holzforschung* **1991**, *45(5)*, 383-388.
9. Flournoy, D.S.; Paul, J.A.; Kirk, T.K.; Highley, T.L. *Holzforschung* in press.
10. Stone, J.E.; Scallan, A.M. *Pulp and Paper Mag. Can.* **1968**, *69*, 288-293.
11. Stone, J.E.; Scallan, A.M. *Cellulose Chem. Technol.* **1968**, *2*, 343-358.
12. Tarkow, H.; Feist, W.C.; Southerland, C.F. *Forest Prod. J.* **1966**, *16(10)*, 61-65.
13. Kerr, A.J.; Goring, D.A.I. *Can. J. Chem.* **1975**, *53*, 952-959.
14. Grethlein, H.E. *Bio/technology* **1985**, *3*, 155-160.
15. Cowling, E.B. *Biotechnol. & Bioeng. Symp. No. 5* **1975**, 163-181.
16. Srebotnik, E.; Messner, K.; Foisner, R. *Appl. Environ. Microbiol.* **1988**, *54(11)*, 2608-2614.
17. Daniel, G.; Nilsson, T.; Pettersson, B. *Appl. Environ. Microbiol.* **1989**, *55(4)*, 871-881.
18. Blanchette, R.A.; Abad, A.R.; Farrell R.L.; Leathers T.D. *Appl. Environ. Microbiol.* **1989**, *55(6)*, 1457-1465.

69

The long term effect of surface treatments on the properties of cellulosic and ligneous museum artefacts

Maureen A. Robson - Studio, 29 Park Avenue, Birmingham, B18 5ND, UK.

ABSTRACT

Cellulose based natural fibres and ligneous composites are a potential resource for the making of ephemeral domestic items especially in developing tropical countries where these fibres are abundant eg grass skirts, woven basketry, wooden drums and decorative spears. For societies lacking written documents, the collection of such material, for studying and display within museums, provide both the anthropologist, and the conservator, a wealth of information for a better understanding of their culture and technology. Many of these items were never intended to last very long and yet in the past, long after their natural life expectancy, these artefacts have been preserved by application of suitable surface treatments to strengthen and support the weakened fibre. In this work, the resulting physical and mechanical properties of these treatments are assessed. Anthropologists and conservators may in the future need to question seriously their professional judgements in justifying unnatural preservation of ephemeral items and the stage at which their natural life cycle should be frozen.

INTRODUCTION

Ethnographic collections contain a diversity of objects composed of wood and related materials. Some of the unusual materials within this category are cork, pith, and treefern. The conservation of cellulose based natural fibres and ligneous composites presents an often difficult choice between applying passive methods - leaving in the natural state which may result in subsequent damage or disintegration, or using interventive conservation methods which will structurally assist the long term preservation of the artefact but may lead to some change in its structure and surface appearance.

Material anthropologists and museums of anthropology collect and study artefacts as a means of understanding the overall culture of a people. For societies lacking written documents, the collection of such material, for study and display within museums, provides both the anthropologist and the conservator a wealth of information for a better understanding of their culture and technology.

Ethnographic conservation appears to be following a parallel course to that of museum anthropology. Throughout this century we have seen the dramatic changes in direction within museums, from an initial emphasis on massive acquisition programmes to one on communication through exhibitions and education, and didactic displays.

The 1960's and 1970's saw the emergence of museum conservation as a powerful discipline, very much in response to the concern with the physical preservation of objects. In the 1980's the recognition of the importance of documentation and information content increased. [5]

DOCUMENTATION

The need to fully consider the ethics and consequences of conservation methods which may cause problems in the future reflects current advances in conservation. Certain treatments may destroy the cultural, historical and technological information associated with the artefact. Thus full documentation without conservation, on selected antiquities or samples taken from them, will safeguard this information and provide a resource for future research. Provided the condition of the antiquities thus selected is continuously monitored under suitable environmental conditions on display or in storage, it will no doubt, pre-empt the necessity for interventive conservation.

The importance for the conservator to maintain adequate records of the variety of conservation and restoration materials and techniques used today, is emphasised as a prerequisite for conservation on any formal training programme. On some occasions it may be necessary to document previous restoration which may be re-restored along with the antiquity as the possibility of greater damage resulting from its removal may not be ignored. Past restoration methods are of historical interest and very often their documentation no longer exists. Detailed records should be made of these early techniques to provide an invaluable resource for future research.

Conservators learn much of anthropological importance from their work with an artefact. Conversely, they depend heavily on the information, particularly technological, available to them from anthropology. It is possible to develop effective working links between conservators and anthropologists with common interests, even though they may be in very different regions and institutions. They must indeed work more closely together. Their overall goals are the same in understanding the human world and helping to preserve its heritage.

The conservator will extend the natural life expectancy of these artefacts, many of which were never intended to last very long and yet their natural wastage would entail the loss of important cultural and technological information about the people that crafted and used them.

PASSIVE CONSERVATION

Sometimes, all that it necessary is the use of passive conservation, by which adequate environmental monitoring will prevent the deterioration of the exhibit. A system with various inbuilt environmental controls according to specific requirements of temperature and relative humidity will ensure the long term maintenance of the exhibits.

All sunlight whether direct or reflected is damaging. Light falling on an object must therefore be reduced. To prevent photodegradation the lighting units within a showcase may be protected where desirable, with ultraviolet absorbent sleeves and diffuse light through a sheet of opal perspex. Direct light sources should be limited to below 50 lux. A relative humidity between 50 - 55 % and temperature 18 - 21°C are recommended for artefacts on display. Slight variations in conservation requirements for environmental controls will naturally occur between collections housed in temperate climates and that in tropical climates.

In the past, before conservators became fully aware of the significance of environmental conditions to assist the long term preservation of vulnerable ethnographic material, the only option was interventive conservation. The natural life expectancy of these artefacts was thus considerably prolonged by application of suitable surface treatments both to strengthen and support the weakened structure.

INTERVENTIVE CONSERVATION

Current cleaning methods seek to ensure safe application, without loss or run of colour or fibre shrinkage on drying, and interventive conservation methods endeavour to guarantee the compatibility of the consolidant with the fabric of the object, and their reversibility both in the short and long term. Although interventive methods structurally assist, they are not conspicuous to the general observer.

Dry brushing, use of dental evacuator, photographic blow brush methods of cleaning when required and depending on degree of the fragility of the fibre, may be recommended.

Wet cleaning methods of localised humidification and misting with deionised water to relax folds or unravel tangled vegetable fibres such as fringes on ceremonial masks need to be carefully monitored to assess long term damage. Indiscriminate use may cause unforeseen damage in the future. It is often sufficient in ambient temperatures to use an inflated polythene bag wrapped in acid free tissue to slowly relax folds in for example a seal skin anorak resulting from bad storage methods without introducing humidification. Ethnographic textiles should never be washed. Very often on testing, dyes may appear to be colourfast but may still run during prolonged immersion. In the same way there is no reliable way of testing for the possibility of fibre shrinkage in a composite garment exhibiting a diverse range of fibres. Dry treatment methods, where appropriate, are to be desired.

Interventive conservation methods for wooden objects present different problems. Additions of natural waxes, synthetic polishes or surface embellishments to the original polished wooden surfaces of such items such as ceremonial clubs and spears should be avoided as they are not indigenous to the country of origin of the objects, may become lodged in ornate incised design and often darken the original wooden surface. In such cases the use of an absorbent dry duster is all that may be required.

Existing worm holes in wooden items such as wooden feast bowls, katamarans and canoes etc, need to be carefully checked in spring during the gestation period of the larvae and for up to at least two years thereafter. Very often remains of past activity eg. orange frass from deep within the holes and thin outer surfaces of the tunnel walls from vertical wood worm shafts near the surface, may be dislodged during transit and found in the tissue or other packing material. This should not be confused with current worm activity which will be evident from observation of the object in situ and will require immediate fumigation to remove infestation. The choice of fumigant may depend on the composite nature of the infected object where a change in colour to the original surface may be undesirable. The long term effect after the use of fumigants containing sodium pentachlorophenol and lindane may be deleterious to the artefact. Phenolic salts may crystallise out on the surface of the artefact during storage, and require removal by dry brushing. Polychrome wooden artefacts with unbound pigment decoration may require localised application of a suitable insecticide in preference to risking salt contamination after fumigation.

Very fragile fibres require some support before they totally disintegrate during storage into minute piles of dust. The general practice of protecting the fibre with bandages during storage very often on removal, leads to further disintegration and subsequent loss of the fragile woven fibre and the problem still needs to be addressed before the item is displayed. Objects cannot be successfully displayed unless fibres are safely supported.

A safe method of bonding fragile fibres on to a support fabric of suitable weight has already been described [3] The use of aqueous emulsions can be recommended. The support fabric is impregnated with the adhesive which is then allowed to dry to form a thin cohesive film. The film is then reactivated by the application of gentle heat from a spatula iron. (This method has been discussed at Symposium '86 "The Care and Preservation of Ethnological Materials" at the Canadian Conservation Institute in Ottawa). Since the samples of adhesives varied in strength, they could be successfully used in supporting different strengths of fibre within the same object, ranging from the severely weakened to the more substantial.

The 1950's saw the emergence of Paraloid B72 (Acrylic polymer) in xylene or toluene which was put to a variety of uses in the field of conservation. Formulations of Paraloid B72 with various solvents and matting agents have been suggested for the repair and conservation of cellulosic and ligneous composite museum artefacts. Surface applications with various solvents and matting agents will attach decorative pigment to polychrome wooden sculpture such as carved totem poles, ancestor figures and ceremonial stools.

By this time long term reversibility was regarded as an important criterion. Consolidation processes provide additional strength to fragile porous archaeological and ethnographic material where a change in colour is undesirable. During the 1950's N-methoxymethyl nylon (soluble nylon) was used for the consolidation of fragile ethnographic material. By the late 1960's, its use was being discouraged because of dirt pick-up and loss of strength and flexibility and insolubility after ageing.

During the late fifties to early sixties a new approach was introduced with the idea of complete immersion in a synthetic resin. For the first time account had to be taken of the possible effect of the conservation process on the external appearance of the porous material, such as a ceremonial mask made from bark cloth (consisting of mulberry fibre pounded with a heavy wooden mallet) over a wooden stick frame. Any testing or sample removal must be carried out prior to treatment or authenticity and dating methods would be lost. The complete artefact was immersed in a suitable solution under vacuum of Bedacryl 122X (polybutyl methacrylate) and later Paraloid B72 (acrylic polymer) in xylene. [4]. Bedacryl 122X (PBMA) has had a chequered career. During 1955-56 it was thought well of by the British Museum's conservation scientists, but fell from favour twenty years later when its long term reversibility was in doubt. Currently, Bedacryl 122X (PBMA) is again in general use and can be purchased in two forms: either as a solid which can be dissolved to form the resin, or in a concentrated solution. Both forms may be diluted with toluene to the required concentration and when applied, dries to a matt finish. These resins were effective as a means of consolidation since they were used to strengthen whole artefacts both on excavation and in the laboratory workshop before polyvinyl alcohol emulsions were available. They were also used on individual fragments which could later be re-assembled and bonded with a suitable adhesive. However, the resin treatment did result in a certain amount of discolouration of the fabric. A vacuum desiccator is used for this purpose and a glass vessel of small dimensions is used as the container, in order is economise in the use of the solvent, this vessel is placed inside the desiccator which is then evacuated. The object is allowed to remain under vacuum until all bubbling has ceased so that consolidant is observed by the porous pot fabric. Air is then admitted when the desiccator reaches atmospheric pressure, the artefact is lifted out of the container and allowed to drain, it can then be dried without acquiring a gloss by leaving it in an atmosphere of the solvent [2]. In the past these resins have been used as an adhesive for both ethnographic textiles and wood.

CONCLUSION

Passive conservation treatments include good house-keeping, safe packing, providing adequate support and appropriate environmental conditions in storage, transit and on display with full documentation of objects and their past and present treatments. These measures may obviate to a large extent the need to introduce interventive conservation methods which will structurally assist in the long term preservation of the item but often lead to some undesirable change in the structure and surface appearance of the ethnographic object.

Very often the consolidant has become irreversible with age and will considerably darken and may embrittle the structure. If this happens it will not be possible to use in the future, any subsequent consolidant that may be developed. Therefore, it is desirable to avoid as far as possible such interventive methods. Although preventative action is normally preferred for painted objects, there are many instances where treatment is needed, to ensure preservation of the paint layer. Resins used to adhere flaking paint, however, tend to be easily absorbed into the paint. The resin tends to migrate to the surface filling the pores as it rises.

Unnatural preservation of ephemeral artefacts may in the future be considered unnecessary provided a readily available source of comparable original artefacts exists. This, however, is not always the case. All these cultures, by their very nature, are dynamic and constantly evolving. In many cases the non-native culture has intervened in the evolution of the art form. It is therefore important to preserve these crucial artefacts. Extensive microbiological and pathogen activity on the vegetable fibres may obviate the need to preserve the artefact. Anthropologists and conservators will need to work more closely together to achieve a balance between the cost of conservation, if required, of the "old" artefact and the acquisition of the new and may need to seriously question during which stage their natural life cycle should be "frozen".

With the skills of modern scientific research, information technology has progressed from written documentation using raw materials such as vegetable fibres from the pulp and paper industry worldwide, to computerisation, where information is permanently stored and more easily retrieved on computerised records. Long term preservation of our heritage today is thus ensured for the future. Conservators will need to face up to these new technological preservation resources which will, no doubt, pre-empt the use of interventive conservation on museum artefacts, providing a safer passive method for the long term storage of our heritage today!

ACKNOWLEDGEMENT

The author wishes to thank the following for their discussion, encouragement and support: David Bailey, Leo Biek, Debbie Bolton, Pat Brown, Andrew Carson, Eileen Collins and Christopher Radbourne, Anne Hudson, Hilde Smith and Pauline Timmins, and to Wendy Firmin, Editorial Critic and Kim Sumner, Secretariat, Kevin and Robert Brown, Directors of Advance Office Services, UK for practical assistance and professional advice.

REFERENCES

1. Horie, C.V. (1987) Materials for Conservation. Butterworths.

2. Plenderleith, H.J. (1956) The Conservation of Antiquities and Works of Art, Oxford University Press.

3. Robson, M.A. (1986) "The Use of Aqueous Polymer Emulsions in the Conservation of Ethnographic Material; A New Method for Supporting Fragile Woven Fibre Decoration, in "The Care and Preservation of Ethnological Materials (eds R. Barley, M Gilberg, J C McCawley and T Stone), Canadian Conservation Institute, Ottawa.

4. Robson, M.A. (1988) "Methods of Restoration and Conservation of Bronze Age Pottery Urns in the British Museum", in Early Advances in Conservation, British Museum Occasional Paper No.65.

5. Robson, M.A. (1989) "Interventive and Passive Conservation for a Travelling Exhibition", Museum Ethnographers Group.

6. Robson, M.A. (1990) "Historical applications of the use of synthetic resins in conservation of cellulosic vegetable fibres and ligneous skin", in The History of Manmade Fibres (ed R. Seymour) Atlanta, Georgia.

7. Robson, M.A. (1990) "Early Advances in the use of Acrylic Resins for the conservation of Antiquities", in The History of Manmade Fibres (ed R. Seymour) Atlanta, Georgia.

Forthcoming Publications

8. Robson, M.A. "A Comparative Study of a range of Glass Adhesives," (IIC)

9. Robson, M.A. "One part Acrylic U.V. Curing Systems with Fillers for Restoration of Porcelain and Enamelware, (UKIC)

70

Protection of wood against biodegradation by chemical modification

R.M. Rowell - USDA, Forest Product Laboratory, One Gifford Pinchot Drive, Madison, WI 53705.

ABSTRACT

It has been known for over 40 years that bonding simple chemicals to wood cell wall polymers greatly reduces or stops attack by many types of microorganisms. Laboratory test results using brown-, white-, and soft-rot fungi and tunneling bacteria indicate that the mechanism of protection is based mainly on the lowering of cell wall moisture. This mechanism does not seem to apply to attack due by termites or marine organisms.

INTRODUCTION

Wood is degraded biologically because organism recognize the carbohydrate polymers (mainly the hemicelluloses) in the cell wall and have very specific enzyme systems capable of hydrolyzing these polymers into digestible units. Biodegradation of the high molecular weight cellulose weakens the wood because crystalline cellulose is primarily responsible for the strength of wood. Strength is lost as the cellulose polymer undergoes degradation through oxidation, hydrolysis, and dehydration reactions.

Microorganism attack of wood has traditionally been controlled through the use of wood preservatives. These are chemicals which control biological attack due to the toxicity of the chemical. It is also possible to greatly reduce or stop microorganism attack by chemically modifying the cell wall polymers. The mechanism of this protection is not known but may be due to the lowering of the equilibrium moisture content below that needed for microorganism attack and by changing the conformation and configuration requirements of the enzyme-substrate reactions.

The purpose of this paper is to review the experimental results on the resistance of chemically modified wood to biological attack and try to draw some conclusions on the mechanism of effectiveness for that resistance.

Most of the experimental research done on biological resistance of chemically modified wood has been done with decay fungi. Table 1 shows a summary of the results of that work [1].

Table 1 - Biological resistance of chemically modified pine wood against brown- and white-rot fungi

Modification Chemical	Weight Percent Gain	Weight Loss After 12 Weeks	
		Brown-rot Fungus	White-rot Fungus
		%	
None	0	68	7
Acetic anhydride	17	<2	<2
Propylene oxide	25	<15	<2
Butylene oxide	22	<3	<1
Methyl isocyanate	20	<3	<1
Formaldehyde	5	<3	<1
β-Propiolactone	25	<2	<2
Acrylonitrile	25	<2	<2

All of the bonded chemicals at a weight percent gain (WPG) over about 20 show good resistance to brown- and white-rot fungi except propylene oxide. Propylene oxide is not effective in preventing attack by brown-rot fungi even though the same number of hydroxyl groups should be modified as were modified by reaction with butylene oxide, methyl isocyanate, β-propiolactone or acrylonitrile [2,3]. This exception of propylene oxide to the protection rule is perhaps the key to understanding the mechanism of the resistance to attack by fungi by chemical modification.

A closer look at fungal attack of propylene oxide modified wood is shown in Table 2. This data would indicate that

Table 2 - Soil block test of propylene oxide-modified southern pine inoculated with two brown-rot fungi and one white-rot fungus

Weight Percent Gain	Weight Loss After 12 Weeks Test		
	Lentinus lepideus	*Gloeophyllum trabeum*	*Coriolus versicolor*
Control	44.2	68.7	3.4
5 - 8	17.5	41.6	4.1
9 - 15	9.8	39.4	4.6
15 - 20	7.4	36.7	4.5
20 - 23	5.8	38.6	2.6
24 - 26	4.8	35.5	1.1
27 - 30	4.6	26.5	0.9

propylene oxide modified wood is more susceptible to white-rot fungi attack at chemical WPG's below about 20 than untreated wood [4]. Complete resistance to attack by *Lentinus lepideus* is not achieved at WPG's up to 30 and attack by *Gloeophyllum trabeum* is only reduced about half compared to untreated wood.

The mechanism of brown-rot fungi attack on lignocellulosics is thought to be as follows [5]:

```
ENZYMES-->HEMICELLULOSES
          [Energy source for
           generation of
           chemical oxidation
           system]------------>ACCESSIBLE CELLULOSE
                                [Strength losses]
                                [Energy source for
                                 generation of beta-
                                 glucosidases]------> WEIGHT
                                                      LOSS
```

This mechanism is consistent with the data that strength losses occur long before weight losses in brown-rot fungi attacked wood [6]. In this mechanism the key to brown-rot fungi resistance lies in the protection of the hemicellulose polymers. If that single component is protected, attack can not proceed. But, the question still remains, why do all chemical modifications except propylene oxide protect wood from brown-rot fungi attack?

Part of the answer may be seen in the data on attack by brown-, white-, and soft-rot fungi and tunneling bacteria in a fungal cellar on acetylated wood flakes shown in Table 3. This

Table 3 - Fungal cellar tests of aspen flakeboards made from control and acetylated flakes[1,2]

Weight Percent Gain	Rating at intervals[3]							
	2 mo	3 mo	4 mo	5 mo	6 mo	12 mo	24 mo	36 mo
0	S/2	S/3	S/3	S/3	S/4	--	--	--
7.3	S/0	S/1	S/1	S/2	S/3	S/4	--	--
11.5	0	0	S/0	S/1	S/2	S/3	S/4	--
13.6	0	0	0	0	S/0	S/1	S/2	S/3
16.3	0	0	0	0	0	0	0	0
17.9	0	0	0	0	0	0	0	0

[1]Nonsterile soil containing brown-, white-, and soft-rot fungi and tunneling bacteria.

[2]Flakeboards bonded with 5% phenol-formaldehyde adhesive.

[3]Rating system: 0 = no attack; 1 = slight attack; 2 = moderate attack; 3 = heavy attack; 4 = destroyed; S = swollen.

data shows that no attack occurs until swelling of the wood occurs [3,4]. This would indicate that the moisture content of the cell wall is critical before attack can take place.

Table 4 shows the equilibrium moisture content (EMC) for several types of chemically modified wood. It can be seen that all bonded chemicals lower the EMC of the wood by about half that of un-reacted wood except propylene oxide. At 85% relative humidity (RH), where biological attack starts to occur, the EMC for propylene oxide is 13.1 and the untreated is 16.4.

Table 4 - Equilibrium moisture content of control and chemically modified pine

Chemical	Weight Percent Gain	Equilibrium Moisture Content at 27 °C			
		0%RH	35%RH	60%RH	85%RH
Control	0	1.0	5.0	8.5	16.4
Acetic anhydride	20.4	0.7	2.4	4.3	8.4
Formaldehyde	3.9	0.4	3.0	4.2	6.2
Propylene oxide	21.9	1.4	3.9	6.1	13.1
Butylene oxide	18.7	1.3	3.5	5.7	10.7

These results would indicate that it is the lowering of the cell wall moisture content that is the key to the mechanism of brown-rot fungi protection.

Chemically modified wood has also been shown to be resistant to attack by termites and marine organisms. Table 5 shows the results of a two week test with the subterranean termite *Reticulitermes flavipes* [3,7]. All bonded chemicals

Table 5 - Wood weight loss in chemically modified southern pine after 2 weeks exposure to *Reticulitermes flavipes*

Chemical	Weight Percent Gain	Wood Weight Loss %
Control	0	31
Propylene oxide	9	21
	17	14
	34	6
Butylene oxide	27	4
	34	3
Acetic anhydride	10.4	9
	17.8	6
	21.6	5

including propylene oxide give about equal protection against termites . This would indicate that cell wall moisture content is not the only critical factor in the protection mechanism for termites.

Table 6 shows the data for chemically modified wood in a marine environment [8]. As with the termite test, all types of chemical modifications of wood help resist attack by marine organisms.

Table 6 - Ratings of chemically modified southern pine exposed to a marine environment

Chemical	Weight Percent Gain	Years of Exposure	Mean rating due to attack by[1]	
			Limnoriid and Teredinid Borers[2]	*Shaeroma terebrans*[3]
Control	0	1	2 - 4	3.4
Propylene	26	11.5	10	
oxide		3		3.8
Butylene	28	8.5	9.9	
oxide		3		8.0
Butyl isocyanate	29	6.5	10	---
Acetic anhydride	22	3	8	8.8

[1]Rating system - 10 = no attack; 9 = slight attack; 7 = some attack; 4 = heavy attack; 0 = destroyed

[2]Installed in Key West, Florida, 1975 to 1987

[3]Installed in Tarpon Springs, Florida, 1984 to 1987

CONCLUSIONS

The data collected to date on the mechanism of protection against biological attack of chemically modified wood would indicate that, for fungi, the critical factor is the moisture content of the cell wall. The moisture content of the hemicellulose component of the cell wall must be lowered to less than about 8% at 85% relative humidity before attack will be stopped.

There is not sufficient data to date to draw any conclusions on the mechanism of protection against termites or marine organisms. Texture, hardness, distribution of bonded chemical, as well as cell wall moisture content may all be important in protecting wood from attack by termites and marine organisms. More research data is needed before any conclusions can be made.

REFERENCES

1. Rowell, R.M. 1983, Commonwealth Forestry Bureau, Oxford, England, 6(12), 363-382.

2. Rowell, R.M.; Esenther, G.R., Nicholas, D.D. and Nilsson, T. 1987, J. Wood Chem. Tech., 7(3), 427-440.
3. Rowell, R.M.; Esenther, G.R.; Youngquist, J.A.; Nicholas, D.D.; Nilsson, T.; Imamura, Y.; Kerner-Gang, W.; Trong, L. and Deon, G. 1988, Proceedings: IUFRO wood protection subject group, Honey Harbor, Ontario, Canada. Canadian Forestry Service, p. 238-266
4. Rowell, R.M. and W.D. Ellis 1984, USDA Forest Service Research Paper FPL 451, Forest Products Laboratory, Madison, WI, 41 pp.
5. Personal cummication from T. Nilsson, Upsalla, Sweden.
6. Rowell, R.M.; Youngquist, J.A. and Imamura, Y. 1988, Wood and Fiber Sci., 20(2), 266-271
7. Rowell, R.M., Hart, S.V. and Esenther, G.R. 1979, Wood Sci, 11(4), 271-274
8. Johnson, B.R. and Rowell, R.M. Material und Organismen 1988, 23(2): 147-156.

71

The preservation of sawn Irish sitka spruce (*Picea sitchensis* (Bong.) Carr.).

Declan C. Cahill*, Martin Buggy and Evan R. Petty - Department of Materials Science and Technology, University of Limerick Ireland, Plassey Technological Park, Limerick, Ireland. *Forest Products Group, EOLAS, Dublin 9, Ireland.

1. INTRODUCTION

A combination of favourable growing conditions (1) and over-production in the agricultural sector (2) has led to increased private and public investment in softwood forestry in Ireland. The State Forestry, operated as Coillte Teo owns 78% of State forest area. In 1989, Sitka spruce represented 45% of area and 54% of volume (inventory) of Irish State forests. Sitka spruce has an annual growth rate of 20-24 m^3/ha in Ireland, compared to 11-14 m^3/ha in Britain and 4 m^3/ha in Sweden. It is estimated (3) that spruce sawlog will account for 80% of total sawlog production in Ireland in 1993-1997.

The primary market for this material is the construction industry, where preservation treatment may enhance service-life. The need to collect basic information on the preservation treatability of Irish grown Sitka spruce is important for both domestic and potential export markets. This will complement other recent work (4,5) on spruce. Two common industrial methods for preservation treatment are double vacuum and CCA (copper/chrome/arsenic) pressure processes. The double vacuum treatment process is generally used to treat softwood joinery and structural timber in Ireland (6). Some structural timber is treated with water-borne CCA but its main use is for farm and roadside timber fencing. The work reported here, examined the double vacuum preservation treatment of common construction timbers sizes of Sitka spruce.

2. EXPERIMENTAL

Sitka spruce trees were selected at random in a number of different forests and sawn to provide samples (7). For comparison, imported whitewood of Finnish and Russian origins was obtained and tested simultaneously with the Irish material.

The following nominal sizes were used in this study.

Tiling batten	:	44 x 35 x 2400 mm
Rafters	:	100 x 35 x 2400 mm
Wall plates	:	100 x 75 x 2400 mm

A total of 718 samples were subjected to double vacuum preservative treatments following BWPDA manual (1986) using Protim Prevac 80 solution and a Protim Prevac plant. Details of the treatment schedules are shown in Table 1.

Gravimetric take up of preservative for each sample was measured. Volumetric uptakes were calculated from relevant densities. The loading of fungicide, tributyl tin oxide (TBTO) in the outer 3 mm lateral zone of samples was measured by atomic absorption spectroscopy (BS 5666; 1980). For a limited number of samples, analysis of 5 - 8 mm zone was also carried out.

TABLE 1

Double Vacuum Schedules

Schedule Code	Initial Vacuum Pressure (Bar)	Initial Vacuum Time (Minutes)	Impregnation Pressure (Bar)	Impregnation Time (Minutes)	Final Vacuum Pressure (Bar)	Final Vacuum Time (Minutes)
DV1	-0.33	3	0	3	-0.67	20
DV2	-0.33	5	1	5	-0.67	20
DV3	-0.17	10	1	40	-0.83	20
DV4	-0.83	10	1	60	-0.83	20

3. RESULTS

Table 2 shows the volumetric uptake (l/m^3) of preservative in samples following double vacuum treatment. The TBTO loading in samples which had received the DVI schedule is shown in Table 3.

4. DISCUSSION

In all treatment schedules, the Irish grown material absorbed more than the control Whitewood. Given the large sample population, a statistical analysis of the effects of variables such as origin, treatment schedule and size on uptake was undertaken using a "Minitab" software package. The effects of timber condition, i.e. moisture content, sapwood, ring width, saw cut, pith and density, were also analysed. Apart from the marked

TABLE 2

Double Vacuum Treatment Preservative Uptakes in Irish Sitka Spruce and Imported Whitewood

Sample Type	No. of Samples	Origin	DV1	DV2	DV3	DV4
Tiling Battens	119	Imported				
		Mean	32.85	46.99	57.28	91.89
		Std. dev.	11.30	24.86	16.84	36.92
	119	Irish				
		Mean	40.35	64.95	95.05	176.53
		Std. dev.	9.64	26.35	32.21	63.44
Rafters	120	Imported				
		Mean	30.66	33.81	44.84	78.27
		Std. dev.	10.32	9.94	11.93	26.29
	120	Irish				
		Mean	40.81	44.99	74.74	129.97
		Std. dev.	15.00	9.48	26.04	44.04
Wall-Plates	120	Imported				
		Mean	17.01	22.32	26.04	47.34
		Std. dev.	2.72	4.74	3.27	16.72
	120	Irish				
		Mean	21.70	28.04	53.20	87.83
		Std. dev.	4.22	5.99	47.70	28.16

TABLE 3

Preservative Loadings (%TBTOm/m) in Double Vacuum treated Irish Sitka Spruce and Imported Whitewood (Schedule DV1)

Component	Irish		Imported	
	0-3mm	5-8mm	0.3mm	5-8mm
Tiling Battens				
Mean :	0.099	0.036	0.122	0.028
Std. dev.:	0.022	0.016	0.091	0.013
Rafters				
Mean :	0.051	0.033	0.120	0.024
Std. dev.:	0.022	0.015	0.065	0.007
Wall-Plates				
Mean :	0.113	0.038	0.078	0.039
Std. dev.:	0.030	0.029	0.023	0.021

effects of origin, size and treatment schedule only density was found to have a statistically valid relationship to uptake for the double vacuum technique.

Considering the results in Table 3, it can be seen that the lateral loadings of TBTO are very similar for Irish and imported material even though the total uptakes are markedly different. This infers that the extra preservative fluid enters the Irish timber by travelling up the end grain of the timber.

An experiment was carried out to give an indication of the effect of end-grain uptake on the overall preservative absorption in Irish Sitka spruce. Twenty wallplate offcuts (465 mm long) were ripped in two to produce matching pairs of samples. One set of the sub-samples was end-sealed with two coats of a PVC (polyvinyl chloride) cement and two coats of PVA (polyvinyl acetate) wood glue. The matching samples of each set were not end-sealed.

Ten matching samples were treated with a double vacuum schedule (DVI) while the other samples received a DV4 schedule. All the samples were weighed before and after treatment and the gravimetric uptake calculated (see Table 4).

TABLE 4

Preservative Treatment of End-sealed Sitka Spruce

		Weight Increase (gm)	
Schedule		End-sealed	Non End-sealed
DV1	Mean:	34.7	55.2
	Std. Dev:	9.7	14.8
DV4	Mean:	58.9	121.6
	Std. Dev:	12.7	26.0

The results clearly indicated the contribution which end-grain absorption made to the overall uptake of preservative solution in Irish Sitka spruce. End-sealing reduced the average preservative uptake achieved with the DV1 schedule by 37%. The reduction caused by end-sealing with the DV4 schedule was 48%. It should be stressed that the above data were based on short samples, which will tend to exagerate the end-grain effect.

To further test this possibility of high end-grain absorption, an investigation of preservative loading analysis was carried out. Although only a small sampling was carried out, the results illustrate this point of different end grain penetration of preservative in Irish and imported spruces. Two wallplates, one Irish and other imported, which had been previously treated with a DV4 schedule, were selected. Cross-sectional samples were removed at different points from the end-grain areas of each sample. Twenty millimetres were removed from each of the faces to reduce the influence of lateral penetration and the remaining core was then analysed for TBTO as before. The results are shown in Table 5.

TABLE 5

End-grain Preservative Analysis

IRISH		IMPORTED	
Distance from end-grain	%TBTO m/m	Distance from end-grain	%TBTO m/m
0 mm	0.3078	0 mm	0.0256
50 mm	0.1371	50 mm	0.0137
100 mm	0.0476	100 mm	0.0120
200 mm	0.0055	200 mm	0.0036
400 mm	0.0040	400 mm	0.0022

These results would support the view that Sitka spruce has a higher preferential end-grain uptake of preservative than imported Whitewood. It is known (8) that movement of liquids is greater in spruce latewood than earlywood. It was found that the latewood membranes in Sitka spruce are very robust and rigid. These tend not to be fully aspirated on drying. Thus it would be expected that higher concentrations of preservative chemicals would be found in the latewood portions of annual rings.

The annual rings from the treated Sitka spruce wallplates referred to above were separated into earlywood and latewood - the boundary was arbitarily visually decided. The two fractions were then quantitatively analysed for TBTO (see Table 6).

These results indicate that the latewood in Sitka spruce had a higher retention of preservative fungicide than the earlywood. This supports the idea that the latewood cells in Sitka spruce allow greater movement of preservative which is mainly due to un-aspirated pits. However, it also shows that preservative was also present even in the earlywood quite deep in the end-grain area.

TABLE 6

Preservative Loadings within Annual Rings

Distance from end-grain		%TBTO m/m
33 mm	earlywood	0.050
	latewood	0.0866
66 mm	earlywood	0.0322
	latewood	0.0449
149 mm	earlywood	0.0105
	latewood	0.0184
335 mm	earlywood	0.0041
	latewood	0.0090

The performance of preserved specimens are currently being evaluated in weathering trials. Analysis of these with the preservative uptake data will provide a basis for the formulation of standards in the preservation of fast growth in timber for construction applications.

5. CONCLUSIONS

The preservation performance of Irish grown sawn Sitka spruce has been quantified using the double vacuum technique. The material absorbed more preservative than control (imported) whitewood. Absorption levels were generally higher than those required for commercial protection. It was found that end grain absorption was an important feature of Sitka spruce behaviour. This may be associated with liquid movement through partially aspirated cells in Sitka spruce latewood.

6. REFERENCES

1. D. O'Brien "A Financial Appraisal of Sitka Spruce". Irish Forestry 43(2), 128 (1986).
2. F.J. Couvery "The Potential of Forestry to Contribute to Rural Development" Agricultural Record September 36 (1981).
3. M. Joyce "Sawmilling - Present and Future" Society of Irish Foresters Symposium. Dublin. April (1989).
4. D. Ronze, M. Mary, A. Zoulalain and W.G. Kauman. "A model of axial flow of organic liquids in two softwoods". Wood Science and Technology, 22, 1 (1988).
5. R.J. Murphy and D.J. Dickinson "Accelerated Diffusion Treatment of Sawn Sitka Spruce Grown in the UK." IRG/WP/3395. Stockholm (1986).
6. D.C. Cahill. "Directory of Preservation Plants in Ireland - 7th Edition" Eolas, Dublin (1989).
7. D.C. Cahill. M. Tech. Thesis, University of Limerick (1990).
8. J.P. Moroney, Ph.D. Thesis, University of Leeds (1977).

72

Pulp and paper mill effluent biobleaching using immobilized *Phanerochaete chrysosporium* BKMF 1767

R.S. Singh , S.S. Marwaha, P.K. Khanna*, and J.F. Kennedy** - Department of Biotechnology, Punjabi University, Patiala, 147002, India. *Department of Microbiology, Punjab Agriculture University, Ludhiana, 141004, India. **Research Laboratory for the Chemistry of Bioactive Carbohydrates and Proteins, Department of Chemistry, University of Birmingham, P.O. Box 363, Birmingham B15 2TT, UK.

ABSTRACT

Studies were conducted on the biobleaching of black liquor using immobilized Phanerochaete chrysosporium BKMF 1767. The physical, chemical and biological characteristics of black liquor of three alkaline extraction stages and combined effluent showed that the black liquor (stage-I) was the most highly polluted. The investigations carried out to optimise the immobilization technique, displayed a 5% (w/v) concentration of alginate matrix optimum for the efficient entrapment of the fungal mycelium. However, the treatment of the beads with different concentrations of aluminium sulphate resulted in the solubilization of the beads. The stationary conditions and 5% (v/v) concentration of the black liquor supported the maximum colour and COD reduction by alginate entrapped white-rot fungi.

INTRODUCTION

Bleached kraft mills usually discharge large volumes of brown coloured effluents as a result of the different processes employed in the processing of wood and pulp. The first alkaline extraction stage bleach plant effluent is the major source of coloured effuents from pulp and paper industries. These effluents are associated with high biochemical oxygen demand (BOD_5), chemical oxygen demand (COD), total solids and organic carbon. The main contributors to the colour are polymeric, chlorinated, heavily oxidized degradation fragments of lignin (Kirk et al., 1983; Royer et al., 1983).

Moreover, their high chlorine contents (3 kg of lignin bound chlorine/tonne of bleached paper, Wainwright, 1990) are highly toxic (Livernoche et al., 1983). These lignin derivatives are highly resistant to microbial attack. As a result, the coloured and toxic materials from bleach plants escape the current biological waste water treatment technologies in an almost unmodified form, going into the receiving water thereby resulting in environmental pollution.

White-rot basidiomycetous fungi possess an active lignolytic system (Paice et al., 1989; Kirk et al., 1978), which is able to degrade protolignin as well as heavily modified lignins, such as kraft and chlorinated lignins (Chang et al., 1983). It is well established that the white-rot fungi can decolourise kraft bleach effluents and exhibit their activity only in the presence of oxygen and an external carbon source (Kirk et al., 1978). The potential of some white-rot fungi in decolourising the effluents of pulp and paper mills has recently been demonstrated (Campbell et al., 1982; Eaton et al., 1982; Chang et al., 1987; Khanna et al., 1990, 1991). However, the literature survey reveals no report on the use of white-rot basidiomycetous fungi for the biobleaching of black liquor from the first extraction stage of pulp and paper mills.

The concept of immobilization of microbial cells on solid support was introduced a number of years ago. Since then, it was developed into an important technique in biotechnology (Jack and Zajic, 1977). The ability of immobilized white-rot fungi to decolourise kraft mill effluents has been demonstrated by Livernoche et al. (1981). Keeping this in view, the advantages of immobilized cell systems over free cells, the present investigations were undertaken to decolourise black liquor from the first extraction stage of pulp and paper mills, using immobilized Phanerochaete chrysosporium BKMF 1767.

MATERIAL AND METHODS

Collection and storage of black liquor and combined effluent: Figure 1 outlines the sites of origin of black liquor and combined effluent in bleached kraft mills of paper manufacture. Black liquor of three alkaline extraction stages (I, II & III) and combined effluent were periodically collected from Shreyans Paper Mills Ltd., Ahmedgarh, Punjab, India, in air tight plastic containers (10L capacity) and stored at $4 \pm 1^{\circ}$C.

Fungal cultures: Four white-rot basidiomycetes namely Polyporus versicolor FRI 165, Polyporus hirsutus FRI 2862 (procured from FRI, Dehradun), Phanerochaete chrysosporium BKMF 1767 (Courtsey Prof. T.K. Kirk, Forest Products Laboratory, US Department of Agriculture, USA) and Sporotrichum pulverulentum MTCC 163 (collectd from IMTECH, Chandigarh) were used for the decolouration of black liquor of pulp and paper mills.

Preparation of starter: The fungal cultures were grown in potato dextrose (PD) broth (50ml contained in 250ml capacity conical flasks) for seven days under stationary conditions at 30°C. The fungal mats were recovered, washed with sterilized distilled water, suspended in unsterilized effluent sample and macerated aseptically using a homogenizer. The mycelial suspensions thus formed were used at 10% (v/v) to screen the cultures for their biobleaching abilities.

Screening of white-rot fungi: The diluted and undiluted black liquor samples, filtered through ordinary filter paper, supplemented with dextrose (1% w/v) and adjusted to pH 5.5 were used to screen the white-rot fungi for their biobleaching abilities. The black liquor samples (45ml) in 250ml conical flasks were inoculated with 5ml of starter culture (10% v/v) and then incubated for 7 days at 30°C. In each case, the samples were analysed for colour and chemical oxygen demand (COD) at 24h intervals.

Immobilization of *Phanerochaete chrysosporium*: The mycelial suspension of *Phanerochaete chrysosporium* made in sterilized distilled water was thoroughly mixed with sodium alginate (5% w/v), unless otherwise specified. The resultant mycelium alginate slurry was extruded as drops through a sterilized glass syringe, equipped with 15 gauge needle, into calcium chloride solution (0.075M) at room temperature. The beads were left suspended in calcium chloride solution (0.075M) for 24h to allow complete gelation. The beads were washed with saline immediately prior to their use to remove excess of calcium ions and untrapped cells.

The alginate beads with entrapped fungal mycelium were treated with different concentrations of aluminium sulphate solution (0.25, 0.50, 0.75 and 1.0M) for 24h at room temperature to effect their strengthening. The influence of different concentrations of sodium alginate (2%, 3%, 4%, 5%, 6% and 7% w/v) on the entrapment capability of the matrix was also studied.

Biobleaching of black liquor with immobilized *P. chrysosporium*: The black liquor filtered through ordinary filter paper was supplemented with dextrose (1% w/v) and adjusted to pH 5.5. It (50ml in 250ml capacity conical flasks) was then inoculated with beads loaded to a known weight of mycelium. The flasks were incubated at 30°C for six days and the samples were analysed for colour and COD at 24h intervals.

The influence of different concentrations (5%, 10%, 15%, 20%, 25%, 50% and 75% v/v) of black liquor and agitation on biobleaching ability of immobilized culture was investigated.

ANALYSES

Colour: CPPA standard method (1974) was used for measuring the colour units of black liquor and combined effluent.

Total dissolved solids (TDS) and total suspended solids (TSS): The procedure of the Indian Standard Institution (ISI, 1966) was followed for the estimation of the black liquor and combined effluent.

Biochemical oxygen demand (BOD_5) and chemical oxygen demand (COD): The BOD_5 (at 20°C for 5 days) and COD of black liquor and combined effluent samples were determined following the procedure of AOAC (1975).

Temperature and pH : The temperature and pH of the black liquor and combined effluent samples were recorded immediately after collection.

Dry fungal biomass : The contents of flasks were filtered using dried and pre weighed Whatman Filter paper No.1 which was then dried in an oven at 60°C to a constant weight to calculate the dry weight of fungal biomass. For the immobilization, a known volume of mycelial suspension was filtered, dried and weighed as described above.

Enumeration of Microflora : The black liquor and combined effluent samples were analysed for their bacterial and fungal counts following the procedure, described by Khanna et al. (1990).

The presence of coliform bacteria in the samples was detected by using single and double strength lactose solution at 37°C. Samples displaying gas and acid production in lactose solution were taken as coliform positive samples.

RESULTS AND DISCUSSION

Characterization and screening of black liquor: The periodic analysis of effluents of three alkaline extraction stages of black liquor generation i.e. stage-I (BL-I), stage-II (BL- II), stage-III (BL-III) and combined effluent (CE) have shown great variations in their biological, physical and chemical characteristics.

The bacterial densities of BL-I, BL-II, BL-III and CE ranged from $11x10^3$-$25x10^3 ml^{-1}$, $18x10^3$-$30x10^3 ml^{-1}$, $32x10^3$- $112x10^3 ml^{-1}$ and $35x10^3$-$105x10^3 ml^{-1}$ respectively. However, fungi were observed only in CE ($4.0x10^3$-$12x10^3 ml^{-1}$). The higher bacterial and fungal populations of combined effluents may be attributed to their self purification capacity. All type of effluents tested were coliform positive. The order of most probable number (MPN) of coliforms was BL- I<BL-II<BL-III<CE.

Similarly, the temperature and pH of waste waters of different stages of pulp and paper making varied markedly. The temperature of BL-I and pH of BL-I, BL-II and BL-III were much higher than BIS permissible limits for the discharge of industrial effluents. The slightly alkaline (CE) to highly alkaline (BL-I, BL-II, BL-III) pH may be attributed to the nature of chemicals used during the bleaching of pulp. The amount of total solids (BL-I: 41,700- 63,000; BL-II: 13,200-17,000; BL-III: 6,400-8,000; CE: 2,670-3,500mgl^{-1}), total dissolved solids (BL-I: 24,000-38,250, BL-II: 9,000-14,250; BL-III: 5,500-6,400; CE: 1,190-1,894mgl^{-1}), total suspended solids (BL-I: 11,600-39,000; BL-II: 2,550-5,500; BL-III: 900-1,900; CE: 1,071- 2,291mgl^{-1}), BOD_5 (BL-I: 11,980-16,550; BL-II: 7,900-10,300; BL-III: 1,285-2,000; CE:

75-140mgl^{-1}) and COD (BL-I: 35,245- 42,250; BL-II: 19,800-22,400; BL-III: 9,325-22,100; CE: 580- 620mgl^{-1}) of tested samples were much higher than the federal permissible limits governing discharge of effluents in India. Similar observations for combined effluents have earlier been made by Khanna et al., 1990.

The variations in the physical and chemical characteristics of the one type of effluents at different time intervals could be attributed to the use of different raw materials in different seasons and also due to variation in total production of paper. These characteristics showed that black liquor of stage-I (Fig.1) is highly polluted as compared to other effluents and was selected for further investigations.

Screening of fungal cultures for decolourization: Among the four strains of white-rot fungi studied Phanerochaete chrysosporium BKMF 1767 showed the highest degree of decolourization (66.29%) and COD reduction (50.20%) of black liquor after seven days of cultivation (Fig.2). With the exception of Polyporus hirsutus FRI 2862, the other strains investigated during the course of this study showed more than 50% decolourization, whereas, all the strains resulted in less than 50% reduction of COD. In batch and continuous processes P. chrysosporium has been reported as an efficient reducer of colour (Chang et al., 1987; Joyce et al., 1983). Moreover, in a continuous process, 50% reduction in COD by this culture has been reported (Eaton et al., 1981). Since P. chrysosporium BKMF 1767 resulted in maximum decolourization and COD reduction in black liquor, it was selected for further studies using immobilized white-rot fungi.

Biobleaching of black liquor with immobilized P. chrysosporium BKMF 1767: The fungal mycelium immobilized to a cell concentration of 185mg dry wt. 50ml^{-1} of undiluted black liquor was employed for the biobleaching of black liquor. No decolourization of black liquor was observed during the treatment time. Moreover, the beads started disintegrating after 24h of incubation. The complete solubilization of beads was observed after 72h of incubation. The dissolution of beads may be attributed to the presence of certain degelatinising ions in concentrated black liquor.

Optimization of immobilization technique:

(i) **Strengthening of alginate beads for entrapment of mycelium:** To overcome the problem of solubilization and maintain the integrity of the calcium alginate beads carrying the entrapped mycelia (185mg dry mycelium 50ml^{-1} of undiluted black liquor), beads were treated with different concentrations of aluminium sulphate for 24h. The treated beads become solubilized within 24h of incubation and resulted in the release of macerated mycelium into the black liquor. In addition during this course no decolourization of the undiluted black liquor was observed. However, the treatment of alginate beads with aluminium sulphate for the strengthening of gel for immobilization of yeast for alcohol production has been recommended (Bajpai et al., 1989).

(ii) **Influence of different concentrations of alginate on the entrapment of P. chrysosporium:** A cell concentration of 138mg dry wt $50ml^{-1}$ of diluted black liquor (5% v/v) was immobilized in different concentrations of sodium alginate. At all the gel concentrations (2-6% w/v) the beads of uniform size and shapes were formed. However, at higher concentration (7% w/v) dissimilar and disformed beads were produced, which may be attributed to high viscosity of the gel and variability of pressure applied during its manual extrusion through glass syringe. Maximum colour (72.13%) and COD (61.75%) reductions of black liquor were observed after six days of incubation at 5% (w/v) gel concentration (Table 1). Livernoche et al. (1981) reported 80% reduction in colour of a Kraft mill combined effluent with immobilized Coriolus versicolor within three days of the treatment.

Optimization of process characteristics for biobleaching of black liquor:

(i) **Influence of different concentrations of black liquor on biobleaching:** A mycelial load of 120mg dry wt. $50ml^{-1}$ of diluted black liquor was employed. The alginate beads were dissolved at 40-70% (v/v) of concentration of black liquor. The maximum declourization (68.08%) and COD reduction (51.84%) were observed in 5% (v/v) black liquor after six days of treatment (Fig.3). However, Livernoche et al. (1983) reported 90% colour reduction with immobilized Coriolus versicolor, after three days of incubation. Since 5% (v/v) black liquor resulted in maximum colour and COD reduction, it was selected for further investigations.

(ii) **Influence of agitation on biobleaching of black liquor:** The mycelial load of 120mg dry wt. $50ml^{-1}$ of diluted black liquor (5% v/v) was used during the course of this experiment. The samples were incubated under stationary conditions in a BOD incubator and shaking conditions at 200rpm on a rotary shaker. The stationary conditions, supported the maximum decolourization (68.08%) and COD reduction (51.84%) after six days of incubation. The corresponding colour and COD reduction under shaking conditions were 50.22% and 42.08%, respectively (Fig.4).

The above experimentation carried out in flask cultures at laboratory scale provides very useful guide lines for further investigations on the biobleacing of black liquor of first stage of alkaline extraction in pulp and paper mills.

ACKNOWLEDGEMENT

The financial support to RSS in the form of Senior Research Fellow from CSIR, New Delhi, is duly acknowledged.

REFERENCES

AOAC. (1975) In : Official Methods for Analysis, Association of Official Analytical Chemists, Washington D.C.

Bajpai, P., Sharma A., Raghuram, N. and Bajpai, P.K. (1989) J. Micro. Biotech. **4**: 87-92.

Campbell, A.G., Gerrad, E.D., Joyce, T.W., Chang, H-m and Kirk, T.K. (1982). In : Proceeding of 1982 Tappi Research and Development Conference, Asheville, North Carolina, TAPPI Press : 209-214.

Chang, H-m, Joyce, T.W., Gerrad, E.D., Huynh, V-b and Kirk, T.K. (1983). In: Recent Advances in Biodegradation Research, Eds. Higuchi, Chang and Kirk, pp.257-268, Uni. Publi. Co., Tokyo.

Chang, H-m, Joyce, T.W., and Kirk, T.K. (1987). US Patent, **4**, 655: 926.

CPPA (1974). Technical Section Standard Method H5P.

Eaton, D.C., Chang, H-m, Joyce, T.W., Jeffries, T.W. and Kirk, T.K. (1981). Annual Meeting Proceeding, TAPPI Press, pp.157-161.

Eaton, D.C., Chang, H-m, Joyce, T.W., Jeffries, T.W. and Kirk, T.K. (1982). TAPPI J. **65**: 89-92.

ISI (1966). Methods of sampling and tests for industrial effluents (Part 1), Indian Standard Institution, New Delhi, **25**: 2488.

Jack, T.R. and Zajic, J.E. (1977). Adv. Biochem. Eng. **5**: 126-154.

Joyce, T.W., Chang, H-m, Campbell, A.G., Gerrad, E.D. Jr. and Kirk, T.K. (1983). In: Biotechnology in Pulp and Paper Industry, London: 30-36.

Khanna, P.K., Mittar, Dev, Marwaha, S.S. and Kennedy, J.F. (1990). Biopaper No.10: **3**: 11. Biopapers J., 16-18.

Khanna, P.K., Mittar, Dev, Marwaha, S.S. and Kennedy, J.F. (1991). J. Chemi. Tech. & Biotech (In Press).

Kirk, T.K., Schultz, E., Connors, W.J., Lorenz, L.F. and Zeikus, J.G. (1978). Arch. Microbiol. **117**: 227-285.

Kirk, T.K., Jeffries, T.W. and Leatham, G.F. (1983). TAPPI J. **66**: 46-51.

Livernoche, D., Jurasek, L., Desrochers, M. and Veliky, I.A. (1981). Biotech. Lett. **3**: 701-706.

Livernoche, D., Jurasek, L., Desrochers, M. and Donca, J. (1983). Biotech. Bioeng. **25**: 2055-2065.

Paice, M.G., Jurasek, L., Ho, C., Boubonnais, R. and Archibald, F. (1989). TAPPI J. **72**: 217-221.

Royer, G., Livernoche, D., Desrochers, M., Jurasek, L., Roulean, D. and Mayer, R.C. (1983). Biotech. Lett. **5**: 321- 326.

Wainwright, M. (1990). Chem. Ind. **1**: 31-34.

73

Kraft mill effluent: biological treatment

N. Durán*, E. Esposito* and V.P. Canhos** - *Instituto de Quimica, Biological Chemistry Laboratory, Universidade Estadual de Campinas, C.P. 6154, Campinas, CEP 13081, S.P. Brazil. **Fundação Tropical, Campinas, S.P., Brazil.

ABSTRACT

We selected four strains with the capacity for the decolorization of kraft effluent from 48 white-rot fungi which degrade eucalyptus wood and 50 white- and brown-rot fungi and one ascomycete which degraded eucalyptus and pinus woods. The selected strains were **Lentinus edodes, Chrysonilia sitophila, Phanerochaete chrysosporium** and **Phaeocoriolellus trabeus**. The decolorization efficiency (ratio of decolorization [%]/mycelial mass [mg]) of these strains in 5 days, in the absence of any external carbon source were the following: 0.97 (L. **edodes**); 0.54 (C. **sitophila**); 0.94 (P. **chrysosporium**) and 0.30 (P. **trabeus**). Chemical analyses and spectroscopic studies, besides molecular weight distribution by gel filtration were indicative that **P. trabeus** modified chlorolignins without any degradation of the polymer. Meanwhile, **L. edodes** exhibited 70% polymer consumption, P. **chrysosporium** and C. **sitophila** by a similar fashion only around 30% was observed. A similar decolorization of E_1 Kraft effluent by **P. chrysosporium** and **L. edodes** in a packed-bed bioreactor using nylon-immobilized mycelium and diffuse aeration was also found.

INTRODUCTION

Attempts have been made to remove the color of effluent from a Kraft mill by **P. chrysosporium** [1]by **Trictoporia sp** [2], by **Aspergillus sp** [3] and by **Coriolus versicolor** [4]. A mycelial color removal process (MyCor) for decolorizing E_1 bleach plant effluent have been published [5,6]. The potential use of white-rot fungi for biopulping and treatment of pulp mill waste effluent has been reviewed [7]. Recently a rapid and efficient method to assay ligninolytic activity and an evaluate the fungal decolorization capacity for Kraft effluent, without any extra carbon sources was devised [8].

EXPERIMENTAL

CULTURE CONDITIONS. Cultures were grown in agitated conditions for ligninase and phenoloxidase production, using malt extract as carbon source, in 125 Erlenmeyer flasks at 25 ^oC, in the dark. A culture medium (50 ml) malt liquid at pH 5.0 was used [8].
ENZYME ASSAYS. Ligninase activities [9] were measured by UV spectroscopy of veratrylaldehyde formed by oxidation of veratryl alcohol (expressed as U/L). Phenoloxidases (Laccase and Peroxidase) were measured using syringaldazine as a substrate in the absence and presence of H_2O_2 respectively [10]. Beta-glucosidase was measured using p-nitrophenyl-beta-D-glucoside(pNPG) as substrate [11]. Mn-Peroxidase was determined by oxidizing phenol red in the presence of H_2O_2 and Mn(II) [12].
EFFLUENT. The effluent (RIGESA, S.C, Brazil) used in this study was the first alkaline extraction stage (pH 9.5 - 13.0) from a bleached kraft pulp (Eucalyptus) and a combination with C_1 (Chlorination effluent) in different proportions.
MEASUREMENT OF COLOR IN TREATED KRAFT EFFLUENT. Color removal by fungi strain during incubation was measured according to the CPPA method [13]. The mycelial pellets (2 g of wet mycelium, 97 % humidity) were inoculated to 25 ml of non-sterile effluent (pH 5.0) at 25^oC and 150 rpm in the absence of any additional carbon or nitrogen source. The effluent was diluted and initial absorbance of 0.44 was used. The carbohydate concentration of E_1 effluent was 0.06%.
DECOLORIZATION WITH DEAD MYCELIUM. The wet mycelium was autoclaved for 30 min at 120 ^oC, and the experiment was carried out as described for live mycelium [14].

COD DETERMINATION. The Chemical Oxygen Demand (COD) was determined by standard methods [15].

MOLECULAR MASS DISTRIBUTION. 0.5 ml of the clean effluent supernatant was applied to 60 x 1.5 cm Sephadex-G-50 column equilibrated with 0.1 N NaOH-LiOH . 2.5 ml were collected (Flow rate of 2 min 15 s/ml) and the chlorolignins were determined at 280 nm [16].

SUGAR DETERMINATION. The carbohydrate determination of kraft effluent was carried out by the 3,5-dinitro salicilic acid (DNS) procedure [17].

PHENOL DETERMINATION. The total phenol content of the effluents was measured by a standard method [18].

DECOLORIZATION OF E_1 KRAFT EFFLUENT IN A PACKED-BED BIOREACTOR. The pre-culture was carried out in plates for **L. edodes:** 1.25% malt extract, 2% agar and wood blocks of E. grandis (5 days). Pre-inoculum in 1.25 % malt extract in 250 ml volume (mycelium from 1 plate) from 5 days in the bioreactor together with 5 g of dried Nylon. The culture was filtered and 300 ml of E_1 Kraft effluent (100%) was added at a pH of 5.0 and at an absorption (465 nm, pH 7.6) of 0.8072. No extra carbon sources was added.For the **P. chrysosporium** the initial pH was 4.5. The pre-inoculum and inoculum were the same as previously reported. Immobilization of **P. chrysosporium** was similar as in **L.edodes,** but the decolorization process was carried out in 1% glucose, with nutrients and minerals with a 50% E_1 effluent.

RESULTS AND DISCUSSIONS

Previously we had shown that **L. edodes** (UEC-2019), **L. edodes** (UEC-2021), **P. trabeus** (UEC-2023), **C. sitophila** (TFB-27441) and **P. chrysosporium** (BKM-F-1767) decolorized efficiently E_1 effluent [8,19]. In TABLE 1 are exhibited the enzymatic activities after 120 h of effluent treatment by the respective mycellium. Laccase, peroxidase, lignin peroxidase, manganese peroxidase and beta-glucosidase were measured. From our observation, beta-glucosidase exerts an important role in the effluent decolorization. The most efficient strain for the decolorization, **L. edodes** (UEC-2019), exhibited a high beta-glucosidase and lignin peroxidase activities. Meanwhile **P. trabeus** (UEC-2023) although it exhibited high Mn-peroxidase and high lignin peroxidase it was extremely inefficient for decolorization, probably due to complete absence of beta-glucosidase activity in the effluent medium.**C. sitophila** [20] together with **P. chrysosporium** exhibited relatively high beta-glucosidase activities, but not high lignin peroxidases

TABLE 1. BIOLOGICAL TREATMENT OF KRAFT EFFLUENT@

Enzymatic Activity (U/L)	Strains BKM-F 1767	TFB 27441	UEC 2019	UEC 2021	UEC 2023
Laccase	0.04	0.00	11.0	2.1	0.8
Peroxidase	0.005	0.00	0.5	3.2	0.4
Ligninase	7.80	7.80	19.3	23.0	260.0
Mn-Peroxidase*	0.04	0.02	0.01	0.01	0.8
β-Glucosidase&	12.0	7.0	14.0	5.0	0.0
Efficiency#	0.94	0.64	0.97	0.6	0.3
COD(%)	26.0	41.0	60.0	61.0	31.0
Decolor.(%)	52.0	41.0	73.0	42.0	24.0

@ 120 hours ; *Absorbance min^{-1} ml^{-1}; &U/ml; #percentage of decolorization/mycelial mass(dry, mg).

activities. Previous results from other studies, mainly with carbon source variation, enphasize the importance of executing the biological treatment under strictly defined conditions. Recently, it was shown that this degradation depends on the transglucosylation activity of glucosidase secreted by the majority of wood-rot fungi [21] and the glucoside formation prevents polymerization and accelerates efficient degradation reactions [22].
Both strains of **L. edodes** were quite efficient in decreasing COD which is perfectly parallel to the molecular mass consumption (Fig.1).

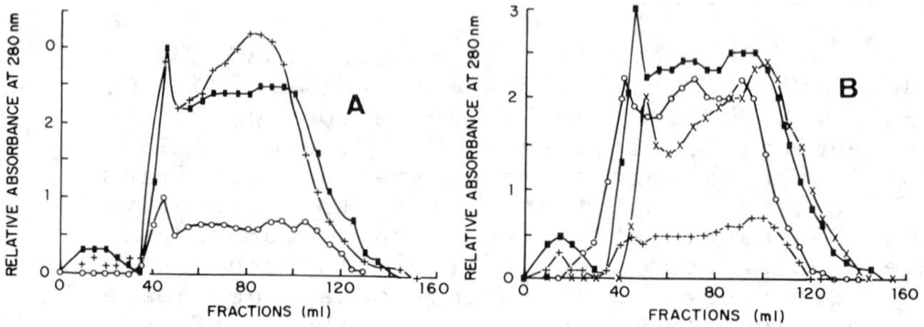

Fig.1. Gel filtration on Sephadex-G-50:a) Control effluent (-■-); **P. trabeus** (-+-).L. **edodes (2021)**. (-○-). b) Control effluent (-■-), P. **chrysosporium** (-×-),L. edodes(2019) (-+-), C. **sitophila** (-○-).

Only 13% of the decolorization was due to mycelial adsorption, then the main process was the mineralization of the chlorolignins.
In a selective study with a packed-bed bioreactor, using nylon-immobilized mycellium and diffuse aeration, L. **edodes** (UEC-2019) was further analyzed

(TABLE 2). The values were similar in both strains, but we have to emphasize that in the case of P. **chrysosporium** the conditions were different. In the latter case a carbon and nitrogen source were added, but not in the case of **L. edodes**.

In summary, the strategy which we have developed seems to us a rapid and efficient way to select new strains for decolorization and decontamination of Kraft effluent. From our results it is clear that lignin peroxidase or other peroxidases are not enough to extend these processes, but also it is necessary the beta-glucosidase activity is determined. The packed-bed bioreactor (300 ml) using immobilized mycellium and diffuse aeration is also a rapid system in order to evaluate the new strains in their capacity for decontamination.

TABLE 2. DECOLORIZATION OF E_1 KRAFT EFFLUENT IN A PACKED-BED BIOREACTOR (300 ML) USING NYLON-IMMOBILIZED MYCELIUM AND DIFFUSE AERATION.

FUNGUS	**L. edodes** (UEC-2019)[&]	**P. chrysosporium** (BKM-F-1767)[#]
PERIOD (h)	120	120
pH (inic. 5.0)	4.1	(inic. 4.5) 4.0
Decolorization(%)	55	67
Decrease of Total Phenols(%)	36	42
Decrease of COD	67	50

[&] See at the Experimental conditions. [#] Ref. [23].

ACKNOWLEDGEMENT: Support from CNPq, FAPESP, FINEP and PADCT Program are acknowledged. We also acknowledged some experimental advised by Mr. M. Barbosa.

REFERENCES

1. Chang, H.-M.; Joyce, T.W.; Vasideravan, B. Intern. Symp. Wood Pulping Chem (Paris) 4th. **1987**, 2, 159.
2. Royer, G.; Desrochers, M.; Jurasek, L.; Rouleau, D.; Mayer, R.C. *J. Chem. Technol. Biotechnol.* **1987**,35B, 14.
3. Dutta, S.A.; Parhad, N.M.; Joshi, S.R. *IAWPC Technol. Ann.* **1985**, 12, 869.
4. Archibald, F.; Paice, M.G.; Jurasek, L. *Enzyme Microb. Technol.* **1990**, 12, 846.

5. Yin, C.-F.; Joyce, T.W.; Chang, H.-M. *J. Biotechnol.*, **1989**, 10, 67.
6. Yin, C.-F.; Joyce, T.W.; Chang, H.-M. *J. Biotechnol.* **1989**, 10, 77.
7. Durán, N.; Ferraz, A.; Mansilla, H. *Arq. Biol. Tecnol.*, **1990**, 33, 295.
8. Esposito, E.; Canhos, V. P.; Durán, N. *Biotechnol. Lett.* **1991**, 13, 571
9. Tien, M.; Kirk,T.K. *Natl. Acad. Sci. USA.* **1984**, 81, 2280.
10. Szklarz, G.D.; Antibus, R.K.; Sinsabaugh, R.L.; Linkins, A. *Mycology,* **1989**, 81, 234.
11. Tan, L.U.L.; Mayers, P.; Saddler, J.N. *Can. J. Microbiol.*1987, 33, 689.
12. Kuwahara, M.; Glenn, J.K.; Morgan, M.A.; Gold, M. *FEBS Lett.* **1984**, 169, 247.
13. CPPA Technical Section Standard Method H5P. (1975).
14. Durán,N.; Erismann, N de M.; Milagres,A.M.F.; Curotto, E.; Esposito,E.; Canhos, V.P. Proc. VIII Biorganic Workshop (Brazil-Chile), Talca, Chile.**1991**, 4-8 .
15. APHA. Standard Methods for Examination of Water and Wastewater. 14 ed. New York (1975).
16. Ferrer, I.; Dezotti, M.; Durán, N. *Biotechnol. Lett.* **1991**, 13, 577.
17. Miller, G.L. *Anal. Chem.*, **1959**, 31, 426.
18. Stand. Meth. for the Examination of Water and Waste-water. 7th Ed. Amer. Publ. Health Ass. 5.68-5.69 (1989).
19. Durán, N. Proc. 2nd Symp. Chem. Lignin Other Wood Components,Campinas, S.P., Brazil **1991**,3, S-06.
20. Ferraz, A.; Durán, N. *Lett. Appl.Microbiol.* **1991**, 13, 82.
21. Kondo, R.; Inamura, H. *Mokuzai Gakkaishi.* **1989**, 35, 1001.
22. Kondo, R.; Imore, T.; Inamura, H.; Nishida, T. *J. Biotechnol.* **1990**, 13, 181.
23. S'Anna G. L. Proc. 2nd Symp. Chem. Lignin Other Wood Components, Campinas,S.P., Brazil **1991**,3, S-04.

74

Mineralization of lignin-bound and free xenobiotics by the white-rot fungus, *Phanerochaete chrysosporium*

H. Sandermann, Jr., D.-H. Pieper[+] and R. Winkler[++] - GSF-Forschungszentrum für Umwelt und Gesundheit GmbH, Institut für Biochemische Pflanzenpathologie, Ingolstädter LandstraBe 1, D-8042 Neuherberg, FRG.

ABSTRACT

The unique capacity of the white-rot fungus, Phanerochaete chrysosporium, to completely mineralize polychlorinated hydrocarbons as well as lignin-bound and free chlorinated anilines has been discovered in 1985. In the cases of DDT and chlorinated anilines, the fermentation conditions for mineralization differed from those needed to induce the extracellular lignin peroxidase complex. Benzo[α]pyrene and chlorophenols were transformed by lignin peroxidase to mutagenic quinoid derivatives that may be intermediates of mineralization. 3,4-Dichloroaniline was transformed to the highly toxic 3,4,3',4'-tetrachloroazobenzene, a structural analogue of 2,3,7,8-TCDD. Mineralization of chloroanilines appears to proceed via an unidentified polar conjugate and a unique N-chlorophenyl-succinimide conjugate. Conjugation provides protective groups that may also play a role in normal lignin mineralization. The presence of toxifying as well as detoxifying reactions makes careful ecotoxicological control mandatory for the use of P. chrysosporium in biological decontamination.

INTRODUCTION

Lignin is a main component of wood and of plant cell walls generally. Due to its complex cross-linked structure, lignin is not degraded in animals. Much of the huge lignin biomass is entering the long-lived pool of soil humus components, but part of the lignin biomass is degraded much faster.

+ Present address: GBF-Bereich Mikrobiologie, D- 3300 Braunschweig, FRG.

++ Present address: GSF-Institut für Bodenökologie, D-8042 Neuherberg, FRG.

In particular, white-rot fungi have long been known to bc specialists in fast lignin degradation [1].

Our interest in lignin degradation arose after we identified lignin as a major binding site in the so-called plant bound residue fractions of pesticides and other xenobiotics. These results have been presented at a previous Cellulose Conference [2]. At that time it was unclear what would happen to the bound xenobiotics in food and decomposer chains. Subsequent animal studies indicated a surprisingly high degree of bioavailability (66%) of lignin-bound chloroanilines in the rat, whereas the mineralization rate in a standard soil system was only about 1% over three months [3]. More recent studies with rats and sheep indicate that the degree of bioavailability depends greatly on the way the lignin metabolites are prepared. An acid-labile (probably, 4-hydroxy-benzylamine) linkage of the chloroanilines appears to be related to high animal bioavailability [4]. The chloroaniline/ lignin metabolites used in the initial animal study [3], as well as the free chloroanilines, have also been examined for mineralization by the white-rot fungus, <u>Phanerochaete chrysosporium</u>. The lignin metabolites used carried a ^{14}C-label in either the bound chloroanilines or in the natural coniferyl alcohol lignin subunits.

The surprising discovery reported in 1985 [5] was that the fungus did not distinguish the ^{14}C-labels in either the natural or the xenobiotic lignin subunit. Moreover, high mineralization of about 50% was obtained in 4 weeks not only for the lignin-bound but also for the free chloroanilines. As in the case of lignin itself [6], mineralization occurred most efficiently in stationary growth phase and under conditions of nitrogen limitation. In 1985, two short communications appeared that also described the unusual capacity of <u>P. chrysosporium</u> for the mineralization of xenobiotics. Bumpus et al. [7] employed 3,4,3',4'-tetrachlorobiphenyl, 2,4,5,2',4',5'-hexachlorobiphenyl, 2,3,7,8-TCDD, DDT, benzo[α]pyrene and lindane. Eaton [8] employed Arochlor 1254 (mixed tetra- to heptachlorobiphenyls). In our subsequent studies, 4-chloroaniline and 3,4-dichloroaniline/lignin metabolites from wheat [9], a 3,4,5,6,-tetrachlorobenzoquinone/lignin metabolite [10], N-succinyl-4-chloroaniline [11] and N-malonyl-4-chloroaniline [12] were mineralized in high yield. Other laboratories have by now successfully mineralized a total of some 20-30 xenobiotics. Reviews on the mineralization capacity of <u>P. chrysosporium</u> have appeared [13,14].

TWO CONTROVERSIES AND THEIR RESOLUTION

Our results on the mineralization of free 3,4-dichloroaniline by <u>P. chrysosporium</u> were attacked in the literature [15]. Mineralization could not be reproduced under fermentation conditions that were similar but not identical to those originally reported. It was concluded that <u>P. chrysosporium</u> had no exceptional metabolic potency but resembled soil microorganisms which have but low mineralization rates for chloroanilines. We found in a re-examination that the fungal strain used in [15] gave high mineralization rates under the original fermentation conditions [5]. The following slight experimental changes completely abolished mineralization: fermentation at 27° C rather than at 39° C and gassing with air rather than oxygen. It became clear that the fungal mineralization system is strictly dependent on fermentation conditions [16]. A general debate that is still going on today concerns the role of the extracellular lignin peroxidase complex in the mineralization of xenobiotics.

Two of the three early studies [7,8] and a subsequent review [13] proclaimed a central role of the lignin peroxidase complex in xenobiotic mineralization, including that of DDT [7,17,18]. However, DDT mineralization could be achieved under conditions where the lignin peroxidase complex was not induced [19]. The controversy was probably decided by the report that DDT is not a substrate for the isolated lignin peroxidase complex [14]. Phenanthrene [20] likewise appeared to be mineralized without involvement of the lignin peroxidase complex. On the other hand, an involvement has been made likely in the cases of benzo[α]pyrene [21,22], pentachlorophenol [23], 2,4-dichlorophenol [24], 2,4-dinitrotoluene [25] and of anthracen [26]. The true CO_2-generating system appears to be still unknown and may be cell-associated. One possible candidate is the cytochrome P-450 system recently discovered in P. chrysosporium for phenanthrene [20].

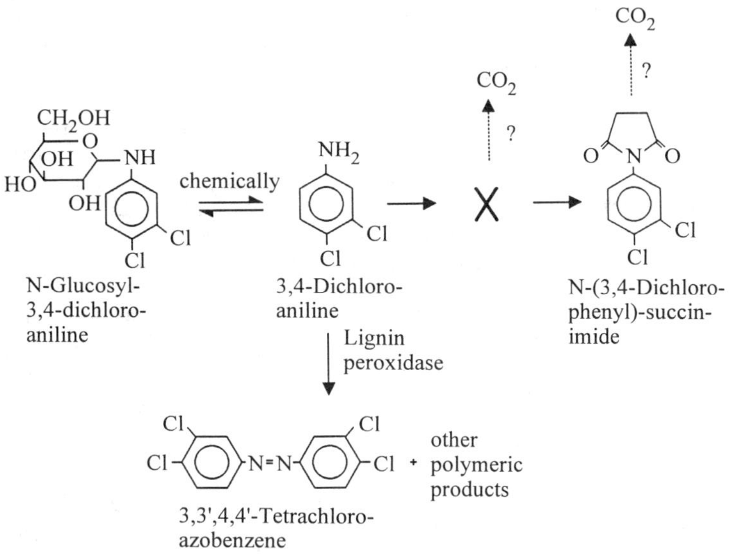

Figure 1. Bioconversion of 3,4-dichloroaniline by Phanerochaete chrysosporium. The spontaneous formation of the N-glucoside and the peroxidase-catalyzed formation of dimer and oligomers are competing with the formation of the polar conjugates "X" and N-3,4-dichlorophenyl-succinimide, and CO_2 development.

MUTAGENIC ACTIVATION BY THE LIGNIN PEROXIDASE COMPLEX

Certain lignin peroxidase products of xenobiotics are recognized mutagens, that will covalently bind to DNA, or cause DNA strand breaks. This applies in particular to the quinoid derivates formed from benzo[α]pyrene [21,22] and pentachlorophenol [23] as well as possibly some of the chlorinated dioxins formed from chlorophenols [27].

In catalyzing these reactions of mutagenic activation, lignin peroxidase behaves like horseradish and certain mammalian peroxidases. The isolated lignin peroxidase complex transformed 3,4-dichloroaniline in 10-15% yield to 3,4,3',4'-tetrachloroazobenzene [28] (see Fig. 1). This compound resembles 2,3,7,8-TCDD in structure and toxicity [29] and therefore constitutes an extreme example for toxification by P. chrysosporium.
The simultaneous presence of toxifying and detoxifying pathways makes a careful ecotoxicological control mandatory for the use of white-rot fungi in biological decontamination processes.

A ROLE FOR PROTECTIVE GROUPS IN MINERALIZATION

Incubation of 3,4-dichloroaniline with the isolated lignin peroxidase complex has led to nonpolar dimeric and oligomeric products [28]. In contrast, the active in-vivo cultures formed polar metabolites in high yield. Nonpolar products were formed only in the inactive controls by an apparent autoxidation process [5,9]. One major type of polar product has been determined to belong to a unique novel class of conjugates, the N-chlorophenyl-succinimides [11] (see Fig. 1).

Recent kinetic studies have shown that 3,4-dichloroaniline disappeared completely within a few days after addition to N-limited agitated cultures. This was accompanied by the appearance of a highly polar and labile conjugate, which has so far not been chemically identified and is termed "X". This product rapidly declined in favour of the appearance of N-3,4-dichlorophenyl-succinimide and $^{14}CO_2$ development.

The N-glucoside conjugate of 3,4-dichloroaniline was observed in the culture medium as a spontaneous product. The mass action law was found to apply to the equilibrium, 3,4-dichloroaniline + D-glucose $\longleftrightarrow$ N-3,4-dichlorophenyl-glucoside, with a dissociation equilibrium constant of 0.1 M. On the basis of the time course study, the N-glucoside was not involved in CO_2 development. All observed metabolic reactions are shown in Fig. 1.

In summary, the fungus behaved like an organic chemist by adding N-protective groups to the oxidation-sensitive amino group prior to ring cleavage and CO_2 development. It is not clear whether "X" or the N-chlorophenyl-succinimide, or both conjugates, act as precursors for CO_2 development. Could O-protective groups be involved in normal lignin mineralization by P. chrysosporium ? Such a mechanism could resolve the paradoxical finding that isolated lignin peroxidase will polymerize rather than depolymerize phenolic compounds [6,31]. Fungi are generally known to form phenolic conjugates [32]. Such substituents could prevent phenoxy-radical formation in favor of ring cleavage, and they could also serve as markers for cellular uptake systems.

REFERENCES

1. Zeikus, J.G. in Alexander, M. (Ed.) Advances in Microbial Ecology. Plenum Publishing Corporation, 1989, 5, 211.
2. Sandermann, H.; Scheel, D.; Trenck, v. d. T. J. Appl. Polymer Sci., Appl. Polymer Symp. **1983**, 37, 407.

3. Sandermann, H.; Arjmand, M.; Gennity, I.; Winkler, R.; Struble, C. B.; Aschbacher. P. J. Agric. Food Chem. **1990**, 38, 1877.
4. Sandermann, H.; Musick, T. J.; Aschbacher, P.J. Agric. Food Chem., Submitted.
5. Arjmand, M.; Sandermann, H. J. Agric. Food Chem. **1985**, 38, 78.
6. Kirk, T.K.; Farrell, R.L. Ann. Rev. Microbiol. **1987**, 6, 465.
7. Bumpus, J.A.; Tien, M.; Wright, D.; Aust, S.D. Science, **1985**, 228, 1434.
8. Eaton, D.C. Enzyme Microbiol. Technol. **1985**, 7, 194.
9. Arjmand, M.; Sandermann, H. Z. Naturforsch. **1986**, 41c, 206.
10. Schäfer, W.; Sandermann, H. J. Agric. Food Chem. **1988**, 36, 370.
11. Arjmand, M.; Sandermann, H.; Pesticide Biochem. Physiol. **1987**, 27, 173.
12. Winkler, R.; Sandermann, H. Pesticide Biochem. Physiol. **1989**, 33, 239.
13. Bumpus, J.A.; Aust, S.D. BioEssays, **1986**, 36, 166.
14. Hammel, K.E. Enzyme Microbiol. Technol. **1989**, 11, 776.
15. Hallinger, S.; Ziegler, W.; Wallnöfer, P.R.; Engelhardt, G. Chemosphere, **1988,** 17, 543.
16 Winkler, R.; Sandermann, H. Seventh Interna-tional Congress of Pesticide Chemistry,IUPAC, Hamburg, Abstract Book, Abstract No.06C-03, **1990**.
17. Bumpus, J. A.; Aust, S. D. Appl. Environ. Microbiol. **1987**, 53, 2001.
18. Fernando, T.; Aust, S. D.; Bumpus, J. A. Chemos-phere **1989**, 19, 1387.
19. Köhler, A.; Jäger, A.; Willershausen, H.; Graf,H. Appl. Microbiol. Biotechnol. **1988**, 29, 618.
20. Sutherland, J.B.; Selby, A.L.; Freeman, J.P.; Evans, F.E.; Cerniglia, C.E. Appl. Environ. Microbiol. **1991**, 57, 3310.
21. Haemmerli, S.D.; Leisola, M.S. A.; Sanglard, D.; Fiechter, A. J. Biol. Chem. **1986**, 261, 6900.
22. Hammel, K.E.; Kalyanaraman, B.; Kirk, T.K. J. Biol. Chem. **1986**, 261, 16948.
23. Mileski, G.J.; Bumpus, J.A.; Jurek, M.A.; Aust, S. D. Appl. Environ. Microbiol. **1988**, 54, 2885.
24. Valli, K.; Gold, M.H. J. Bacteriol. **1991**, 173, 345.
25. Valli, K.; Brock, B.J.; Joshi, D.K.; Gold, M.H. Appl. Environ. Microbiol. **1992**, 58, 221.
26. Hammel, K.E.; Green, B.; Gai, W.Z. Proc. Natl. Acad. Sci. **1991**, 88, 10605.
27. Svenson, A.; Kjeller, L.O.; Rappe, C. Environ. Sci. Technol. **1989**, 23, 900.
28. Pieper, D. H.; Winkler, R.; Sandermann, H. Angew. Chem. **1992**, 104, 60; Int. Ed. **1992**, 31, 68.
29. Poland, A.; Knutson, J. C. Ann. Rev. Pharmacol. Toxicol. **1982**, 22, 54.
31. Haemmerli, S. D.; Leisola, M. S. A.; Fiechter, A. FEMS Microbiol. Lett. **1986**, 35, 33.
32. Turner, W. B.; Aldridge, D. C. Fungal Metabolites II, Academic Press, New York, **1983**.

Index

acellulolytic strains (C-), 415-420
acetylation, 474-477
acid hydrolysis, 74, 83, 86
  of cellulose, 357
activation, 400
adsorption, 393
air quality,
  effect on wood, 329-334
air-lift fermenter, 431
alcohol, 17
alkali metal hydroxides, 172
alkali, 17
alkalisation, 124
amine oxide solvent, 309
anaerobic digestion, 75-80, 365-370
ascomycete, 493
Aspen, 38, 45, 51-58
Aspergillus fumigatus, 409, 432

bacterial cellulose, 167
bagasse, 32, 123-128
banana tree fibres, 32
basidiomycetes, 371, 373
beating, 103
biobleaching, 485-492
biogas, 365-370
biomethanation, 365-370
biotechnology, 135-139
black liquor, 485-492
BOD_5, 485-492
bond cleavage, 347-354
bound water, 236
Brazilian lignocellulosics, 31-36
breaking elongation, 179
breaking twist angle, 261-266
brightness, 90, 315-319
brown-rot, 457-462
bubble column, 431
bursting strength, 223
butanetetracarboxylic acid, 207, 284
butylene oxide, 474-477
byssinosis, 161

carbohydrates, 69, 107
carboxyl content, 317-320
carboxymethyl cellulase (endoglucanase), 377-382
carboxymethylation, 232, 233
catalysts poisons, 283
  for delignification, 21
cationic cotton, 249
cell wall pore volume, 457-462
cellulase treatment, 189, 191
cellulases, 371-376, 384, 409, 432
cellulolytic enzymes, 373
cellulose, 153, 157, 255-258, 457-462
  conservation of, 465-471
  hydrolysis of, 355-361
cellulose acetate, 237-242
cellulose crystallinity, 141-146, 326
cellulose dissolution, 130, 131
cellulose fibres, 129-133
cellulose oxidation, 451-456
cellulose production,
  environmental aspects, 303-310
cellulose structural models, 169
cellulose triacetate fibres, 243-247
Ceriporiopsis, 428
chemical impregnation, 51-56
chemical modification, 474-477

chemical pretreatment, 57, 58
chemimechanical pulping, 45-50, 51-56
chemithermomechanical pulping, 45-50
chlorine bleaching, 97
chlorine dioxide bleaching, 97, 98
chlorine free bleaching, 25-28
choline chloride, 250
citric acid, 284
CMP, 37-39
COD, 485-492
colour strength, 255-260
condensation, 70
contact ratio, 113
conversion of lignocellulosics, 356
convolution angles,183
cooking,
  high temperature, 51-56
core yarn, 213-218, 220
cotton, 205-210, 213-218, 219, 249, 283
cotton fabrics, 289-294
cotton fibre, 136, 188, 190, 199, 200, 231, 261-266
cotton linters, 167
cotton/PET blends, 226, 228
cotton processing dust, 161
cotton stalks, 75-80
cresol, 70
crosslinking, 249
crystallinity, 11-16
crystallite size, 153-158
CTMP, 37-39

DDT,
  mineralization of, 501
degree of polymerisation, 12, 13, 326
delignification, 5-9, 11-16
diaper environmental impact analysis, 299
3,4-dichloroaniline,
  mineralisation of, 501, 502
dihydroxypropylation, 232, 233
dimethoxyphenol, 347-354
dimethyloldihydroxy ethyleneurea 207, 208
direct liquefaction, 341-346
discoloration, 330-331
dissolving pulp, 83-87
double vacuum treatment, 480
durable press, 205-208, 213-218, 226-228, 250, 261-266, 283

effect of Cr(III) on bleaching, 98
effluents, 485-492
electron microscopy, 393, 394
elemental map, 162
elongation, 315-319
environmental aspects,
  of cellulose production, 303-310
environmental conditions, 315-319
environmentally acceptable technology, 362
enzymatic activity, 377-382
enzyme, 187-190
enzyme-phenolic complexes, 404
enzumic hydrolysis, 399
  of cellulose, 357
equilibrium moisture content, 476, 477
ESCA, 39, 40
esterification, 283
expert systems, 300, 301
explosion pulp, 38-41
extention orientation curve, 183
extinction ratio,

fading rate constant, 255-260
Fenton's reagent, 451-456
fibre, 111-116, 141-146
  from bagasse pulp, 123-128
  cotton, 261-266
  derivatised, 261-266
  Raman spectra of, 243-247
fibre breaking tenacity, 178
fibre characteristics,51-56
fibre curl, 113
fibre distribution, 38, 40
fibre length, 111-114
fibre propereties, 65, 132, 133, 304
fibre quality, 136, 137
fibre thickness, 111-116
fibre width, 111-114
fibrets, 237-242
fibrillar orientation, 199-201
fibrillar structure, 237-242
flame retardant, 213-218
flax, 117-121
free radicals, 451-456
FTIR, 141-451
fungally degraded wood, 445-451
fungi, 474, 475
G. resinaceum, 377-382
garment dyeing, 249
gene regulation, 138, 139
genetic engineering, 394 395
genetic improvement, 415-420
glass filament, 213-218
β-glucosidase, 377-382, 493-497
glycosans, 361
glycoside hydrolases, 384
Gossypium, 135

hemicellulases, 371-376, 409
Hermans orientation factor, 177
historic linen textiles, 153, 157
HPLC, 103
hydration, 196
hydrogen peroxide, 451-456
hydrothermolysis, 63-68
hygroscopicity, 187

iactonic acid, 289-294
immobilisation, 485-492
impure substrate, 410
inhibition, 400
intimate-blend, 205-208
irradiated wood, 330-332
irradiation, 83-86
isocyanates, 474-477

Kawabata evaluation, 225
Kraft pulp, 104
Kraft effluent, 493-497
Kraft paper pulp, 96

latent acid catalyst, 255, 256
lateral order distribution, 171
lattice distortions, 167, 168
LCR, 172
leaching, 108
life cycle analysis, 297-302
lightfastness, 255-260
lignin, 5-9, 69, 107, 141-146, 335-340, 457-462
lignin degradation, 67, 419, 420
lignin peroxidase, 422, 501, 502
lignin recondensation, 67
ligninases, 371-376, 493-497
lignocellulases, 409
linear regression analysis, 51-56
linen, 323
    short staple, 117-121
linseed stalks, 75-80

magnefite pulp, 104
maleic acid,
    in crosslinking, 289-294
man-made fibre, 219
manganese peroxidase, 422
marine organisms, 477
mechanical behaviour,
    of wood, 445-451
mechanical strength, 45-50
mercerising agent, 199, 204
microfibres, 237-242
microfibril dimension, 167
mineralisation of Xenobiotics, 500
molecular modelling, 393, 394
molecular weight distribution, 83-88
museum artefacts, 465-471
mutagenic activation, 502

N-methylmorphine N-oxide, 309
n-methylol (DMA) resin, 255, 256
NIRS, 141-146
NMR, 141-146
non-cellulosic polysaccharides, 141-146
nylon-filament, 213-218
nylon-immobilized, 493-497

oxidation, 325
oxygen, 17
oxygen-alkali pulping, 25-28
oxygen mass transfer, 431

packing density, 199, 204
paper, 111-116
paper making, 75-80
paper properties, 51-56
pentachlorophenol, 421-429
Phanerochaete chrysosporium, 457-462, 485-492, 499
Phanerochaete, 421, 426
1,10-phenanthroline, 21
phenolation, 73
phenolic compounds, 404
phosphate, 284
phosphite, 284
photodegradation, 315-319, 323
photofading, 255
photoirradiation, 332
physical strength, 39
phytomass refining, 69
Picea glauca, 90
plant wastes, 372-376
pleurotus ostreatus, 415, 416
polycarboxylic acids, 261-266, 283
polyester 205-210
polyester filament, 213-218
polyethylene, 221
Poplar wood, 63-68
pore size,
    of cellulose fibre, 197
Postia placenta, 457-462
prediction models, 51-56
preservation, 480

preservation uptake, 481
pretreatment, 25, 28, 365-370
propylene oxide, 474-477
pulp,
  bagasse, 123-128
  dissolving, 83-87
  Kraft, 96, 104
  magnefite, 104
  steam explosion, 51-58
  ultrahigh yield, 51-56
pulp and paper properties, 57-61
pulp preparation,75-80
pulp properties, 45-50
pulp use, 308
pulping, 5-9, 17
  chemimechanical, 45-56
pulsation apparatus, 25
pulsation intensity, 26

quantitative traits, 139

radiation,
  effect on wood, 329-334
Raman specta,
  of fibres, 243-247
ramie, 167
rayon, 168
reactive dye, 256
reactivity, 83-87
recycled solvent, 341, 342
resist treatments, 250
respiratory toxicity, 164
restricted diffusion, 196-198
ring spinning, 213-218

scanning electron microscopy,
  of cellulose fibres, 34
Scherrer technique, 153, 154
secondary cell wall, 199
semiconductors, 90
Sitka Spruce, 479
softwood Kraft pulp, 194
solvent-spun cellulose production, 303-310
solid state fermentation, 412
solvation, 70
solvent system, 129
solvolysis pulp, 11-16
specific refining energy, 57-61
spectrophotometric measurements, 315
spectroscopy, 335
spinning, 125, 221
Spruce Kraft pulp, 90
Spruce wood structure, 483
staple fibres, 126
steam explosion pulp, 45-50, 51-56, 57-61
steam explosion treatment, 117-120
stone finishing, 251
straw, 63-68
strength uniformity, 199-202
structural probe, 172
structure,
  fibrillar, 237-242
substrate specificity, 386
sugar cane bagasse, 32, 63-68
sugar cane, 123
sulfonate, 39, 40
sulfuric acid, 70
sunlight, 90
surface analysis, 39
surface chemistry, 330
surface treatment, 465-471
swelling, 70

temperature stabilites, 410
tenacity, 178, 199-203
tensile properties, 261-266
tensile strength, 222, 315-319, 335-339
tents, 213-218
termites, 476
tetraploids and diploids, 183
textile properties, 235
textiles, 153, 157
thermal reaction pathways, 347-354
thermolysis, 347-354
thermolytical hydrolysis of cellulose, 360
titanium dioxide, 90
total organic carbon, 187-191
toxic inorganic elements, 161
Trichoderma viride, 188, 384
triethanolamine, 286

ultra-high-yield pulp, 37, 38, 45-50, 51-56
ultraviolet light, 92
uniforms, 213-218
urea treatment of fibres, 32

valonia cellulose, 167
viscose, 83-87
viscose process, 125
viscose rayon manufacture, 307

water absorbability, 187
water diffusion, 193-197
water hyacinth, 365-370

water retention value, 261-266
water/2-propanol mixture, 341-346
weathering, 335-340
wheat straw, 63-68
white-rot fungi, 421-429, 451-456, 457-462, 485-492, 493
wood composition, 307
wood, 341-346
    effect of air quality on, 329-334
    delignification, 5-9
    fungally degraded, 445-451
    effect of radiation on, 329-334
wrinkle resistance, 289-294

X-ray diffraction, 153-155
xanthation, 124
XRFS, 161
xylanolytic enzymes, 373

yarn core, 206-210

zinc oxide, 90